THE NEW TESTAMENT

A HISTORICAL INTRODUCTION TO THE EARLY CHRISTIAN WRITINGS

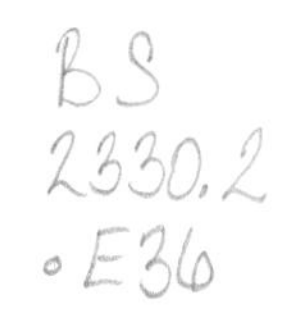

THE NEW TESTAMENT

A HISTORICAL INTRODUCTION TO THE EARLY CHRISTIAN WRITINGS

Bart D. Ehrman

New York Oxford
Oxford University Press
1997

Oxford University Press

Oxford New York
Athens Auckland Bangkok Bogota
Bombay Buenos Aires Calcutta
Cape Town Dar es Salaam Delhi
Florence Hong Kong Istanbul Karachi
Kuala Lumpur Madras Madrid Melbourne
Mexico City Nairobi Paris Singapore
Taipei Tokyo Toronto

and associated companies in
Berlin Ibadan

Published by Oxford University Press, Inc.
198 Madison Avenue, New York, New York 10016

Library of Congress Cataloging-in-Publication Data

Ehrman, Bart D.

The New Testament: A Historical Introduction to the Early Christian Writings / Bart D. Ehrman.

p.cm.

Includes biographical references.

ISBN 0-19-508481-0

1. Bible. N.T.—Introductions 2. Bible. N.T.—History of Biblical events. 3. Bible. N.T.—History of contemporary events. 4. Church history—Primitive and early church, ca.30–600. 5. Christian literature, Early. I. Title

BS2330.2.E36 1997

255.6'7—dc20 95–42342 CIP

1 3 5 7 9 8 6 4 2

Printed in the United States of America
on acid-free paper

To David R. Adams

Brief Table of Contents

Preface xix

Acknowledgments xxi

Credits xxiii

Chapter 1 What Is the New Testament: The Early Christians and Their Literature 1

Chapter 2 The World of Early Christian Traditions 16

Chapter 3 The Traditions of Jesus in Their Greco-Roman Context 40

Chapter 4 The Christian Gospels: A Literary and Historical Introduction 51

Chapter 5 Jesus, the Suffering Son of God: The Gospel according to Mark 56

Chapter 6 The Synoptic Problem and Its Significance for Interpretation 72

Chapter 7 Jesus, the Jewish Messiah: The Gospel according to Matthew 79

Chapter 8 Jesus, the Savior of the World: The Gospel according to Luke 96

Chapter 9 Luke's Second Volume: The Acts of the Apostles 115

Chapter 10 Jesus, the Man Sent from Heaven: The Gospel according to John 133

Chapter 11 From John's Jesus to the Gnostic Christ 154

Chapter 12 Jesus from Different Perspectives: Other Gospels in Early Christianity 171

Chapter 13 The Historical Jesus: Sources, Problems, and Methods 185

Chapter 14 The Historian and the Problem of Miracle 198

Chapter 15 Jesus, the Apocalyptic Prophet 203

Chapter 16 From Jesus to the Gospels 233

Chapter 17 Paul the Apostle: The Man and His Mission 241

Chapter 18 Paul and His Apostolic Mission: Thessalonians as a Test Case 257

Chapter 19 Paul and the Crises of His Churches: 1 and 2 Corinthians, Galatians, Philippians, and Philemon 271

Chapter 20 The Gospel according to Paul: The Letter to the Romans 299

Chapter 21 Does the Tradition Miscarry? Paul in Relation to Jesus, James, Thecla, and Theudas 311

Chapter 22 In the Wake of the Apostle: The Deutero-Pauline and Pastoral Epistles 320

Chapter 23 From Paul's Female Colleagues to the Pastor's Intimidated Women: The Oppression of Women in Early Christianity 341

Chapter 24 Christians and Jews: Hebrews, Barnabas, and Later Anti-Jewish Literature 351

Chapter 25 Christians and Pagans: 1 Peter, the Letters of Ignatius, the Martyrdom of Polycarp, and Later Apologetic Literature 366

Chapter 26 Christians and Christians: James, the Didache, Polycarp, 1 Clement, Jude, and 2 Peter 383

Chapter 27 Christians and the Cosmos: The Revelation of John, The Shepherd of Hermas, and the Apocalypse of Peter 398

Chapter 28 Epilogue: Do We Have the Original New Testament? 414

Glossary of Terms 423

Index 431

CONTENTS

Preface . . . xix
Acknowledgments . . . xxi
Credits . . . xxiii

Chapter 1 What Is the New Testament? The Early Christians and Their Literature . . . 1

Box 1.1 The Hebrew Bible and the Christian Old Testament . . . 2

* The Diversity of Early Christianity . . . 3

Box 1.2 The Canon of Scripture . . . 4

* The New Testament Canon of Scripture . . . 7

Box 1.3 The Common Era and Before the Common Era . . . 8

Box 1.4 The Layout of the New Testament . . . 9

* Summary and Implications for our Study . . . 13

* Some Additional Reflections: The Historian and the Believer . . . 13

Chapter 2 The World of Early Christian Traditions . . . 16

* The Problem of Beginnings . . . 16

* One Remarkable Life . . . 17

Box 2.1 Pagan and Gentile . . . 18

Box 2.2 The Greco-Roman World . . . 19

* The Environment of the New Testament: Religions in the Greco-Roman World . . . 20

Box 2.3 Divine Rulers as Savior Gods . . . 26

Box 2.4 The Roman Empire . . . 27

Box 2.5 Christianity as a Mystery Cult . . . 31

* The Environment of the New Testament: Judaism as a Greco-Roman Religion . . . 32

Chapter 3 The Traditions of Jesus in Their Greco-Roman Context 40

* Oral Traditions behind the Gospels 40

Box 3.1 Orality and Literacy in the Ancient World 45

Box 3.2 Differences between Mark and John on the Time of Jesus' Death 48

* Some Additional Reflections: The Authors of the Gospels 49

Chapter 4 The Christian Gospels: A Literary and Historical Introduction 51

* The Question of Genre 51

* Biography as a Greco-Roman Genre 52

Box 4.1 Plutarch on Biography 53

* The Gospels as Ancient Biographies 54

Chapter 5 Jesus, the Suffering Son of God: The Gospel according to Mark 56

* The Beginning of the Gospel: Jesus the Messiah, the Son of God Who Fulfills the Scripture 56

Box 5.1 The Jewish Messiah 57

* Jesus the Authoritative Son of God 59

* Jesus the Opposed Son of God 59

Box 5.2 Jewish Scribes, Pharisees, Sadducees, Herodians, and Chief Priests 60

* Jesus the Misunderstood Son of God 63

* Jesus the Acknowledged Son of God 63

Box 5.3 The Messianic Secret in Mark 64

* Jesus the Suffering Son of God 65

Box 5.4 Son of God and Son of Man 66

Box 5.5 The Charge of Blasphemy according to Mark .67

* Jesus' Death as the Son of God .67

* Jesus the Vindicated Son of God .68

* Conclusion: Mark and His Readers .70

Chapter 6 The Synoptic Problem and Its Significance for Interpretation .72

* Methods for Studying the Gospels .72

* The Synoptic Problem .72

Box 6.1 Ironing Out the Problems: One Potential Difficulty in Mark's Account .76

Box 6.2 The Contents of Q .72

* The Methodological Significance of the Four-source Hypothesis .78

Chapter 7 Jesus, the Jewish Messiah: The Gospel according to Matthew .79

* The Importance of Beginnings: Jesus the Jewish Messiah in Fulfillment of the Jewish Scriptures .80

Box 7.1 Matthew's Scheme of Fourteen .82

* Jesus and His Forerunner from Matthew's Perspective .85

* The Portrayal of Jesus in Matthew: The Sermon on the Mount as a Springboard .86

Box 7.2 The Golden Rule .89

* Jesus and the Jewish Cultic Practices Prescribed by the Law .90

* Jesus Rejected by the Jewish Leaders .91

Box 7.3 Gentiles in Matthew's Community .92

* Matthew and His Readers .93

Box 7.4 Was Matthew a Jew? .95

Chapter 8 Jesus, the Savior of the World: The Gospel according to Luke 96

* The Comparative Method and the Gospel of Luke 96

* A Comparative Overview of the Gospel 98

* The Preface to Luke's Gospel 98

Box 8.1 Apologetic Literature in Early Christianity 99

* Luke's Birth Narrative in Comparative Perspective 101

Box 8.2 Historical Problems with Luke's Birth Narrative 102

Box 8.3 The Virgin Birth in Matthew and Luke 103

* From Jew to Gentile: Luke's Portrayal of Jesus the Rejected Prophet 103

* Luke's Distinctive Emphases throughout His Gospel 107

Box 8.4 Jesus' Bloody Sweat in Luke 108

Box 8.5 Jesus as a Righteous Martyr 109

Box 8.6 The Institution of the Lord's Supper in Luke 112

* Conclusion: Luke in Comparative Perspective 113

Chapter 9 Luke's Second Volume: The Acts of the Apostles 115

* The Genre of Acts and Its Significance 115

Box 9.1 The Book of Acts: An Ancient Novel? 117

* The Thematic Approach to Acts 117

* From Gospel to Acts: The Opening Transition 118

Box 9.2 Luke's Mysterious Two Men 120

* Luke's Artistry as a Storyteller 121

* Themes in the Speeches in Acts 122

Box 9.3 The Death of Judas 123

* In Sum: Prominent Themes of Luke-Acts 128

* The Author of Luke-Acts and His Audience 129

Chapter 10 Jesus, the Man Sent from Heaven: The Gospel according to John 133

* The Gospel of John from a Literary-Historical Perspective 134

Box 10.1 Jesus' Signs in the Fourth Gospel 135

Box 10.2 "The Jews" in the Fourth Gospel 136

* The Gospel of John from a Thematic Perspective 136

* The Gospel of John from a Comparative Perspective 137

Box 10.3 Signs and Faith in the Fourth Gospel 140

Box 10.4 Jesus and the "I Am" Sayings in John 141

* The Gospel of John from a Redactional Perspective 142

* The Socio-Historical Method 145

Box 10.5 The Death of the Beloved Disciple in the Johannine Community 146

* The Gospel of John from a Socio-Historical Perspective 147

Box 10.6 John's De-Apocalypticized Gospel 152

* The Author of the Fourth Gospel 152

Chapter 11 From John's Jesus to the Gnostic Christ 154

* The Questions of Genre and Author 154

Box 11.1 A Letter from Greco-Roman Egypt 155

* The New Testament Epistolary Literature and the Contextual Method ... 156

Box 11.2 The Gospel and Epistles of John: Some Thematic Similarities .. 157

* The Johannine Epistles from a Contextual Perspective 158

* Reflections on the Contextual Method 160

* Beyond the Johannine Community: The Rise of Christian Gnosticism . . .161

Box 11.3 House Churches in Early Christianity 162

Box 11.4 How Do You Know a Gnostic When You See One? 165

Box 11.5 Gnostics and the Jewish Scriptures 166

* Gnostics and the Johannine Community 169

Chapter 12 Jesus from Different Perspectives: Other Gospels in Early Christianity 171

* Narrative Gospels 172

Box 12.1 The Gospel of the Ebionites and Early Gospel Harmonies 173

* Sayings Gospels 177

Box 12.2 Judas Thomas as Jesus' Twin Brother 178

* Infancy Gospels 181

Box 12.3 The Older Sayings of the Gospel of Thomas 182

* Conclusion: The Other Gospels 183

Chapter 13 The Historical Jesus: Sources, Problems, and Methods 185

* Problems with Sources 185

* Non-Christian Sources 186

Box 13.1 Christianity as a Superstition in the Roman World 187

* Christian Sources 188

Box 13.2 The Testimony of Flavius Josephus 189

* The Historical Approach to the Gospels: Criteria
for Reconstructing the Life of Jesus 192

Box 13.3 Aramaisms as a Criterion of Authenticity 193

* Conclusion: Reconstructing the Life of Jesus 196

Chapter 14 The Historian and the Problem of Miracle 198

* Miracles in the Modern World and in Antiquity .198

* The Historian and Historical Method 200

Chapter 15 Jesus, the Apocalyptic Prophet 203

* Political Crises in Palestine and Their Ramifications 203

* The Formation of Jewish Sects 205

Box 15.1 Divine Revelation in the Dead Sea Scrolls 209

* Popular Modes of Resistance to Oppression 211

* An Ideology of Resistance 215

* Jesus as a Jewish Apocalyptic Prophet 217

* The Apocalyptic Deeds of Jesus 219

Box 15.2 Was Jesus a Cynic Philosopher? 220

Box 15.3 The Temple Incident as an Enacted Parable 221

* The Apocalyptic Teachings of Jesus 224

Box 15.4 The Cosmic Deliverer of Israel 226

* The Apocalyptic Death of Jesus 228

Box 15.5 Jesus and Judas, the Betrayer 230

Chapter 16 From Jesus to the Gospels 233

* The Beginning of Christianity 233

* Jesus' Resurrection from an Apocalyptic Perspective 234

Box 16.1 Jesus, the Messiah, and the Resurrection 235

* Jesus' Death, according to the Scriptures 236

* The Emergency of Different Understandings of Jesus 237

Box 16.2 Vicarious Suffering in Jewish Martyrologies and Other Greco-Roman Literature 238

Chapter 17 Paul the Apostle: The Man and His Mission 241

* The Study of Paul: Methodological Difficulties 242

Box 17.1 The Pauline Corpus 243

Box 17.2 Other Sources for the Life of Paul 246

* The Life of Paul 247

Box 17.3 Paul on the Road to Damascus 250

Chapter 18 Paul and His Apostolic Mission: Thessalonians as a Test Case 257

* The Founding of the Church in Thessalonica 257

* The Beginnings of the Thessalonian Church: A Socio-Historical Perspective 263

Box 18.1 Rules for a Private Association 264

* The Church as Thessalonica after Paul's Departure 265

Box 18.2 Charges against the Christians 268

Box 18.3 The Thessalonians' Perplexity 269

* Conclusion: Paul the Apostle 269

Chapter 19 Paul and the Crises of His Churches: 1 and 2 Corinthians, Galatians, Philippians, and Philemon 271

* 1 Corinthians 271

Box 19.1 Overview of 1 Corinthians 276

Box 19.2 Possibilities of Existence in the Afterlife 277

* 2 Corinthians 280

Box 19.3 The Partitioning of 2 Corinthians 283

* Galatians 285

Box 19.4 The Logic of the Opponents' Position in Galatia 286

Box 19.5 Cephas and Peter 288

* Philippians 292

Box 19.6 The Christ Hymn of Philippians 295

* Philemon 296

Chapter 20 The Gospel according to Paul: The Letter to the Romans 299

* The Occasion and Purpose of the Letter 299

Box 20.1 The Beginnings of the Roman Church 300

* The Theme of the Epistle 302

Box 20.2 Paul's Gospel to the Romans 303

* Pauline Models for Salvation 304

Box 20.3 Judicial and Participationist Models of Salvation in Paul 307

* The Flow of Paul's Argument 307

Box 20.4 Other Models of Salvation in Paul 309

* Conclusion: Paul and the Romans 309

Chapter 21 Does the Tradition Miscarry?
Paul in Relation to Jesus, James, Thecla, and Theudas 311

* Paul in Relation to What Came Before 311

* Paul in Relation to What Came After 314

Box 21.1 Jesus and Paul: Some of the Differences 315

* Conclusion: Pauline Christianities 318

Chapter 22 In the Wake of the Apostle: The Deutero-Pauline and Pastoral Epistles 320

* Pseudonymity in the Ancient World 320

Box 22.1 Paul's Third Letter to the Corinthians 322

* The Deutero-Pauline Epistles 323

Box 22.2 The Resurrection of Believers in Paul and Colossians 328

Box 22.3 The Vocabulary of Salvation in Paul and Ephesians 331

* The Pastoral Epistles 332

* The Historical Situation and Authorship of the Pastoral Epistles 335

Box 22.4 Church Hierarchy in Ignatius 338

* Conclusion: The Post-Pauline Pastoral Epistles 339

Chapter 23 From Paul's Female Colleagues to the Pastor's Intimidated Women: The Oppression of Women in Early Christianity 341

* Women in Paul's Churches 341

* Women Associated with Jesus 342

* Paul's Understanding of Women in the Church 344

* Women in the Aftermath of Paul 345

Box 23.1 Similarities between 1 Tim 2:11-15 and 1 Cor 14:34-35 346

* Ancient Ideologies of Gender 347

* Gender Ideology and the Pauline Churches 349

Chapter 24 Christians and Jews: Hebrews, Barnabas, and Later Anti-Jewish Literature 351

* Early Christian Self-Definition 351

Box 24.1 Disparate Views on Christians and Jews 353

* Continuity and Superiority: The Epistle to the Hebrews 354

Box 24.2 Divergent Views of Christ in Hebrews 356

* Discontinuity and Supremacy: The Epistle of Barnabas 359

Box 24.3 Gematria in Early Christianity 362

* Conclusion: The Rise of Christian Anti-Semitism 362

Box 24.4 Melito's Passover Sermon 364

Chapter 25 Christians and Pagans: 1 Peter, the Letters of Ignatius, the Martyrdom of Polycarp, and Later Apologetic Literature 366

* The Persecution of the Early Christians 366

Box 25.1 The Christian Disruption of the Family: The Case of Perpetua .369

* Christians in a Hostile World: The Letter of 1 Peter 371

* Christians Sentenced to Death: The Letters of Ignatius 374

* Christians before the Tribunal: The Martyrdom of Polycarp 377

Box 25.2 An Alternative View of Christian Martyrdom 378

* Christians on the Defense: The Later Apologetic Literature 380

Chapter 26 Christians and Christians: James, the Didache, Polycarp, 1 Clement, Jude, and 2 Peter 383

* The Epistle of James 384

* The Didache 385

Box 26.1 The Development of the Lord's Prayer 387

* Polycarp's Letter to the Philippians 388

* 1 Clement 389

Box 26.2 Polycarp and the Early Christian Tradition 390

Box 26.3 Other Problems in the Corinthian Church 392

*Jude 393

*`2 Peter 394

Box 26.4 Peter, the Smoked Tuna, and the Flying Heretic 395

* Conclusion: Conflicts within the Early Christian Communities 396

Chapter 27 Christians and the Cosmos: The Revelation of John, The Shepherd of Hermas, and the Apocalypse of Peter 398

* Introduction: The End of the World and the Revelation of John 398

* The Content and Structure of the Book of Revelation 399

* The Book of Revelation from a Historical Perspective 400

* Apocalyptic Worldviews and Apocalypse Genre 401

Box 27.1 The Book of Revelation as Underground Literature 403

* The Revelation of John in Historical Context 404

Box 27.2 Contemporary Interpretations of the Book of Revelation 407

* The Shepherd of Hermas 409

Box 27.3 The Shepherd of Hermas and the Muratorian Canon 410

* The Apocalypse of Peter 412

Chapter 28 Epilogue: Do We Have the Original New Testament? 414

* The Manuscripts of the New Testament 414

* Changes in the New Testament Text 415

* The Criteria for Establishing the Original Text 418

* Conclusion: The Original Text of the New Testament 420

Glossary of Terms 423

Index 431

PREFACE

With so many textbooks on the New Testament from which to choose, it seems only fair to begin this one by indicating some of its distinctive features. While there are several outstanding introductory texts, most of them approach the New Testament from a theological or literary perspective. I have no trouble with these vantage points per se; they do not, however, happen to be mine. In this text, I am first and foremost interested in questions that pertain to the history of early Christianity and to the early Christian writings as they reflect that history and as they helped to shape it.

I am interested, for example, in the life of the historical Jesus (a matter surprisingly left untouched in a number of introductory treatments), in the history of the traditions that circulated about him, in the ways that the authors of our New Testament documents agreed and disagreed with one another (which I treat as a historical question), in the missionary practices of the apostle Paul and others like him, in the ways early Christians differed from their Jewish and pagan neighbors, in the rise of Christian anti-Semitism, in the social opposition evoked by the earliest Christians, in the role of women in the early church, and in a wide range of other questions that are more the province of the historian than of the theologian or literary critic. My historical orientation has led me to situate the early Christian literature more firmly than is normally done in the social, cultural, and literary world of the early Roman Empire. Thus, for example, I do not discuss Greco-Roman religion, the sociopolitical history of Palestine, and other related issues merely as background (for instance, in a kind of introductory appendix, as is commonly done). I have instead evoked the context of the early Christian writings at critical junctures throughout the book, as a way to help beginning students unpack the meaning and significance of these writings. Thus, for example, the discussion of religion in the Greco-Roman world sets the stage for reflections on the traditions about Jesus that were being circulated and sometimes modified within that world. The discussion of the social history of Palestine is reserved for a later chapter on the historical Jesus, since knowing about first-century Palestinian Judaism is presumably of greatest relevance for understanding a first-century Palestinian Jew. Reflections on the philosophical schools appear (principally) in the discussion of the missionary activities of Paul, for which they are particularly apropos. Justifications for these and other decisions are made en route.

Four other features of the presentation derive more or less from its fundamentally historical orientation. First, since the books of the New Testament represent only some of the writings produced by the earliest Christians, I have taken pains to situate them within their broader literary context. Thus, students are introduced, at least briefly, to other surviving pieces of early Christian literature through the early second century (e.g., the Apostolic Fathers and some of the Gnostic texts from Nag Hammadi).

Second, I have taken a rigorously comparative approach to all of these texts. The discussions focus not only on who wrote the various books of the New Testament and on what they have to say but also on how these authors relate to one another. For example, do Mark, John, and Thomas understand the significance of Jesus in the same way? Do Matthew, Paul, and Barnabas see eye to eye on the Jewish Law? Do the authors of 1 Corinthians, 1 Timothy, and Revelation share the same views of the end times? Do Jesus, Paul, and Luke all represent fundamentally the same religion?

Third, unlike most authors of introductions to the New Testament, I not only mention a variety of methods for the study of ancient literature, I actually model them. Students typically have difficulty understanding how genre analysis and redaction criticism, to pick just two examples, actually work. I introduce and apply these and several other methods, explaining what I am doing along the way, to show not only what we can know about these ancient Christian writings but also *how* we can know what we know.

Finally, rather than simply state what scholars have said about various critical issues involved in the study of early Christian literature and history (an approach that never makes for the most scintillating reading), I have tried to engage the reader by showing *why* scholars say what they say. In other words, I provide the evidence and mount the arguments that strike scholars as compelling and allow readers to decide for themselves whether or not they agree.

Teaching should engage students and reading should stimulate them. Yet most textbooks, in most fields, are so dreadfully boring. I hope that this, in particular, will not be among the faults of the present book.

Notes on Suggestions for Further Reading

The bibliographical suggestions at the end of each chapter are meant to guide beginning students who are interested in pursuing one or more of the issues raised in this book. To avoid overwhelming the student with the enormous quantity of literature in the field, for most chapters I have limited myself to seven or eight entries (more for longer chapters, fewer for shorter ones). All of the entries are books, rather than articles, and each is briefly annotated. Some of the entries are more suitable for advanced students, and these are indicated as such. For most chapters I have included at least one work that introduces or embraces a markedly different perspective from the one that I present. I have not included any biblical commentaries in the lists, although students should be urged to consult these, either one-volume works such as the *Harper's Bible Commentary* and the *Jerome Biblical Commentary* or commentaries on individual books, as found in the Anchor Bible, Hermeneia, Interpretation, and the New International Commentary series.

For some of the issues that I discuss, there are no adequate full-length treatments for beginning-level students to turn to, but there are excellent discussions of virtually everything having to do with the New Testament in Bible dictionaries that are readily available in most college libraries. Students should browse through the articles in such one-volume works as the *Harper's Bible Dictionary* and the *Mercer Dictionary of the Bible*. In particular, they should become intimately familiar with the impressive six-volume *Anchor Bible Dictionary*, which is destined to be a major resource for students at all levels for years to come. (Just with respect to Chapter 1 of this text, for example, the *Anchor Bible Dictionary* presents full-length treatments, with bibliographies, of early Christianity, Christology, the Ebionites, Marcion, Gnosticism, Nag Hammadi, heresy and orthodoxy, and the New Testament canon.)

ACKNOWLEDGMENTS

I have incurred piles of moral debt while writing this book and here would like to acknowledge my chief creditors. First and foremost are my bright and interesting undergraduate students in New Testament and Early Christianity at Rutgers University and the University of North Carolina at Chapel Hill. They have kept my teaching a challenging and lively experience. I am particularly grateful, in ways they may not know, to my gifted and energetic graduate students at UNC and Duke, especially those who assisted me directly in my research on this project: Judy Ellis, Mark Given, and Kim Haines-Eitzen, three scholars from whom you will be hearing more.

I am indebted to my erudite friends and colleagues in the field, who have taught me a great deal and have always been eager to teach me a great deal more. Along the way I have talked with a slew of scholars about this project, and here must beg their forbearance in not divulging all of their names, lest I inadvertently leave off one or two. I would, however, like to make special mention of three of my closest friends and dialogue partners: Beth Johnson of New Brunswick Theological Seminary, Joel Marcus of the University of Glasgow, and Dale Martin of cross-town rival Duke. These learned and insightful New Testament scholars have read every word of my manuscript and insisted that I change most of them. Two other New Testament scholars have selflessly provided advice, assistance, and support: Paul W. Meyer, my former teacher at Princeton Seminary and now colleague and resident mentor at UNC, and Jeff Siker, my good friend and erstwhile racquetball victim, backgammon foe, and confidant.

I would also like to acknowledge my wife, Cindy, who suffered through a careful reading of a preliminary draft of the manuscript and who went above and beyond the call of conjugal duty in making a number of helpful suggestions and useful comments.

I am grateful to the University of North Carolina at Chapel Hill for a semester research leave that allowed me to work full-time on the project, and to my colleagues in the Department of Religious Studies who have always been supportive in the extreme.

I am indebted to my two Oxford editors: Cynthia Read, who suggested the project in the first place and cajoled me into taking it on, and Robert Miller, who assumed editorial duties midstream and with uncommon skill made the passage home extraordinarily smooth.

I have dedicated the book to my teacher, David R. Adams, a great New Testament scholar who infected me, and all of his graduate students, with a passion for teaching and who, above all else, taught us how to think.

In addition to these personal notes, I would like to acknowledge my gratitude to previous scholars whose labors make such introductory textbooks possible.

Most of the quotations of the Bible, including the Apocrypha, are drawn from the New Revised Standard Version. Some, however, represent my own translations. The quotation from Plutarch in Chapter 4 is from Louis Ropes Loomis, *Plutarch: Selected Lives and Essays* (Roslyn: N.Y.: Walter J. Black, 1951). Quotations from the Gospel of Peter and the Gospel of Thomas in Chapter 12 are taken from David R. Cartlidge and David L. Dungan, *Documents for the Study of the Gospels* (Philadelphia: Fortress, 1980). Quotations from Tacitus in Chapter 13 are from Henry Bettenson, ed., *Documents of the Christian Church*, 2d ed. (New York: Oxford University Press, 1963). The reconstruction of the Testimonium Flavium in Chapter 13 comes from John Meier, *A Marginal Jew: Rethinking the Historical Jesus*, vol. 1, Anchor Bible Reference Library (New York: Doubleday, 1991), p. 61. Quotations of the Dead Sea Scrolls in Chapter 15 are drawn from Geza Vermes, *The Dead Sea Scrolls in English*, 2d ed. (New York: Penguin, 1975).

The correspondence between Paul and Seneca in Chapter 17 is taken from Edgar Hennecke, *New Testament Apocrypha*, ed. Wilhelm Schneemelcher, trans. R. McL. Wilson, vol. 2 (Philadelphia: Westminster Press, 1965). The material from Fronto in Chapter 18 comes from *The Octavius of Marcus Minucius Felix*, ed. and trans. G. W. Clark (Mahway, N.J.: Newman, 1974); the inscription from the Latvian burial society, also in Chapter 18, comes from N. Lewis and M. Rheinhold, Roman Civilization, vol. 2 (New York: Columbia University Press, 1955). The quotations from Melito of Sardis in Chapter 24 are from Gerald Hawthorne, "A New English Translation of Melito's Paschal Homily," in *Current Issues in Biblical and Patristic Interpretation*, ed. G. Hawthorn (Grand Rapids, Mich.: Eerdmans, 1975). Quotations from Tertullian in Chapter 25 are taken from Alexander Roberts and James Donaldson, *The Ante-Nicene Fathers*, revised by A. Cleveland Coxe, vol. 3 (reprint, Edinburgh: T & T Clark; Grand Rapids, Mich.: Eerdmans, 1989). In Chapters 26-27 some of the translations of Polycarp, Ignatius, and the *Didache* are from Cyril C. Richardson, ed., *Early Christian Fathers* (New York: Macmillan, 1978); others are my own.

CREDITS

Fig. 1.1: British Library. Fig. 2.1: Numismatic Museum, Athens/Hellenic Republic Ministry of Culture. Fig. 2.4: Forum, Pompeii/Erich Lessing/Art Resource, NY. Fig. 2.5: Louvre/Alinari/Art Resource, NY. Fig. 2.8: Ritmeyer Archaeological Design, England. Fig. 3.3: Archaeological Museum, Piraeus/Foto Marburg/Art Resource, NY. Fig. 4.1: Cathedral Treasury, Aachen/Foto Marburg/Art Resource, NY. Fig. 5.1: Alinari/Art Resource, NY. Fig. 5.3: British Museum. Fig. 7.1: Museo Lateranense, Vatican Museums/Alinari/Art Resource, NY. Fig. 8.1: Staatsbibliothek, Munich/Foto Marburg/Art Resource, NY. Fig. 8.2: Bart Ehrman. Fig. 8.3: Hirmer Verlag München. Fig. 9.2: André Held. Fig. 10.1 Robert Miller. Fig. 10.2 André Held. Fig. 10.4a: Museo Lateranense, Vatican Museums/Alinari/Art Resource, NY. Fig. 10.4b: Robert Miller. Figs. 11.2 and 12.1: Institute for Antiquity and Christianity, Claremont, CA. Figs. 13.1 and 13.2: André Held. Fig. 14.1: Museo Nazionale Romano delle Terme, Rome/Alinari/Art Resource, NY. Fig. 14.2: Corinth excavations, American School of Classical Studies, Athens; photos by I. Ioannidou and L. Bartziotou. Fig. 15.1: British Museum. Fig. 15.3: Israel Museum, Jerusalem. Fig. 15.4: Robert Miller. Fig. 15.6: André Held. Fig. 15.7: Grotte, St. Peter's Basilica, Vatican City/Alinari/Art Resource, NY. Fig. 16.1: Vatican Museums/Alinari/Art Resource, NY. Fig. 17.1: University of Michigan. Fig. 18.2: Mostra Augustea, Rome/Alinari/Art Resource, NY. Fig. 18.3: Alinari/Art Resource, NY. Fig. 18.4: C. M. Dixon. Fig. 18.5: Museo Archeologico Nazionale, Naples/Alinari/Art Resource, NY. Fig. 19.1: André Held. Fig. 19.2: Alinari/Art Resource, NY. Fig. 19.3: André Held. Fig. 19.4: Israel Museum, Jerusalem. Fig. 20.1: Mostra Augustea, Rome/Alinari/Art Resource, NY. Fig. 22.1: British Library. Fig. 22.2: André Held. Fig. 23.1: Robert Miller. Fig. 23.2: Catacomb of Priscilla, Rome/Scala/Art Resource, NY. Fig. 24.1: British Museum. Fig. 25.2: Roger Wood. Fig. 26.1: British Library. Fig. 27.2: Hirmer Verlag München. Fig. 27.3: André Held. Fig. 28.1: Photo courtesy of Bruce Metzger: Manuscripts of the Greek Bible, Oxford University Press, 1981. Fig. 28.2: Rylands Library, University of Manchester.

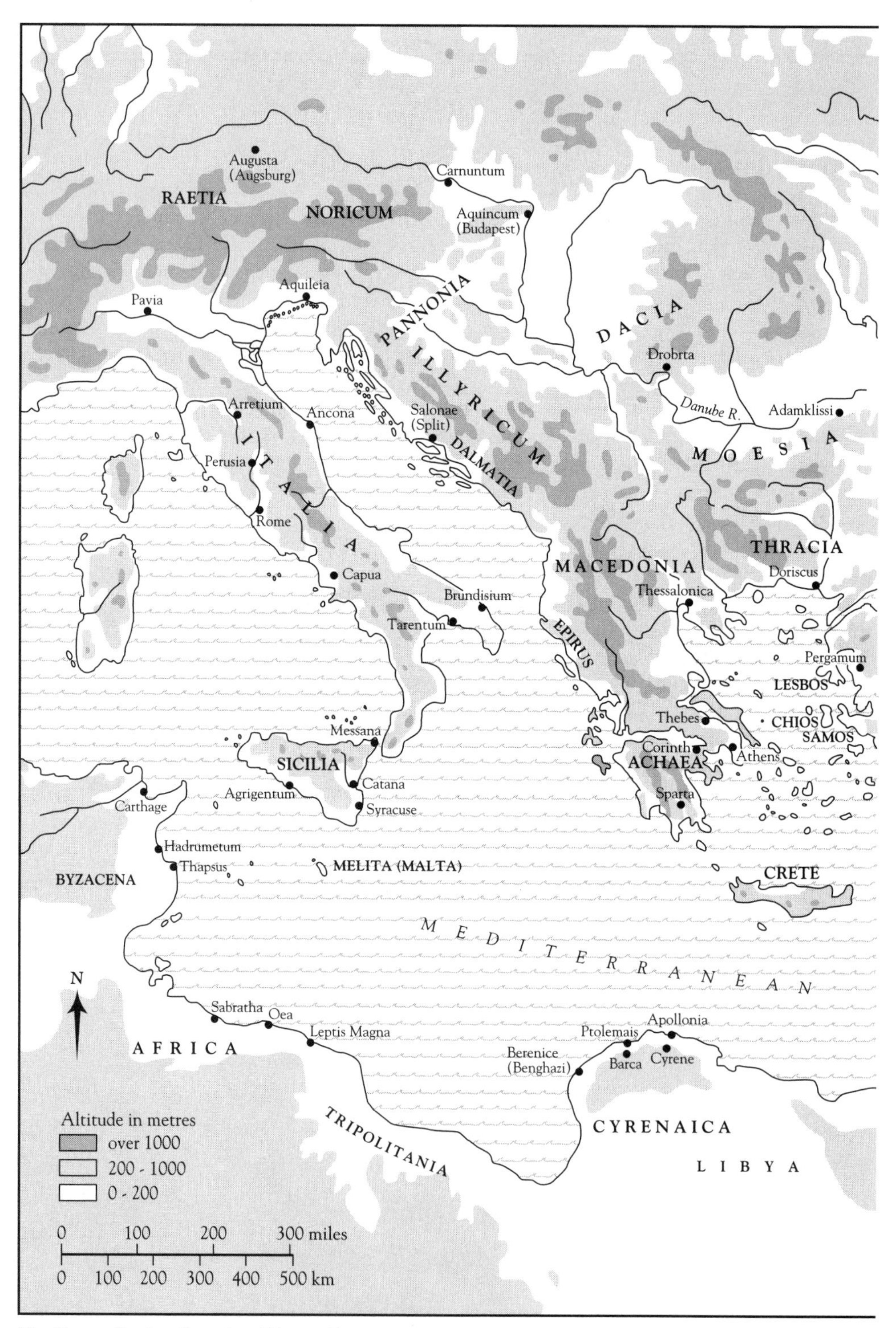

The Roman Empire: Central and Eastern Provinces.

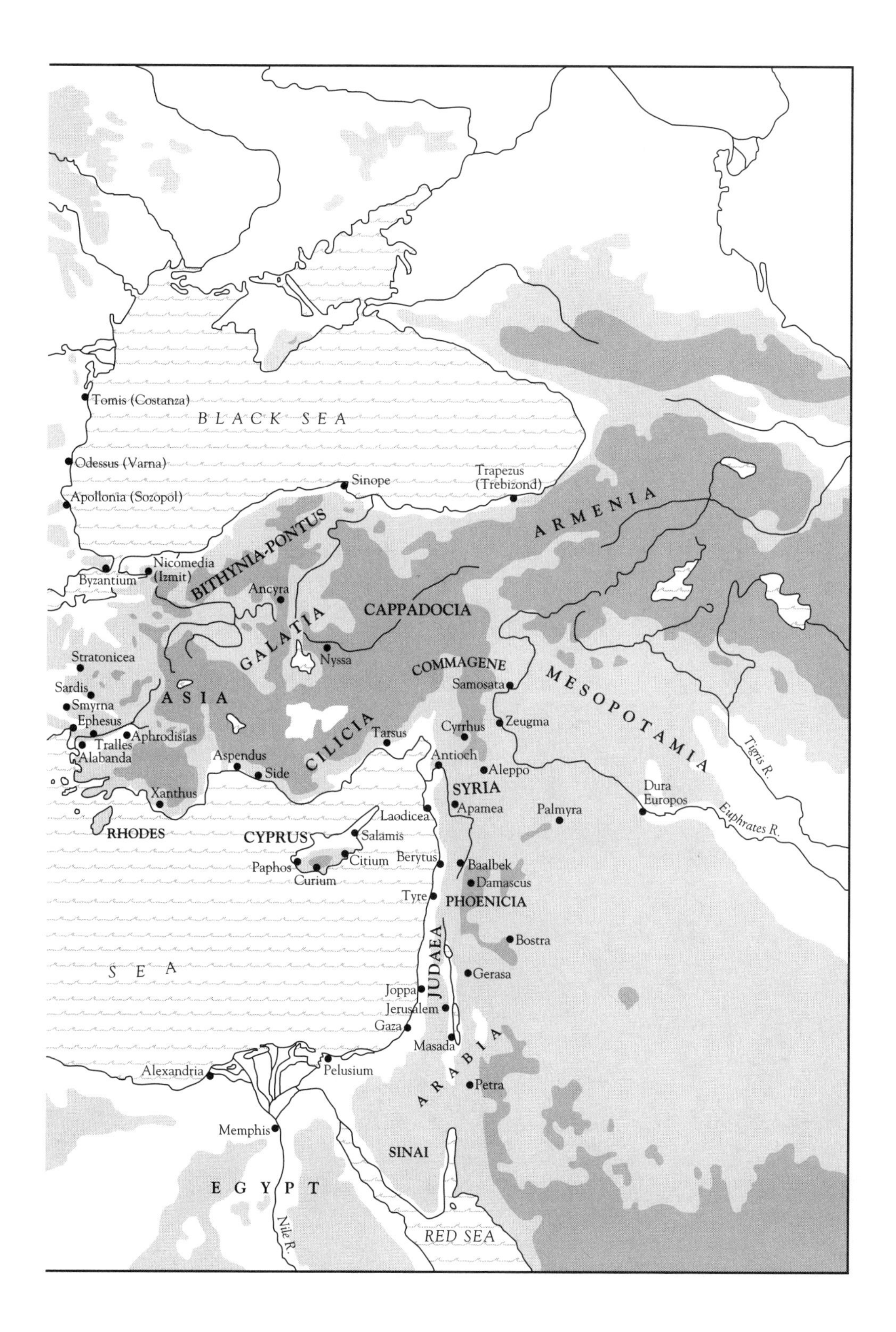

Tomis (Costanza)
BLACK SEA
Odessus (Varna)
Sinope
Trapezus
(Trebizond)
Apollonia (Sozopol)
ARMENIA
BITHYNIA-PONTUS
Nicomedia
(Izmit)
Byzantium
Ancyra
CAPPADOCIA
GALATIA
Nyssa
Stratonicea
COMMAGENE
MESOPOTAMIA
Samosata
Sardis
ASIA
Smyrna
Ephesus
Zeugma
Cyrrhus
Tarsus
CILICIA
Tralles
Aphrodisias
Alabanda
Aspendus
Side
Antioch
Aleppo
Tigris R.
SYRIA
Xanthus
Apamea
Dura
Europos
Palmyra
Laodicea
Euphrates R.
RHODES
CYPRUS
Salamis
Citium
Berytus
Baalbek
Paphos
Curium
Damascus
Tyre
PHOENICIA
Bostra
SEA
JUDAEA
Gerasa
Joppa
Jerusalem
Gaza
Masada
ARABIA
Alexandria
Pelusium
Petra
Memphis
SINAI
EGYPT
Nile R.
RED SEA

CHAPTER 1

What Is the New Testament? The Early Christians and Their Literature

Christianity in the modern world is a richly diverse phenomenon. Ask any Pentecostal preacher who has attended a Roman Catholic mass, or a Greek Orthodox monk who has happened upon a Baptist tent revival, or an Episcopalian nun who has visited a Jehovah's Witness prayer meeting. There is, to be sure, common ground among many Christian groups, but when you compare the beliefs and practices of an Appalachian snake handler with those of a New England Presbyterian, you may be more struck by the differences than the similarities.

Is this kind of rich diversity a modern development? Many people appear to think so. For them, Christianity was originally a solid unity, but with the passing of time (especially since the Protestant Reformation) this unity became fractured and fragmented. Historians, however, recognize that in some ways Christian differences today pale in comparison with those that existed among believers in the distant past. If we turn the clock back 1,800 years to the end of the second century, we find people calling themselves Christian who subscribe to beliefs that no modern eye has seen or ear heard, Christians who believe that there are 2 different gods, or 32, or 365, Christians who claim that the Old Testament is an evil book inspired by an evil deity, Christians who say that God did not create the world and has never had any involvement with it, Christians who maintain that Jesus did not have a human body, or that he did not have a human soul, or that he was never born, or that he never died.

Of course, many people today would argue that such views could not be Christian. What is striking to the historian, though, is that people who claimed to be Christian believed these things. Moreover, these believers invariably maintained that their ideas were taught by Jesus himself. In many instances, they could appeal to written proof, that is, documents allegedly penned by Jesus' own apostles.

The New Testament also contains books that were thought to have been written by Jesus' own apostles. These books, however, do not teach that there are several gods, or that the creator of the world is evil, or that Jesus did not have a real body. Are there historical grounds for thinking that the New Testament books actually were written by Jesus' apostles and that books supporting contrary views were forgeries? Indeed, how is it that some books claiming to be written by the apostles were included in the New Testament, but others were not? Moreover, even if the books that made it into the New Testament agree on certain fundamental points (for example, that there is only one God), is it possible that they might disagree on others (such as who Jesus is)? That is to say, if Christians in the second century, a hundred fifty years or so after Jesus, held such a wide range of beliefs, is it not possible that Christians of the first century (when the books of the New Testament were being written) did as well? Did all of the early Christians agree on the fundamental points of their religion?

SOME MORE INFORMATION

Box 1.1 The Hebrew Bible and the Christian Old Testament

The terms "Jewish Scriptures" and "Hebrew Bible" both refer to the collection of books considered sacred in the religion of Judaism, books that were written almost entirely in Hebrew. Many of these writings were regarded as holy even before Jesus' day, especially the first five books of Moses, known as the Torah or Law.

About a century after Jesus, the collection of books into the Hebrew Scriptures was more or less fixed. Altogether, the collection comprised twenty-four different books. Because of a different way of counting them, they number thirty-nine books in English translation (the twelve minor prophets in English Bibles, for example, count as only one book in the Hebrew Bible).

Christians have long referred to these books as the "Old Testament," to set them apart from the books of the "New Testament" (the new set of books that reveal God's will to his people). Throughout our study, I will use the term "Old Testament" only when referring explicitly to Christian views; otherwise, I will call these books the Jewish Scriptures or Hebrew Bible.

The Hebrew Bible (5 books)

The Torah (5 books)
- Genesis
- Exodus
- Leviticus
- Numbers
- Deuteronomy

The Prophets (8 books)
- Former Prophets
  - Joshua
  - Judges
  - Samuel (counts as 1 book)
  - Kings (counts as 1 book)
- Later Prophets
  - Isaiah
  - Jeremiah
  - Ezekiel
  - The Twelve (count as 1 book)
    - Hosea
    - Joel
    - Amos
    - Obadiah
    - Jonah
    - Micah
    - Nahum
    - Habakkuk
    - Zephaniah
    - Haggai
    - Zechariah
    - Malachi

The Christian "Old Testament"

The Pentateuch (5 books)
- Genesis
- Exodus
- Leviticus
- Numbers
- Deuteronomy

Historical Books (12 books)
- Joshua
- Judges
- Ruth
- 1 and 2 Samuel
- 1 and 2 Kings
- 1 and 2 Chronicles
- Ezra
- Nehemiah
- Esther

Poetry and Wisdom Books (5 books)
- Job
- Psalms
- Proverbs
- Ecclesiastes
- Song of Solomon

Prophetic Books (17 books)
- Major Prophets
  - Isaiah
  - Jeremiah
  - Lamentations
  - Ezekiel
  - Daniel

The Writings (11 books)	Minor Prophets
Job	Hosea
Psalms	Joel
Proverbs	Amos
Ruth	Obadiah
Song of Solomon	Jonah
Ecclesiastes	Micah
Lamentations	Nahum
Esther	Habakkuk
Daniel	Zephaniah
Ezra-Nehemiah (1 book)	Haggai
Chronicles (1 book)	Zechariah
	Malachi

These are some of the issues that we will consider as we begin to examine the earliest Christian writings. They are not, of course, the only issues. There is an extraordinarily broad range of important and intriguing questions that readers bring to the New Testament—about where it came from, who its authors were, what their messages were—and many of these will occupy us at considerable length in the pages that follow. But the issue of Christian diversity is a good place for us to begin our investigation. It cannot only provide a useful entrée into important questions about the early stages of the Christian religion, starting with the teachings of Jesus, but it can also enlighten us about the nature of the New Testament itself, specifically about how and why these various books came to be gathered together into one volume and accepted by Christians as their sacred scripture (see box 1.1).

THE DIVERSITY OF EARLY CHRISTIANITY

As I have intimated, Christian diversity is somewhat easier to document in the second century, after the books of the New Testament were written, than in the first. This is because, quite simply, there are more documents that date to this period. The only Christian writings that can be reliably dated to the first century are found in the New Testament itself, although we know that other Christian books were produced at this time. We begin our investigation, then, by examining several examples of later forms of Christianity, before seeing how these are relevant to the study of the New Testament itself.

Jewish-Christian Adoptionists

Consider first the form of religion embraced by a group of second-century Jewish Christians known to be living in Palestine, east of the Jordan River. These believers maintained that Jesus was a remarkable man, more righteous in the Jewish Law than any other, a man chosen by God to be his son. Jesus received his adoption to sonship at his baptism; when he emerged from the waters of the Jordan, he saw the heavens open up and the Spirit of God descend upon him as a dove, while a voice from heaven proclaimed, "You are my son, today I have begotten you."

According to these Christians, Jesus was empowered by God's Spirit to do remarkable miracles and to teach the truth of God. Then, at the end of his life, he fulfilled his divine commission by dying as a willing sacrifice on the cross for the sins of the world, a sacrifice that put an end to all sacrifices. Afterwards God raised him from the dead. Jesus then ascended into heaven, where he presently reigns.

There may seem to be little that is remarkable about these beliefs—until, that is, one probes a bit further into the details. For even though Jesus was chosen by God, according to these Christians, he was not himself divine. He was a righteous man but nothing more than a man. In their view, Jesus was not born of a virgin, he did not exist prior to his birth, and he was not God. He was adopted by God to be his son, the savior of the world. Hence the name bestowed upon this group by others: they were "adoptionists." For them, to call Jesus God was a blasphemous lie. For if Jesus were God, and his Father were also God, there would be two Gods. But the Jewish Scriptures emphatically state otherwise: "Hear O Israel, the Lord our God, the Lord is one" (Deut 6:4).

According to these Christians, this one God chose Israel and gave it his Law (in the Jewish Scriptures). Furthermore, Jesus taught that his followers must continue to obey the entire Law (except the law that required animal sacrifice) in all its details—and not just the Ten Commandments! Those who were not born Jews must first become Jews in order to follow Jesus. For men, this meant being circumcised; for men and women, it meant observing the sabbath and keeping kosher food laws.

On what grounds did these Christians advance this understanding of the faith? They had a sacred book written in Hebrew which they claimed contained the teachings of Jesus himself, a book that was similar to what we today know as the Gospel of Matthew (without the first two chapters). What about the other books of the New Testament, the other Gospels and Acts, the epistles, and Revelation? Odd as it might seem, these Jewish Christians had never heard of some of these books, and rejected others of them outright. In particular, they considered Paul, one of the most prominent authors of our New Testament, to be an arch-heretic rather than an apostle. Since, in their opinion, Paul blasphemously taught that Christ brought an end to the Jewish Law, his writings were to be rejected as heretical. In short, these second-century Christians did not have our New Testament canon (see box 1.2).

Marcionite Christians

The Jewish-Christian adoptionists were by no means unique in not having our New Testament. Consider another Christian group, this one scattered throughout much of the Mediterranean in the mid to late second century, with large numbers of congregations flourishing especially in Asia Minor (modern-day Turkey). Their opponents called them "Marcionites" because they subscribed to the form of Christianity advanced by the second-century scholar and evangelist Marcion, who himself claimed to have uncovered the true teachings of Christianity in the writings of Paul. In sharp contrast to the Jewish Christians east of the Jordan, Marcion maintained that Paul was the true apostle, to whom Christ had especially appeared after his resurrection to impart the truth of the gospel. Paul, according to Marcion, had begun as a

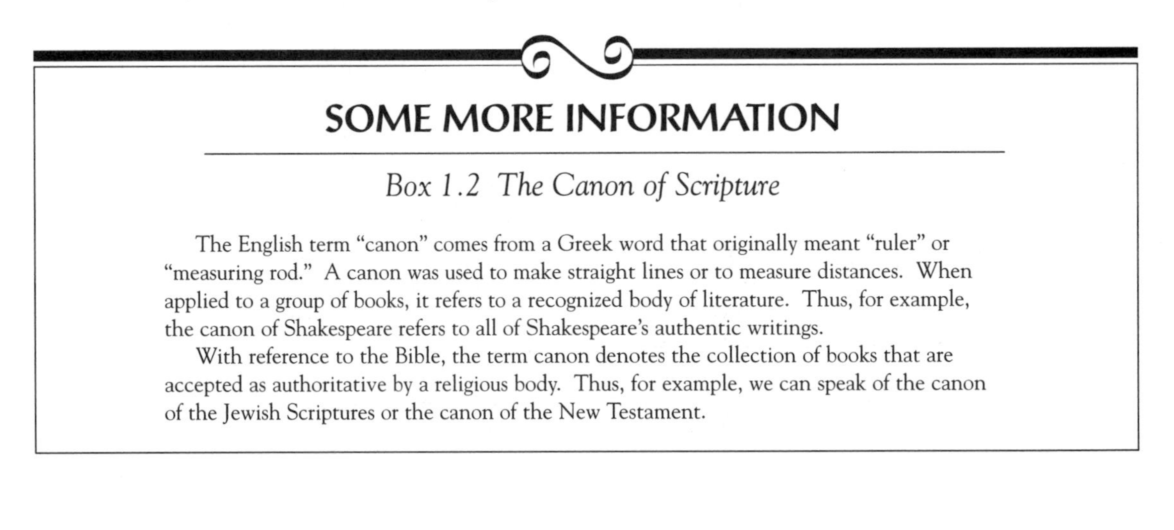

SOME MORE INFORMATION

Box 1.2 The Canon of Scripture

The English term "canon" comes from a Greek word that originally meant "ruler" or "measuring rod." A canon was used to make straight lines or to measure distances. When applied to a group of books, it refers to a recognized body of literature. Thus, for example, the canon of Shakespeare refers to all of Shakespeare's authentic writings.

With reference to the Bible, the term canon denotes the collection of books that are accepted as authoritative by a religious body. Thus, for example, we can speak of the canon of the Jewish Scriptures or the canon of the New Testament.

good Jew intent on obeying the Law to the utmost, but the revelation of Christ showed him beyond doubt that the Jewish Law played no part in the divine plan of redemption. For him, Christ himself was the only way of salvation. Marcion argued that Paul's writings effectively set the gospel of Christ over and against the Law of the Jews, and that the apostle had urged Christians to abandon the Jewish Law altogether.

For Marcion and his followers, the differences between the religion preached by Jesus (and his apostle, Paul) and that found in the Jewish Scriptures were plain to see. Whereas the Jewish God punishes those who disobey, they claimed, the God of Jesus extends mercy and forgiveness; whereas the God of the Jews says "an eye for an eye and a tooth for a tooth," the God of Jesus says to "turn the other cheek"; and whereas the Old Testament God tells the Israelites to conquer Jericho by slaughtering its entire population—men, women, and children—the God of Jesus says to love your enemies. What do these two Gods have in common? According to the Marcionites, nothing. For them, there are two separate and unrelated Gods, the God of the Jews and the God of Jesus.

Marcionite Christians maintained that Jesus did not belong to the wrathful and just God of the Jews, the God who created the world and chose Israel to be his special people. In fact, Jesus came to save people from this God. Moreover, since Jesus had no part in the Creator, he could have no real ties to the material world that the Creator-God made. Jesus therefore was not actually born and did not have a real flesh-and-blood body. How, then, did Jesus get hungry and thirsty, how did he bleed and die? According to Marcionites, it was all an appearance: Jesus only seemed to be human. As the one true God himself, come to earth to deliver people from the vengeful God of the Jews, Jesus was never born, never got hungry or thirsty or tired, never bled or died. Jesus' body was only a phantasm.

The contrasts between the Jewish Christians and the Marcionites are stark. One group said that Jesus was totally human and not divine, the other said that he was totally divine and not human. One group staunchly maintained that there was only one God, the other asserted that there were in fact two. One said that the true God created the world, called Israel to be his people, and gave them the Law, the other said that the true God had never had any dealings with the world or with Israel. One group urged that believers must follow the Law, the other argued that they should reject it altogether. Both groups considered themselves to be the true Christians.

Most significantly for our purposes here, these groups did not appeal to the same authorities for their views. On the contrary, whereas the Jewish Christians rejected Paul as a heretic, the Marcionites followed him as the greatest of the apostles. Moreover, instead of adhering to a version of Matthew's Gospel, the Marcionites used a truncated version of something like our Gospel of Luke, along with ten of Paul's letters (all of those found in the New Testament, with the exceptions of 1 and 2 Timothy and Titus). Even these were not exactly the letters as we have them today, however. Marcion believed that earlier heretics had willfully modified these books by inserting positive references to the God of the Jews, and his creation, and his Scriptures; accordingly, he excised these passages, giving his followers a form of the New Testament strikingly different from that which we now read: eleven books, all of them shortened, and no Old Testament.

Gnostic Christians

The Jewish-Christian adoptionists and the Marcionites were not the only two Christian groups vying for converts in the second century. In fact, there were many other groups supporting a wide range of other beliefs on the basis of a wide range of other authorities as well. Some of the best known in modern times are the various sects of Christian Gnostics, so named because of their claim that special "gnosis" (Greek for "knowledge") is necessary for salvation.

We know that Gnostic Christians were located in major urban areas throughout much of the Mediterranean during the second and third centuries, especially in Egypt, Syria, Asia Minor, Rome, and Gaul. Gnostics were themselves wildly diverse, with different groups believing radically different things (see chapter 11). Some Gnostics agreed with Marcion that Jesus was totally divine and not at all human, and for much the same reason that he did: Jesus represented a different God from the one who created this world. Others, however, claimed that Jesus Christ represented two distinct beings, the human Jesus and the divine Christ. These Gnostics agreed with the Jewish-Christian adoptionists that Jesus was the most righteous man on earth and that something special had happened at his baptism. They did not think, though, that God adopted him to be his son; instead, they maintained that his baptism was the moment at which the divine being, the Christ, came into the man Jesus, empowering him for his healing and, especially, teaching ministry. At the end of Jesus' life, immediately before his death, the Christ then departed from him once again to return to heaven. This is why Jesus cried out in such anguish on the cross, "My God, my God, why have you left me behind?" (cf. Mark 15:34).

Who, though, was this divine Christ? For many Gnostics, he was one of the deities that made up the divine realm. Unlike the Jewish Christians who were strict monotheists (believing in only one God) or the Marcionites who were strict ditheists (believing in two), Gnostics were polytheists (believing in many). In some of the Gnostic systems that we know about there were 32 different gods; in others as many as 365. Moreover, for all of these systems, the true God was not the God of the Old Testament. Unlike Marcion, however, Gnostics did not believe that the Old Testament God was simply vengeful and righteous, a God who had high standards (the Law) and little patience with those who did not meet them. For many of them, the creator God of the Old Testament was inherently evil, as was this material world that he created.

Gnostics felt a sense of alienation from this world and knew that they did not belong here. They were spiritual beings from the divine realm who had become entrapped in the realm of matter by the evil God and his subordinates. Salvation meant escaping from this material world. Thus a god from the divine realm entered into the man Jesus, and left him prior to his death, so that he could impart to the imprisoned spirits the knowledge (gnosis!) that is necessary for escape.

This was secret knowledge not divulged to the masses, not even to the mass of Christians. It was meant only for the chosen, the elect, the Gnostics themselves. They did not deny that Jesus taught the crowds publicly, but they believed he reserved the secret teachings that led to salvation only for the elect who were able to act upon them. The Gnostics passed on this teaching by word of mouth and claimed that it could be discovered through a careful reading of the writings of the apostles. It lay there hidden beneath the surface. Thus, for the Gnostic, the literal meaning of these texts was not what mattered; the truth necessary for salvation could be found only in the secret meaning, a meaning exclusively available to Gnostic interpreters, those "in the know."

Since Gnostic Christians were not tied to the literal meaning of their texts, they were not as compulsive as other Christians about collecting a group of books and ascribing special authority to them (in contrast, for example, to the Marcionites). Various Gnostics nonetheless did have their own favorites. We know that many of them were especially drawn to the Gospel of John and that others cherished Gospels that most modern people have never heard of: the *Gospel of Thomas*, the *Gospel of Mary*, the *Gospel of Philip*, and the *Gospel of Truth*. Some of these books have only recently been discovered by archaeologists. Each of them was thought to convey the true teachings of Jesus and his apostles.

How is it that most of these books cannot be found in our own New Testament? Or for that matter, how is it that the versions of Matthew, Luke, and Paul read by Jewish-Christian adoptionists and Marcionites were not included? Why do the views of these other groups not have equal representation in the Christian Scriptures? The answer can be found by examining the story of one other group of second-century Christians.

"Proto-Orthodox" Christians

The "proto-orthodox" Christians represent the forerunners (hence the prefix "proto") of the group that became the dominant form of Christianity in later centuries. When this group later acquired more converts than any of the others and stifled its opposition, it claimed that its views had always been the majority position and that its rivals were, and always had been, "heretics," who willfully "chose" (the Greek root of the word "heresy") to reject the "true belief" (the literal meaning of "orthodoxy").

We ourselves can use the term "proto-orthodox" only in retrospect, since the adherents of this position did not actually know that their views would become dominant, nor did they think of themselves as forerunners of believers to come later; like all the other groups of their day, they simply saw themselves as the true Christians. The story of their victory over their opponents is fascinating, but aspects of it are hotly debated among modern-day scholars. Some think that the proto-orthodox beliefs were original to Christianity, others that they developed over time. Some scholars claim that the proto-orthodox had always been in the majority throughout Christendom, others that other forms of Christianity were predominant in many parts of the Mediterranean (e.g., Jewish Christians in parts of Palestine, Gnostics in parts of Egypt and Syria, Marcionites in Asia Minor). Fortunately, we do not need to resolve these thorny problems here.

But there are aspects of the proto-orthodox struggle for dominance that are directly germane to our study of the New Testament. To begin with, we can consider what these Christians believed in contrast to the other groups we have discussed.

Proto-orthodox Christians agreed with the Jewish Christians when they said that Jesus was fully human, but disagreed when these people denied that he was divine. They agreed with the Marcionites when they said that Jesus was fully divine, but disagreed when they denied that he was human. They agreed with the Gnostics when they said that Jesus Christ taught the way of salvation, but disagreed when they said that he was two beings rather than one and when they claimed that his true teachings had been secret, accessible only to the elect few. In short, proto-orthodox Christians argued that Jesus Christ was both divine and human, that he was one being instead of two, and that he had taught his disciples the truth. They claimed that the apostles had written the teachings of Jesus down and that, when interpreted in a straightforward and literal fashion, the books that were passed on from the apostles to their followers revealed the truth necessary for salvation. These views may sound familiar to readers who have had any involvement with Christianity, for the side that held these views won the debate and determined the shape of Christianity up to the present day.

The proto-orthodox position, then, attempted to counteract the claims of the groups that they opposed. In part, this meant that the proto-orthodox group had to reject some documents that claimed to be written by apostles but that advanced beliefs contrary to their own, for example, the *Gospel of Peter*, the *Gospel of Philip*, or the *Gospel of Thomas*, all of which appeared to support Gnostic perspectives. Some of the writings used by the opposing groups, though, were quite popular among the proto-orthodox Christians as well. For example, the Gospel of Matthew was well-loved by Jewish Christians, and the Gospel of John was a favorite of many Gnostics. Indeed, by accepting and ascribing authority to both of these Gospels, the proto-orthodox believers were able to balance the "heretical" claims that could be made when only one of them was taken to be the ultimate authority.

In other words, if Jesus appears to be fully human in one Gospel and fully divine in another, by accepting both authorities as Scripture the proto-orthodox were able to claim that both perspectives were right, and that an exclusive emphasis on Jesus as only human, or purely divine, was a perversion of the truth. The development of the canon of Scripture within proto-orthodox circles is in large part an attempt to define what true Christians should believe by eliminating or compromising the views of other groups.

Because the proto-orthodox group represented the party that eventually became dominant in Christianity (by at least the fourth century), Christians of all later generations inherited their canon of Scripture, rather than the canons supported by their opponents.

THE NEW TESTAMENT CANON OF SCRIPTURE

The purpose of this sketch is not to give a complete account of Christianity in the second century but instead to indicate how early Christianity was extremely diverse and to show how this diver-

SOME MORE INFORMATION

Box 1.3 The Common Era and Before the Common Era

Most students will be accustomed to dating ancient events as either A.D. ("anno domini," Latin for "year of our Lord") or B.C. ("before Christ"). This terminology may make sense for Christians, for whom A.D. 1996 is indeed "the year of our Lord 1996." It makes less sense, though, for Jews, Muslims, and others for whom Jesus is not the "Lord" or the "Christ." Scholars have therefore begun to use a different set of abbreviations as more inclusive of others outside the Christian tradition. In this book I will follow the alternative designations of C.E. ("the Common Era," meaning common to people of all faiths who utilize the traditional Western calendar) and B.C.E. ("before the Common Era"). In terms of the older abbreviations, then, C.E. corresponds to A.D., and B.C.E. to B.C.

sity led to the collection of books into a sacred canon. The Christian Scriptures did not drop from the sky one day in April the year Jesus died. They were written by individual authors at different points of time, in different countries, to different communities, with different concerns; they were later read by an even wider range of Christians and were eventually collected together into what we now call the New Testament. Before launching into a study of these various books, we should reflect further on how and when they (and not others) came to be placed in the canon. We can begin with some preliminary observations concerning the shape of the canon as we now have it.

The New Testament: Some Basic Information

The New Testament contains twenty-seven books, written in Greek, by fifteen or sixteen different authors, who were addressing other Christian individuals or communities between the years 50 and 120 C.E. (see box 1.4). As we will see, it is difficult to know whether any of these books was written by Jesus' own disciples.

The first four books are "Gospels," a term that literally means "good news." The four Gospels of the New Testament proclaim the good news by telling stories about the life and death of Jesus—his birth, ministry, miracles, teaching, last days, crucifixion, and resurrection. These books are traditionally ascribed to Matthew, Mark, Luke, and John. Proto-orthodox Christians of the second century claimed that two of these authors were disciples of Jesus: Matthew, the tax collector mentioned in the First Gospel (Matt 9:9), and John, the beloved disciple who appears in the Fourth (e.g., John 19:26). The other two were reportedly written by associates of famous apostles: Mark, the secretary of Peter, and Luke, the traveling companion of Paul. This second-century tradition does not go back to the Gospels themselves; the titles in our Bibles (e.g., "The Gospel according to Matthew") were not found in the original texts of these books. Instead, their authors chose to remain anonymous.

The next book in the New Testament is the Acts of the Apostles, written by the same author as the Third Gospel (whom modern scholars continue to call Luke even though we are not certain of his identity). This book is a sequel to the Gospel in that it describes the history of early Christianity beginning with events immediately after Jesus' death; it is chiefly concerned to show how the religion was disseminated throughout parts of the Roman Empire, among Gentiles as well as Jews, principally through the missionary

SOME MORE INFORMATION

Box 1.4 The Layout of the New Testament

Gospels: The Beginnings of Christianity (4 books)
- Matthew
- Mark
- Luke
- John

Acts: The Spread of Christianity (1 book)
- The Acts of the Apostles

Epistles: The Beliefs, Practices, and Ethics of Christianity (21 books)
- Pauline Epistles
  - Romans
  - 1 and 2 Corinthians
  - Galatians
  - Ephesians
  - Philippians
  - Colossians
  - 1 and 2 Thessalonians
  - 1 and 2 Timothy
  - Titus
  - Philemon
- General Epistles
  - Hebrews
  - James
  - 1 and 2 Peter
  - 1, 2, and 3 John
  - Jude

Apocalypse: The Culmination of Christianity (1 book)
- The Revelation of John

This schematic arrangement is somewhat simplified. All of the New Testament books, for example (not just the epistles), are concerned with Christian beliefs, practices, and ethics, and Paul's epistles are in some ways more reflective of Christian beginnings than the Gospels. Nonetheless, this basic orientation to the New Testament writings can at least get us started in our understanding of the early Christian literature.

labors of the apostle Paul. Thus, whereas the Gospels portray the *beginnings* of Christianity (through the life and death of Jesus), the book of Acts portrays the *spread* of Christianity (through the work of his apostles).

The next section of the New Testament comprises twenty-one "epistles," that is, letters written by Christian leaders to various communities and individuals. Not all of these epistles are, strictly speaking, items of personal correspondence. The book of Hebrews, for example, appears to be an early Christian sermon, and the epistle of 1 John is a kind of Christian tractate. Nonetheless, all twenty-one of these books are traditionally called epistles. Thirteen of them claim to be written by the apostle Paul; in some cases, scholars have come to question this claim. In any event, most of these letters, whether by Paul or others, address theological or practical problems that have arisen in the Christian communities they address. Thus, whereas the Gospels describe the beginnings of Christianity and the book of Acts its spread, the epistles are more directly focused on Christian beliefs, practices, and ethics.

Finally, the New Testament concludes with the Book of Revelation, the first surviving instance of a Christian apocalypse. This book was written by a prophet named John, who describes the course of future events leading up to the destruction of this world and the appearance of the world to come. As such, it is principally concerned with the culmination of Christianity.

Other Early Christian Writings

The books I have just described were not the only writings of the early Christians, nor were they originally collected into a body of literature called the "New Testament." We know of other Christian writings that have not survived from antiquity. For example, the apostle Paul in the first letter to the Corinthians, refers to an earlier writing that he had sent them (1 Cor 5:9) and alludes to a letter that they themselves had sent him (7:1). Unfortunately, this correspondence is lost.

Other noncanonical writings, however, have survived. The best known of these are by authors collectively called the "Apostolic Fathers." These were Christians living in the early second century, whose writings were considered authoritative in some proto-orthodox circles, on a par with the writings of the Gospels or Paul. In fact, some of our ancient manuscripts of the New Testament include writings of the Apostolic Fathers as if they belonged to the canon. Other, previously unknown, Christian writings have been discovered only within the present century. Some of these writings clearly stand at odds with those within the New Testament; some of them appear to have been used as sacred scripture by certain groups of Christians. A number of them claim to be written by apostles. The most spectacular find occurred in the mid-1940s near the town of Nag Hammadi, Egypt, where a bedouin digging for fertilizer accidentally uncovered a jar containing thirteen fragmentary books in leather bindings. The books contain anthologies of literature, some fifty-two treatises altogether, written in the ancient Egyptian dialect called Coptic. Whereas the books themselves were manufactured in the mid–fourth century C.E. (we know this because some of the bindings were strengthened with pieces of scratch paper that were dated), the treatises that they contain are much older: some of them are mentioned by name by authors living in the second century. Before this discovery, we knew that these books existed, but we didn't know what was in them.

What kind of books are they? I earlier indicated that Gnostic Christians appealed to written authorities that did not make it into the New Testament, some of them allegedly written by apostles. These are some of those books. Included in the collection are epistles, apocalypses, and collections of secret teachings. Yet more intriguing are the several Gospels that it contains, including one allegedly written by the apostle Philip and another attributed to Didymus Judas Thomas, thought by some early Christians to be Jesus' twin brother (see box 12.2).

These books were used by groups of Christian Gnostics during the struggles of the second, third, and fourth centuries, but they were rejected as heretical by proto-orthodox Christians. Why were they rejected? The question takes us back to the issues raised earlier concerning how Christians went about deciding which books to include in the New Testament and when their decisions went into effect.

The Development of the Christian Canon

Proto-orthodox Christians did not invent the idea of collecting authoritative writings together into a sacred canon of Scripture. In this they had a precedent. For even though most of the other religions in the Roman Empire did not use written documents as authorities for their religious beliefs and practices, Judaism did.

Jesus and his followers were themselves Jews who were conversant with the ancient writings that were eventually canonized into the Hebrew Scriptures. Although most scholars now think that a hard and fast canon of Jewish Scripture did not yet exist in Jesus' own day, it appears that most Jews did subscribe to the special authority of the Torah (see box 1.1). Also, many Jews accepted the authority of the Prophets as well. These writings include the books of Joshua through 2 Kings in our English Bibles, as well as the more familiar prophets Isaiah, Jeremiah, Ezekiel, and the twelve minor prophets. According to our earliest accounts, Jesus himself quoted from some of these books; we can assume that he accepted them as authoritative.

Thus Christianity had its beginning in the proclamation of a Jewish teacher, who ascribed authority to written documents. Moreover, we know that Jesus' followers considered his own teachings to be authoritative. Near the end of the first century, Christians were citing Jesus' words and calling them "Scripture" (e.g., 1 Tim 5:18). It is striking that in some early Christian circles the correct interpretation of Jesus' teachings was thought to be the key to eternal life (e.g., see John 6:68 and *Gosp. Thom.* 1). Furthermore, some of Jesus' followers, such as the apostle Paul, understood themselves to be authoritative spokespersons for the truth. Other Christians granted them this claim. The book of 2 Peter, for example, includes Paul's own letters among the "Scriptures" (2 Pet 3:16).

Thus by the beginning of the second century some Christians were ascribing authority to the words of Jesus and the writings of his apostles. There were nonetheless heated debates concerning which apostles were true to Jesus' own teachings (cf. Marcion and the Jewish Christians on Paul), and a number of writings that claimed to be written by apostles were thought by some Christians to be forgeries. It is interesting to reflect on how our present New Testament emerged from this conflict, for, in fact, the first person to establish a fixed canon of Scripture appears to have been none other than Marcion. Marcion's insistence that his sacred books (a form of Luke and ten truncated letters of Paul) made up the Christian Bible evidently led other Christians to affirm a larger canon, which included other Gospels (Matthew, Mark, and John) and other epistles (the Pastorals and the eight general epistles) as well as the books of Acts and Revelation.

It appears then that our New Testament emerged out of the conflicts among Christian groups, and that the dominance of the proto-orthodox position was what led to the development of the Christian canon as we have it. It is no accident that Gospels that were deemed heretical—for instance, the *Gospel of Peter* or the *Gospel of Philip*—did not make it into the New Testament. This is not to say, however, that the canon of Scripture was firmly set by the end of the second century. Indeed, it is a striking fact of history that even though the four Gospels were widely considered authoritative by proto-orthodox Christians then—along with Acts, most of the Pauline epistles, and several of the longer general epistles—the collection of our twenty-seven books was not finalized until much later. For throughout the second, third, and fourth centuries proto-orthodox Christians continued to debate the acceptability of some of the other books. The arguments centered around (*a*) whether the books in question were ancient (some Christians wanted to include *The Shepherd* of Hermas, for example; others insisted that it was penned after the age of the apostles); (*b*) whether they were written by apostles (some wanted to include Hebrews on the grounds that Paul wrote it; others insisted that he did not); and (*c*) whether they were widely accepted among proto-orthodox congregations as containing correct Christian teaching (many Christians, for example, disputed the doctrine of the end times found in the book of Revelation).

Contrary to what one might expect, it was not until the year 367 C.E., almost two and a half centuries after the last New Testament book was written, that any Christian of record named our current twenty-seven books as the authoritative canon of Scripture. The author of this list was Athanasius, the powerful bishop of Alexandria, Egypt. Some scholars believe that this pronouncement on his part, and his accompanying proscription of heretical books, led monks of a nearby monastery to hide the Gnostic writings discovered 1,600 years later by the bedouin near Nag Hammadi, Egypt.

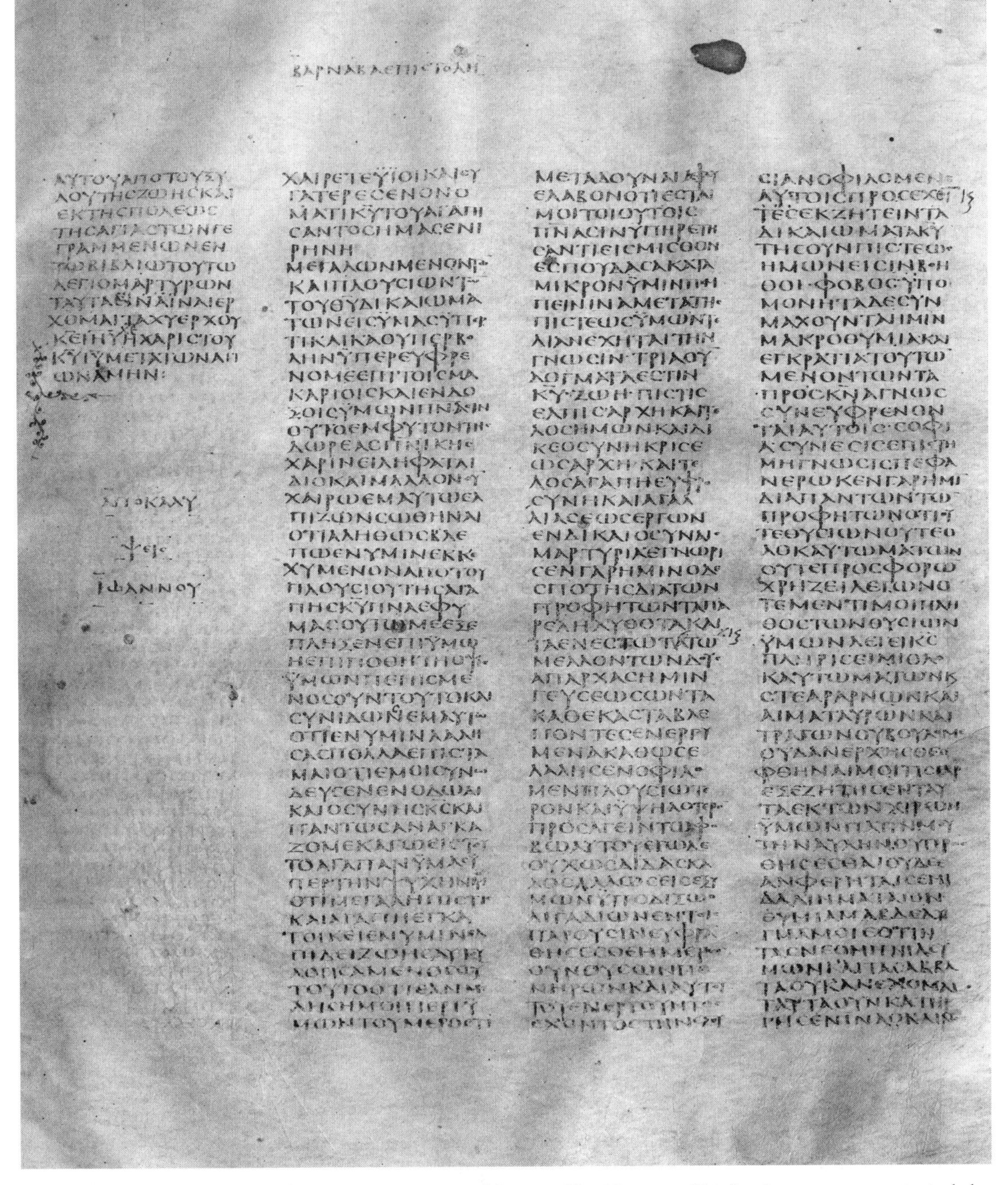

Figure 1.1 Codex Sinaiticus, the oldest surviving manuscript of the entire New Testament. This fourth-century manuscript includes *The Shepherd* of Hermas and the *Epistle of Barnabas* (the first page of which is pictured here), books that were considered part of the New Testament by some Christians for several centuries.

SUMMARY AND IMPLICATIONS FOR OUR STUDY

The question of the canonization of the New Testament books is important for the study of the New Testament. In explaining how and why later Christians decided to include some books rather than others in the canon, this chapter has highlighted the following points about the early Christians and their literature.

1. Early Christianity was extremely diverse. It was not the unified monolith that modern people sometimes assume.
2. This diversity was manifest in a wide range of writings, only some of which have come down to us in the New Testament.
3. The New Testament canon was formed by proto-orthodox Christians who wanted to show that their views were grounded in the writings of Jesus' own apostles.
4. Whether these writings actually represented the views of Jesus' own apostles, however, was in some instances debated for decades, even centuries.

Perhaps the most important aspect of the canon is that the various books of the New Testament are typically read as standing in essential harmony with one another. But do the books of the New Testament agree in every major way? Or are they only thought to agree because they have been placed together, side by side, in an authoritative collection that is venerated as sacred Scripture? Is it possible that when these books are taken out of their canonical context they stand at real tension with one another?

These are among the most difficult and controversial issues that we will address in our study of the New Testament writings. In order to anticipate my approach, I might simply point out that historians who have carefully examined the New Testament have found that its authors do, in fact, embody remarkably diverse points of view. These scholars have concluded that the most fruitful way to interpret the New Testament authors is to read them individually rather than collectively. Each author should be allowed to have his own say,* and should not be too quickly reconciled with the point of view of another. For example, we should not assume that Paul would always say exactly what Matthew would, or that Matthew would agree in every particular with John, and so on. Following this principle, scholars have been struck by the rich diversity represented within the pages of the New Testament. This point cannot be stressed enough. The diversity of Christianity did not begin in the modern period, as some people unreflectively assume, nor did it begin in the second century, in the fragmented forms of Christianity discussed earlier in this chapter. The diversity of Christianity is already evident in the earliest writings that have survived from the Christians of antiquity, most of which are preserved within the canon of the New Testament.

In this book, we will approach the writings of the New Testament from this historical perspective, looking at each author's work individually, rather than allowing the shape of the later Christian canon to determine the meaning of all of its constituent parts.

SOME ADDITIONAL REFLECTIONS: THE HISTORIAN AND THE BELIEVER

Most of the people interested in the New Testament, at least in modern American culture, are Christians who have been taught that it is the inspired word of God. If you yourself belong to this camp, then you may find the historical perspective that I have mapped out in this chapter

* Throughout this book I will be using the masculine pronoun to refer to the authors of the early Christian literature, simply because I think all of them were males. For discussion of some of the relevant issues, see chapter 23 and box 3.1.

somewhat difficult to accept, in that it may seem to stand at odds with what you have been taught to believe. If so, then it is for you in particular that I want to provide these brief additional reflections.

Here is the question: how can a Christian who is committed to the Bible affirm that its authors have a wide range of perspectives, and that they sometimes disagree with one another? I can address the question by stressing that this book is a historical introduction to the early Christian writings, principally those found in the New Testament, rather than a confessional one. This is an important distinction because the New Testament has always been much more than a book for Christian believers. It is also an important cultural artifact, a collection of writings that stands at the foundation of much of our Western civilization and heritage. These books came into existence at a distant point in time and have been transmitted through the ages until today. In other words, in addition to being documents of faith, these books are rooted in history; they were written in particular historical contexts and always read within particular historical contexts. For this reason, they can be studied not only by believers for their theological significance but also by historians (whether they happen to be believers or not) for their historical significance.

Historians deal with past events that are matters of the public record. The public record consists of human actions and world events—things that anyone can see or experience. Historians try to reconstruct what probably happened in the past on the basis of data that can be examined and evaluated by every interested observer of every persuasion. Access to these data does not depend on presuppositions or beliefs about God. This means that historians, as historians, have no privileged access to what happens in the supernatural realm; they have access only to what happens in this, our natural world. The historian's conclusions should, in theory, be accessible and acceptable to everyone, whether the person is a Hindu, a Buddhist, a Muslim, a Jew, a Christian, an atheist, a pagan, or anything else.

To illustrate the point: historians can tell you the similarities and differences between the worldviews of Mohandas Gandhi and Martin Luther King Jr., but they cannot use their historical knowledge to tell you that Gandhi's belief in God was wrong or that Martin Luther King's was right. This judgment is not part of the public record and depends on theological assumptions and personal beliefs that are not shared by everyone conducting the investigation. Historians can describe to you what happened during the conflicts between Catholics and Lutherans in sixteenth-century Germany; they cannot use their historical knowledge to tell you which side God was on. Likewise, historians can explain what probably happened at Jesus' crucifixion; but they cannot use their historical knowledge to tell you that he was crucified for the sins of the world.

Does that mean that historians cannot be believers? No, it means that if historians tell you that Martin Luther King Jr. had a better theology than Gandhi, or that God was on the side of the Protestants instead of the Catholics, or that Jesus was crucified for the sins of the world, they are telling you this not in their capacity as historians but in their capacity as believers. Believers are interested in knowing about God, about how to behave, about what to believe, about the ultimate meaning of life. The historical disciplines cannot supply them with this kind of information. Historians who work within the constraints of this discipline are limited to describing, to the best of their abilities, what probably happened in the past.

This book approaches the New Testament from the historical rather than the confessional perspective. I am not going to convince you either to believe or to disbelieve the Gospel of John; I will describe how it probably came into existence and discuss what its message was. I am not going to persuade you that Jesus really was or was not the Son of God; I will try to establish what he said and did based on the historical data that are available. I am not going to discuss whether the Bible is or is not the inspired word of God; I will show how we got this collection of books and indicate what they say and reflect on how scholars have interpreted them. This kind of information may or may not be of some use to the reader who happens to be a believer, but it will certainly be useful to one who is interested in history, especially the history of early Christianity and its literature.

SUGGESTIONS FOR FURTHER READING

Bauer, Walter. *Orthodoxy and Heresy in Earliest Christianity*. Trans. Robert Kraft et al. Ed. Robert Kraft and Gerhard Krodel. Philadelphia: Fortress, 1971. The classic study of the wide-ranging diversity of second- and third-century Christianity, suitable only for more advanced students.

Dunn, James D. G. *Unity and Diversity in the New Testament: An Inquiry into the Character of Earliest Christianity*, 2d ed. London: SCM Press, 1990. A very informative discussion that applies Bauer's view of early Christian diversity to the New Testament itself; highly recommended for students who have already completed a course in New Testament.

Ehrman, Bart D. *The Orthodox Corruption of Scripture: The Effect of Early Christological Controversies on the Text of the New Testament*. New York: Oxford University Press, 1993. Chapter 1, "The Text of Scripture in an Age of Dissent," explores the diversity of early Christianity on a more introductory level than Bauer.

Gamble, Harry. *The New Testament Canon: Its Making and Meaning*. Philadelphia: Fortress, 1985. A clearly written and informative overview of the formation of the NT canon.

von Harnack, Adolph. *Marcion: The Gospel of the Alien God*. Trans. John E. Steely and Lyle D. Bierma. Durham, N.C.: Labyrinth Press, 1990. The classic study of the life and teachings of Marcion.

Metzger, Bruce M. *The Canon of the New Testament: Its Origin, Development and Significance*. Oxford: Clarendon Press, 1987. The authoritative discussion of the formation of the canon, for advanced students.

Pagels, Elaine. *The Gnostic Gospels*. New York: Random House, 1976. An enormously popular and provocative account of the views of some of the early Gnostics in relation to emerging Christian orthodoxy.

CHAPTER 2

The World of Early Christian Traditions

THE PROBLEM OF BEGINNINGS

Where does one begin a study of the New Testament? One might be inclined to begin with the Gospel of Matthew. This, however, is probably not the best choice: even though Matthew is the first book in the canon, it was not the first to be written. Indeed, as we will see later, it was probably not even the first Gospel to be written.

The first New Testament book to be written was probably 1 Thessalonians, one of the letters penned by the apostle Paul. For this reason, some teachers begin their courses on the New Testament with the life and writings of Paul. While this choice makes better sense than beginning with Matthew, it has problems of its own. Paul lived after Jesus and based many of his teachings on his belief in Jesus' death and resurrection. Would it not make better sense, then, to begin with the life and teachings of Jesus?

The problem with beginning with Jesus is that we do not have any writings from him, and the Gospels that record his words and deeds were written long after the fact, indeed, even after Paul. To be sure, during Paul's lifetime Christians were talking—and some perhaps even writing—about Jesus, telling what he said and did, recounting his conflicts and explaining his fate. Unfortunately, we do not have direct access to these older traditions. We know them only insofar as they were written down later, especially in the Gospels. This means, somewhat ironically, that if we want to begin with the earliest and most important figure in the New Testament, we have to start with documents that were written relatively late.

But this is not the only problem with beginning our study with the traditions about Jesus. What is even more problematic is that these first-century traditions do not "translate" easily into the twentieth century, where our commonsense assumptions, worldviews, values, and priorities are quite different from those shared by the early followers of Jesus. Contrary to what many people think, it is very difficult for us today to understand the original meanings of the sayings of Jesus and the stories about him. This is one reason that modern people have such deeply rooted disagreements over how to interpret the New Testament. It comes from a different world. And many of the ideas and attitudes and values that we take for granted today as common sense would have made no sense in that world; that is, they would have been "nonsense."

In the early Christian world, there was no such thing as a middle class as we know it, let alone a Protestant work ethic, with all of its promises of education and prosperity for those who labor hard. In that world, only a few persons belonged to the upper class and nearly everyone else was in the lower. Few people had any hope for social mobility, slaves made up perhaps a third of the total population in major urban areas, and many of the poor were worse off than the enslaved. There were no cures for most diseases. Many babies died, and adult women had to bear, on average, five children simply to keep the population constant. Most people were uneducated and ninety percent could not read. Travel was slow and dangerous and long trips were rare; most people never ventured far from home during their lives. In the world of early Christianity, everyone, except most Jews, believed in a multiplicity of gods; they knew that divine beings of all sorts were constantly involved with their everyday lives, bringing rain, health, and peace—or their opposites.

People living in the ancient world would have understood the stories about Jesus in light of these realities. This applies not only to how they reacted to these stories and integrated them into their own worldviews but even to how, on the very basic level, they understood what the stories meant. For you can understand something only in light of what you already know.

This point can be illustrated through a modern example. When I was in college in the 1970s, I drove an Austin Healy Sprite. Today this fact does not impress most of my students, who have never heard of an Austin Healy Sprite. If I want to explain to them what it was, I have to do so in terms that they already know. I usually begin by telling them that the Sprite was the same car as the MG Midget. What if they have never heard of a Midget? I tell them that it was a 1970s version of the Mazda Miata. This is a car they generally know. If they don't, I might tell them that the Sprite was a sports car. What if they don't know what that is? I explain: it's a small two-seat convertible that sits low to the ground and is generally considered sporty. What if they don't know what a convertible is, or a two-seater? What if they don't know what a car is? "Well, a car is like a horseless carriage." My explanation, though, assumes that they know what carriages are and what relation horses generally have to them. And if they don't?

My point is that we can understand something only in light of what we already know. Imagine how you yourself might explain an elephant or a roller coaster or a kumquat to someone who had never seen one. What, though, has any of this to do with the New Testament? For one thing, it explains why I think that the most sensible place to begin our study is with the life of a famous man who lived nearly 2,000 years ago in a remote part of the Roman empire.

ONE REMARKABLE LIFE

From the beginning his mother knew that he was no ordinary person. Prior to his birth, a heavenly figure appeared to her, announcing that her son would not be a mere mortal but would himself be divine. This prophecy was confirmed by the miraculous character of his birth, a birth accompanied by supernatural signs. The boy was already recognized as a spiritual authority in his youth; his discussions with recognized experts showed his superior knowledge of all things religious. As an adult he left home to engage in an itinerant preaching ministry. He went from village to town with his message of good news, proclaiming that people should forgo their concerns for the material things of this life, such as how they should dress and what they should eat. They should instead be concerned with their eternal souls.

He gathered around him a number of disciples who were amazed by his teaching and his flawless character. They became convinced that he was no ordinary man but was the Son of God. Their faith received striking confirmation in the miraculous things that he did. He could reportedly predict the future, heal the sick, cast out demons, and raise the dead. Not everyone proved friendly, however. At the end of his life, his enemies trumped up charges against him, and he was placed on trial before Roman authorities for crimes against the state.

Even after he departed this realm, however, he did not forsake his devoted followers. Some claimed that he had ascended bodily into heaven; others said that he had appeared to them, alive, afterwards, that they had talked with him and touched him and become convinced that he could not be bound by death. A number of his followers spread the good news about this man, recounting what they had seen him say and do. Eventually some of these accounts came to be written down in books that circulated throughout the empire.

But I doubt that you have ever read them. In fact, I suspect you have never heard the name of this miracle-working "Son of God." The man I have been referring to is the great neo-Pythagorean teacher and pagan holy man of the first century C.E., Apollonius of Tyana, a worship-

SOME MORE INFORMATION

Box 2.1 Pagan and Gentile

Throughout our discussions I will be using the terms "pagan" and "Gentile." When historians use the term "pagan," they do not assign negative connotations to it (as you may when you use it in reference, say, to your roommate or next-door neighbor). When used of the Greco-Roman world, the term simply designates a person who subscribed to any of the polytheistic religions, that is, anyone who was neither a Jew nor a Christian. The term "paganism," then, refers to the wide range of ancient polytheistic religions outside of Judaism and Christianity. The term "Gentile" designates someone who is not a Jew, whether the person is pagan or Christian. It too carries no negative connotations.

per of the Roman gods, whose life and teachings are recorded in the writings of his later follower Philostratus, in his book *The Life of Apollonius*.

Apollonius lived at about the time of Jesus. Even though they never met, the reports about their lives were in many ways similar. At a later time, Jesus' followers argued that Jesus was the miracle-working Son of God, and that Apollonius was an impostor, a magician, and a fraud. Perhaps not surprisingly, Apollonius's followers made just the opposite claim, asserting that he was the miracle-working Son of God, and that Jesus was a fraud.

What is remarkable is that these were not the only two persons in the Greco-Roman world who were thought to have been supernaturally endowed as teachers and miracle workers. In fact, we know from the tantalizing but fragmentary records that have survived that numerous other persons were also said to have performed miracles, to have calmed the storm and multiplied the loaves, to have told the future and healed the sick, to have cast out demons and raised the dead, to have been supernaturally born and taken up into heaven at the end of their life. Even though Jesus may be the only miracle-working Son of God that we know about in our world, he was one of many talked about in the first century.

Clearly, then, if we want to study the early traditions told about Jesus, traditions that are our only access to the man himself, we have to begin by situating them in their original context in the Greco-Roman world (see box 2.2). The stories about Jesus were told among people who could make sense of them, and the sense they made of them in a world populated with divine beings may have been different from the sense that we make of them in our foreign world. These stories may have had a commonsensical meaning for people in antiquity that they do not have for us.

Figure 2.1 A Roman coin from around the time of Jesus, with the likeness of Caesar Augustus and a Latin inscription, "Augustus, Son of the Divinized Caesar." If Julius Caesar, the adopted father of Augustus, was a god, what does that make Augustus?

SOME MORE INFORMATION

Box 2.2 The Greco-Roman World

The "Greco-Roman world" is a term that historians use to describe the lands surrounding the Mediterranean from the time of Alexander the Great through the first three or four centuries of the Roman Empire (see box 2.4).

Alexander was arguably the most significant world conqueror in the history of Western civilization. Born in 356 B.C.E., he succeeded to the throne of Macedonia as a twenty-year old when his father, King Philip II, was assassinated. Alexander was single-minded in his desire to conquer the lands of the Eastern Mediterranean. A brilliant military strategist, he quickly and boldly—some would say ruthlessly—overran Greece to the South and drove his armies along the coastal regions of Asia Minor (modern-day Turkey) to the East, into Palestine and then Egypt. He finally marched into the heart of the Persian Empire, overthrowing the Persian monarch Darius, and extending his territories as far away as modern-day India.

Alexander is particularly significant in the history of Western civilization because of his decision to impress a kind of cultural unity upon the conquered lands of the eastern Mediterranean. In his youth he had been trained in Greece by the great philosopher Aristotle and became convinced that Greek culture was superior to all others. As a conqueror he actively promoted the use of the Greek language throughout his domain and built Greek-style cities, with gymnasiums, theaters, and public baths, to serve as administrative and commercial centers. Moreover, he generally encouraged the adoption of Greek culture and religion throughout all his cities, especially among the upper classes. Historians have named this cultural process "Hellenization," after the Greek word for Greece, *Hellas*.

Upon Alexander's untimely death at the age of thirty-three (323 B.C.E.), his realm was divided among his leading generals. During their reigns and those of their successors, Hellenism (i.e., Greek culture) continued to flourish in major urban centers around the eastern Mediterranean (less so in rural areas). Throughout this period, as political boundaries shifted and kings and kingdoms came and went, a person could travel from one part of Alexander's former domain to the other and still communicate with the local inhabitants by speaking the lingua franca of the day, Greek. Moreover, such a person could feel relatively at home in most major cities, amidst Greek customs, institutions, traditions, and religions. Thus, more than at any time in previous history, the eastern Mediterranean that emerged in Alexander's wake experienced a form of cultural unity and cosmopolitanism (a "cosmopolite" is a "citizen of the world," as opposed to a person who belongs only to one locality).

The Roman empire arose in the context of the Hellenistic world and took full advantage of its unity, promoting the use of the Greek language, accepting aspects of Greek culture, and even taking over features of the Greek religion, to the point that the Greek and Roman gods came to be thought of as the same, only with different names. This complex unity achieved culturally through Hellenization and politically through the conquests of Rome (see box 2.4) are summed up by the term, Greco-Roman world.

We will begin our reflections by discussing ancient "pagan" religions (see box 2.1), since it was primarily among pagans that Christians told most of their stories and acquired most of their converts when the books of the New Testament were being written. We will then turn to consider early Judaism, one of the distinctive religions of the Greco-Roman world, the religion of the earliest Christians and of Jesus himself.

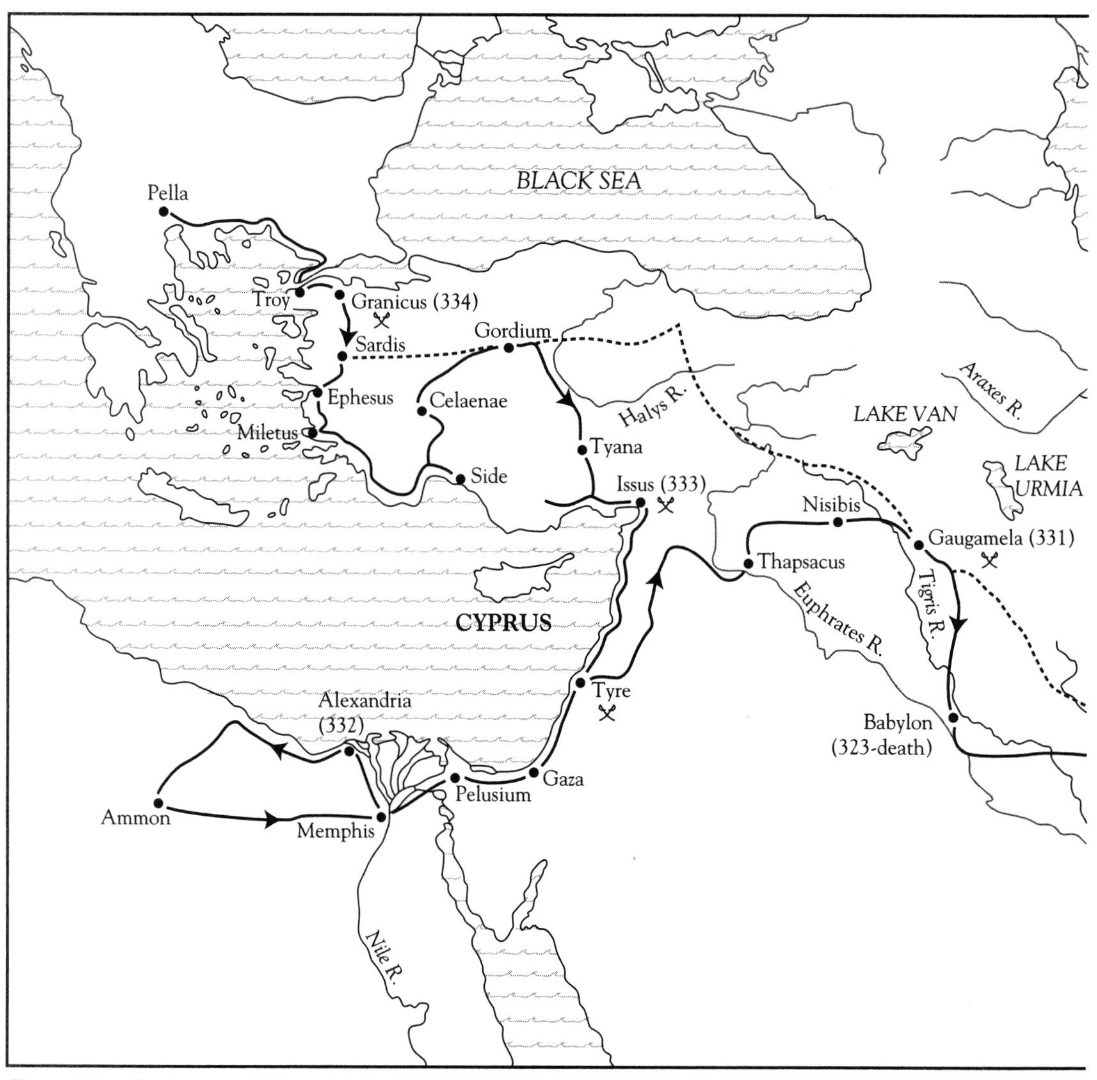

Figure 2.2 The Journeys of Alexander the Great (334-323 B.C.E.).

THE ENVIRONMENT OF THE NEW TESTAMENT: RELIGIONS IN THE GRECO-ROMAN WORLD

Greco-Roman Religiosity: A Basic Sketch

Odd as it may seem, to understand the nature and function of religion in the Greco-Roman world, we have to abandon almost all of our own notions about religion today. What do twentieth-century Americans think about when they think about organized religion? The following list is by no means exhaustive, but it does include a number of popular notions held by many people in our society (though not by all people, of course, for our world is fantastically diverse):

1. Religious organization and hierarchy (e.g., the Christian denominations and their leaders, whether a pope, a Methodist bishop, or the leader of the Southern Baptist convention)
2. Doctrinal statements (e.g., the creeds said in churches, the basic beliefs endorsed by all believers)

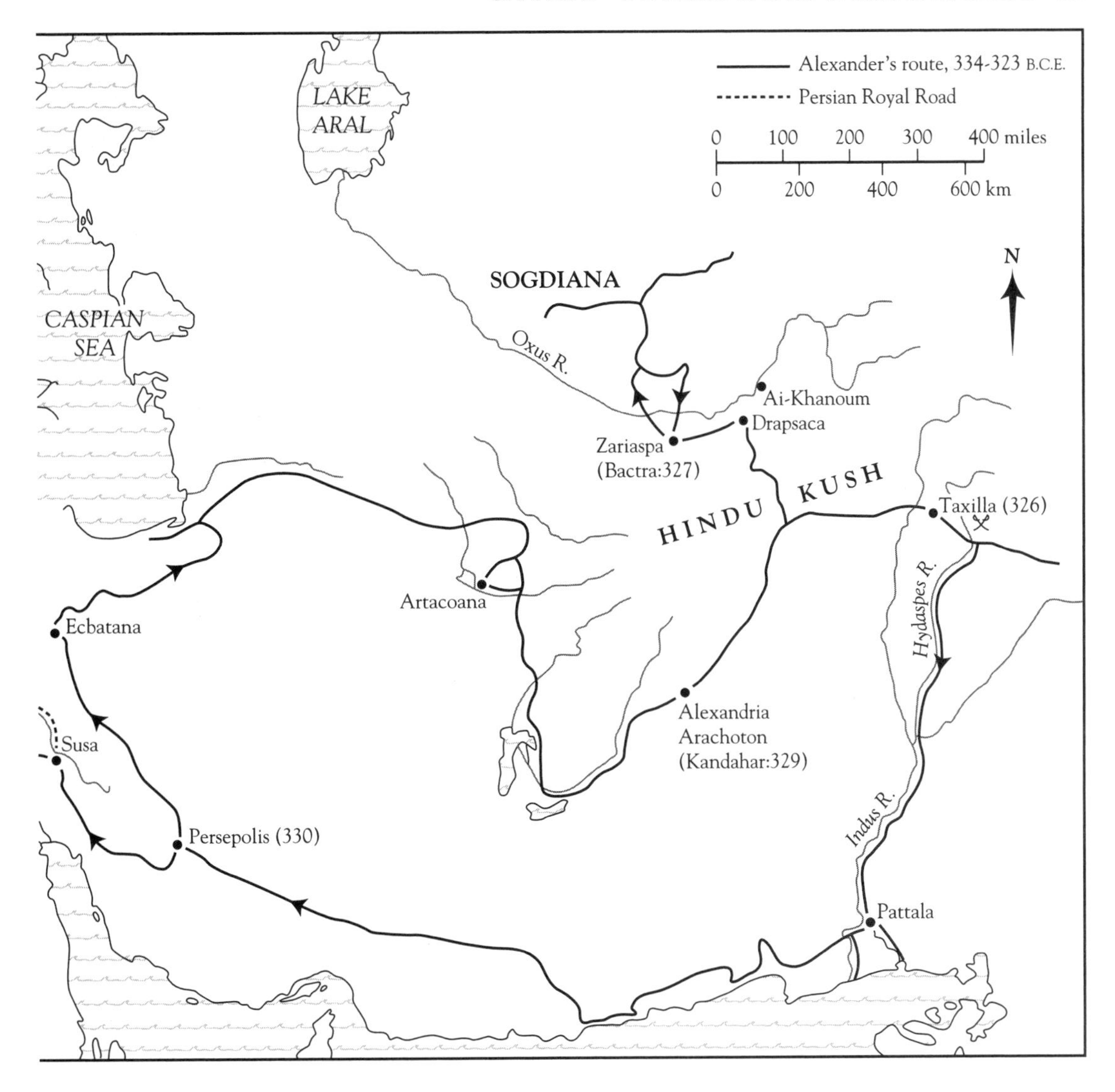

3. Ethical commitments (i.e., religiously motivated guidelines for conducting one's daily interactions with others)
4. Beliefs about the afterlife (which for some people in our time is *the* reason for being religious)
5. Sacred written authorities (e.g., the Hebrew Bible or the New Testament or the Koran)
6. The separation of church and state (an important element in American politics *and* religion)
7. Exclusive commitments (e.g., a member of a Baptist church cannot also be a Hare Krishna, just as a practicing Jew cannot be a Mormon).

One of the most striking and startling aspects of ancient religion is that outside of Judaism, none of these features applies. In the so-called pagan religions of the Roman empire, there were no national or international religious organizations with elected or appointed leaders who had jurisdiction over the various local cults. There were no creedal statements or, indeed, any necessary articles of faith whatsoever for devotees. Whereas ethics were generally as important to people then as they are today, daily ethical demands played virtually no role in the practice of religion itself. Many people evidently did not hold a firm belief in life

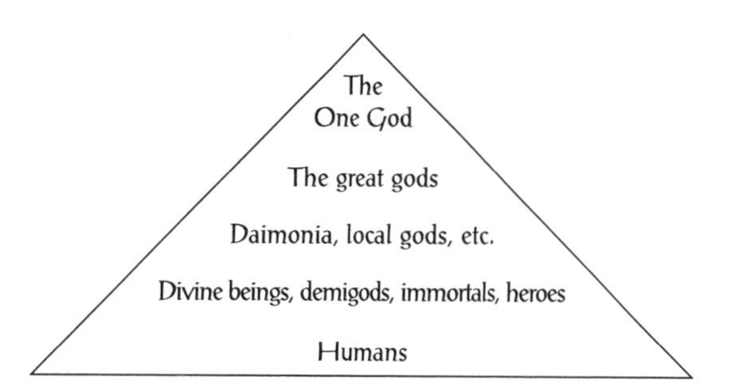

Figure 2.3 The Divine Pyramid as Understood in Greco-Roman Religions.

after death; those who did, so far as we can tell, did not generally become more religious as a result. Pagan religions were never centered on sacred writings to guide the individual's beliefs and practices. And there was no such thing as separation of church and state; on the contrary, since the gods made the state great, the state responded by encouraging and sponsoring the worship of the gods. Finally, virtually no one in the pagan world argued that if you worshipped one god, you could not also worship another: exclusive adherence to one cult was practically unknown.

How can we fathom a set of religions so different from our own? Since we can only understand something in light of what we already know, we can begin by considering a series of contrasts between modern and ancient religions, somewhat along the lines I have already laid out.

Polytheism Instead of Monotheism. Modern religions in the West (Judaism, Christianity, and Islam) are monotheistic, advocating belief in one Divine Being. For most modern Westerners, it is simply common sense to think that there is only one God. For persons in the ancient world, however, this was non-sense. Everyone knew that there were many gods, of all sorts and descriptions, of all functions and locations: gods of the field and forest, gods of the rivers and streams, gods of the household and courtyard, gods of the crops and weather, gods of healing, gods of fertility, gods of war, gods of love.

The belief in many gods came down from prehistoric times; in the Greco-Roman world, nearly everybody took their existence for granted. Not that everybody worshipped the same gods. On the contrary, many gods were localized deities of a certain place or a certain family. With the conquest of villages, towns, and countries by other villages, towns, and countries, local gods sometimes spread to other regions, occasionally becoming national or international. Sometimes conquered peoples would accept the gods of their conquerors, either by substituting them for their own (since the gods of the victors were, after all, demonstrably more powerful), or by using the new names for their old gods (which is simply another mode of substitution), or by adding the new gods to those that they already worshipped.

There were of course the "great gods" who were worshipped throughout different portions of the Mediterranean. These included the gods mentioned by the ancient poets Homer and Hesiod. The writings of these ancients, for example, Homer's *Iliad* and *Odyssey*, were not considered to be some kind of Scriptural authority in the way the Bible was for Jews and later for Christians, but they were good stories that people told and enjoyed hearing, even if they did sometimes portray the gods in a somewhat unfavorable light as conducting themselves in wild and capricious ways.

How did the average person understand the relationship of the great gods to those of their own locality? Recent scholarship has shown that most people in the Greco-Roman world conceived of the divine realm as a kind of pyramid of power, with the few but mightiest god(s) at the top and the more numerous but less powerful deities at the bottom (see figure 2.3). Some of the most highly educated thinkers, for example, philosophers and their students, maintained that at the very peak of the pyramid was one almighty God, whether understood to be the Greek Zeus, the Roman Jupiter, or some unknown and unknowable God, so powerful as to be beyond human comprehension. This God was ultimately responsible for the world and for all that happens in it; ironically, though, he was so powerful that he was all but inaccessible to mere mortals.

The pyramid's next tier represented the powerful gods worshipped in different localities throughout the empire. Among Greek people, these would include Poseidon, Hera, Aphrodite, Artemis, Dionysus, and others of Greek myth and

Figure 2.4 Many inhabitants of the Roman empire offered worship to the genius (ruling spirit) of the emperor as god, as seen in this depiction of a sacrifice taken from an altar before the temple of the emperor Vespasian in Pompei. Notice the priest on the right holding a sledgehammer with which to stun the sacrificial bull before another priest slices its throat.

legend; in Roman circles these would be identified by their Latin names: Neptune, Juno, Venus, Diana, and Bacchus. These gods were thought to be incredibly powerful and altogether worthy of worship and praise. Many of them were associated with significant functions of human society. For example, Ares (Latin Mars) was the god of war, Aphrodite (Venus) the goddess of love, and Dionysus (Bacchus) the god of wine.

Below this tier was another inhabited by lesser gods, including the local deities who had limited powers (although they were still far beyond anything humans could imagine) but who were in more direct contact with human affairs. Included on this tier were the *daimonia*. This Greek term is hard to translate into English. The cognate term "demons" carries the wrong connotation altogether, for the *daimonia* were not evil fallen angels who temporarily inhabited human bodies, forcing them to do all sorts of nasty things. To be sure, some of them were dangerous, but for the most part they were relatively indifferent to human activities and so had to be persuaded, through cultic acts, to behave in ways that would lead to benefit rather than harm.

Finally, on the bottom level of the divine pyramid was a range of divine beings who more or less bridged the gap between mortals and the gods. Included here were humans who, at their deaths, had been divinized (i.e., made immortal, like the gods). These were typically great men, philosophers or warriors, whose extraordinary deeds won them special favors from the gods at death as well as in life. Also found here were demigods, individuals said to have been born to the union of a god or goddess with a mortal, as found, for instance, in a number of Greek and Roman myths and folktales. This final category is of particular interest for us because it included select human beings who were widely believed to have been far more than human, including great philosophers like Pythagoras, whose wisdom was thought by some to be inexplicable if merely human, powerful athletes like Heracles, whose strength was far beyond the mere mortal, and great rulers like Alexander of Macedonia, whose power to affect human lives was nearly divine.

Some people considered the Roman emperor to be this kind of divine being. He was not the one God, or even one of the Olympians. Indeed, from the divine perspective he was very much a subordinate. But from the human point of view, he was fantastically powerful, himself divine, and for some inhabitants of the empire worthy of worship and praise. Also included among such beings were Apollonius of Tyana and other so called sons of God, whose supernatural teachings and miraculous deeds demonstrated their divine lineage.

Pagans who heard stories about Jesus and his miracles would have had no difficulty understanding what they meant. Among other things they meant that Jesus was himself divine, a divine man come to earth.

Present Life instead of Afterlife. Many people in the modern world are motivated in their religious commitments by a belief in the afterlife. Fearing eternal torment or longing for eternal bliss, they turn to religion as a way of securing happiness after death.

This view would have made little sense to most people in the ancient world. Recent studies of ancient gravestone inscriptions, in fact, suggest that whereas some people subscribed to a notion of the afterlife (as we will see later when we consider the mystery cults), the majority did not. Moreover, of those who did, most believed that it involved some kind of vague shadowy existence that was to be postponed as long as possible at all costs, a netherworld to which all people were destined, whether moral or immoral, faithful or unfaithful. And yet nearly everyone in the ancient world believed in the gods and participated in religion.

For most ancient persons, religion was not the way to guarantee an afterlife; it was a way to secure life in the here and now. For the majority of people in the ancient world, life was constantly lived on the edge. There was nothing like modern medication to prevent and cure disease; a tooth abscess would frequently prove fatal. There were no modern surgical methods and only primitive forms of anesthesia; women often died in childbirth, and simple operations could be hellish nightmares. There were no modern methods of agriculture and limited possibilities for irrigation; a minor drought one year could lead to a poor village's starvation the next. There were no modern modes of transportation: in rural areas, food distribution was limited at best. War, famine, disease, poverty—the eternal blights of the human race—were constant and perennial concerns of ancient persons. And, of course, all the anxieties of interpersonal relations were still very much alive then as well; they knew the tragic loss of a child or friend, fear for personal safety, unrequited love.

In a world that is helpless against the elements, the gods play a major role. They supply rain for the crops, fertility for the animals, children for the family. They bring victory in war and prosperity in peace. They heal the sick and comfort the downtrodden. They provide security and hope and love. These are things beyond the control of mere mortals; they can come only from the gods.

Cultic Acts rather than Doctrine. But how could the powerful and immortal gods be influenced to provide what was needed in this life? The gods were not impressed by anyone's beliefs about them nor did they require people to say the proper creed or acknowledge the proper "truths." Odd as this may seem to us moderns, doctrine played virtually no role in these religions: it scarcely mattered what people believed. What mattered was how people showed their devotion to the gods. The gods wanted to be worshipped through proper cultic acts.

The English term "cult" derives from the Latin term for "care." The ancient concept of *cultus deorum* thus referred to the "care of the gods" (cf. the English word "agriculture," meaning the "care of the fields"). How, though, did one "care" for the gods? How did one attend to them so as to secure their favor? For the ancient person the answer was simple: through prayer and sacrifice. Local and family deities had their own established cults. Daily cultic acts might involve pouring out a little wine before a meal in honor of one of the family gods or saying a prayer for favor. Periodic festivals

Figure 2.5 A depiction of the practice of extispicy from an ancient altar. Notice the priest who is stooped over to examine the entrails of the recently sacrificed bull to discern whether the sacrifice has been acceptable to the gods.

would be celebrated in which a group of worshippers would sacrifice an animal, or have a local priest do so, while set prayers were spoken. The inedible parts of the animal would be burned to the god, the rest would be prepared and eaten by the participants in a picnic-like atmosphere.

Throughout the empire, special festival days were set aside for the worship of the state gods. These were the powerful gods who had shown favor to Rome and made it great. People worshipped them to secure their continued favor and patronage. Great celebrations in the capital city itself would follow standard rituals by priests trained in the sacred traditions, they would perform the required sacrifices and say the established prayers in precisely the same way year after year. The Romans generally assumed that if religious practices worked they must be right and must be retained. That they did work was plain for all to see in the grandeur and power of Rome itself.

Moreover, it was possible to know for certain whether a particular cultic act had proved acceptable to the gods, for the gods would say so. One of the standard religious practices of the Romans that seems most bizarre to modern persons involved the art of "extispicy"—the reading of a sacrificial animal's entrails (Latin *exta*) by a specially trained priest (a "haruspex") to determine whether the god(s) had accepted the sacrifice. If the entrails were not perfect—for example, if they were not healthy, or the proper size, or in the proper place—then the rite was to be performed again.

The practice of extispicy shows that Roman religion was not simply a one-way street in which

SOME MORE INFORMATION

Box 2.3 Divine Rulers as Savior Gods

With respect to the homage paid to the Roman emperor as a divine being, the "Savior" of the human race, consider the following inscription set up in honor of Gaius Julius Caesar Germanicus, otherwise known to history as the emperor Caligula, by the city council of Ephesus in Asia Minor, around 48 C.E.

> The council and the people (of the Ephesians and other Greek) cities, which dwell in Asia and the nations (acknowledge) Gaius Julius, the son of Gaius Caesar, as High Priest and Absolute Ruler, ... the God Visible who is born of (the Gods) Ares and Aphrodite, the shared Savior of human life.

the worshipper tried to placate the gods. The gods had ways of communicating with humans as well. They did so through various modes of "divination" (ways of discerning the divine will). Roman priests called augurs, for instance, were trained in interpreting the flights or eating habits of birds ("taking the auspices") to determine whether the gods were in favor of a projected action on the part of the state, such as a military expedition. For private direction from the god, there were sacred places called "oracles," where people perplexed about their own future could come to address a question to a god, whose priestess would enter into a trance, become filled with the divine spirit, and deliver a response, sometimes written down by an attendant, often in poetic verse. Sometimes the gods communicated by more natural means, for example, by sending a thunderclap or a dream as a sign.

There was close interaction between the divine and human realms in the ancient world. The gods spoke to humans through dreams and oracles and physical signs, and humans served the gods, securing their favor through prayers and sacrifices.

Church and State Together instead of Separated. In the Greco-Roman world there was no separation between the function of the state and the performance of religion. Quite the contrary, government and religion both functioned, theoretically, to secure the same ends of making life prosperous, meaningful, and happy. The gods brought peace and prosperity and made the state great. In turn, the state sponsored and encouraged the worship of the gods. For this reason, state priesthoods in the Roman empire were (to use our modern terminology) political appointments. The priests of the leading priestly "colleges" in Rome were senators and other leading officials. Temples were dedicated to the gods because of great military victories, the temple staff was supplied by the state, and celebrations were overseen by the government.

The emperor encouraged the cult of the gods, and in some parts of the empire (although not in the city of Rome itself) he himself was recognized as divine. At first, emperors were worshipped only after they had died and were proclaimed by the Senate to have become divinized. Outside of Rome, however, even during the New Testament period, living emperors came to be worshipped as the divine "Savior" of the empire. These divine men had brought deliverance from the evils that threatened the well-being of the state. Some of the emperors discouraged this practice, but officials in the provinces sometimes promoted it (see box 2.3). Thus, local cults devoted to the emperor existed throughout much of Asia Minor when the Apostle Paul arrived with his word of the Savior Jesus. By the second century, cities throughout the empire held celebrations in which sacrifices were made on behalf of the emperor or his "genius," that is, the divine spirit that ruled over his family.

SOME MORE INFORMATION

Box 2.4 The Roman Empire

The traditional date for the founding of Rome is 753 B.C.E. It began as a small farming village which grew over time into a city spread over a large area that included the "seven hills of Rome." For nearly 250 years Rome was ruled by local kings, whose abuses led to their ouster in 510 B.C.E. For nearly half a millenium thereafter, Rome was a republic governed by an aristocratic oligarchy called the Senate, which was made up of the wealthiest and most influential members of its highest class.

As it refined its political and legislative systems, Rome also grew strong militarily, eventually conquering and colonizing the entire Italian peninsula and then, after three protracted wars against the city of Carthage in North Africa, known as the Punic (Wars 264-241 B.C.E., 218-202 B.C.E., and 149-146 B.C.E.), acquiring control of the entire Mediterranean region.

The late republic period saw an increasing number of internal struggles for power, many of them violent, as prominent generals and politicians attempted to seize the reins of power. When Julius Caesar tried to become a dictator, he was assassinated in 44 B.C.E. The Republic (ruled by the Senate) was not finally transformed into an Empire (ruled by an emperor) until Caesar's great-nephew and adopted son Octavian, a wealthy aristocrat and Rome's most successful general, brought a bloody end to the civil wars that had racked the city and assumed full control in the year 27 B.C.E.

Even after this time, the Senate continued to exist and to oversee aspects of the immense Roman bureaucracy, which included the governance of provinces that eventually stretched from Spain to Syria. Official posts were sometimes delegated to members of the "equestrian" class as well. These had a lower rank and less wealth than senators, but they were nonetheless members of the landed aristocracy. But with the inauguration of the reign of Octavian, who soon assumed the name Caesar Augustus (roughly meaning "the most revered emperor") there was one ultimate ruler over Rome, an emperor who wielded virtually supreme power. Emperors who succeeded Caesar Augustus after his death in 14 C.E. were of varying temperaments and abilities. For the period of our study, they include the following:

Tiberius (14–37 C.E.)
Caligula (37–41 C.E.)
Claudius (41–54 C.E.)
Nero (54–68 C.E.)
Four different emperors in the tumultuous year of 68–69 C.E. including, finally,
Vespasian (69–79 C.E.)
Titus (79–81 C.E.)
Domitian (81–96 C.E.)
Nerva (96–98 C.E.)
Trajan (98–117 C.E.)
Hadrian (117–138 C.E.)

The political implications of this kind of worship may seem clear to us, living so many centuries later. The belief that the gods were directly involved in the Roman state surely helped to secure the peace of the empire. One might rebel against a powerful mortal, but who would take up arms against a god?

Tolerance instead of Intolerance. Because of the ill-fated experience of the early Christians, who were occasionally persecuted by the Roman authorities, many people today assume that Romans were by and large intolerant when it came to religion. Nothing could be further from the truth. Certainly, refusing to perform a sacrifice to the gods on behalf of the emperor, or refusing to throw some incense on the altar to his genius, might cause trouble. This refusal would be seen as a political statement (again, to use our modern terms), as a vote of no confidence or, even worse, open defiance of the power of the state and the even greater power of the gods who made it great. Moreover, since everyone knew that there were lots of gods, all of whom deserved worship, it made little sense to refuse to take part in cultic acts.

Basic tolerance was one of the central aspects of ancient Greco-Roman religion. Unlike some forms of Christianity that eventually arose in its midst, the empire's other religions were altogether forebearing of one another (see chapter 25). There was no reason that everyone should worship the same gods any more than everyone should have the same friends. All the gods deserved to be worshipped in ways appropriate to them. Thus when people visited or relocated to a new place, they would typically begin to worship the gods who were known there; sometimes they would continue to worship their own gods as well. The various religious rites were by and large tolerated; local practices were honored, and those who worshipped the state gods did not try to drive out their opposition. There was no sense of exclusivity in Greco-Roman religions, no sense that my gods are real and yours are false, that you must convert to my gods or be punished.

Magic and Mystery in Greco-Roman Religion

Magic was big business in the Roman empire. This should come as no shock, given what we have already seen about the religions of the period. If the function of religion was to perform cultic acts in order to sway the gods to act on your behalf, what was one to do if the established religion didn't work? Many people in the Greco-Roman world (even people actively involved in "religion") opted to go an alternative route, resorting to what was known even then as "magic."

Older scholarship understood magic to be the superstitious manipulation of divine powers, that is, the performance of incantations and ritual acts in such a way as to compel supernatural forces to grant a person's desires. It does indeed appear that something like this was widely practiced throughout the Roman world. We not only have ancient literary texts in which such practices are described, we also have discovered a number of magical texts, that is, documents that were used for magical purposes. These include long recipes for potions with exotic ingredients (the ancient equivalents of the eye of newt and hair of a bat), mystical incantations with repetitions of meaningless syllables (analogous to "abracadabra" but sometimes going on for paragraphs), and tablets that invoke curses on an enemy (a kind of ancient voodoo). These devices were "guaranteed" to produce the desired results, for example, the death of an enemy or the unbridled passion of an alluring neighbor.

The problem for scholars today, though, is deciding how these practices differ substantially from what we call religion. If Greco-Roman religion involved rituals and fixed prayers that had to be performed in certain set ways in order to secure the favor of the gods, how is that so different from what we term magic? In fact, it appears not to be so different. Ancient religion and ancient magic involved similar actions and anticipated similar (divine) results. Ultimately, of course, neither could provide absolute guarantees. Why, then, did the ancients themselves refer to some practices as magical?

Anthropological studies of the phenomenon suggest that when a society at large approves of a cultic practice (or at least when its elite members do), it is labeled "religious," whereas similar practices that are not approved are viewed suspiciously and called "magical." Magic, then, can be seen as the dark side of religion; it is mysterious and secretive and socially marginal. This is why two ancient miracle workers producing similar results might be perceived differently, the one as a son of God (a term of approbation) the other as a magician (disapprobation). The former is on the side of the good and the sanctioned; the latter has used dark powers and unapproved methods.

This is not to say that ancient Greco-Roman society altogether disapproved of secrecy and mystery in religion. On the contrary, sanctioned forms of mystery existed in certain local cults, and some of these came to enjoy an international reputation. Modern scholars commonly refer to these forms of religion as the "mystery cults." In some respects the

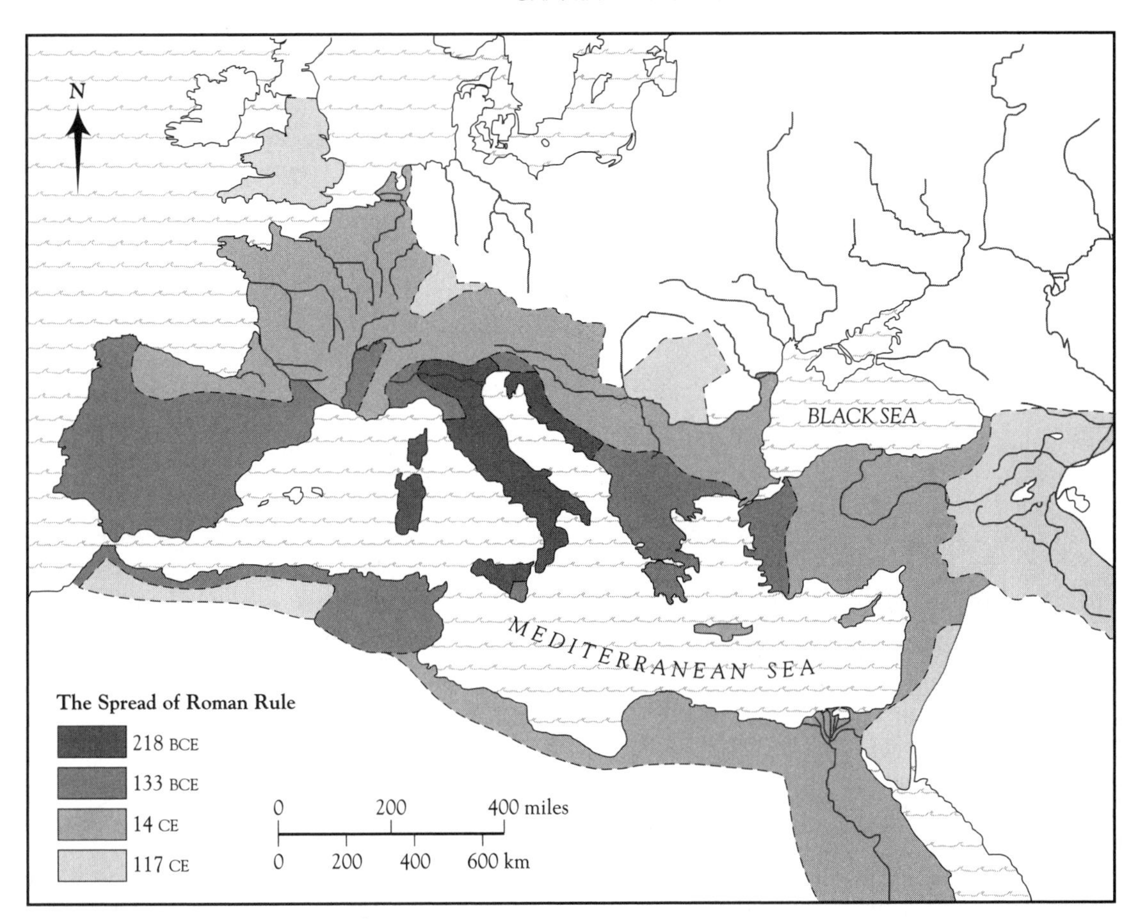

Figure 2.6 The Spread of Roman Rule.

mystery cults stand out as exceptional in the religious climate of the Greco-Roman world; quite possibly, it was precisely their atypical character that made them so sought after. Regrettably, despite their popularity, we are remarkably ill-informed concerning these cults. Indeed, they are called mysteries, in part, because participants could not divulge what happened during their sacred rituals. As a consequence, our evidence has to be pieced together from isolated comments and fragmentary remains.

From this evidence, though, we can get some idea about what most of the mysteries were like and how they differed from both the state and other local cults. We have seen that most religions in the period were concerned with both individual and community needs (e.g., rain, fertility, victory, peace, prosperity). The mystery cults were relatively distinct in focusing chiefly on the well-being of the individual. Moreover, whereas almost all other religions were centered on life in the here and now, mystery cults appear to have placed some emphasis (older scholarship believed it was exclusive emphasis) on providing a happy existence in the life after death. Finally, even though there was wide tolerance of different religions in the Greco-Roman world, and no general sense of exclusive attachment to one deity over another, within the mysteries we find individuals who are principally devoted to one god or goddess for life. Even these, however, do not appear to have claimed that theirs was the only true god or goddess; instead, theirs was the only one *for them*.

The mysteries, it appears, met personal, individual needs and resonated with many persons in the Greco-Roman world who did not find existential fulfillment (to use a modern phrase) in the local and state cults in which they participated. Each of the mystery cults was different; each had its own special location and its own customs and rituals. Many of them evidently centered around a mythology of the death and resurrection of a god or goddess, a mythology ultimately rooted in ancient fertility religion, in which the death of winter gives way to the new life of spring. Moreover, the periodic ritual of these cults apparently celebrated this mythology in a way that enabled the participants to become part of the entire transformative process of new life. That is to say, the enacted myth about the gods was transmuted into reality for the devotees, who believed they would live again, happily, after death. For those who had been found worthy to be a follower of the mystery's god or goddess, there was promised not only a more satisfying existence now but also a more blissful afterlife.

Not just anyone could walk in off the streets to join one of these mystery cults. Each of them appears to have emphasized rituals of initiation for membership. Those who wished to join were typically put through a period of ceremonial cleansing (involving fastings, prayers, and sometimes ritual washings) and instruction prior to being admitted to the ranks of the devotees. We have evidence to suggest that those who experienced the initiation, who could then join in the ceremonies when they were periodically celebrated, felt at greater peace with themselves and the world.

Among the better known mysteries in the ancient world were those involving the Greek goddesses Demeter and her daughter Kore (sometimes called Persephone) at the town of Eleusis in Greece, the goddess Isis and her husband Osiris from Egypt, the Greek God Dionysus (also known as Bacchus), and the Persian God Mithras. Despite the occasional instance of a devotee being committed to only one or the other of these mysteries, we know of many instances in which persons were initiated into several of them. Furthermore, initiation did not at all preclude worship of the local and state gods; some of the Roman emperors were themselves initiates.

Philosophy and Religion in the Greco-Roman World

There is one final aspect of the Greco-Roman world to consider before turning our attention to the place that Judaism occupied within it. I have already mentioned that Greco-Roman cults did not overly concern themselves with doctrines about the gods or with the moral behavior of their devotees. This does not mean, however, that there was no room for reflection on the meaning of life, the nature of personal happiness, and the need for ethical behavior. This kind of thinking, though, lay largely outside the province of cult and within the province of philosophy.

Philosophy and religion were not thought to be irreconcilable entities; indeed, some of the best known philosophers were priests in pagan temples. They nonetheless represented two different spheres of activity with two different sets of concerns. Greco-Roman philosophy was not concerned with placating the gods or petitioning their involvement in the affairs of the community. It was instead concerned with showing how a person could attain well-being in this world, a world that is at best filled with meaninglessness and boredom, and at worst wracked with pain and misery.

Professional philosophers were a relatively rare breed in the Greco-Roman world, whose preindustrial society had scant resources to support large numbers of people who did little but think and teach others to do likewise. Moreover, few people had the time or ability to read philosophical treatises; indeed, most people were illiterate and couldn't read *anything* (see box 3.2). Nonetheless, philosophical ideas were widely known, due in large part to their typical mode of communication. On street corners and thoroughfares of major urban areas throughout the empire, philosophers of all stripes could be found proclaiming their views and urging others to adopt them in their own lives, rather like street preachers in some places today.

Of the important philosophical schools during the first century of the Common Era, three stood out as prominent: the Stoics, the Platonists, and the Epicureans. Each of these traditions traced its roots back over three hundred years, and the dif-

SOMETHING TO THINK ABOUT

Box 2.5 Christianity and the Mystery Cults

Scholars in the earlier part of this century were struck by how similar the ancient descriptions of the mysteries were to what we know about Christianity, a secret society whose members worshipped a divine being who died and was raised from the dead, and who could bring peace on earth and eternal life after death. Initiates into the society went through a period of ritual purification (baptism) and instruction, and members, according to this view, periodically celebrated the myths of the cults beginning (in the Lord's Supper).

Recent scholarship, however, has been less inclined to call Christianity a mystery cult, or to claim that it simply borrowed its characteristic ideas and practices from previously existing religions. In part this is because we do not know very much about what exactly happened during the mystery rituals, especially in the period when Christianity began. For example, did they typically partake of a meal, commemorating the death of their savior god? We simply don't know.

All the same, the broad parallels between Christianity and these other religions do remain intriguing and worthy of reflection. Maybe the question scholars have asked should be posed differently: would non-Christian outsiders have looked upon Christianity as a kind of mystery cult, analogous to others that they knew?

ferences between them ran wide and deep, but for our study their common features are more important than their differences.

All three philosophies tried to show how an individual could achieve personal well-being in the midst of a harsh and sometimes capricious world. Each group defined well-being in a somewhat different way, but they all generally portrayed it as a kind of inner peace that comes from living in conformity with nature. For the Stoics, for example, this meant living in harmony with the world as it was structured by the divine; for the Epicureans it meant realizing that the divine realm has nothing to do with this world and locating personal peace of mind in the simple pleasures of daily existence. For all the philosophies, though, the attainment of well-being involved an exercise of reason, a mental effort of reconfiguring one's understanding of the world and the nature of reality. Only an exercise of the mind could provide a person with the tools necessary to live a full life internally and protect one from hardships that strike externally.

Thus philosophers put a high premium on both education and discipline; or, to put the matter slightly differently, they were concerned with doctrines (what to think) and ethics (how to live). These emphases explain one further aspect of philosophy that contrasted it with religion. As I've indicated, cults throughout the Roman world were by and large tolerant of one another; there was scarcely any reason to convert others away from one set of gods to another. The same could not be said, though, of philosophy, for here was an area in which if one person was right, the others were wrong. For this reason, proponents of various philosophical schools tended to insist on the validity of their own views and to be somewhat intolerant of the views of others (even though they freely borrowed their ideas from one another, making it sometimes difficult to discern their differences). In other words, unlike the religions of the Greco-Roman world, the philosophies worked to convert people to their points of view. These were, in short, missionary movements.

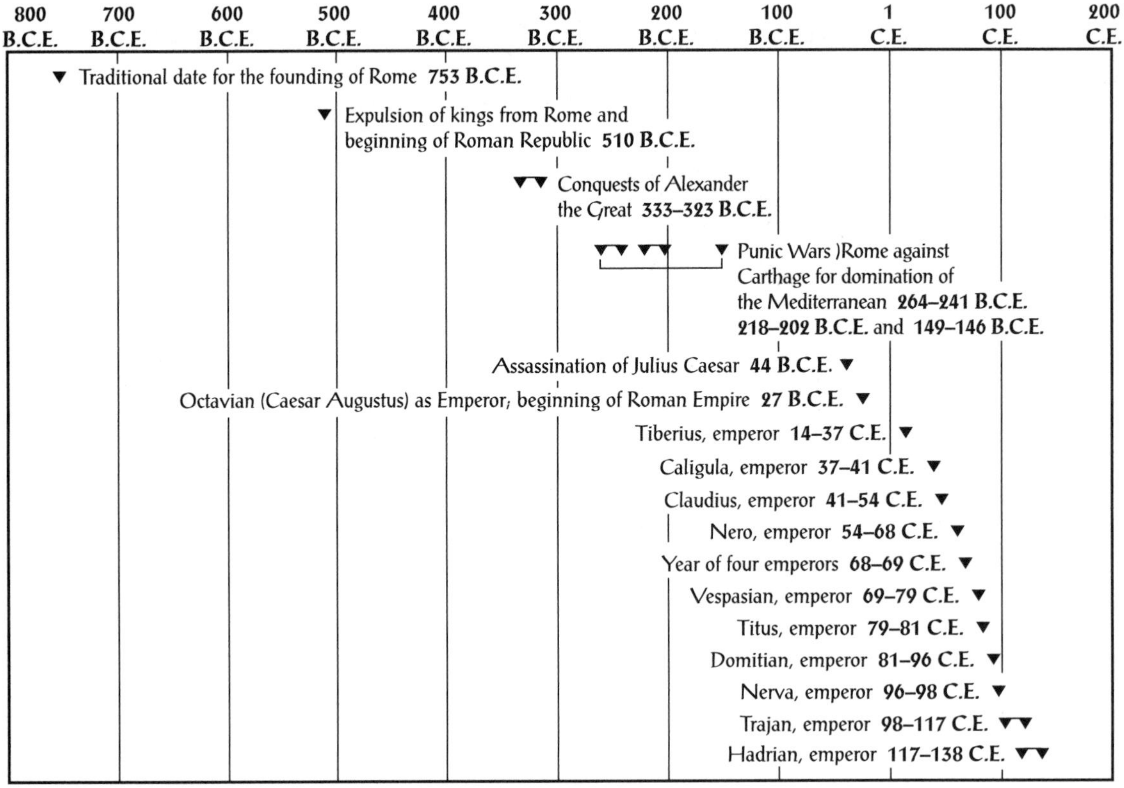

800 B.C.E. | 700 B.C.E. | 600 B.C.E. | 500 B.C.E. | 400 B.C.E. | 300 B.C.E. | 200 B.C.E. | 100 B.C.E. | 1 C.E. | 100 C.E. | 200 C.E.

▼ Traditional date for the founding of Rome **753 B.C.E.**

▼ Expulsion of kings from Rome and beginning of Roman Republic **510 B.C.E.**

▼▼ Conquests of Alexander the Great **333–323 B.C.E.**

▼▼ ▼▼ ▼ Punic Wars)Rome against Carthage for domination of the Mediterranean **264–241 B.C.E. 218–202 B.C.E.** and **149–146 B.C.E.**

Assassination of Julius Caesar **44 B.C.E.** ▼

Octavian (Caesar Augustus) as Emperor; beginning of Roman Empire **27 B.C.E.** ▼

Tiberius, emperor **14–37 C.E.** ▼

Caligula, emperor **37–41 C.E.** ▼

Claudius, emperor **41–54 C.E.** ▼

Nero, emperor **54–68 C.E.** ▼

Year of four emperors **68–69 C.E.** ▼

Vespasian, emperor **69–79 C.E.** ▼

Titus, emperor **79–81 C.E.** ▼

Domitian, emperor **81–96 C.E.** ▼

Nerva, emperor **96–98 C.E.** ▼

Trajan, emperor **98–117 C.E.** ▼▼

Hadrian, emperor **117–138 C.E.** ▼▼

Figure 2.7 Time Line of Key Events in Hellenistic and Roman Times.

JUDAISM AS A GRECO-ROMAN RELIGION

It is reasonable to think that Judaism is the most important religion of the Greco-Roman world for understanding Jesus and emerging Christianity. Jesus was himself a Jew, as were his earliest followers. He was born to Jewish parents and raised in a Jewish culture; he worshipped the Jewish God, learned the Jewish Scriptures, kept Jewish customs, became a Jewish teacher, and preached to Jewish crowds. He was executed for allegedly claiming to be the Jewish king. What did it mean to be a Jew in the first century of the Roman empire?

I will postpone discussion of specific aspects of Judaism in Jesus' homeland of Palestine to a later chapter, where we will take up issues of the relationship of Jesus to his own environment (see chapter 15). There we will consider known Jewish groups (e.g. the Pharisees, Sadducees, and Essenes), the significance of the Dead Sea Scrolls, the political and social upheavals in Palestine during the first century, and so on. We will also see how the rich diversity of early Christianity and of Greco-Roman religion were matched by that of early Judaism. Some scholars have been so struck by this diversity that they opt to speak of early Judaisms rather than early Judaism. Even with this diversity, however, people in the ancient world appear to have meant something in particular when they called somebody a Jew. What might that have been?

Judaism was everywhere understood to be one of the religions of the Roman empire. Notwithstanding the caricatures that one sometimes reads, in which Judaism is said to have been absolutely unique and unlike other Greco-Roman religions, most people in the ancient world recognized it to be an ancient form of cultic devotion similar to others in many ways. Of course there were distinctive features, but every religion, not just Judaism, was distinctive.

Like other Greco-Roman religions, Judaism included the belief in a higher realm in which there was a powerful deity who could benefit humans and who showed special favor to those who worshipped him in ways prescribed from antiquity. The principal cultic acts of this religion involved animal sacrifice and prayer. Sacrifices were performed in a sacred temple (located in Jerusalem) by specially appointed priests. Portions of the animal, for most sacrifices, would be burned in honor of the deity. The priest would skin, prepare, and sometimes cook the carcass; the worshipper would then take it home to eat with his family and friends as a feast. Prayers were an important part of the worship of the Jewish God, usually addressing personal and communal needs (e.g., peace, fertility, prosperity, health.) In many fundamental respects, then, Judaism was comparable to other Greco-Roman religions. In other important ways, though, it was different.

Monotheism: The Belief in the One True God

As we have seen, virtually all of the religions in the empire were polytheistic. Before Christianity, Judaism alone was committed to the notion that there was one and only one true God who was to be worshipped and praised. To be sure, the difference between Jews and pagans on this score should not be blown out of proportion, as if they were absolutely dissimilar. We have already observed that some pagans, chiefly some philosophers and their followers, also believed that there was one chief deity who was ultimately responsible for the world and what happens within it, whether Zeus, Jupiter, or whoever else was thought to occupy the peak of the divine pyramid. The other gods, including the *daimonia* and the demigods, were of less power and eminence. Jews, too, believed that there were immortal beings, far greater in power than humans, who existed somewhere between them and the true God. In the modern world we might call these beings angels and archangels; for ancient Jews they also included such beings as the "cherubim" and "seraphim."

The key difference between Jews and persons of other religions, then, was not that Jews denied the existence of a hierarchy of supernatural beings; the difference was that Jews as a rule insisted that only the one Creator God, the supreme deity himself, was to be worshipped. Moreover, this one God was not the unknown and unknowable deity of some philosophers, nor was he the Greek Zeus or the Roman Jupiter. He was the God of the Jews, who was so holy—so far removed from anything that anyone could think or say—that even his name was not to be pronounced. Originally, this deity, like many others in the Greco-Roman world, was a local god who was worshipped in the land of Judea (or Judah, as it was earlier called). Those who worshipped this God were the people who lived there, the Judeans, whence we get the term "Jew."

About 550 years before Jesus, a large number of the Judeans were forced to leave their homeland because of a military, political, and economic crisis spawned by the invasion of the Babylonians. Many of those who relocated in places like Babylonia and Egypt retained their belief in the God of their homeland and continued to worship him in the ancient ways, maintaining the various customs followed in Judea—except, of course, that they could not worship in the Temple in Jerusalem (neither, though, could the Jerusalemites themselves for the better part of a century, as the building lay in ruin). Hence, by the Greco-Roman period being a Jew meant worshipping the God of the Judeans, that is, the God of Israel. Jews scattered throughout the world, away from Judea, were said to live in the "Diaspora," a term that literally means "dispersion." By the time of Jesus, there were far more Jews in the Diaspora than in Palestine. By some estimates, Jews comprised 7 percent of the total population of the Roman empire, which is usually set at around 60 million in the first century. Only a fraction of these lived in the Jewish homeland. Some scholars calculate that in the days of Jesus, twice as many Jews lived in Egypt as in all of Palestine itself.

Most of the Jews in the Diaspora stopped speaking Hebrew, the ancient tongue of Judea. By the second century before Jesus, many Jews read (or heard) their Scriptures only in Greek translation (see box 2.2), the so-called Septuagint translation.

Thus a distinctive feature of Jews around the world was that they did not worship a god of their own locality but the one God of their distant homeland, the God of Israel and no other. Moreover, they claimed that this God had shown them special favor. For most non-Jews this was an audacious claim (even though Romans, as we have seen, made similar claims about their own gods). Jews nonetheless maintained that the one God, the creator of heaven and earth, was uniquely their God. Hence, the second distinctive aspect of Judaism: their belief in the pact that God had made with Israel, or, using their own term, the covenant.

The Covenant: Israel's Pact with Its God

Most Jews were committed to the belief that the one true God had entered into a special relationship with them in the ancient past. God had chosen Israel from among all the other nations of the earth to be his special people. As part of his agreement with them, he promised that he, the creator and sustainer of all things, would protect and defend them in all their adversities.

Jews had ancient stories that told how God had fulfilled this promise. The most important were stories connected with the Exodus of the children of Israel from their slavery in Egypt, stories that eventually came to be embodied in the Jewish Scriptures. According to the ancient accounts, Israel had been maliciously subjected to forced labor for 400 years. God heard their cries and sent a savior, Moses, whose miraculous deeds compelled the king of Egypt to release them from bondage. Thus God delivered his people from slavery, destroying the powerful Egyptian army in the process, and brought them through trial and tribulation to the Promised Land. After they did battle with the nations who possessed the land, they entered in and became a great nation.

In light of God's actions on their behalf, Jews maintained that he had chosen them and made a covenant with them to be their God. That was his side of the agreement. In exchange, Jews were to obey his laws, laws pertaining to how they were to worship him and behave toward one another. As we will see, Jews as a rule did not consider this Law of God an onerous burden. Quite the contrary, the Law was God's greatest gift to his people. The existence of this divinely given Law, and the Jews' commitment to follow it, is a third distinctive aspect of this religion.

The Law: Israel's Covenantal Obligations

The English word "law" is a rather wooden translation of the Hebrew term "Torah," which is perhaps better rendered "guidance" or "direction." Ancient Jews sometimes used the word to refer to the set of laws that Moses received on Mount Sinai, as recorded in the books of Exodus, Leviticus, Numbers, and Deuteronomy. It was also used, though, to refer to these books themselves, along with their companion volume Genesis. These are the heart and soul of the Jewish Scriptures; today they are also sometimes called the "Pentateuch" (meaning "the five scrolls"). These books record the Jewish traditions of creation and primeval history, including the stories about Adam and Eve, Noah's ark, and the Tower of Babel, as well as the stories surrounding the Jewish Patriarchs and Matriarchs: Abraham and Sarah, Isaac and Rebeccah, Jacob and Leah and Rachel, and the twelve fathers of the twelve tribes of Israel, that is, Judah and his brothers. In addition, they narrate the traditions about Moses, the Exodus from Egypt, and the wanderings in the wilderness prior to the entry into the Promised Land. In particular, they contain the actual laws that God is said to have delivered to Moses on Mount Sinai after the Exodus from Egypt, laws that were to govern the worship of the Jews and their actions within their community, including, for example, the Ten Commandments.

Christians in the modern period frequently misunderstand the intent and purpose of this Jewish Law. It is not the case that ancient Jews (or modern ones, for that matter) generally thought that they had to keep all of the laws in order to earn God's favor. This was not a religion of works in the sense that one had to follow a long list of do's and don'ts in order to find salvation. Quite the contrary, as recent scholars have increasingly realized, ancient Jews were committed to following the Law because they had already been shown favor by God. The Jews were chosen to be God's special people, and the Law was given to show them how to live up to this calling. For this reason, keeping the Law was not a dreaded task that everyone hated; Jews typically considered the Law a great joy to uphold.

The Law consisted of rules pertaining to both cultic and communal life. There were laws on how to worship God properly and on how to live with one's neighbor. In the context of the first century, most of these laws would not have seemed out of the ordinary. Jews were not to commit murder or steal or bear false witness, they were to make restitution when they or something they owned did damage to a neighbor, and they were to perform sacrifices to God, following certain set practices. Even though other cults did not have written rules and regulations governing ethical behavior, there was nothing unusual in people wanting to encourage such activities. Other Jewish laws, however, did strike outsiders as peculiar. Jews, for example, were commanded to circumcise their baby boys—an act that they interpreted as the "sign of the covenant," for it showed that they (or at least the males among them) were distinct from all other nations as God's chosen people. Even though several other peoples (such as Egyptians) also practised circumcision, Jews in the empire were occasionally maligned for it, as it seemed to most outsiders to involve nothing short of forced mutilation.

Jews were also commanded not to work on the seventh day of the week, the Sabbath, but to keep it holy. Even though pagans observed periodic festivals in honor of their own gods, it was otherwise unheard of to take a weekly vacation from work. For the Jews this was a great good: for one day in seven they could relax from their labors with family and friends, enjoy a special meal, and join in a communal service of worship to their God. To some pagan observers, however, the custom showed that Jews were naturally lazy. Other laws that led to widespread derision involved the Jews' dietary restrictions. God had for some mysterious reason commanded Jews not to eat certain kinds of food, including pork and shellfish, common foods among other peoples in the Mediterranean region. This struck many outsiders as bizarre and superstitious.

Most Jews did not consider these laws (even the dietary ones) to be picayune requirements that few people wanted to follow and that nobody could. For comparison, consider the ancient Jewish legal code in light of our own. We too, for example, have laws against consuming certain edible substances (especially certain liquids, powders, and tablets). And our own legal system is far more complicated than anything available to the ancient Jew, indeed far more complicated than the average citizen can possibly understand (just consider our tax laws!). By comparison with modern law, the law embodied in the Jewish Torah was not particularly harsh or onerous or complicated. And for ancient Jews it was not the law of political bureaucrats; it was the law of God. Keeping it was a great joy, because doing so showed that the Jews were the elect people of God.

Temple and Synagogue: Israel's Places of Worship

There were two particularly important institutions for Jewish worship in the first century: the Temple in Jerusalem, where the animal sacrifices so central to the prescriptions of the Torah were to be performed, and the local synagogues, where Jews throughout the empire could worship God by studying and discussing the Law in the context of communal gathering and prayers.

The Jewish Temple. Jewish practices of animal sacrifice do not appear to have been so different from those of other ancient religions. Moreover, the Jewish Temple itself was not unlike other temples, it was a sacred structure in which the deity was believed to dwell, where worshippers could come to perform cultic acts in his honor and in hopes of receiving divine benefits as a result. At the same time, the Jewish Temple was known to be one of the grandest in the world of antiquity, spoken of with praise and admiration even by those who were not among its devotees. In the days of Jesus, the Temple complex encompassed an area roughly 500 yards by 325 yards, large enough, as one modern scholar has pointed out, to enclose twelve soccer fields (Sanders 1992). From the outside, its stone walls rose 100 feet from the street, as high as a modern ten-story building. No mortar had been used in its construction; instead, the

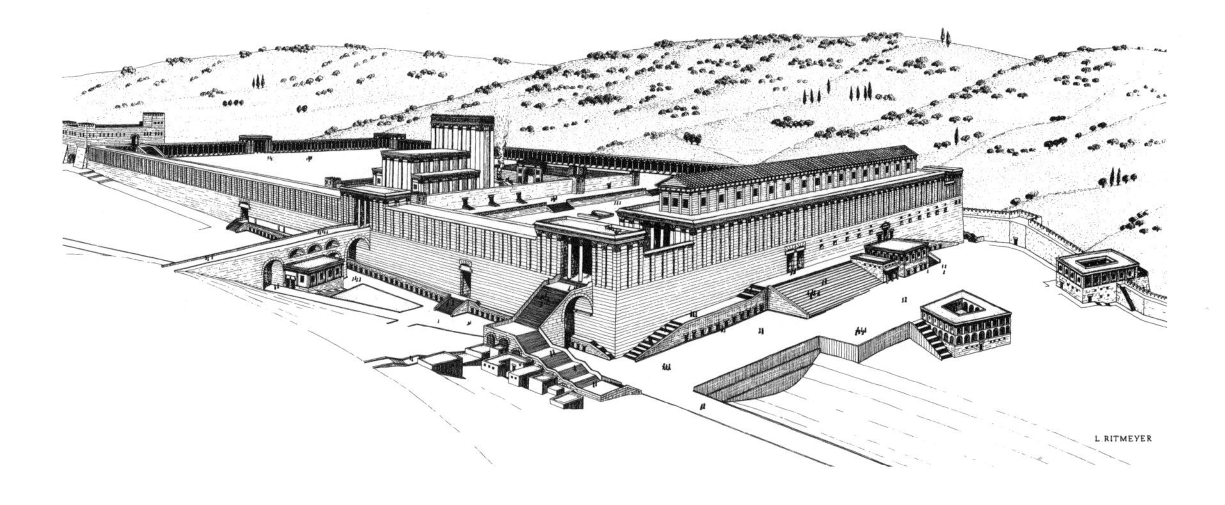

Figure 2.8 A pictorial reconstruction of the Jewish Temple in Jerusalem.

stones, some of them 50 yards in length, had been carefully cut to fit together neatly. The gates into the temple were 45 feet high by 44 feet wide (with two doors, 22 feet wide, in each); one ancient source indicates that 200 men were required to close them each evening. From all of our ancient descriptions, the Temple complex appears to have been a fantastically beautiful set of buildings made with the best materials money could buy, including gold, which overlaid extensive portions of the structures. As you might imagine, its construction was an immense feat; when it was completed in 63 C.E., 18,000 local workers were reportedly left unemployed. It was destroyed just seven years later at the climax of the Jewish war against Rome, never to be built again.

One of the things that made the Jerusalem Temple unique in the Greco-Roman world is that in the opinion of most Jews of the period, it was to be the only temple for the God of Israel. Whereas numerous temples could be devoted to any of the pagan gods, this God would receive sacrifices only in the Temple in Jerusalem. Jews from around the world, even those who never set foot inside, paid an annual tax to help defray the costs of its upkeep and administration. In no small measure, this special reverence for the place derived from the belief that God himself dwelt in the Temple, in a special room called the Holy of Holies. The belief that a god might actually be present in a holy place was widespread throughout antiquity. In most ancient temples, however, the deity was present in the cult image, or "idol," kept in a sacred room. The sacred room in the Jerusalem Temple, on the other hand, was completely empty. Since the Jewish God was so holy, unlike all else that is, he explicitly forbade any images to be made of him.

No one could enter this holiest of rooms except the Jewish high priest, and he did so only once a year on the Day of Atonement (Yom Kippur), when he performed a sacrifice for the sins of the people. The Holy of Holies was thus the most sacred spot in the Temple and the rest of the building complex was structured so as to emphasize the holiness that emanated from its center. Before the Holy of Holies was the sanctuary, into which only certain priests could go; around it was the court of the priests, which allowed only priests and their assistants, the Levites. Farther out was the court of the Israelites, into which only Jewish men could go to bring their offerings to the priests. Beyond that was the court of (Jewish) women, who were not allowed any nearer to the inner sanctum (Jewish men could assemble there as well), and finally beyond that came the court of the Gentiles, where even non-Jews could congregate.

Thus, the idea of a temple and the activities of prayer and sacrifice that transpired there were not so different from what one could find in other religions in the empire. Apart from the details of the cultic ceremonies (which, of course, differed to some degree in all ancient religions), what made this Temple unlike others was the fact that, according to its adherents, it was the only one to be built to their God, who dwelt there in holiness apart from any sacred image.

The Synagogue. Despite the fact that Jews from around the world paid an annual tax to support the Temple, most could not worship there on a regular basis. Indeed, many could not afford to make a pilgrimage there, ever. For this reason, apparently, centuries before Jesus—scholars debate when, exactly—Jews in the Diaspora devised an alternative mode of worship, one that did not involve sacrifice of animals but focused instead on discussing the sacred traditions of the Torah and praying to the God of Israel. These activities took place in the community, as Jews came together on the Sabbath in either a home or a separate meeting place, sometimes a freestanding building, usually under the leadership of the more highly educated and literate of their members. The Scriptures were read and discussed and set prayers were said. These gatherings were called "synagogues," from the Greek word for "gathering together," a term that eventually came to refer to the building in which the meetings took place.

By the time of Jesus there were synagogues wherever there were communities of Jews in the empire, both in Palestine and abroad. In many respects these were not unlike the gathering places of like-minded individuals among non-Jews, where certain religious activities occurred and prayers were said. Greco-Roman "associations" were commonly organized, for example, for workers of the same trade in a locale, who might share a range of common interests. It was not unusual to find other associations organized for the purpose of periodic social gatherings, where members would pool their funds to provide ample food and drink and, perhaps strangely to the modern observer, provide, through a reserve, a proper burial for their deceased members.

Rarely, though, would such organizations, whether trade associations or funeral clubs, include men, women, and children; rarely would they meet together every week; and rarely would they devote themselves principally to the purposes of prayer and discussion of sacred traditions. To this extent, Jewish synagogues were distinctive.

The Jewish Context for the Traditions about Jesus

Although Judaism had some features that set it apart from other cults, Jews around the world did not all agree on every major aspect of their religion or engage in the same religious practices. As we shall see, this religion was richly diverse in the first century of the Common Era. At the same time, Judaism was not altogether unlike other religions of the empire. As we have seen, for example, even some pagans could accept the notion of monotheism. They also accepted that the gods had made special provisions for certain people (for example, the state gods of Rome), that they had given certain commandments (such as how to worship them), and that they were to be honored in certain places (temples) in certain ways, including prescribed prayers and sacrifices. Thus Judaism should be seen as one of the Greco-Roman religions, distinct and yet similar to the others, just as all religions of that world were distinct and yet similar to one another.

There is one further similarity between Judaism and the pagan religions of its environment, a similarity of particular importance to the traditions about Jesus that circulated throughout this world. Just as Judaism shared with other religions the notion that there were other divine beings of lesser majesty and power than the one true God, so too it maintained that these other divine beings sometimes appeared to people in human form. There are records of such appearances in the Jewish Scriptures, as when angels came and spoke to humans, imparting a divine revelation or performing a spectacular miracle. Moreover, there are accounts in Judaism of human beings who appeared to be far more than human. For example, Moses was said in the Hebrew Scriptures to have performed miracles through the power of God (e.g., sending the plagues against Egypt), the prophet named Elisha reportedly healed the blind and multiplied loaves for the hungry, and Elijah overwhelmed his opponents through the power of God, supplied food and drink to those in need, and even raised the dead.

Outside of the Hebrew Scriptures we know of Jews who were thought to stand in a special relation with God. These Jewish holy men, sometimes called the sons of God, reportedly could heal the sick and calm the storm. Some Jews believed that God spoke directly and intimately to them. The later rabbis sometimes told stories of such holy men, some of whom lived near the time of Jesus, also in Galilee. For example, Hanina ben Dosa and Honi the "circle-drawer" were famous among the rabbis for their memorable teachings and miraculous deeds. Thus the stories about Jesus, the miracle-working Son of God, would have made sense not only to pagans, who were familiar with accounts of divine men, but to Jews as well, whether in Palestine or the diaspora.

SUGGESTIONS FOR FURTHER READING

Anthologies of Texts

Barrett, C. K., ed. *The New Testament Background: Selected Documents*. 2d. ed. New York: Harper & Row, 1989. A standard collection of Jewish and pagan texts relevant to the study of the New Testament.

Cartlidge, David R., and David L. Dungan, eds. *Documents for the Study of the Gospels*. 2d ed. Philadelphia: Fortress, 1994. Presents a valuable selection of ancient literary texts that are closely parallel to the New Testament Gospels, including portions of Philostratus's Life of Apollonius.

Charlesworth, James H., ed., *The Old Testament Pseudepigrapha*. 2 vols. Garden City, N.Y.: Doubleday, 1983, 1985. The most complete collection of noncanonical writings of early Judaism from before and around the time of the New Testament, with full and informative introductions.

Lane, Eugene, and Ramsey MacMullen, eds. *Paganism and Christianity: 100–425 C.E.: A Sourcebook*. Philadelphia: Fortress, 1992. A handy anthology of ancient texts that deal specifically with religion in the Greco-Roman world.

Lefkowitz, Mary R., and Maureen B. Fant, eds. *Women's Lives in Greece and Rome: A Source Book in Translation*. Baltimore: Johns Hopkins University Press, 1982. A superb collection of ancient texts illuminating all the major aspects of women's lives in the Greco-Roman world.

Meyer, Marvin, ed. *The Ancient Mysteries: A Sourcebook*. San Francisco: Harper & Row, 1987. An anthology of ancient literary texts that discuss the mystery cults, with helpful introductions.

Shelton, Jo-Ann, ed. *As the Romans Did: A Source Book in Roman Social History*. New York: Oxford University Press, 1988. A very useful anthology of ancient texts dealing with every major aspect of life in the Roman world, including religion.

Vermes, Geza, ed. *The Dead Sea Scrolls in English*. 3d ed. Baltimore: Penguin Books, 1987. The most accessible collection and translation of the Dead Sea Scrolls available in English, with a clear introduction.

Studies of the Greco-Roman World

"Pagan" Religions

Ferguson, John. *Religions in the Roman Empire*. Ithaca, N.Y.: Cornell University, 1970. An introductory overview of the wide variety of Roman religions, with some emphasis on archaeological and other nonliterary sources.

Howatson, M. C., ed. *Oxford Companion to Classical Literature*. 2d ed. Oxford: Oxford University Press, 1989. For quick reference to names, myths, literary works, events, and other aspects of the ancient Greek and Roman worlds, this work is an indispensable tool for beginning students.

Lane Fox, Robin. *Pagans and Christians*. New York: Alfred A. Knopf, 1987. A long but fascinating discussion of the relationship of pagans and Christians during the first centuries of Christianity, valuable especially for its brilliant sketch of what it meant to be a pagan in the second and third centuries of the Common Era.

MacMullen, Ramsey. *Paganism in the Roman Empire*. New Haven, Conn.: Yale University, 1981. An authoritative discussion of the nature of Roman religion, for somewhat more advanced students.

Early Judaism

Cohen, Shaye. *From the Maccabbees to the Mishnah*. Philadelphia: Westminster Press, 1987. Perhaps the best place for beginning students to turn for a clear overview of early Judaism.

Kraft, Robert, and George Nicklesburg, eds. *Early Judaism and its Modern Interpreters*. Philadelphia: Fortress Atlanta: Scholars, 1986. A collection of significant essays on major aspects of early Judaism, for more advanced students.

Sanders, E. P. *Judaism Practice and Belief, 63 B.C.E.–66 C.E.* London: SCM Press; Philadelphia: Trinity Press International, 1992. A full, detailed, and authoritative account of what it meant to be a Jew immediately before and during the time of the New Testament.

Sandmel, Samuel. *Judaism and Christian Beginnings*. New York: Oxford University Press, 1978. An insightful introductory sketch of early Judaism by a prominent Jewish scholar.

The Social World of Early Christianity

Malherbe, Abraham. *Social Aspects of Early Christianity*. 2d ed. Philadelphia: Fortress, 1983. A clear and interesting treament of early Christianity from a socio-historical, rather than literary or theological, perspective; ideal for beginning students.

Stambaugh, J. E. and D. L. Balch. *The New Testament in Its Social Environment*. Philadelphia: Westminster, 1986. A nice overview of major aspects of the social world of early Christianity, including discussions of ancient modes of communication and transportation, ancient economies, social classes, education, and urban life.

CHAPTER 3

Where It All Began: The Traditions of Jesus in Their Greco-Roman Context

We have already touched on one of the ironies involved in the historical study of the New Testament. If we choose to begin our study not with the earliest New Testament author, Paul, but with the person on whom his religion is in some sense based, Jesus, then we are compelled to begin by examining books that were written *after* Paul, Indeed, some of these books were among the last New Testament books to be produced. To reach the beginning, we have to start near the end.

At the same time, even though the Gospels themselves were written relatively late, they preserve traditions about Jesus that existed much earlier, many of them circulating among Christians long before Paul wrote his letters. Now that we have discussed several important aspects of the Greco-Roman environment within which the Christian religion was born and grew, we can examine the traditions themselves, as embodied near the end of the first century in the Gospels of Matthew, Mark, Luke, and John, and somewhat later in the Gospels ascribed to Peter and Thomas. How did these various authors acquire their traditions about Jesus?

ORAL TRADITIONS BEHIND THE GOSPELS

For the moment, we will leave aside the question of who these authors were (see "Some Additional Reflections" at the end of the chapter), except to point out that all of the New Testament Gospels are anonymous; their authors did not sign their names. Our principal concern at present involves a different issue, namely how and where these anonymous authors acquired their stories about Jesus. Here we are in the fortunate position of having some definite information, for one of these authors deals directly with this matter. Luke (we do not know his real name) begins his Gospel by mentioning earlier written accounts of Jesus' life and by indicating that both he and his predecessors acquired their information from Christians who had told stories about him (Luke 1:1–4). That is to say, these writings were based to some extent on oral traditions, stories that had circulated among Christians from the time Jesus died to the moment the Gospel writers put pen to paper. How much of an interval, exactly, was this?

No one knows for certain when Jesus died, but scholars agree that it was sometime around 30 C.E. In addition, most historians think that Mark was the first of our Gospels to be written, sometime between the mid 60s to early 70s. Matthew and Luke were probably produced some ten or fifteen years later, perhaps around 80 or 85. John was written perhaps ten years after that, in 90 or 95. These are necessarily rough estimates, but almost all scholars agree within a few years.

Perhaps the most striking thing about these dates for the historian is the long interval between Jesus' death and the earliest accounts of his life. Our first written narratives of Jesus (i.e., the Gospels) appear to date from thirty-five to sixty-five years after the fact. This may not seem like a

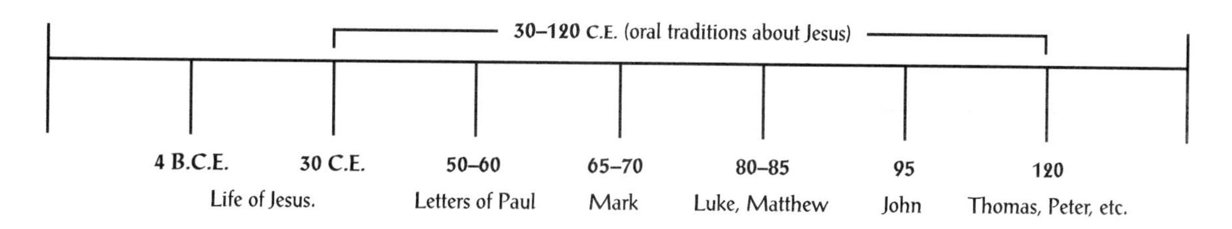

Figure 3.1 Time Line of the Early Christian Movement.

long time, but think about it in modern terms. For the shortest interval, (the gap between Jesus and Mark), this would be like having the first written record of Eisenhower's presidency appear today. For the longest interval (between Jesus and John), it would be like having stories about a famous preacher from the early years of the Great Depression show up in print for the first time this week. We should not assume that the Gospel accounts are necessarily unreliable simply because they are late, but the dates should give us pause. What was happening over these thirty, forty, fifty, or sixty years between Jesus' death and the writing of the Gospels?

Without a doubt, the most important thing that was happening for early Christianity was the spread of the religion from its inauspicious beginnings as a tiny sect of Jesus' Jewish followers in Jerusalem—the Gospels indicate that there were eleven men and several women who remained faithful to him after his crucifixion, say a total of fifteen or twenty people altogether—to its status as a world religion enthusiastically supported by Christian believers in major urban areas throughout the Roman Empire. Missionaries like Paul actively propagated the faith, converting Jews and Gentiles to faith in Christ as the Son of God, who was crucified for the sins of the world and then raised by God from the dead.

We know that this tiny group of Jesus' disciples had so multiplied by the end of the first century that there were believing communities in cities of Judea and Samaria and Galilee, probably in the region East of Jordan; in Syria, Cilicia, and Asia Minor; in Macedonia and Achaia (modern day Greece); in Italy; and possibly in Spain. By this time Christian churches may have sprung up in the Southern Mediterranean, probably in Egypt and possibly in North Africa as well.

To be sure the Christians did not take the world by storm. As we will see later in Chapter 25, Roman officials in the provinces appear to have taken little notice of the Christians until the second century; strikingly, there is not a single reference to Jesus or his followers in pagan literature of any kind during the first century of the Common Era. Nonetheless, the Christian religion quietly and persistently spread, not converting millions of people, but almost certainly converting thousands in numerous locations throughout the entire Mediterranean.

What did Christians tell people in order to convert them? Our evidence here is frustratingly sparse: examples of missionary sermons in the book of Acts and some intimations of Paul's preaching in his own letters (e.g. 1 Thess 1:9–10). We cannot tell how representative these are. Moreover, there are good reasons for thinking that most of the Christian mission was conducted not through public preaching, say on a crowded street corner, but privately, as individuals who had come to believe that Jesus was the Son of God told others about their newfound faith and tried to convince them to adopt it as well.

Since we know that in the Greco-Roman world religion was a way of securing the favor of the gods, we are probably not too far afield to think that if faith in Jesus were known to produce beneficial, or even miraculous, results, then people might be persuaded to convert. If a Christian testified, for example, that praying to Jesus, or through Jesus to God, had healed her daughter, or that a representative of Jesus had cast out an evil spirit, or that the God of Jesus had miraculously provided food for a starving family, this might spark interest in

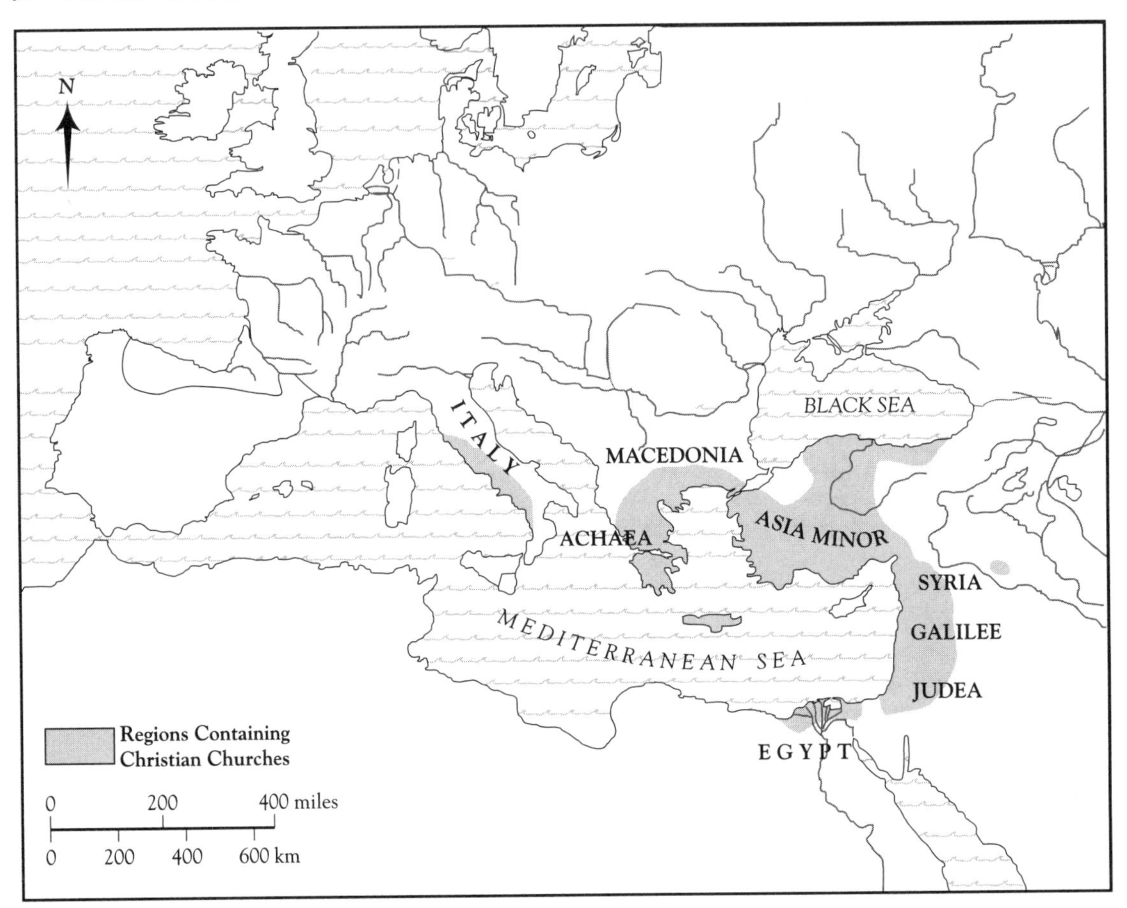

Figure 3.2 Christian Churches in Existence by 100 C.E.

her neighbor or co-worker. Those with an interest in Jesus would want to learn more about him. Who was he? When did he live? What did he do? How did he die? The Christian, in turn, would be both compelled and gratified to tell stories about Jesus to anyone interested.

Such opportunities to tell stories about Jesus must have presented themselves throughout the major urban areas of the Mediterranean for decades prior to the writing of the Gospels. Otherwise there is no way to account for the spread of the religion in an age that did not enjoy the benefits of telecommunication. When people had heard enough (however much that might have been), they might have decided to believe in Jesus. This would have involved, among other things, adopting aspects of Jesus' own religion, which for non-Jews meant accepting the Jewish God and abandoning their own, since Jews maintained that this One alone was the true God. Once the converts did so, they could join the Christian community by being baptized and receiving some rudimentary instruction. Presumably it was the leaders of the Christian congregation who performed the baptisms and taught the converts. These leaders would have been the earliest people to adopt the new religion in the locality or people with special gifts for leadership, possibly the more highly educated among them, who were therefore best suited to giving instruction.

We do not know exactly what the leaders would have told new converts, but we can imagine that they would have imparted some of the essentials of the faith: information about the one true God, his creation, and his son Jesus. To some extent, this would have involved telling yet other stories about who Jesus was, about how he came into the world, about what he taught, what he did, why he suffered, and how he died. Stories about Jesus were thus being told throughout the Mediterranean for decades, both to win people to faith and to edify those who had been brought in. They were told in evangelism, in instruction, and probably in services of worship.

We do not know the precise identity of those who were telling the stories about Jesus. Was every story told by one of the apostles? Impossible. The mission goes on for years and years and years all over the map. Were the stories told by other eyewitnesses? Equally impossible. They must have been told, then, for the most part, by people who had not been there to see them happen, who had heard them from other people, who also had not been there to see them happen. The stories were passed on by word of mouth from one convert to the next. They were told in different countries, in Egypt, Judea, Galilee, Syria, and Cilicia, throughout Asia Minor, Macedonia, Achaia, Italy, and Spain. They were told in different contexts, for different reasons, at different times. They were told in a language other than Jesus' own (he spoke Aramaic, while most of the converts spoke Greek), often by people who were not Jews, almost always by people who were not eyewitnesses and had never met an eyewitness.

Let me illustrate the process with a hypothetical example. Suppose I am a Greek- speaking worshipper of the goddess Artemis from Ephesus. I listen to a stranger passing through town, who tells of the wonders of Jesus, of his miracles and supernatural wisdom. I become intrigued. When I hear that this wandering stranger has performed miracles in Jesus' name— my neighbor's son was ill, but two days after the stranger prayed over him, he became well—I decide to inquire further. He tells of how Jesus performed great miracles and of how, even though wrongly accused by the Romans for sedition and crucified, he was raised by God from the dead. Based on everything I've heard, I decide to forego my devotion to Artemis. I put my faith in Jesus, get baptized, and join the local community.

I take a trip for business to nearby Smyrna. While there, I tell friends about my new faith and the stories I've learned about my new Lord. Three of them join me in becoming Christian. They begin to discuss these things with their neighbors and friends. Mostly they are rejected, but they

Figure 3.3 Stories of the power of the gods to heal the sick were widespread in the Greco-Roman world. Here we see a relief from the temple of the healing god Asclepius in the city of Piraeus, showing the god and his female assistant (on the right) curing a sleeping patient.

acquire several converts, enough to come together once a week for worship, to discuss their faith, and to tell more stories. These new converts tell their own families the stories, converting some of them, who then take the word yet further afield.

And so it goes. As the new converts tell the stories, the religion grows, and most of the people telling the stories are not eyewitnesses. Indeed they have never laid eyes on an eyewitness or anyone else who has.

This example does not imply that if we had accounts based on eyewitnesses, they would necessarily be accurate. Even the testimonies of eyewitnesses can, and often do, conflict. But the scenario I have painted does help to explain why there are so many differences in the stories about Jesus that have survived from the early years of Christianity. These stories were circulated year after year after year, primarily by people who had believed their entire lives that the gods were sometimes present on earth, who knew of miracle workers who had appeared to benefit the human race, who had themselves heard fantastic stories about this Jewish holy man Jesus, and who were trying to convert others to their faith or to edify those who had already been converted. Furthermore, nearly all of these story tellers had no independent knowledge of what really happened. It takes little imagination to realize what happened to the stories.

You are probably familiar with the old birthday party game "telephone." A group of kids sits in a circle, the first tells a brief story to the one sitting next to her, who tells it to the next, and to the next, and so on, until it comes back full circle to the one who started it. Invariably, the story has changed so much in the process of retelling that everyone gets a good laugh. Imagine this same activity taking place, not in a solitary living room with ten kids on one afternoon, but over the expanse of the Roman Empire (some 2,500 miles across), with thousands of participants—from different backgrounds, with different concerns, and in different contexts—some of whom have to translate the stories into different languages (see box 3.1).

The situation, in fact, was even more complicated than that. People in the Christian communities that sprang up around the Mediterranean encountered severe difficulties in living their daily lives and thus sought help and direction from on high. The traditions about Jesus were part of the bedrock of these communities; his actions were a model that Christians tried to emulate; his words were teachings they obeyed. Given this context, is it conceivable that Christians could have made up a story that proved useful in a particular situation? Creating a story is not far removed from changing one, and presumably people would have good reasons for doing both.

Christians would not have to be deceitful or malicious to invent a story about something that Jesus said or did; they would not even have to be conscious of doing so. All sorts of stories about people are made up without ill intent, and sometimes stories are told about persons that we know are not historically accurate: ask any well-known person who is widely talked about, a politician, religious leader, or university professor.

The Nature of the Gospel Traditions

It does not appear that the authors of the early Gospels were eyewitnesses to the events that they narrate. But they must have gotten their stories from somewhere. Indeed, one of them acknowledges that he has heard stories about Jesus and read earlier accounts (Luke 1:1–4). In the opinion of most New Testament scholars, it is possible that in addition to preserving genuine historical recollections about what Jesus actually said and did, these authors also narrated stories that had been modified, or even invented, in the process of retelling.

The notion that the Gospels contain at least some stories that had been changed over the years is not pure speculation; in fact, we have hard evidence of this preserved in the Gospels themselves (we will examine some of this evidence in a moment). We also have reason to think that early Christians were not particularly concerned that stories about Jesus were being changed. Odd as it may seem to us, most believers appear to have been less concerned than we are about what we would call the facts of history. Even though we as twentieth-century persons tend to think that something cannot be true unless it happened, ancient Christians, along with a lot of other ancient people, did not think this way. For them, something could be true whether or not it happened. What mattered more than historical fact was what we might call religious or moral truth.

On one level, even modern people consider "moral truth" to be more important than historical fact, that is, they will occasionally concede that something can be true even if it didn't happen. Let me illustrate. Every second grader in the country has heard the story of George Washington and the cherry tree. As a young lad, George takes the axe to

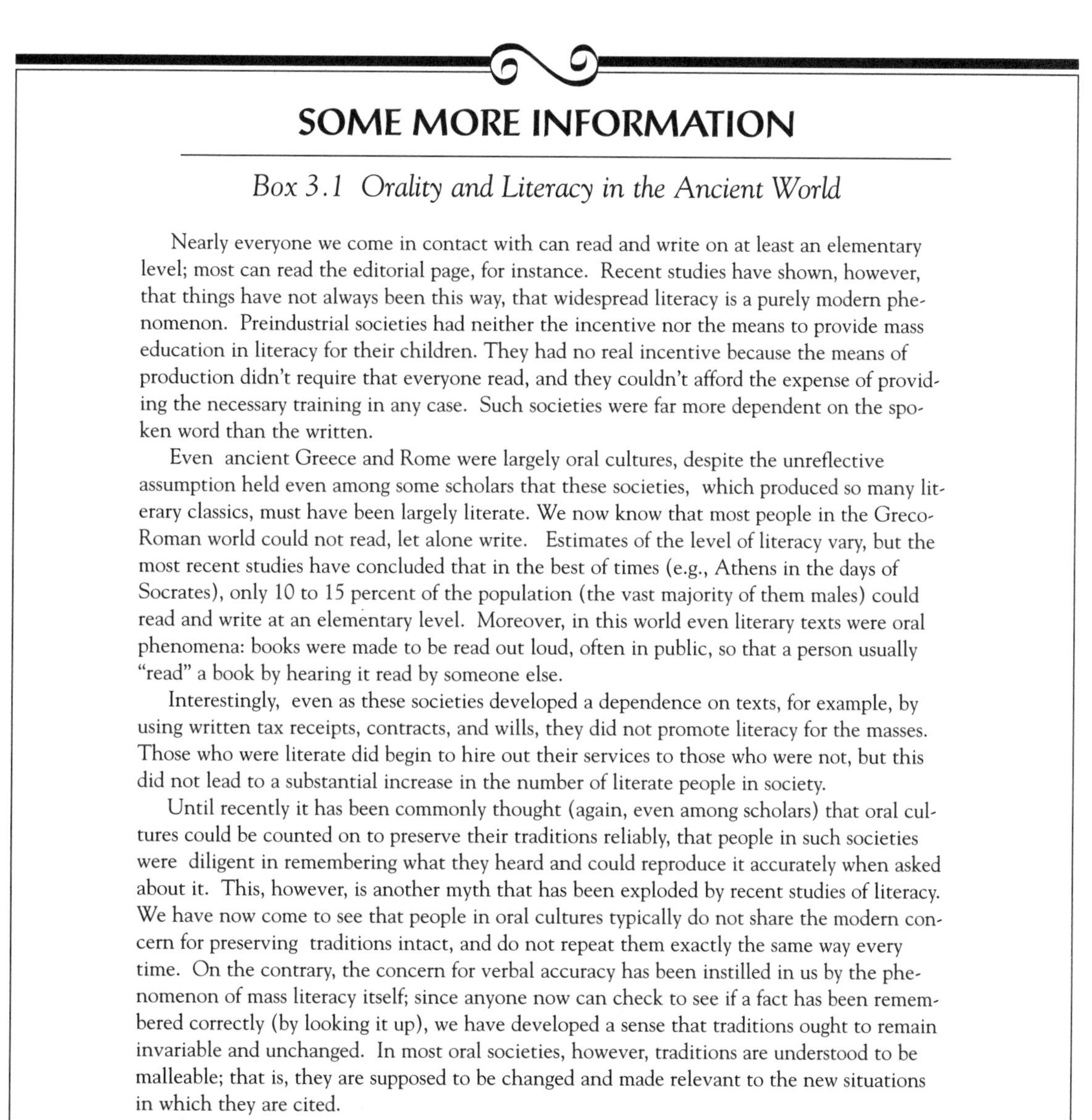

SOME MORE INFORMATION

Box 3.1 Orality and Literacy in the Ancient World

Nearly everyone we come in contact with can read and write on at least an elementary level; most can read the editorial page, for instance. Recent studies have shown, however, that things have not always been this way, that widespread literacy is a purely modern phenomenon. Preindustrial societies had neither the incentive nor the means to provide mass education in literacy for their children. They had no real incentive because the means of production didn't require that everyone read, and they couldn't afford the expense of providing the necessary training in any case. Such societies were far more dependent on the spoken word than the written.

Even ancient Greece and Rome were largely oral cultures, despite the unreflective assumption held even among some scholars that these societies, which produced so many literary classics, must have been largely literate. We now know that most people in the Greco-Roman world could not read, let alone write. Estimates of the level of literacy vary, but the most recent studies have concluded that in the best of times (e.g., Athens in the days of Socrates), only 10 to 15 percent of the population (the vast majority of them males) could read and write at an elementary level. Moreover, in this world even literary texts were oral phenomena: books were made to be read out loud, often in public, so that a person usually "read" a book by hearing it read by someone else.

Interestingly, even as these societies developed a dependence on texts, for example, by using written tax receipts, contracts, and wills, they did not promote literacy for the masses. Those who were literate did begin to hire out their services to those who were not, but this did not lead to a substantial increase in the number of literate people in society.

Until recently it has been commonly thought (again, even among scholars) that oral cultures could be counted on to preserve their traditions reliably, that people in such societies were diligent in remembering what they heard and could reproduce it accurately when asked about it. This, however, is another myth that has been exploded by recent studies of literacy. We have now come to see that people in oral cultures typically do not share the modern concern for preserving traditions intact, and do not repeat them exactly the same way every time. On the contrary, the concern for verbal accuracy has been instilled in us by the phenomenon of mass literacy itself; since anyone now can check to see if a fact has been remembered correctly (by looking it up), we have developed a sense that traditions ought to remain invariable and unchanged. In most oral societies, however, traditions are understood to be malleable; that is, they are supposed to be changed and made relevant to the new situations in which they are cited.

The importance of these new studies should be obvious, as we begin to reflect on the fate of the traditions about Jesus as they spread by word of mouth throughout the largely illiterate Greco-Roman world.

the tree in his father's front yard. When his father comes home and asks, "Who cut down my cherry tree?" George confesses, "I cannot tell a lie. I did it."

Historians know that this never happened. In fact, the Christian minister who fabricated the story later admitted to having done so. Why then do we tell the story? For one thing, the story stresses one of the ultimate values that we claim as a country. We use the story to teach children that our country is rooted in integrity. Who was George Washington? He was the father of our nation. What kind of man was he? He was an honest man, a man of integrity! Really? How honest was he? Well, one time when he was a boy. . . .

The account of George Washington and the cherry tree is told for at least one other reason as well, relating not so much to national image as to personal ethics. We tell this story to children because we want them to know that they should not lie under any circumstances. Even if they've done something bad, something harmful, they should not try to deceive others about it. It is better to come clean and deal with the consequences than to distort the truth and make things worse. So we tell the story, not because it really happened, but because in some sense we think it is true.

The stories about Jesus in the early church may have been similar. To be sure, many of them are accounts of things that really did happen (part of our task will be figuring out which ones did). Others are historical reminiscences that have been changed, sometimes a little, sometimes a lot, in the retelling. Others were made up by Christians, possibly well-meaning Christians, at some point prior to the writing of the Gospels. But they all are meant to convey the truth, as the storyteller saw it, about Jesus.

The evidence that stories about Jesus were changed (or made up) in the process or retelling can be found in the stories themselves as they have come down to us in the Gospels. In numerous instances different Gospels tell the same story, but the stories differ in significant ways. Sometimes these differences represent simple shifts in emphasis. At other times, however, they represent irreconcilable conflicts. What is striking is that whether the changes are reconcilable or not, they often point to an attempt by some early Christian storyteller to convey an important idea about Jesus. Here we will look at just one example; dozens could easily be cited. The point is that many of the earliest Christians appear to have been willing to change a historical fact to make a theological point.

The illustration I have chosen concerns a small detail with profound implications—the day and time of Jesus' death, which are described differently in the Gospels. All four Gospels of the New Testament indicate that Jesus was crucified sometime during Passover week, in Jerusalem, on orders of the Roman governor, Pontius Pilate, but there is a slight discrepancy in the accounts. To understand it, you will need some background information.

In the days of Jesus, Passover was the most important Jewish festival. It commemorated the exodus of the children of Israel from their bondage in Egypt. The Hebrew Scriptures narrate the commemorative event itself (Exod 7–12). According to the ancient accounts, God raised up Moses to deliver his people and through him brought ten plagues on the land of Egypt to convince the Pharaoh to set his people free. The tenth plague was by far the worst: the death of every first-born human and animal in the land. In preparation for the onslaught, God instructed Moses to have every family of the Israelites sacrifice a lamb and spread its blood on the lintels and doorposts of their houses. In that way, when the angel of death came to bring destruction, he would see the blood on the doors of the Israelites and "pass over" them to go to the homes of the Egyptians.

The children of Israel were told to eat a quick meal in preparation for their escape. There was not time even to allow the bread to rise; they were therefore to eat it unleavened. The Israelites did as they were told; the angel of death came and went. The Pharaoh pleaded with the children of Israel to leave, they fled to the Red Sea, where they made their final escape through the parted waters.

The Israelites were instructed through Moses to commemorate this event annually. Hundreds of years later, in the days of Jesus, the Passover celebration brought large numbers of pilgrims to Jerusalem, where they would participate in sacrifices in the Temple and eat a sacred meal of symbolic foods, including a lamb, bitter herbs to recall their bitter hardship in Egypt, unleavened bread, and several cups of wine. The sequence of events was typically as follows. Lambs would be brought to the Temple, or purchased there, for sacrifice with the assistance of a priest. They would then be prepared for the

Passover meal by being skinned, drained of their blood, and possibly butchered. Each person or family who brought a lamb would then take it home and prepare the meal. That evening was the Passover feast, which inaugurated the weeklong celebration called the Feast of Unleavened Bread.

As you may know, in Jewish reckoning, a new day begins when it gets dark (that is why the Jewish Sabbath begins on Friday evening). So the lambs would be prepared for the Passover meal on the afternoon of the day before the meal would actually be eaten. When it got dark, the new day started, and the meal could begin.

This now takes us to the dating of Jesus' execution. The Gospel of Mark, probably our earliest account, clearly indicates when Jesus was put on trial. On the preceding day, according to Mark 14:12, the disciples ask Jesus where he would have them "prepare" the Passover. This is said to happen on the day when the priests "sacrifice the passover lamb," or the day of Preparation for the Passover (the afternoon before the Passover meal). Jesus gives them their instructions and they make the preparations. That evening—the start of the next day for them—they celebrate the meal together (14:17–25).

At this special occasion, Jesus takes the symbolic foods of the meal and endows them with additional symbolic meaning, saying, "This is my body . . . this is my blood of the covenant" (14:22–24). Afterwards, he goes with his disciples to the Garden of Gethsemane, where he is betrayed by Judas Iscariot and arrested (14:32, 43). He is immediately put on trial before the Jewish Council, the Sanhedrin (14:53). He spends the night in jail; early in the morning the Sanhedrin delivers him over to Pilate (15:1). After a short trial, Pilate condemns him to death. He is led off to be crucified, and is nailed to the cross at 9:00 a.m. (15:25). Thus, in the Gospel of Mark, Jesus is executed the day after the Preparation of the Passover, that is, on the morning after the Passover meal had been eaten.

Our latest canonical account of this event is in the Gospel of John. Many of the details here are similar to Mark: the same persons are involved and many of the same stories are told. There are differences, though, and some of these are significant. John's account of the trial before Pilate, for example, is much more elaborate (18:28–19:16). In part, this is because in his version the Jewish leaders refuse to enter Pilate's place of residence and send Jesus in to face Pilate alone. As a result, Pilate has to conduct the trial by going back and forth between the prosecution and the defendant, engaging in relatively lengthy conversations with both before pronouncing his verdict. What is particularly striking, and significant for our investigation here, is that we are told exactly when the trial comes to an end with Pilate's verdict: "Now it was the day of Preparation for the Passover, and it was about 12:00 noon" (John 19:14). Jesus is immediately sent off to be crucified (19:16).

The day of Preparation for the Passover? This is the day before the Passover meal was eaten, the day the priests began to sacrifice the lambs at noon. How could this be? In Mark, Jesus had his disciples prepare the Passover on that day, and then he ate the meal with them in the evening after it became dark, only to be arrested afterwards.

If you read John's account carefully, you will notice other indications that Jesus is said to be executed on a different day than he is in Mark. John 18:28, for example, gives the reason that the Jewish leaders refuse to enter into Pilate's place of residence for Jesus' trial. It is because they do not want to become ritually defiled, and thereby prevented from eating the Passover meal that evening (recall, in Mark, they would have eaten the meal the evening before the trial). This difference in dating explains another interesting feature of John's Gospel. In this account Jesus never instructs his disciples to prepare for the Passover, and he evidently does not eat a Passover meal during his last evening with them (he does not, for example, take the symbolic foods and say, "This is my body" and "This is my blood"). The reason for these differences should by now be clear: in John's Gospel, Jesus was already in his tomb by the time of this meal.

We seem to be left with a difference that is difficult to reconcile. Both Mark and John indicate the day and hour of Jesus' death, but they disagree. In John's account, he is executed on the day on which preparations were being made to eat the Passover meal, sometime after noon. In Mark's account he is killed the following day, the morning after the passover meal had been eaten, sometime around 9:00 a.m. If we grant that there is a difference, how do we explain it?

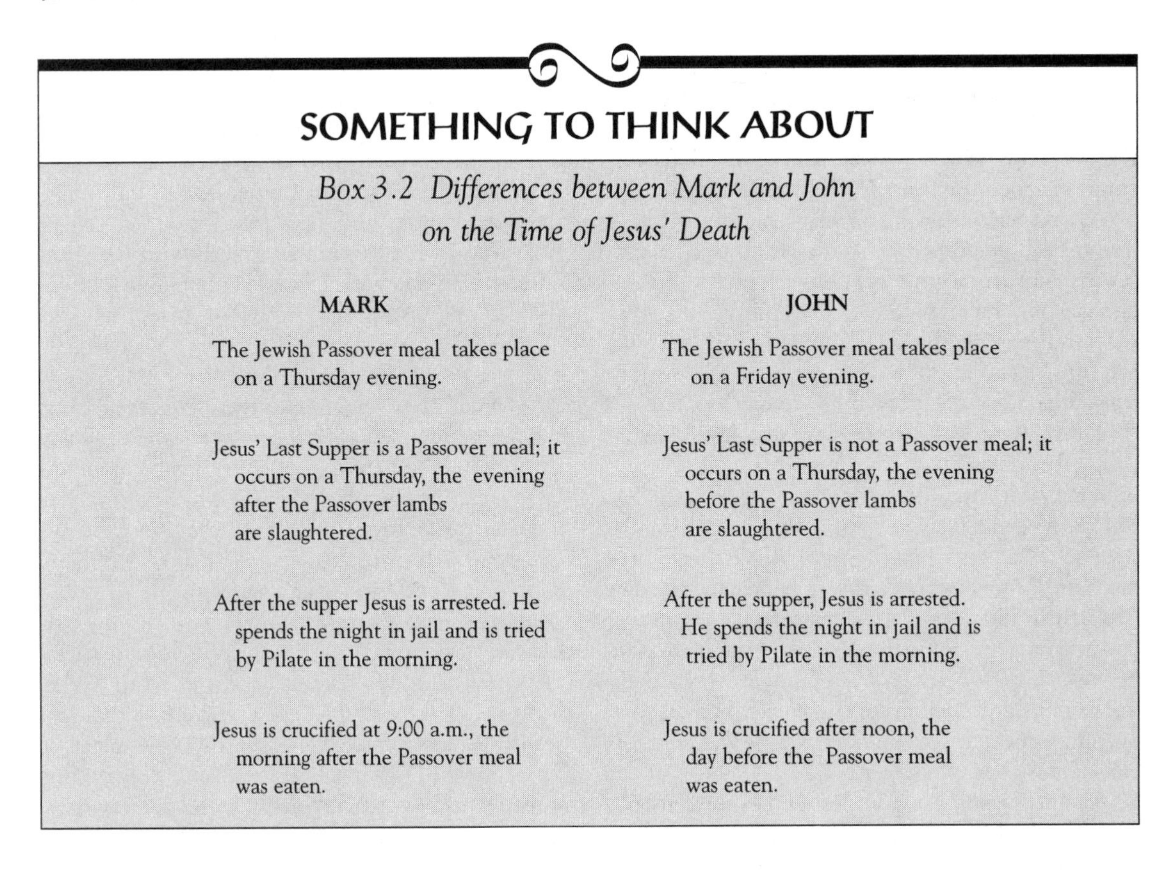

SOMETHING TO THINK ABOUT

Box 3.2 Differences between Mark and John on the Time of Jesus' Death

MARK	JOHN
The Jewish Passover meal takes place on a Thursday evening.	The Jewish Passover meal takes place on a Friday evening.
Jesus' Last Supper is a Passover meal; it occurs on a Thursday, the evening after the Passover lambs are slaughtered.	Jesus' Last Supper is not a Passover meal; it occurs on a Thursday, the evening before the Passover lambs are slaughtered.
After the supper Jesus is arrested. He spends the night in jail and is tried by Pilate in the morning.	After the supper, Jesus is arrested. He spends the night in jail and is tried by Pilate in the morning.
Jesus is crucified at 9:00 a.m., the morning after the Passover meal was eaten.	Jesus is crucified after noon, the day before the Passover meal was eaten.

Some scholars have argued that John's account is more accurate historically, since it coincides better with Jewish sources that describe how criminal trials were to be conducted by the Sanhedrin. If these scholars are right, then Mark or one of his sources may have changed the day on which Jesus was killed in order to promote the idea that Jesus himself had instituted the Lord's Supper during the Passover meal. This is possible, but may not be the best explanation. The Jewish sources that describe the procedures of the Sanhedrin were written nearly 200 years after this event, and thus are probably not our best guide.

If we concede that the later account (John's) is on general principle less likely to be accurate, since so many more years and so many more storytellers would have intervened between the account and the events it narrates, an intriguing possibility arises to explain why John, or his source, may have changed the detail concerning Jesus' death. John is the only Gospel in which Jesus is actually identified as "the lamb of God who takes away the sins of the world." Indeed, he is called this at the very start of the Gospel, by his forerunner, John the Baptist (1:29; cf. 1:36). In the Fourth Gospel, Jesus' death represents the salvation of God, just as the sacrifice of the lamb represented salvation for the ancient Israelites during the first Passover. Perhaps John (or his source) made a change in the day and hour of Jesus' death precisely to reinforce this theological point. In this Gospel, Jesus dies on the same day as the Passover lamb, at the same hour (just after noon)—to show that Jesus really is the lamb of God.

This analysis gives just one example of how historical facts may have been changed to convey theological "truths." We could easily examine other examples pertaining to such key events in the Gospels as Jesus' birth, his baptism, his miracles, his teachings, and his resurrection. The main point is that the stories that Christians told and retold about Jesus were not meant to be objective history lessons for students interested in key events of Roman imperial times. They were meant to convince people that Jesus was the miracle-working Son of God whose death brought salvation to the world, and to edify and instruct those who already believed. Sometimes the stories were modified to express a theological truth. For the early Christians who passed along the stories we now have in the Gospels, it was sometimes legitimate and necessary to change a historical fact in order to make a theological point. These are the stories that the Gospel writers inherited.

This conclusion has some profound implications for our investigation of the Gospels. The first concerns the Gospels as pieces of early Christian literature. Just as the Gospel writers inherited stories that try to make a point, they themselves have attempted to produce coherent accounts of Jesus' life and death to make certain points. Each Gospel author may have had his own points to make, and these may not have been the same in every case. Mark's point may not have been John's point in his story of Jesus' crucifixion. It is important then—indeed, absolutely crucial—that we allow each author to have his own say, rather than assume that they are all trying to say the same thing. We need to study each account for its own emphases.

The second implication concerns the Gospels as historical sources for what happened during the life of Jesus. If the Gospels have differences in historical detail, and each Gospel preserves traditions that have been changed, then it is impossible for the historian simply to take these stories at face value and uncritically assume that they provide historically accurate information. We will therefore need to develop some criteria for deciding which features of the Gospels represent Christianizations of the tradition and which represent the life of Jesus as it can be historically reconstructed.

Over the course of the next five chapters we will devote our attention to the first aspect of our study, the literary emphasis of each Gospel. Once we understand in greater detail where the Gospels came from and what each one has to say, we will then be equipped to address the second issue, asking broader historical questions in an attempt to establish what actually happened in the life of Jesus.

SOME ADDITIONAL REFLECTIONS: THE AUTHORS OF THE GOSPELS

Proto-orthodox Christians of the second century, some decades after most of the New Testament books had been written, claimed that their favorite Gospels had been penned by two of Jesus' disciples—Matthew, the taxcollector, and John, the beloved disciple—and by two friends of the apostles—Mark, the secretary of Peter, and Luke, the traveling companion of Paul. Scholars today, however, find it difficult to accept this tradition for several reasons.

First of all, none of these Gospels makes any such claim about itself. All four authors chose to keep their identities anonymous. Would they have done so if they had been eyewitnesses? This certainly would have been possible, but one would at least have expected an eyewitness or a friend of an eyewitness to authenticate his account by appealing to personal knowledge, for example, by narrating the stories in the first person singular ("On the day that Jesus and I went up to Jerusalem. . .").

Moreover, we know something about the backgrounds of the people who accompanied Jesus during most of his ministry. The disciples appear to have been uneducated peasants from Galilee. Both Simon Peter and John the son of Zebedee, for example, are said to have been peasant fishermen (Mark 1:16–20) who were "uneducated," that is, literally, unable to read and write (Acts 4:13). Now it is true that the Gospels do not represent the most elegant literature from antiquity, but their authors were at least relatively well educated; they write, for the most part, correct Greek. Could two of them have been disciples?

Again, it is possible. Jesus and his apostles, however, appear to have spoken Aramaic, the common language of the Jews in Palestine. Whether they could also have spoken Greek as a second language is something that scholars have long debated, but at the very least it is clear that Greek was not their native tongue. The authors of the Gospels, on the other hand, are absolutely fluent in Greek. Did the apostles go back to school after Jesus died, overcome years of illiteracy by learning how to read and write at a relatively high level, become skilled in foreign composition, and then later pen the Gospels? Most scholars consider it somewhat unlikely.

Perhaps an even more important aspect of the authorship of the Gospels is the evidence that they appear to preserve stories that were in circulation for a long period. This observation certainly applies to narratives for which no eyewitnesses were evidently present. For example, if Pilate and Jesus were alone at the trial in John 18:28–19:16, and Jesus was immediately executed, who told the Fourth Evangelist what Jesus actually said? An early Christian must have come up with words that seemed appropriate to the occasion. The same principle applies to the other accounts of the Gospels as well. All of them appear to have circulated by word of mouth among Christian converts throughout the Mediterranean world.

One of our four authors, Luke, explicitly tells us that he used oral and written sources for his narrative (Luke 1:1–4), and he claims that some of these sources were drawn ultimately from eyewitnesses. This circumstance raises another interesting question. Is it likely that authors who extensively used earlier sources for their accounts were themselves eyewitnesses? Suppose, for example, that Matthew actually was a disciple who accompanied Jesus and witnessed the things he said and did. Why then would he take almost all of his stories, sometimes word for word, from someone else (as we will see in Chapter 6)?

In short, it appears that the Gospels have inherited traditions from both written and oral sources, as Luke himself acknowledges, and that these sources drew from traditions that had been circulating for years, decades even, among Christian communities throughout the Mediterranean world.

SUGGESTIONS FOR FURTHER READING

Dibelius, Martin. *From Tradition to Gospel.* Trans. B. L. Woolf. New York: Scribner, 1934. This ground-breaking study deals with the character of the traditions about Jesus in circulation orally prior to being written down in the Gospels.

Gerhardsson,Birger. *Manuscript and Memory: Oral Tradition and Written Transmission in Rabbinic Judaism and Early Christianity.* Lund, Sweden: Gleerup, 1961. One of the most influential studies to maintain, contrary to the present chapter, that the traditions about Jesus in the New Testament Gospels were not changed, for the most part, in the process of being retold; for advanced students.

Harris, William V. *Ancient Literacy.* Cambridge, Mass: Harvard University Press, 1989. A brilliant analysis by a major classicist who seeks to determine how many people could read and write in the ancient world and what their reasons were for doing so; for advanced students.

McKnight, Edgar V. *What is Form Criticism?* Philadelphia: Fortress, 1969. A basic introduction to the study of how oral traditions about Jesus were modified and formed prior to the writing of the Gospels.

Macmullen, Ramsey. *Christianizing the Roman Empire A.D. 100–400.* New Haven, Conn.: Yale University Press, 1984. A concise and insightful account of the spread of Christianity through the Roman world, including discussion on how Christians engaged in their mission and the reasons for their success.

Ong W. J. *Orality and Literacy.* London: Routledge, 1982. An intriguing discussion of the social and psychological differences between oral and written cultures (between cultures in which traditions are typically heard and those in which they are typically read); for more advanced students.

CHAPTER 4

The Christian Gospels: A Literary and Historical Introduction

Now that we have learned something about the traditions of Jesus that were circulating throughout the Roman world during the middle decades of the first century, we are in a position to consider the early Christian Gospels that eventually came to embody them. There are more Gospels than the ones found in the New Testament, of course, and we will take account of such early documents as the *Gospel of Thomas* and the *Gospel of Peter*. Since our principal concern, however, is with the earliest Christian writings, most of our attention will be focused on the canonical four.

We have already learned significant bits of information about these books. They were written thirty-five to sixty-five years after Jesus' death by authors who did not know him, authors living in different countries who were writing at different times to different communities with different problems and concerns. The authors all wrote in Greek and they all used sources for the stories they narrate. Luke explicitly indicates that his sources were both written and oral. These sources appear to have recounted the words and deeds of Jesus that had been circulating among Christian congregations throughout the Mediterranean world. At a later stage we will consider the question of the historical reliability of these stories. Here we are interested in the Gospels as pieces of early Christian literature.

The first thing to observe is that just as the oral traditions functioned to meet certain needs of the early Christians (e.g., evangelism, instruction, edification), so too the Gospels were penned for certain reasons. Unfortunately, even though these reasons may have been clear to their authors, and perhaps to their first readers as well, they can only be inferred by us, living so many centuries later. It will nonetheless be one of our goals to examine each of the early surviving Gospels to ascertain, insofar as possible, its own orientation, or "take" on the life and death of Jesus. Before examining the Gospels individually, however, we should say a few words about them as a group.

THE QUESTION OF GENRE

Readers bring different sets of expectations to different kinds of literature. When we read a short story, we have a different set of expectations than when we read a newspaper editorial. As educated readers, we know how short stories and editorials "work," and we expect certain features in the one but not the other. The editorial, for example, will not contain character development, plot conflict, plot resolution, and so on. So too we expect different things from a science fiction novel and a science textbook, from a clever limerick and a salacious Harlequin Romance.

These expectations have a profound effect on the way we read literature. Suppose you were to read about a breakthrough in genetic research that could potentially save the human race from some of its worst diseases. At present, however, the research is highly dangerous. If artificially manipulated gene specimens were to escape the laboratory, they could mutate beyond control and bring worldwide ruin and despair. If this were in a science fiction novel, you might be intrigued and recommend the book to a friend. If it were on the front page of the *New York Times*, you might be appalled and write your senator.

We know what to expect from a piece of literature, in part, because we have become accustomed to certain literary conventions that characterize different kinds of writing. Pieces of writing that share a range of conventions are classified together as a genre. The conventions involve *(a)* form (Is the work poetry or prose? long or short? narrative or descriptive?) *(b)* content (Is it about nature or society? a twelfth-century philosopher or a twenty-second century space traveler?) and *(c)* function (Does the work aim to entertain? inform? persuade? a little of each?).

What kind of literature is a Gospel? Or, to put it somewhat differently, when ancient persons read or heard one of these books, what kinds of expectations did they have? Until recently, modern scholars generally agreed that the New Testament Gospels were unlike anything else in all of literature, that they were an entirely new genre invented by the Christians, and represented by only four surviving works. The Gospels were obviously about the man Jesus and thus were somewhat like biographies, but compared to modern biographies they appeared altogether anomalous. In one respect this older view seems reasonable; as we will see in some detail momentarily, the Gospels do indeed differ from modern biographies. Scholars have nonetheless come to reject the idea that they are totally unlike anything else. There is probably no such thing as a kind of literature that is absolutely unique; if there were, no one would have any idea how to read it or know what to make of it. If people in antiquity could read the Gospels and make sense of them, then we have to assume that these books were not in fact completely foreign to them.

This question of how people in antiquity would understand a book should itself give us pause. While it may be true that the Gospels differ from modern genres like biography, they may not have differed from ancient genres. In fact, scholars of ancient literature have found significant parallels between the Gospels and several ancient genres. Some of these investigations have plausibly suggested that the Gospels are best seen as a kind of Greco-Roman (as opposed to modern) biography.

BIOGRAPHY AS A GRECO-ROMAN GENRE

We have numerous examples of Greco-Roman biographies, many of them written by some of the most famous authors of Roman antiquity, such as Plutarch, Suetonius, and Tacitus. One of the ways to understand how this genre worked is to contrast it with modern biographies, following the principle that we can learn something only in light of what we already know. In doing so, we must constantly bear in mind that literary genres are highly flexible; just think of all the different kinds of novels or short stories you have read.

Most modern biographies are full of data—names, dates, places, and events—all of which show a concern for factual accuracy. A modern biography, of course, can deal with the whole of a person's life or with only a segment of it. Typically it is concerned with both public and private life and with how the subject both reacts to what happens and is changed by it. In other words, the inner life of the person, his or her psychological development based on events and experiences, is quite often a central component and is used to explain why the character behaves and reacts in certain ways. Thus modern biographies tend not only to inform but also to explain. They also entertain, of course, and often propagandize as well, especially when they concern political or religious figures.

Most ancient biographies were less concerned with giving complete factual data about an individual's life, or a chosen period of it. Research methods were necessarily different, with few surviving documents to go on, and (by our standards) inadequate tools for record-keeping and data recovery. Biographers often relied heavily on oral information that had circulated for long periods of time. Indeed, many of them expressed a preference for oral sources; these at least could be interrogated! Modern biographers are somewhat more leery of hearsay. Yet more significantly, most ancient biographers were less interested in showing what actually happened in their subjects' lives than in portraying their essential character and personality traits (see box 4.1). This is a key difference between ancient and modern biographies: in the ancient world, prior to the formulation of modern notions of human psychology that have arisen since the Enlightenment, there was little sense that the human personality developed in light of its experiences and encounters with other people. Thus Greco-Roman biography does not generally deal with the inner life, and especially does not do so in the sense of what we would call character formation.

For the ancient biographer, character traits were thought to be relatively constant throughout a person's life. A person's experiences were opportunities to demonstrate what those traits were, rather than occasions for these traits to develop. Therefore, when an ancient biographer employed a chronological framework to organize an individual's life, it was strictly for organizational purposes; it was not to show how the person became who he or she was. Great persons were who they were, and everyone else could try to model themselves on the positive aspects of their characters while avoiding their pitfalls. Biographies were usually

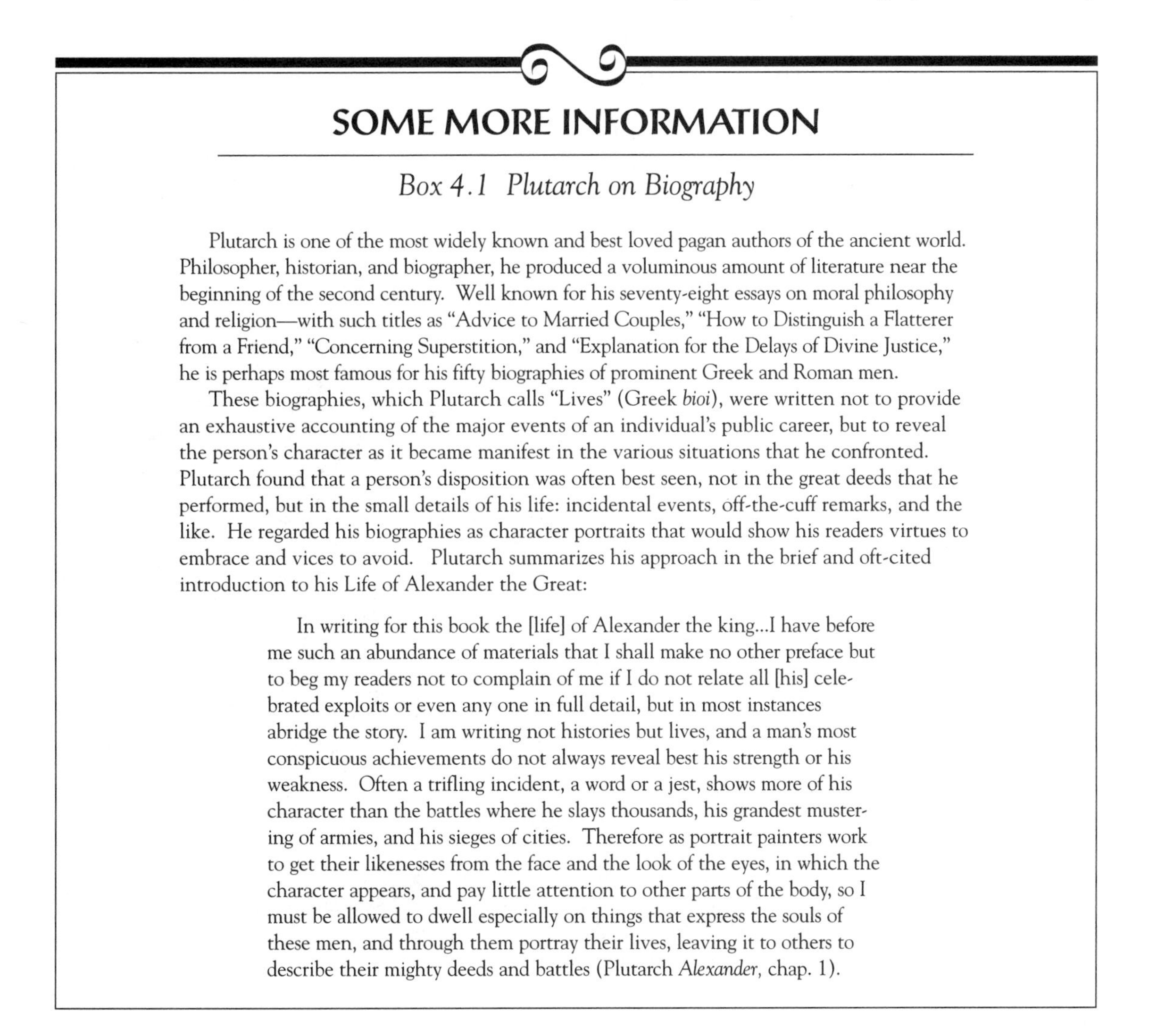

SOME MORE INFORMATION

Box 4.1 Plutarch on Biography

Plutarch is one of the most widely known and best loved pagan authors of the ancient world. Philosopher, historian, and biographer, he produced a voluminous amount of literature near the beginning of the second century. Well known for his seventy-eight essays on moral philosophy and religion—with such titles as "Advice to Married Couples," "How to Distinguish a Flatterer from a Friend," "Concerning Superstition," and "Explanation for the Delays of Divine Justice," he is perhaps most famous for his fifty biographies of prominent Greek and Roman men.

These biographies, which Plutarch calls "Lives" (Greek *bioi*), were written not to provide an exhaustive accounting of the major events of an individual's public career, but to reveal the person's character as it became manifest in the various situations that he confronted. Plutarch found that a person's disposition was often best seen, not in the great deeds that he performed, but in the small details of his life: incidental events, off-the-cuff remarks, and the like. He regarded his biographies as character portraits that would show his readers virtues to embrace and vices to avoid. Plutarch summarizes his approach in the brief and oft-cited introduction to his Life of Alexander the Great:

> In writing for this book the [life] of Alexander the king...I have before me such an abundance of materials that I shall make no other preface but to beg my readers not to complain of me if I do not relate all [his] celebrated exploits or even any one in full detail, but in most instances abridge the story. I am writing not histories but lives, and a man's most conspicuous achievements do not always reveal best his strength or his weakness. Often a trifling incident, a word or a jest, shows more of his character than the battles where he slays thousands, his grandest mustering of armies, and his sieges of cities. Therefore as portrait painters work to get their likenesses from the face and the look of the eyes, in which the character appears, and pay little attention to other parts of the body, so I must be allowed to dwell especially on things that express the souls of these men, and through them portray their lives, leaving it to others to describe their mighty deeds and battles (Plutarch *Alexander*, chap. 1).

meant to highlight those various aspects, not so much for the sake of providing history lessons, as for giving instruction in proper behavior. Personal qualities could be conveyed by a variety of stories about the person. Many of these stories were drawn from narratives that an author inherited from oral traditions, such as sayings, speeches, anecdotes, and stories about conflicts.

As I have already mentioned, there was considerable flexibility in how an ancient biography might portray a person's life, depending in good measure on what kind of public figure he or she was: a military person, a political ruler, a philosopher, a religious leader. The genre could encompass any of these kinds of figures, and different subgenres developed accordingly, each with its own sets of expectations. The role of the miraculous, for example, was typically pronounced in the life of a religious figure (e.g., Philostratus's biography of Apollonius of Tyana): miraculous signs might accompany his birth, he might manifest divine power in his own miracles and inspired teachings, and he might be glorified after his death through an ascension into heaven or through receiving cult from those whose lives he had touched.

If I were to attempt a definition of the Greco-Roman biography, then, it might be something like this: ancient biography was a prose narrative recounting an individual's life, often within a chronological framework, employing numerous subgenres (such as sayings, speeches, anecdotes, and conflict stories) so as to reflect important aspects of his or her character, principally for purposes of instruction (to inform about what kind of person he or she was), exhortation (to urge others to act similarly), or propaganda (to show his or her superiority to rivals).

THE GOSPELS AS ANCIENT BIOGRAPHIES

Many recent scholars have come to recognize that the New Testament Gospels are a kind of ancient biography. Of course, the Gospels have distinctive features of their own, but this is what we would expect, since numerous subtypes typically develop within broader genres of literature. And every individual book has distinctive features as well.

Most of the distinctive features of the Gospels relate directly to their Christian character. They are the only biographies written by Christians about the man they worship as the Son of God who died for the salvation of the world. As we will see, for example, the New Testament Gospels put an inordinate amount of emphasis on the death of the main character, something highly unusual for ancient biography. The stress on Jesus' death, however, is determined by the distinctive emphasis of these works and is not out of bounds for the genre. Instead, it shows that the Gospels are a kind of sub-subgenre, or one type of ancient religious biography. Moreover, the Gospels differ in some ways not only from other Greco-Roman biographies, but also from each other.

We began with the question of how an ancient person might have understood the form of the Gospels. It appears that ancient readers, whether they actually read the words off the page or heard someone else do so, would have recognized them

Figure 4.1 Picture of the Four Evangelists associated with their traditional symbols (John the Eagle, Luke the Ox, Mark the Lion, Matthew the Man) from an eighth-century manuscript of the Gospels.

as biographies of a religious leader. How did this understanding affect the way ancient persons read these books? Ancient readers and hearers of books like these would probably expect to find that the main character was an important religious figure and that all of the action of the narrative revolved around him. They might anticipate a miraculous beginning to his life and a miraculous ending. They might look forward to descriptions of his divinely inspired teachings and superhuman deeds. They would not expect to see anything like what we might call "character development." Instead, they would look for how the character acted and reacted to the various challenges with which he was confronted, demonstrating who he was through his carefully crafted words and impressive deeds. Moreover, they would expect to be able to discern important aspects of his character and identity at the outset of the narrative, in the opening scenes of the action. We ourselves can benefit from reading the Gospels with these expectations in mind.

SUGGESTIONS FOR FURTHER READING

Aune, David. *The New Testament in its Literary Environment*. Philadelphia: Westminster, 1987. A superb introduction to the genres of the New Testament writings in relation to other literature of the Greco-Roman world.

Burridge, Richard. *What Are the Gospels? A Comparison with Greco-Roman Biography*. Cambridge: Cambridge University Press, 1992. A thorough study that emphatically argues that the Gospels are best understood as a kind of ancient biography.

Cartlidge, David R. and David L. Dungan, eds. *Documents for the Study of the Gospels*. 2d ed. Philadelphia: Fortress, 1994. Presents an excellent selection of ancient literary texts that are closely parallel to the New Testament Gospels, including selections from Philostratus's *Life of Apollonius* and Philo's *Life of Moses*.

Talbert, Charles. *What Is a Gospel? The Genre of the Canonical Gospels*. Philadelphia: Fortress, 1977. One of the earliest recent attempts to situate the Gospels within the context of Greco-Roman literature.

CHAPTER 5

Jesus, the Suffering Son of God: The Gospel According to Mark

We begin our study of the Gospels with Mark, the shortest of the four in the New Testament. We do not know who the author was, only that he was a Greek-speaking Christian, presumably living outside of Palestine, who had heard a number of stories about Jesus. Mark (as I will continue to call him since we do not know his real name) penned an extended account of Jesus' life beginning with his appearance as an adult to be baptized by John and ending with the report of his resurrection. In addition to stories that he had heard, Mark may also have used some written sources for portions of his narrative. If so, these sources no longer survive. Of the full-length Gospels that do survive, Mark appears to have been the first written. As we will see, this Gospel was itself used by the authors of Matthew and Luke for many of their stories about Jesus (see chapter 6).

An introductory text such as this cannot provide an exhaustive analysis of Mark (or the other Gospels). My purpose here is simply to provide some guidance for your own interpretation of the book, by supplying you with important keys for unlocking its meaning. My working assumption, throughout our discussions, is that you have already familiarized yourself with the contents of the book by reading it carefully all the way through a couple of times.

There are a number of ways we could approach this investigation. Indeed, we will be taking different approaches to each of the Gospels that we examine. We will study Mark, however, in light of the issues discussed in the previous chapter. Let's assume that we are informed readers of this text, conversant with the genre and knowledgeable about the world within which it was written. Knowing that Mark is a kind of Greco-Roman biography about Jesus, we can ask, who was Jesus, according to this literary portrayal, and what did he do? And how is this message conveyed through the shape of the narrative?

THE BEGINNING OF THE GOSPEL: JESUS THE MESSIAH, THE SON OF GOD WHO FULFILLS SCRIPTURE

One of the first things that strikes the informed reader of Mark's Gospel is how thoroughly its traditions are rooted in a Jewish worldview. The book begins, as do many other ancient biographies, by naming its subject: "The Beginning of the Gospel of Jesus Christ" (1:1). Readers living in the Greco-Roman world would not recognize "Christ" as a name; for most of them it was not even a meaningful title. The word comes from the verb "anoint" and typically referred to someone who had just had a rubdown (with oil). "Christ" *was* a title in Jewish circles, however, as the Greek equivalent of the Hebrew word "messiah." Mark, then, is a book about Jesus the messiah.

SOME MORE INFORMATION

Box 5.1 The Jewish Messiah

The term "messiah" comes from a Hebrew word that means "anointed one," the exact equivalent of the Greek term *christos* (thus "messiah" and "Christ" mean the same thing). In the Hebrew Bible the term is applied to the Jewish King, who was anointed with oil at his inauguration ceremony as a symbolic expression of God's favor; he was called "the Lord's anointed" (see 1 Sam 10:1; Ps 2:2).

The term came to refer to a future deliverer of Israel only after the Babylonians overthrew the nation of Judea in 587 B.C.E. and removed the Jewish king from the throne. From that time on, there was no anointed one (messiah) to rule for several centuries (until the Harmonean rulers). But some Jews recalled a tradition in which God had told David, his favorite king, that he would always have a descendant on the throne (2 Sam 7:14–16). This is probably the origin of the idea that there would be a future messiah to fulfill God's promises, a future king like David who would rule the people of God once again as a sovereign nation in the Promised Land.

By the time of the New Testament, different Jews had different understandings of what this future ruler would be like. Some expected a warrior-king like David, others a more supernatural cosmic judge of the earth, and still others (such as the community that produced the Dead Sea Scrolls) a priestly ruler who would provide the authoritative interpretations of God's law for his people (see Chapter 15). All of these figures are designated "messiah" in the ancient Jewish sources.

In no source prior to the writing of the New Testament, however, is there any reference to a future messiah who is to suffer and die for the sins of the people. This notion appears to be a Christian creation, as we will see more fully in Chapter 16. It may represent a combination of the belief in a future messianic deliverer with the notion that the one who is truly righteous suffers, a notion expressed in such biblical passages as Psalms 22 and 69, and Isaiah 53. Surprisingly for many Christian readers, the term "messiah" never occurs in these passages.

Jews in the first century could have meant a range of things by the title messiah, as scholars have come to realize (see box 5.1). Many of these meanings, however, can be subsumed under two major rubrics (which are not necessarily mutually exclusive). For some Jews, the messiah was the future king of Israel, who would deliver God's people from their oppressors and establish a sovereign state in Israel through God's power. For others, he was a cosmic deliverer from heaven, who would engage in supernatural warfare with the enemies of the Jews and bring a divine victory over their oppressors. Both notions had been around for some time by the first century; both, obviously, were designations of grandeur and power.

Mark begins his Gospel by calling Jesus the messiah. But as we will see—and as everyone who read the book probably already knew—Jesus did not conform to either of the general conceptions of this title. He neither overthrew the Romans in battle nor arrived on the clouds of heaven in judgment. Instead, he was unceremoniously executed for treason against the state. What in the world could it mean to call *him* the messiah? This is one of the puzzles that Mark's Gospel will attempt to solve.

The Jewishness of the Gospel becomes yet more evident in the verses that follow. First there is a tantalizing statement that the story, or at least the first part of it, is a fulfillment of an ancient prophecy recorded in the Jewish Scriptures (it is quoted, of course, in the Greek translation, the

Septuagint; 1:2–3). Then there is the appearance of a prophet, John the Baptist, proclaiming a Jewish rite of baptism for the forgiveness of sins. John's dress and diet (1:6) are reminiscent of another Jewish prophet, Elijah, also described in the Hebrew Scriptures (cf. 2 Kings 1:8). This John not only practices baptism, he also preaches of one who is to come who is mightier than he. Mightier than a prophet of God? Who could be mightier than a prophet?

Jesus himself then appears, coming from the northern part of the land, from the region of Galilee and the village of Nazareth. He is baptized by John, and upon emerging from the waters, he sees the heavens split open and the Spirit of God descend upon him like a dove. He then hears a voice call out from heaven: "You are my beloved Son, in you I am well pleased" (1:11). The proclamation appears to have serious implications: Jesus is immediately thrust out into the wilderness to confront the forces of evil (he is "tempted by Satan," 1:13). He returns, victorious through the power of God ("the angels" have "ministered to him" 1:13), and begins to make his proclamation that God's kingdom is soon to appear (1:14–15).

Figure 5.1 Picture of Jesus' Baptism by John and the Descent of the Dove, from a Vault Mosaic in Ravenna.

Here, then, is a Gospel that begins by describing the forerunner of Jesus, the Son of God, and the miraculous proclamation of his own Sonship. Up to this point a Gentile reader may have recognized the Jewish character of the account but the designation "Son of God" would no doubt have struck a familiar chord. When Jesus was proclaimed the Son of God (by God himself no less), most readers in the Greco-Roman world would probably have taken this to mean that he was like other sons of God—divinely inspired teachers or rulers whose miraculous deeds benefitted the human race. But given the Jewishness of the rest of the beginning, perhaps we should inquire what a Jewish reader would make of the title Son of God.

Even within Jewish circles there were thought to be special persons endowed with divine power to do miracles and to deliver inspired teachings (see Chapter 2). Two of them we know by name Hanina ben Dosa and Honi the "circle-drawer." These men, living roughly at the time of Jesus, were understood to have a particularly intimate relationship with God, and as a result were thought to have been endowed with special powers. Accounts of their fantastic deeds and marvelous teachings are recorded in later Jewish sources. What made these persons special was their unique relationship with the one God of Israel. The notion that mere mortals could have such a relationship was itself quite ancient, as shown by the Jewish Scriptures themselves, where an individual was sometimes called "the son of God." The king of Israel, for example, was thought to mediate between God and humans and so stand in a special relationship with God as a child does to a parent. Even kings with dubious public records were sometimes called "the son of God" (e.g., 2 Sam 7:14; Ps 2:7–9). And others receive the title as well: occasionally the entire nation of Israel, through whom God worked his will on earth (Hos 11:1), and sometimes God's heavenly servants, beings that we might call angels (Job 1:6; 2:1). In all of these instances in Jewish circles, "the son of God" referred to someone who had a particularly intimate relationship with God, who was chosen by God to perform a task, and who thereby mediated God's will to people on earth. Sometimes these sons of God were associated with the miraculous.

What, then, does Mark mean by beginning his account with the declaration, by God himself, that Jesus (this one who was to be executed as a criminal) is his son? We can begin our quest for an answer by examining key incidents in the Gospel's opening chapter, recalling that ancient biographies tended to set the character of their subjects in the early scenes.

JESUS THE AUTHORITATIVE SON OF GOD

The reader is immediately struck by the way in which Jesus is portrayed as supremely authoritative. At the outset of his ministry, he sees fishermen plying their trade. He calls to them and without further ado they leave their boats and family and hapless co-workers to follow him (1:16–20). Jesus is an authoritative leader; when he speaks, people obey.

Jesus enters the synagogue to teach and astonishes those who hear. Mark tells us why: "He taught them as one who had authority, and not as the scribes" (1:22). Jesus is an authoritative teacher; when he gives instruction, people hang onto his every word.

He immediately encounters a man possessed by an unclean spirit, who recognizes him as "the Holy One of God" (1:24). Jesus rebukes the spirit and by his word alone drives it out from the man. Those who witness the deed declare its significance: "With authority he commands even the unclean spirits, and they obey him" (1:27). Not only does he drive out the evil spirits, the embodiments of the opposition to God, he also heals the sick, both relatives of his followers (1:29–31) and unknown townsfolk (1:32–34). Soon he is seen healing all who come, both the ill and the possessed. Jesus is an authoritative healer; when he commands the forces of evil, they listen and obey.

This portrayal of Jesus as an authoritative Son of God sets the stage for the rest of the Gospel. Throughout his public ministry, Jesus goes about doing good, healing the sick, casting out demons, even raising the dead (5:1–43). His fame spreads far and wide as rumors of his fantastic abilities reach the villages and towns of Galilee (1:28; 1:32–34; and 1:45). Moreover, he attracts the crowds by his inspired and challenging teaching, especially when he tells parables, brief stories of everyday, mundane affairs that he endows with deeper spiritual significance. Interestingly, most of those who hear his words do not understand what they mean (4:10–13).

Given the incredible following that Jesus amasses, the amazing teachings that he delivers, and the miraculous deeds that he performs, one would think that he would become immediately and widely acknowledged for who he is, a man specially endowed by God, the Son of God who provides divine assistance for those in need. Ironically, as the careful reader of the Gospel begins to realize, nothing of the sort is destined to happen. Jesus, this authoritative Son of God, is almost universally misunderstood by those with whom he comes in closest contact. Even worse, despite his clear concern to help others and to deliver the good news of God, he becomes hated and opposed by the religious leaders of his people. Both of these characteristics are major aspects of Mark's portrayal of Jesus. He is the opposed and misunderstood Son of God.

JESUS THE OPPOSED SON OF GOD

A good deal of Mark's Gospel shows that despite Jesus' fantastic deeds the leaders of his people oppose him from the outset; and their antagonism escalates until the very end, where it results in the catastrophe of his execution. Despite this hostility between Jesus and the leaders of Israel, Mark does not portray Jesus as standing in opposition to the religion of Judaism (at least as Mark sees it). Recall that Jesus is said to be the Son of the Jewish God, the Jewish messiah, come in fulfillment of the Jewish Scriptures and preceded by a Jewish prophet. He teaches in the Jewish synagogue and works among the Jewish people. Later we will find him teaching in the Temple, observing the Jewish Passover, and discussing fine points of the Jewish Law with Jewish scholars. Indeed, even though Jesus' understanding of the Law will come to be challenged, Mark maintains that he was himself faithful to the Law. Consider the account of the leper in one of the opening stories (1:40–44). After Jesus heals the man, he instructs him to show himself to a Jewish priest and to make an offering on behalf of his cleansing "as Moses commanded" (1:44). Jesus is scarcely bent on subverting the Jewish religion.

Why, then, do the Jewish leaders, the scribes and Pharisees in Galilee and the chief priests in Jerusalem—oppose him (see box 5.2)? Do they not recognize who he is? In fact, they do not recognize him, as we will see momentarily. Even more seriously, they are gravely offended by the things that he says and does. This is evident in the accounts recorded in 2:1–3:6, a group of conflict stories that show a crescendo in the tension between Jesus and the Jewish leaders, the scribes and Pharisees. At first these leaders merely question his actions (2:7), they then take offense at some of his associations (2:16) and his activities (2:18), then protest the actions of his followers (2:24), and finally take serious exception to his own actions and decide to find a way to put him to death (3:6).

In particular, these authorities take umbrage at Jesus' refusal to follow their own practices of purity. He eats with the unrighteous and with sinners, those thought to be unclean and to pollute the pure. For Jesus, these are the ones who need his help (2:15–17). Nor does he follow the Pharisees'

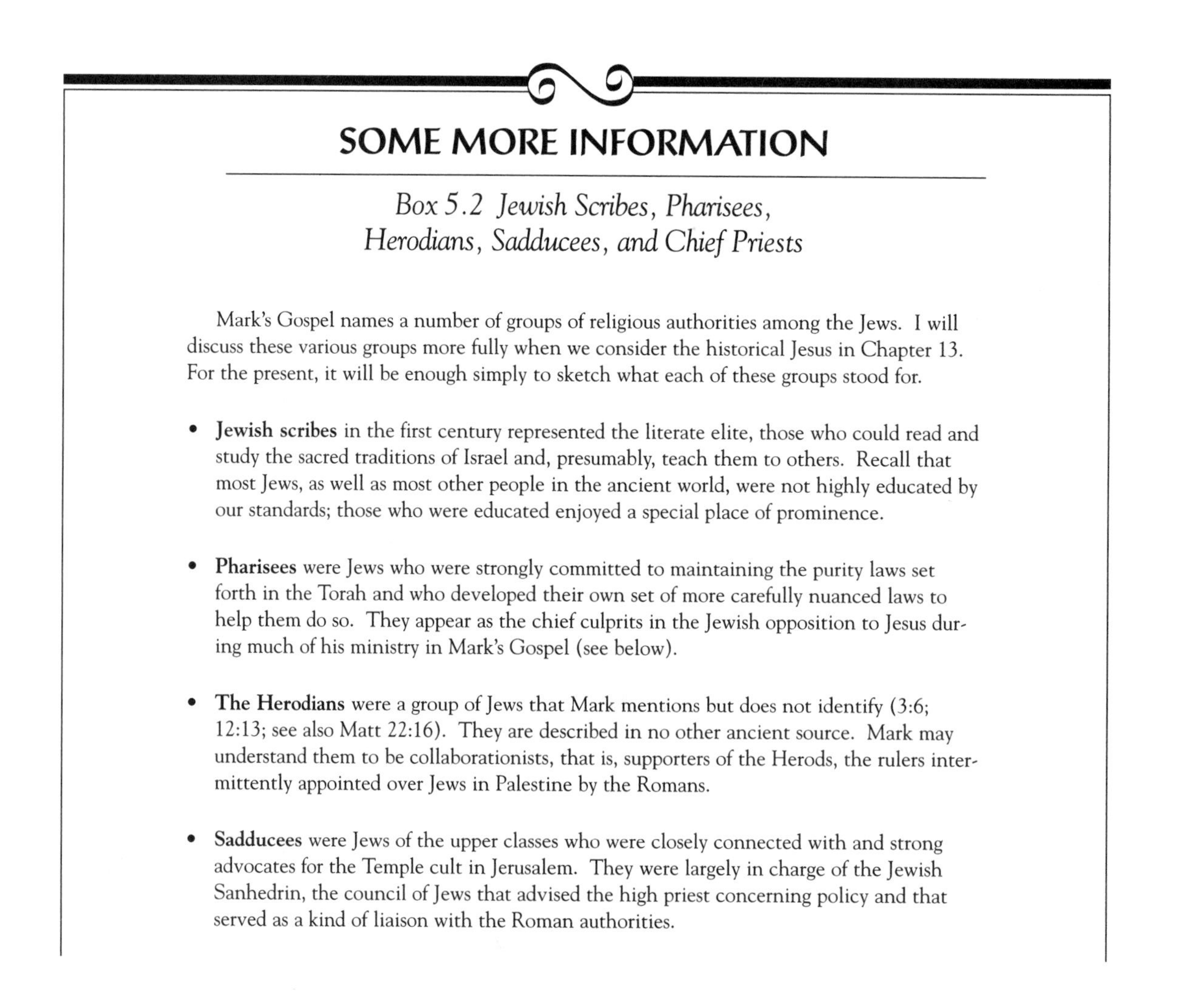

SOME MORE INFORMATION

Box 5.2 Jewish Scribes, Pharisees, Herodians, Sadducees, and Chief Priests

Mark's Gospel names a number of groups of religious authorities among the Jews. I will discuss these various groups more fully when we consider the historical Jesus in Chapter 13. For the present, it will be enough simply to sketch what each of these groups stood for.

- **Jewish scribes** in the first century represented the literate elite, those who could read and study the sacred traditions of Israel and, presumably, teach them to others. Recall that most Jews, as well as most other people in the ancient world, were not highly educated by our standards; those who were educated enjoyed a special place of prominence.

- **Pharisees** were Jews who were strongly committed to maintaining the purity laws set forth in the Torah and who developed their own set of more carefully nuanced laws to help them do so. They appear as the chief culprits in the Jewish opposition to Jesus during much of his ministry in Mark's Gospel (see below).

- **The Herodians** were a group of Jews that Mark mentions but does not identify (3:6; 12:13; see also Matt 22:16). They are described in no other ancient source. Mark may understand them to be collaborationists, that is, supporters of the Herods, the rulers intermittently appointed over Jews in Palestine by the Romans.

- **Sadducees** were Jews of the upper classes who were closely connected with and strong advocates for the Temple cult in Jerusalem. They were largely in charge of the Jewish Sanhedrin, the council of Jews that advised the high priest concerning policy and that served as a kind of liaison with the Roman authorities.

- **Chief priests** were the upper classes of the Jewish priesthood who operated the Temple and oversaw its sacrifices. They would have been closely connected with the Sadducees (presumably a number of Sadducees were among them) and would have been the real power players in Jesus' day, the ones with the ear of the Roman governor in Jerusalem and the ones responsible for regulating the lives of the Jewish people in Judea. Their leader, the high priest, was the ultimate authority over civil and religious affairs when there was no king in Judea.

This basic information about Jewish groups should make us curious about certain aspects of Mark's Gospel. We know from other sources that the Pharisees were not numerous in the days of Jesus; there certainly were not enough to stand at every wheat field to spy out itinerant preachers on the Sabbath (see Chapter 15). Nor, evidently, were they influential in the politics of Palestine at the time, or even concerned that everyone else (i.e., non-Pharisaic Jews) conform to their own rules and regulations for purity. And yet they appear as Jesus' chief adversaries in Mark's narrative, constantly hounding him and attacking him for failing to conform to their views. Can this be historically accurate?

Scholars have long known that some decades after Jesus' death, nearer the end of the first century, the Pharisees did become more prominent in Palestinian life. After the destruction of Jerusalem in 70 C.E. they were given authority by the Romans to run the civil affairs of Palestinian Jews. Moreover, we know that Pharisees interacted frequently with Christian churches after the death of Jesus. Indeed, the one Jewish persecutor of the church about whom we are best informed was Paul, a self-proclaimed Pharisee.

Is it possible that the opposition leveled against the church by Pharisees after Jesus' death affected the ways that Christians told stories about his life? That is to say, because of their own clashes with the Pharisees, could Christians have narrated stories in which Jesus himself disputed with them (usually putting them to shame), even though such disputes would have happened only rarely during his own lifetime?

prescriptions for keeping the seventh day holy (2:23–3:6); he puts human needs above the requirement to rest on the Sabbath. In Jesus' view, the Sabbath was made for the sake of humans and not humans for the Sabbath; it is therefore legitimate to prepare food or heal a person in need on this day (2:27; 3:4). From the Pharisees' perspective (as portrayed by Mark), these are not honest disagreements over matters of policy. They are dangerous perversions of their religion, and Jesus needs to be silenced. The Pharisees immediately take counsel with their sworn enemies the Herodians (see box 5.2) and decide to have him killed (3:6).

After these opening stories of conflict, Jewish authorities are constantly on the attack. In virtually every instance they are the ones who initiate the dispute, even though Mark consistently portrays Jesus as getting the better of them in dialogue (see esp. 11:27–12:40). In the end, however, the chief priests triumph, convincing the Roman governor that Jesus has to die. Why, ultimately, do they do so? The short answer is that they find Jesus threatening because of his popularity and find his words against their Temple cult offensive, as shown in his violent and disruptive actions in the Temple itself (11:18). But in the larger picture painted by Mark's Gospel, the Jewish authorities do not seek Jesus' death merely because they are jealous or because they disagree with him over legal, theological, or cultic matters. They oppose Jesus because he is God's unique representative on earth—God's authoritative Son—and they, the leaders of Israel, cannot understand who he is or what he says. In this, however, they are not alone, for virtually no one else in Mark's narrative can understand who he is either.

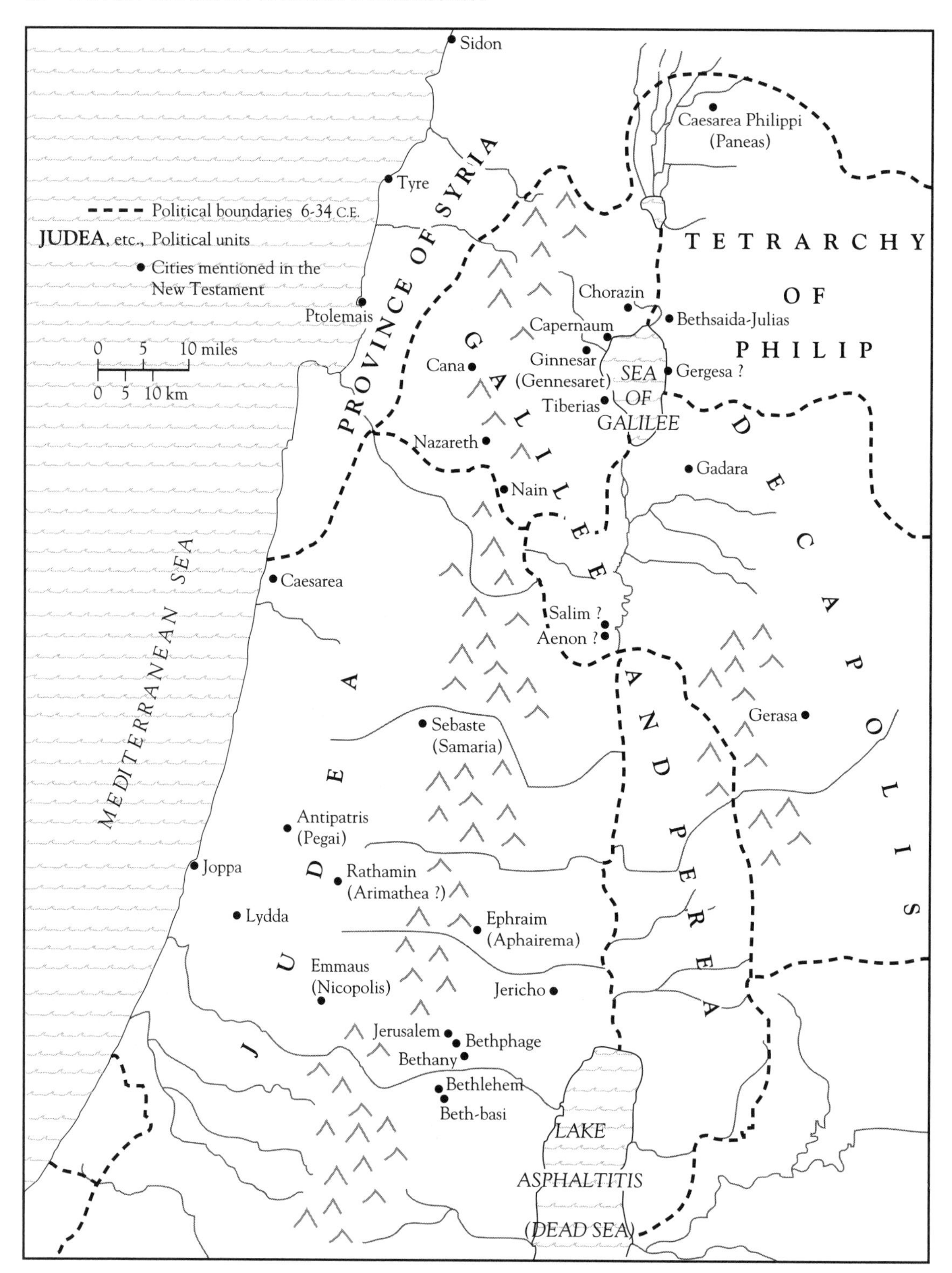

Figure 5.2 Palestine in New Testament Times.

JESUS THE MISUNDERSTOOD SON OF GOD

One way to establish misunderstanding as a Markan theme is to read carefully through the first half of the Gospel and ask, who realizes that Jesus is the Son of God? The answer may come as a bit of a surprise. Clearly God knows that Jesus is his Son, because he himself declares it at the baptism (1:11). And since this declaration comes directly to Jesus ("You are my beloved Son"), the reader can assume that he knows it as well. In addition, the evil demons recognize Jesus as the Son of God; on several instances they scream it out when they encounter him (3:11, cf. 1:24). Who else knows? Only two other persons: the author of the Gospel, who recounts these various tales, and you the reader.

Through the first half of this Gospel, no one else recognizes Jesus' identity, including even those who are closest to him. Early on, when he comes to his home town, his family tries to snatch him from the public eye because they think that he has gone crazy (3:21). Jesus' own townspeople neither understand nor trust him. When he teaches in their synagogue, they take offense at his words and wonder how he has the ability to do such miraculous deeds, since he is a mere carpenter whose (unremarkable) family they know (6:1–6). The Jewish scholars think they know the source of his power. Refusing to acknowledge the divine authority behind Jesus' words and deeds—how could one so profane come from God (2:7)?—they claim that he is possessed by Beelzebul, the prince of the demons, and so does miracles through the power of the Devil (3:22).

Perhaps most striking of all, Jesus' own disciples fail to understand who he is, even though he has specially chosen them to follow him (3:13–19) and given them private instruction (e.g., 4:10–20). When they watch him calm a violent storm at sea with a word, their question is genuine: "Who then is this, that even wind and sea obey him?" (4:41). When they later behold Jesus walking upon the water, they continue to be mystified: "For they did not understand . . . but their hearts were hardened" (6:51–52). When, later still, Jesus warns them "to beware of the leaven of the scribes and Pharisees" (8:15), they mistake his meaning, thinking he is angry because they have forgotten to bring bread, even though they had seen him miraculously feed thousands of hungry people on two different occasions. Now Jesus expresses his own exasperation: "Do you not yet understand?" (8:21). No, they do not. But they will begin to have an inkling, right here at the midpoint of the Gospel.

JESUS THE ACKNOWLEDGED SON OF GOD

One of the keys to understanding Mark's portrayal of Jesus lies in the sequence of stories that begins immediately after Jesus' exasperated question of 8:21. The sequence begins with perhaps the most significant healing story of the Gospel, an account that Mark appears to have invested with special symbolic meaning. This is a story of a blind man who gradually regains his sight (8:22–26).

It is striking that the healing takes place in stages. Indeed, it is the only miracle in the Gospel that Jesus does not perform immediately and effortlessly. When he is asked to heal the blind man, he takes him by the hand, leads him out of the village, spits on his eyes, and asks if he can see. The man replies that he can, but only vaguely: people appear like walking trees. Jesus then lays his hands upon his eyes and looks intently at him, and the man begins to see clearly.

A perceptive reader will recognize the symbolism of the account in light of its immediate context. In the very next story, the disciples themselves, who until now have been blind to Jesus' identity (cf. 8:21), gradually begin to see who he is, in stages. It starts with a question from Jesus: "Who do people say that I am?" (8:27). The disciples reply that some think he is John the Baptist, others Elijah, and yet others a prophet raised from the dead. He then turns the question on them: "But who do you say that I am?" (8:29). Peter, as spokesperson for the group, replies, "You are the Christ."

This is a climactic moment in the narrative. Up to this point, Jesus has been misunderstood by everyone, by family, neighbors, religious leaders, and followers, and now, halfway through the account, someone finally realizes who he is, at least in part. (The reader knows that Peter's confession is correct to some extent, because for Mark Jesus is the messiah: recall how he identifies him in the very first verse of the narrative as "Jesus the

Christ.") Rather than rejecting or repudiating Peter's confession, Jesus orders the disciples not to spread the word: "And he sternly ordered them not to tell anyone about him" (8:30; see box 5.3).

Still, Peter's identification of Jesus as the messiah is correct *only* in part. That is to say, Peter has begun to see who Jesus is, but still perceives him only dimly. The reader know this because of what happens next. Jesus begins to teach that he "must suffer many things, and be rejected by the elders and the chief priests and the scribes, and be killed, and after three days rise from the dead" (8:31). Jesus is the messiah, but he is the messiah who has to suffer and die. And this makes no sense to Peter. He takes Jesus aside and begins to rebuke him.

But why would Peter reject Jesus' message of his approaching "Passion" (a term that comes from the Greek word for "suffering")? Evidently he understands the role of the messiah quite differently from the way Jesus (and Mark) does. The author never delineates Peter's view for us, but perhaps it is not so difficult to figure out. If Peter uses the term "messiah" in the way most other first-century Jews did, then he understands Jesus to be the future deliverer of Israel, a man of grandeur and power

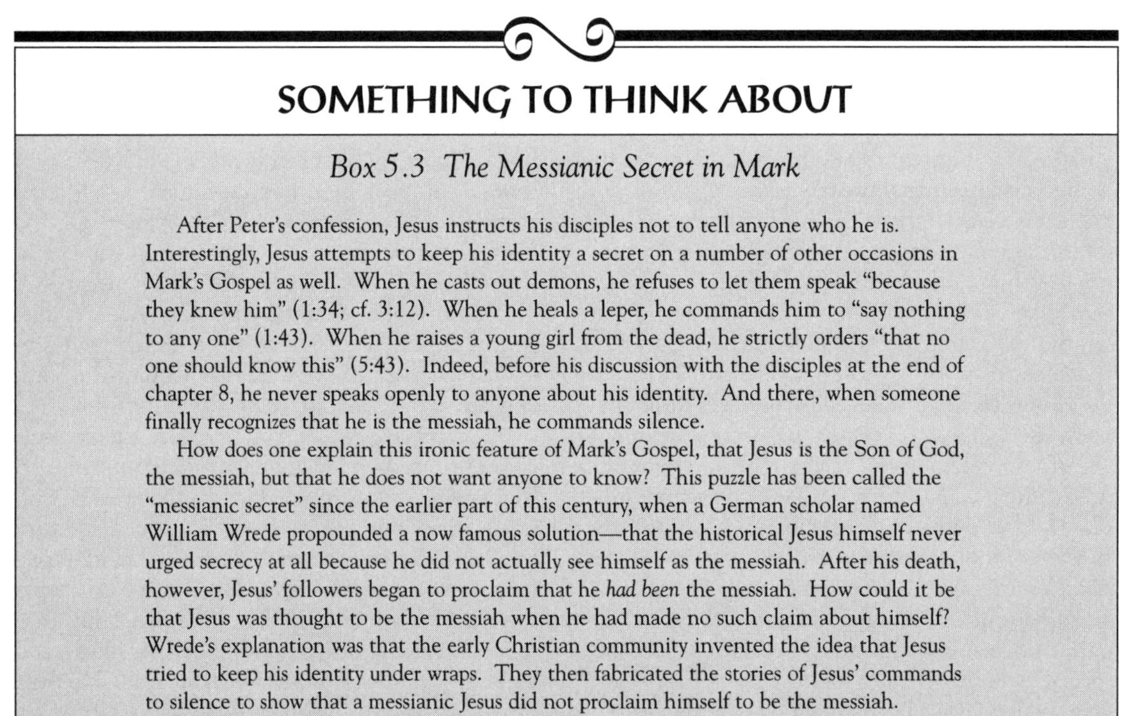

SOMETHING TO THINK ABOUT

Box 5.3 The Messianic Secret in Mark

After Peter's confession, Jesus instructs his disciples not to tell anyone who he is. Interestingly, Jesus attempts to keep his identity a secret on a number of other occasions in Mark's Gospel as well. When he casts out demons, he refuses to let them speak "because they knew him" (1:34; cf. 3:12). When he heals a leper, he commands him to "say nothing to any one" (1:43). When he raises a young girl from the dead, he strictly orders "that no one should know this" (5:43). Indeed, before his discussion with the disciples at the end of chapter 8, he never speaks openly to anyone about his identity. And there, when someone finally recognizes that he is the messiah, he commands silence.

How does one explain this ironic feature of Mark's Gospel, that Jesus is the Son of God, the messiah, but that he does not want anyone to know? This puzzle has been called the "messianic secret" since the earlier part of this century, when a German scholar named William Wrede propounded a now famous solution—that the historical Jesus himself never urged secrecy at all because he did not actually see himself as the messiah. After his death, however, Jesus' followers began to proclaim that he *had been* the messiah. How could it be that Jesus was thought to be the messiah when he had made no such claim about himself? Wrede's explanation was that the early Christian community invented the idea that Jesus tried to keep his identity under wraps. They then fabricated the stories of Jesus' commands to silence to show that a messianic Jesus did not proclaim himself to be the messiah.

Different scholars evaluate the merits of this solution differently, and we will have occasion to return to it when we take up questions pertaining to the historical Jesus in chapter 15. In the present chapter we are interested in how the messianic secret functions literarily in the context of Mark's story of Jesus. Here Jesus is clearly the messiah (cf. 1:1); but just as clearly he is not the great king or cosmic warrior that many Jews may have anticipated. Why then the commands to silence? One explanation is that Jesus in Mark's Gospel does not want people to have the wrong idea about him, for example, by thinking that he is the kind of messiah they have anticipated. For Mark, the title "messiah" does not signify earthly grandeur and power but just the opposite. As messiah, Jesus was the Son of God who had to suffer and die.

who will usher in God's kingdom in a mighty way (whether as a warrior-king or as a cosmic judge of the earth; see box 5.1). But for Mark, this is only a partial truth, a dim perception of who Jesus is. For him, Jesus is the messiah who must suffer and die to bring about salvation for the world.

Peter's failure to perceive this truth forces Jesus to turn the rebuke back on him: "Get behind me Satan! For you are setting your mind not on divine things but on human things" (8:33). The idea that the messiah had to suffer may have appeared totally anomalous to most Jews of the first century, including Jesus' own disciples; but in Mark's view, to understand Jesus in any other way is to succumb to the temptations of the devil. Thus Peter has begun to see, but not yet clearly; he is like a blind man who has partially recovered his sight. Perhaps this is better than being totally blind, but in another sense it is worse, because partial perception can lead to misperception: people seem to be trees and Jesus appears to be the messiah of popular expectation. For Mark, however, Jesus is the suffering Son of God.

JESUS THE SUFFERING SON OF GOD

Throughout the early portions of Mark's Gospel the reader is given several indications that Jesus will have to die (e.g., 2:20; 3:6). After Peter's confession, however, Jesus begins to be quite explicit about it. Even though he is the Christ, the Son of God—or rather because he is—he must suffer death. Three times Jesus predicts his own impending passion in Jerusalem: he is to be rejected by the Jewish leaders, killed, and then raised from the dead. Strikingly, after each of these "Passion predictions" Mark has placed stories to show that the disciples never do understand what Jesus is talking about.

We have already seen the first prediction in 8:31. When Jesus declares that he must be rejected and killed, Peter, who has just declared Jesus to be the messiah, not understanding fully what this means, takes him aside to rebuke him (8:32). Jesus turns the rebuke back on him and begins to teach that suffering is to be not only his lot but that of his followers as well: "Whoever would come after me must take up the cross and follow me." Being a disciple means affliction and pain, not power and prestige; it means giving up one's life in order to gain the world. Those who reject these words will have no part of Christ at the end of the age (8:34–38).

The next prediction occurs a chapter later, after Jesus' hidden glory is revealed on the Mount of Transfiguration to three of the disciples, who even then fail to understand what they have seen (9:2–13; especially vv. 6, 10). In nearly the same terms as before, Jesus predicts his coming death, and Mark states that the disciples do not know what he means (9:30–31). Immediately afterwards, they begin to argue over who is the greatest among them (9:33–34). Jesus again tells them that being his disciple means a life of lowly servitude rather than grand eminence.

The final prediction occurs in the chapter that follows (10:33–34). In this instance, the details are somewhat more graphic, but the response of the disciples is remarkably similar. James and John, two of his closest followers, request positions of prominence when Jesus enters into his glorious kingdom. Jesus has to tell them, yet again, that following him means certain death, and that if they want to be great they must become the slaves of all. This, in fact, is what he has done himself: "For the Son of Man came not to be served, but to serve, and to give his life a ransom for many" (10:45).

From this point on, the narrative marches inexorably towards Jesus' death, as Mark recounts the familiar stories of the "Passion narrative." Jesus triumphally enters Jerusalem to shouts of acclamation from the crowds, who appear to accept the disciples' notion of what it means for Jesus to be the messiah (11:1–10). He enters the Temple and drives out those who are in business there, incurring yet further opposition from the Jewish leadership (11:15–19). He teaches in the Temple, and engages in disputes with his opponents among the leaders, who try to trap him and stir up the crowds against him (11:28–12:40). He launches into a lengthy description of the imminent destruction of the Temple, when the end of time comes and the cosmic judge, the Son of Man, appears to bring judgment to the earth and salvation to the follow-

SOME MORE INFORMATION

Box 5.4 Son of God and Son of Man

The way that most people understand the terms "Son of God" and "Son of Man" today is probably at odds with how they would have been understood by many Jews in the first century. In our way of thinking, a "son of God" would be a god (or God) and a "son of man" would be a man. Thus, "Son of God" refers to Jesus' divinity and "Son of Man" to his humanity. But this is just the opposite of what the terms meant for many first-century Jews, for whom "son of God" commonly referred to a human (e.g., King Solomon; cf. 2 Sam 7:14) and "son of man" to someone divine (cf. Dan 7:13–14).

In the New Testament Gospels, Jesus uses the term "son of man" in three different ways. On some occasions he uses it simply as a circumlocution for himself; that is, rather than referring directly to himself, Jesus sometimes speaks obliquely of "the son of man" (e.g., Matt 8:20). In a related way, he sometimes uses it to speak of his impending suffering (Mark 8:31). Finally, he occasionally uses the term with reference to a cosmic figure who is coming to bring the judgment of God at the end of time (Mark 8:38), a judgment that Mark's Gospel expects to be imminent (9:1; 13:30). For Mark himself, of course, the passages that speak of the coming Son of Man refer to Jesus, the one who is returning soon as the judge of the earth. As we will see later, scholars debate which, if any, of these three uses of the term can be ascribed to the historical Jesus.

ers of Jesus (13:1–36). He assures his hearers that this apocalyptic drama will unfold soon, within their own generation (13:30).

Finally we reach the account of the Passion itself. Jesus is anointed with oil by an unknown woman, evidently the only person in the entire narrative who knows what is about to happen to him (14:1–9; she may, however, simply be performing a kind deed that Jesus himself explains as a preparation for his burial). He celebrates his Last Supper with his disciples (14:12–26) and then goes out with them to the Garden of Gethsemane to pray that he not be required to suffer his imminent ordeal (14:26–42). God, however, is silent. Jesus is arrested (14:43–52) and put on trial before the Jewish Council, the Sanhedrin, where he is confronted with witnesses who accuse him of opposing the Temple (14:53–65). The false witnesses on the inside are matched by the false disciples on the outside: while Jesus is being tried, Peter, as predicted, denies him three times (14:66–72).

Jesus is finally questioned directly by the high priest concerning his identity: "Are you the Christ, the Son of the Blessed One?" The reader, of course, already knows the answer: Jesus is the messiah, the Son of God, but not in any way that these Jewish authorities would recognize. Jesus now confesses to his identity and again predicts that the Son of Man, the cosmic judge from heaven, will soon arrive on the clouds of heaven (14:61–62; see box 5.4). The Sanhedrin charges him with blasphemy and finds him worthy of death (see box 5.5). The next morning they deliver him over to Pilate, who tries him on the charge of claiming to be King of the Jews (15:1–15). When Jesus refuses to answer his accusers, Pilate condemns him to execution for treason against Rome. Pilate gives the Jewish crowds the option of releasing Jesus or a Jewish insurgent, Barabbas (15:6–15). They prefer Barabbas. Jesus is flogged, mocked, and beaten. They take him off and crucify him at 9:00 a.m (15:25).

JESUS' DEATH AS THE SON OF GOD

It is clear from Mark's Gospel that Jesus' disciples never do come to understand who he is. As we have seen, he is betrayed to the Jewish authorities by one of them, Judas Iscariot. On the night of his arrest, he is denied three times by another, his closest disciple, Peter. All the others scatter, unwilling to stand up for him in the hour of his distress. Perhaps Mark wants his readers to understand that the disciples were shocked when their hopes concerning Jesus as messiah were thoroughly dashed: Jesus did not bring victory over the Romans or restore the kingdom to Israel. For Mark, of course, these hopes were misplaced. Jesus was the Son of God, but he was the Son of God who had to suffer. Until the very end, when Jesus was actually crucified, there is nobody in the Gospel who fully understands this.

Mark's narrative may even intimate that at the end Jesus himself was in doubt. In Gethsemane he prays three times not to have to undergo his fate, suggesting perhaps that he thinks there could be another way. When he finally succumbs to his destiny, he appears yet more uncertain, and with good reason. Deserted by his own followers, condemned by his own leaders, rejected by his own

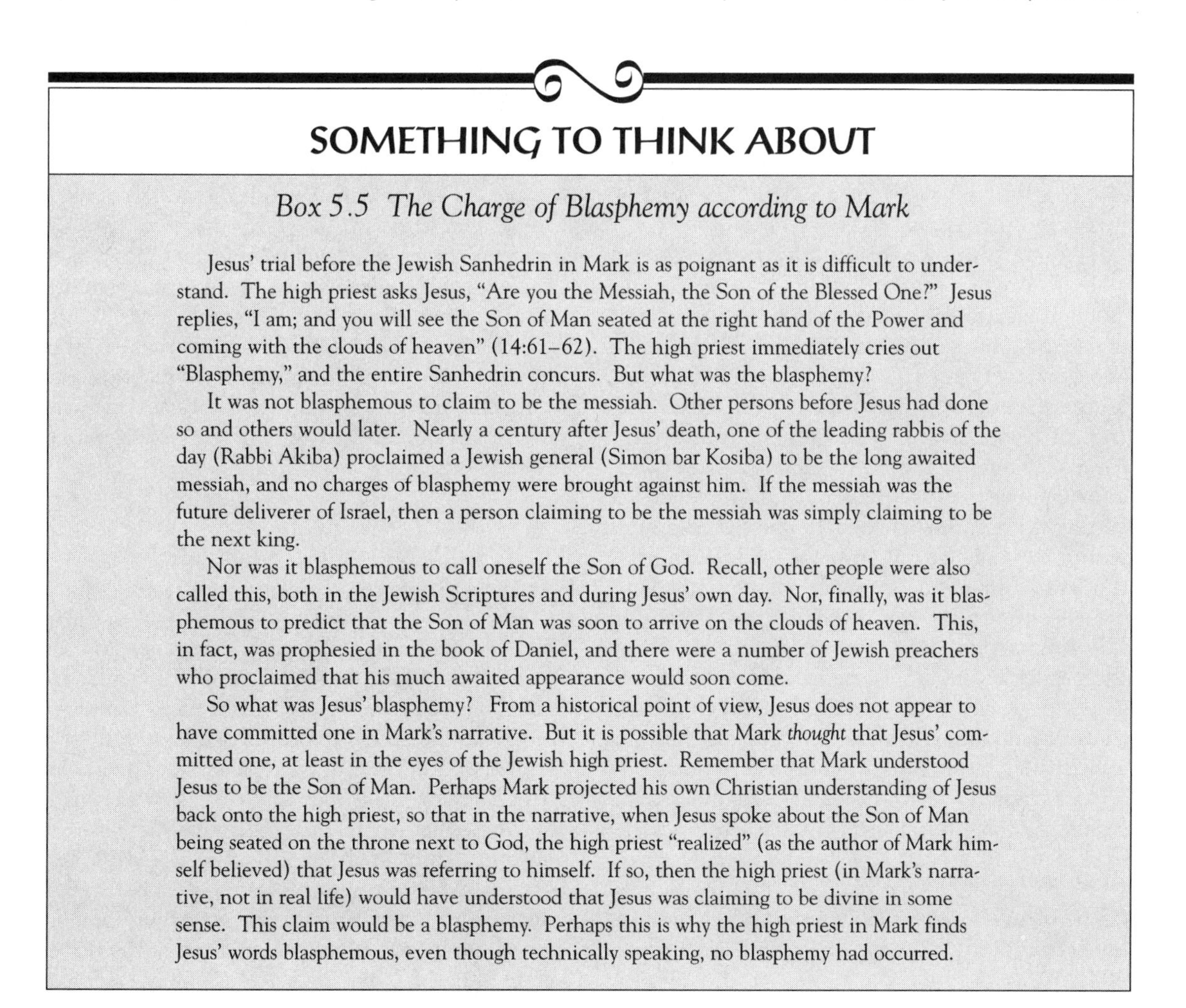

SOMETHING TO THINK ABOUT

Box 5.5 The Charge of Blasphemy according to Mark

Jesus' trial before the Jewish Sanhedrin in Mark is as poignant as it is difficult to understand. The high priest asks Jesus, "Are you the Messiah, the Son of the Blessed One?" Jesus replies, "I am; and you will see the Son of Man seated at the right hand of the Power and coming with the clouds of heaven" (14:61–62). The high priest immediately cries out "Blasphemy," and the entire Sanhedrin concurs. But what was the blasphemy?

It was not blasphemous to claim to be the messiah. Other persons before Jesus had done so and others would later. Nearly a century after Jesus' death, one of the leading rabbis of the day (Rabbi Akiba) proclaimed a Jewish general (Simon bar Kosiba) to be the long awaited messiah, and no charges of blasphemy were brought against him. If the messiah was the future deliverer of Israel, then a person claiming to be the messiah was simply claiming to be the next king.

Nor was it blasphemous to call oneself the Son of God. Recall, other people were also called this, both in the Jewish Scriptures and during Jesus' own day. Nor, finally, was it blasphemous to predict that the Son of Man was soon to arrive on the clouds of heaven. This, in fact, was prophesied in the book of Daniel, and there were a number of Jewish preachers who proclaimed that his much awaited appearance would soon come.

So what was Jesus' blasphemy? From a historical point of view, Jesus does not appear to have committed one in Mark's narrative. But it is possible that Mark *thought* that Jesus' committed one, at least in the eyes of the Jewish high priest. Remember that Mark understood Jesus to be the Son of Man. Perhaps Mark projected his own Christian understanding of Jesus back onto the high priest, so that in the narrative, when Jesus spoke about the Son of Man being seated on the throne next to God, the high priest "realized" (as the author of Mark himself believed) that Jesus was referring to himself. If so, then the high priest (in Mark's narrative, not in real life) would have understood that Jesus was claiming to be divine in some sense. This claim would be a blasphemy. Perhaps this is why the high priest in Mark finds Jesus' words blasphemous, even though technically speaking, no blasphemy had occurred.

people, he is publically humiliated, beaten, spat upon, flogged. He is nailed to the cross, and even there he is mocked by passersby, Jewish leaders, and the two criminals who are crucified along with him. He suffers throughout this entire ordeal in silence, until the very end, when he cries out the words of Scripture: "My God, My God, why have you forsaken me?" (15:34; cf. Ps. 22:2). He then utters a loud cry and dies.

Is this a genuine question of the dying Jesus? Does he truly feel forsaken in the end even by God? Does he not fully understand the reason for his death? These are questions on which readers of the account may disagree. On one point, however, there can be no disagreement. Even though no one else in the Gospel appears to know the significance of Jesus' death, the reader knows. Mark reveals it by narrating two events that transpire immediately after Jesus breathes his last: the curtain in the Temple is torn in half from top to bottom (15:38), and the Roman centurion confesses Jesus to be the Son of God (15:39).

Without posing the historical question of what really happened to the curtain in the Temple (there is no reference in any non-Christian source of it being torn or damaged in any way), one might ask how the reader is supposed to understand Mark's claim that it was ripped asunder. Most ancient Jews ascribed a particular holiness to the Temple as the one place in which sacrifices could be offered up to God. This was a sacred place to be revered and respected. The most sacred area within the holy Temple was the Holy of Holies, the square room in whose darkness God's very presence was thought to dwell. This room was so holy that no one could enter, except on one day of the year, the Day of Atonement (Yom Kippur), when the Jewish high priest could go behind the thick curtain into the presence of God to perform a sacrifice to atone for the sins of the people.

Mark indicates that when Jesus died, the curtain separating this holiest of places from the outside world was torn in half. The event appears to signify that God is no longer removed from his people; his holiness is now available to all. No longer do his people need to rely on the Jewish high priest and his sacrifice for their sins on the Day of Atonement. The ultimate sacrifice has been made, voiding the necessity of all others. Jesus, the Son of God, has "given his life as a ransom for many" (10:45). People now have direct access to God, who comes to them in the death of Jesus.

The second event cited by Mark is equally significant. No one throughout the Gospel has fully understood that Jesus is the Son of God who has to suffer. Until now. Strikingly, it is not one of Jesus' family or followers who understands. It is the Roman centurion who has presided over his crucifixion. This pagan soldier, seeing Jesus die, proclaims, "Surely this man was God's Son" (15:39). This brings the recognition of Jesus' true identity full circle. It was proclaimed at his baptism at the beginning of the Gospel (from heaven); it is now proclaimed at his crucifixion at the end (on earth). Moreover, it is significant who makes the proclamation: a pagan soldier, one who had not been Jesus' follower. This in itself may intimate what will happen to the proclamation of Jesus through the years until the time when Mark pens his account. The proclamation will not find fertile soil among Jews, either those who had known Jesus or those who had not. It will be embraced principally by those outside of Judaism, by Gentiles as represented by this Roman centurion. Jesus is the Son of God, rejected by his own people but acknowledged by the Gentiles, and it is this confession of the suffering and death of the Son of God, Mark reveals, that has brought salvation to the world. This, however, is not the end of the story.

JESUS THE VINDICATED SON OF GOD

One of the most fascinating aspects of Mark's Gospel is the way in which he chose to conclude it. Jesus is buried by a respected leader among the Jews, Joseph of Arimathea (indicating, perhaps, that not all Jews, or even all prominent Jews, were bound to reject him; 15:42–47, cf. 12:28–34). Two women see where he is placed. The next day is the Sabbath. Early in the morning on the day after Sabbath, Mary Magdalene, Mary the mother of James, and Salome come to provide a more decent burial for the body, but they discover that the stone before the tomb has been rolled away. Going inside, they find a young man in a white robe who

Figure 5.3 One of the earliest surviving portrayals of Jesus' crucifixion, from an ancient ivory panel.

tells them that Jesus has risen. He instructs them to tell the disciples and Peter that Jesus is going ahead of them to Galilee and that they are to go there to see him (16:1–7). Then comes the breathtaking conclusion. The women flee the tomb and tell nobody anything, "for they were afraid" (16:8).

Christian readers from time immemorial have been shocked and dismayed by this conclusion. How could it end without the disciples hearing that Jesus has been raised? How could they remain in their ignorance? Surely the women must have told someone. In the early church, some copyists of this Gospel were so put off by the ending that they added one of their own, appending twelve additional verses that describe some of Jesus' appearances to his disciples. Modern scholars are unified, however, in recognizing this ending as secondary (see Chapter 28). Some have proposed, in its stead, that we assume that the final page of the Gospel somehow got lost (which makes the questionable assumption, among other things, that the Gospel was written on separate pages instead of on a scroll).

These various explanations for Mark's ending, however, may be unnecessary. Mark devoted considerable effort to demonstrating that the disciples never could understand what Jesus meant when he talked about dying and rising again. They never do understand, to the very end. Mark's readers, however, understand. In fact, they understand a lot of things—about who Jesus really is, about how he was thoroughly misunderstood, about how his message was to go to the Gentiles, and about what it means for those who believe in him to be his disciples.

CONCLUSION: MARK AND HIS READERS

Can we decide who the original readers of this Gospel probably were? It is impossible, of course, to learn very much about them. Our only evidence comes from the Gospel itself, and conclusions drawn on these slim grounds will necessarily be tentative. But there are a few intimations both about the first readers and about Mark's overarching concerns for them, and I will conclude this discussion by considering them.

The first readers of this Gospel appear to have been the Christians of Mark's community, most of whom would have been illiterate,and thus "read" the Gospel by hearing it read (see box 3.1). They evidently resided outside of Palestine and had Greek as their primary language. There are clues in the Gospel that most of them had not converted to Christianity from Judaism, the most striking of which comes in 7:3–4, where Mark has to explain the Pharisaic custom of washing hands before eating for ceremonial cleansing. Presumably, if his audience were Jewish, they would already know this custom, and Mark would not have to explain it. What is even more intriguing is the fact that Mark appears to misunderstand the practice: he claims that it was followed by "all the Jews." We know from ancient Jewish writings that this is simply not true. For this reason, many scholars have concluded that Mark himself was not Jewish.

Many of Mark's traditions, however, are concerned with showing the Jewishness of Jesus and appear to presuppose strictly Jewish beliefs and practices. How can we explain this? Many of the oral traditions found in this Gospel must go back to the earliest Jewish followers of Jesus, who embodied their own beliefs and concerns in them. As the stories were passed along, their Jewish character was preserved. Mark and many people in his congregation (*some* of them Jewish?) converted to faith in Jesus, which necessarily involved converting to Jesus' religion, Judaism. They too came to worship the Jewish God and saw in Jesus the Jewish messiah, whose death brought about salvation not only for Jews but for the whole world.

It may be that this community continued to face opposition from a local Jewish synagogue that actively rejected these Christian claims about Jesus. And it may be that this opposition at times turned ugly. This would explain why Mark emphasizes that Jewish leaders, especially Pharisees, failed to understand Jesus and that following him involves a high cost. For Mark, following Jesus is not a ticket to glory, it is the path to suffering; being a disciple does not bring exaltation but humiliation and pain.

Mark stresses, however, that the suffering would not last forever. In fact, it would not last long. Just as Jesus was vindicated, so too will be his faithful followers. And the end was near (9:1). This may have been suggested to Mark by current events: many scholars believe that the Gospel was written during the early stages of the Jewish War (66–70 C.E.), at the conclusion of which the Temple itself was destroyed. Does this war mark the beginning of the end, predicted by Jesus as certain to occur during the lifetime of some of his disciples (see 8:38–9:1 and all of chap. 13)? Indeed, for the Markan community, the Son of Man was at the gate, ready to make his appearance. Those who were ashamed of Jesus' words would be put to shame when the Son of Man arrived; those who accepted his words and became his followers would then enter into glory. Just as Mark's Jesus may not have fully understood the meaning of his own crucifixion, so too the Christian community currently experiencing suffering may not fathom its full meaning, but ultimately their pain will lead to redemption. This is just one of the paradoxical claims of Mark's Gospel.

Mark's story of Jesus is replete with such paradoxes: the glorious messiah is one who suffers an ignominious death; exaltation comes in pain, salvation through crucifixion; to gain one's life one must lose it; the greatest are the most humble; the most powerful are the slaves; prosperity is not a blessing but a hindrance; leaving one's home or field or family brings a hundredfold homes and fields and families; the first will be last and the last first. These lessons provide hope for a community that is in the throes of suffering, experiencing the social disruptions of persecution. They make particular sense for a community that knows that its messiah, the Son of God, was rejected and mocked and killed, only to be vindicated by God, who raised him from the dead.

SUGGESTIONS FOR FURTHER READING

Hooker, Morna *The Message of Mark*. London: Epworth, 1983. A very nice overview of the most significant features of Mark's Gospel; ideal for beginning students.

Kingsbury, Jack D. *The Christology of Mark's Gospel*. Philadelphia: Fortress, 1983. A useful discussion of Mark's view of Jesus from a literary-critical perspective which looks for clues to the meaning of the text in the flow of the narrative.

Matera, Frank J. *What Are They Saying about Mark?* New York: Paulist, 1987. An excellent overview of modern scholarship on Mark, for beginning students.

Nickle, Keith. *The Synoptic Gospels: Conflict and Consensus*. Atlanta: John Knox, 1980. One of the best introductory discussions of the background and message of the three Synoptic Gospels.

Sanders, E. P. and Margaret Davies. *Studying the Synoptic Gospels*. Philadelphia: Trinity Press International, 1989. A detailed and thorough discussion of the Synoptic problem and of the major scholarly approaches to each of the three Synoptic Gospels; for advanced students.

Tolbert, M. A. *Sowing the Gospel: Mark's World in Literary-Historical Perspective*. Minneapolis, Minn.: Fortress, 1989. An insightful and provocative study of Mark's literary technique and method, which tries to understand the entire narrative of Mark's Gospel in light of the literary conventions of its first-century milieu; best suited for advanced students.

Wrede, William. *The Messianic Secret*. Trans. J. C. G. Greig. Cambridge: Clarke, 1971. The classic study of Mark's literary technique and theological agenda.

CHAPTER 6

The Synoptic Problem and Its Significance for Interpretation

METHODS FOR STUDYING THE GOSPELS

Now that we have studied one of the early Christian Gospels, we can take a step back and reflect on what we have done. In analyzing Mark, I began by establishing the genre of the book, arguing that it was a kind of Greco-Roman biography, and then asked how an informed reader might understand its message. This hypothetical reader was one who knew how the genre of the book works and who had all of the background information of the first-century world that the author appears to presuppose.

A literary theorist would identify this approach as one kind of "reader-response criticism." For our purposes, however, since the method focuses on a text's literary genre within its historical context, I will call it the "literary-historical method." It is by no means self-evident that the literary-historical method is the best way to approach a text from antiquity. Indeed, most readers of the New Testament have never used it! But in many respects it is superior to other ways of reading the text; it is better, for example, than thinking that the historical context of what an author says, or the literary genre that the author uses, are of no importance to the message. At the same time, there are other ways besides the literary-historical to study a text. In this chapter I will establish the theoretical grounds for using another method that has enjoyed enormous popularity among scholars of the Gospels. It has traditionally been called "redaction criticism."

A "redactor" is someone who edits a text; "redaction criticism" is the study of how authors have created a literary work by modifying or editing their sources of information. The underlying theory behind the method is simple. An author will modify a source of information only for a reason—why change what a source has to say if it is acceptable the way it is? If enough changes point in the same direction, we may be able to uncover the redactor's principal concerns and emphases.

We can subject the Gospels to a redactional analysis because we are convinced that their authors used actual sources in constructing their narratives; that is, they didn't make up most of their stories themselves. Moreover, we are relatively certain that at least one of these sources still survives. To put the matter baldly: most scholars believe that Matthew and Luke used the Gospel of Mark as a source for many of their stories about Jesus. By seeing how they edited these stories, we are able to determine their distinctive emphases. To justify the method, we must obviously begin by demonstrating that Matthew and Luke used Mark as a source.

THE SYNOPTIC PROBLEM

Matthew, Mark, and Luke are often called the "Synoptic Gospels." This is because they have so many stories in common that they can be placed side by side in columns and "seen together" (the literal meaning of the word "synoptic"). Indeed, not only do these Gospels tell many of the same stories, they often do so using the very same words.

This phenomenon is virtually inexplicable unless the stories are derived from a common literary source. To illustrate, consider a modern–day parallel. You have no doubt noticed over the years that when newspapers, magazines, and books all describe the same event, they do so differently. Take any three newspapers from yesterday and compare their treatment of the same news item. At no point will they contain entire paragraphs that are word for word the same, unless they happen to be quoting the same source, for example, an interview or a speech. These differences occur because every author wants to emphasize certain things and has his or her own way of writing. When you do find that two papers have exactly the same account, you know that they have simply reproduced a feature from somewhere else. This happens, for example, when two newspapers pick up the same news story from the Associated Press.

We have a similar situation with the Gospels. There are passages shared by Matthew, Mark, and Luke that are verbatim the same. This can scarcely be explained unless all three of them drew these accounts from a common source. But what was it? The question is complicated by the fact that the Synoptics not only agree extensively with one another, they also disagree. There are some stories found in all three Gospels, others found in only two of the three, and yet others found in only one. Moreover, when all three Gospels share the same story, they sometimes give it in precisely the same wording and sometimes word it differently. And sometimes two of them will word it the same way and the third will word it differently. The problem of how to explain the wide-ranging agreements and disagreements among these three Gospels is called the "Synoptic Problem."

Scholars have propounded a number of theories over the years to solve the Synoptic Problem. Many of the theories are extraordinarily complex and entirely implausible. For an introduction to the problem, we do not need to concern ourselves with all of these solutions. We will instead focus on the one that most scholars have come to accept as the least problematic. This explanation is sometimes called the "four-source hypothesis." According to this hypothesis, Mark was the first Gospel to be written. It was used by both

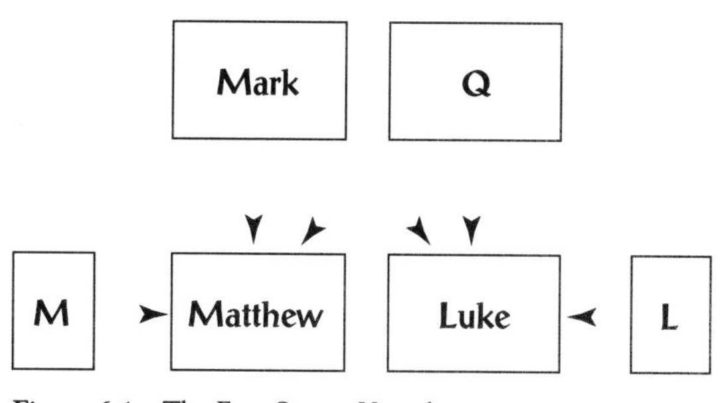

Figure 6.1 The Four-Source Hypothesis.

Matthew and Luke. In addition, both of these other Gospels had access to another source, called Q (from the German word for "source," *Quelle*). Q provided Matthew and Luke with the stories that they have in common that are not, however, found in Mark. Moreover, Matthew had a source (or group of sources) of his own, from which he drew stories found in neither of the other Gospels. Scholars have simply labeled this source (or sources) M (for Matthew's special source). Likewise, Luke had a source (or group of sources) for stories that he alone tells; not surprisingly, this is called L (Luke's special source). Hence, according to this hypothesis, four sources lie behind our three Synoptic Gospels: Mark, Q, M, and L (see figure 6.1).

The cornerstone of this hypothesis is the theory that Matthew and Luke both used Mark. We therefore begin our discussion by considering the arguments that scholars have generally found to be persuasive of "Markan priority."

Arguments for Markan Priority

For the past century or so, three arguments have proved widely convincing for establishing Mark's priority to Matthew and Luke:

Patterns of Agreement. Since the main reason for thinking that the Gospels share a common source is their verbatim agreements, it makes sense to examine the nature of these agreements in order to decide which of the books was used by the other two. If you were to make a detailed comparison of the word-for-word agreements among these Gospels, an interesting pattern would emerge. Sometimes all three of the Gospels tell a story in precisely the same way. This can easily be accounted for; it would happen whenever two of the authors borrowed their account from the earliest one, and neither of them changed it. Sometimes all three Gospels differ. This would happen whenever the two authors who borrowed the story each changed it, in different ways. Finally, sometimes two of the three are exactly alike, but the third differs. This would occur when both of the later authors borrowed the story but only one of them changed it; in this case one of the redactors would agree with the wording of his source, and the other would not.

In this final kind of situation, certain patterns of agreement typically occur among the Synoptic Gospels. Sometimes Matthew and Mark share the wording of a story when Luke differs, and sometimes Mark and Luke share the wording when Matthew differs. But it is extremely rare to find Matthew and Luke sharing the wording of a story also found in Mark when Mark differs. Why would this be?

If Matthew were the source for Mark and Luke, or if Luke were the source for Matthew and Mark, you would probably not get this pattern. Consider these examples. If both Matthew and Luke used Mark, then sometimes they would both reproduce the same wording. That's why all three sometimes agree. Sometimes they would both change the wording for reasons of their own. That's why all three sometimes differ. Sometimes Matthew would change Mark's account when Luke left it the same. That's why Mark and Luke sometimes agree against Matthew. And sometimes Luke would change Mark's account when Matthew left it the same. That's why Matthew and Mark sometimes agree against Luke.

The reason then that Matthew and Luke rarely agree against Mark in the wording of stories found in all three is that Mark is the source for these stories. Unless Matthew and Luke accidentally happen to make precisely the same changes in their source (which does happen on occasion, but not commonly and not in major ways), they cannot *both* differ from the source *and* agree with one another. The fact that they rarely do differ from Mark while agreeing with one another indicates that Mark must have been their source.

You may be relieved to recall that we are not going to worry about the complexities of the problem.

The Sequence of Narrative. One of the most striking aspects of the Synoptic Problem is that even though Matthew and Luke do not often agree together against Mark in the wording of stories that all three of them share, they do extensively agree in the wording of stories that are not found in Mark. For example, both Matthew and Luke have versions of the Lord's Prayer and the Beatitudes. Most, but not all, of these other stories are sayings of Jesus. Later we will pursue the evidence that suggests that Matthew and Luke must have gotten these stories from the source scholars call Q. What is significant at the present juncture is that even the stories not found in Mark suggest that Mark was a source for Matthew and Luke. This conclusion is based on the *sequence* of the stories found in these other two Gospels.

Matthew and Luke often present the stories of their Gospels in the same sequence (Jesus did this, then he did that, then he said this, and so on). What is odd is that when they do preserve the same sequence, it is almost always with stories that are also found in Mark. The other stories that the two Gospels share—that is, those *not* found in Mark—are in virtually every instance located in different places of their narratives.

The best explanation for the different story sequences is that Matthew and Luke each used Mark as one of their sources and also had a different source that they plugged into the narrative framework of Mark at different places. That is to say, not having any indication from Mark's Gospel where traditions like the Lord's Prayer or the Beatitudes would have fit into the life of Jesus, each author put them in wherever he saw fit. Almost never, though, did the stories go in at the same places.

This curiosity of sequence can scarcely be explained if Mark were not one of the sources for Matthew and Luke. Imagine for a moment a different scenario, that Matthew were the source for Mark and Luke. In this hypothetical case, Mark must have decided to remove some of Matthew's stories (since his Gospel is much shorter than Matthew's). Many of these Matthean stories that

Mark omitted, however, were retained by Luke. But when Luke copied Matthew, why would he have rearranged precisely these stories? That is to say, why would Luke have rearranged only those stories that Mark did not bother to copy, while keeping the stories that Mark did copy in the same sequence?

It is almost impossible to think that Luke worked this way (or Matthew, if Luke were the source for both him and Mark). Therefore, the additional stories of Matthew and Luke that occur in different places in their narrative must indicate that Mark was one of their sources, into which they both inserted these other stories.

Characteristics of the Changes. One final argument that is typically advanced for Markan priority is that the kinds of differences in wording that one finds among the three Gospels suggests that Matthew and Luke used Mark as a source. Some of these arguments, again, get rather technical; here I will simply explain the issues in general terms.

Sometimes Mark uses a Greek style of writing that is somewhat awkward or not aesthetically pleasing, sometimes he uses unusual words or phrases, and sometimes he presents difficult ideas. In many instances, however, these problems are not found when Matthew or Luke narrates these accounts. This difference suggests that Mark was the earliest of the three to be written. It is difficult to understand why Mark would introduce awkward grammar or a strange word or a difficult idea into a passage that originally posed no problem, but it is easy to understand why Matthew or Luke might have wanted to eliminate such problems. It is more likely, therefore, that Mark was first and that it was later modified by one or both of the other authors (see box 6.1).

A final and related point is that Mark is the shortest of the three Synoptics. If the author had used one of the others as his source, why would he have eliminated so many good stories? Did he want to produce a shorter version of the life of Jesus? This may sound plausible, but a close examination of the Gospel texts shows that it can't be right: in almost every instance that Mark and Matthew tell the same story, Mark's is longer. Mark doesn't appear, then, to be the work of a condenser. The conclusion that most scholars have reached, therefore, is that Mark's Gospel is the first to have been written, and that it was used independently by both Matthew and Luke.

The Q Source

Once Mark is established as prior to Matthew and Luke, the Q hypothesis naturally suggests itself. Matthew and Luke have stories not found in Mark, and in these stories they sometimes agree word for word. Whence do these stories come?

It is unlikely that one of the authors used Mark, added several stories of his own, and that his account then served as the source for the other. If this were the case, we would not be able to explain the phenomenon noted above, that these stories found in Matthew and Luke but not in Mark are almost always inserted by these other authors into a different sequence of Mark's narrative. Why would an author follow the sequence of one of his sources, except for stories that are not found in his other one? It is more likely that these stories were drawn from another source that no longer exists, the source that scholars have designated as Q.

Notwithstanding the extravagant claims of some scholars, we simply do not know the full extent or character of Q (see box 6.2). It is probably best for methodological purposes to define it strictly as material shared by Matthew and Luke that is not also found in Mark. It is indeed striking that almost all of this material comprises sayings of Jesus. But there are at least two narratives involved: the full story of Jesus' three temptations in the wilderness (Matt 4:1–11; Luke 4:1–13; Mark has only a brief mention of the Temptation, Mark 1:12–-13) and the story of the healing of the centurion's servant (Matt 8:5-10; Luke 7:1–10).

Most scholars think that Q must have been a written document; otherwise it is difficult to explain such long stretches of verbatim agreement between Matthew and Luke. It is not certain, however, that Matthew and Luke had Q in precisely the same form: they may have had it in slightly different editions. The same could be true of their other source, the Gospel of Mark.

SOMETHING TO THINK ABOUT

Box 6.1 Ironing Out the Problems: One Potential Difficulty in Mark's Account

For a simple illustration of a potential difficulty in Mark's narrative that comes to be straightened out by one of the other Synoptics, consider the opening of the story of the rich young ruler. (Actually, the story is somewhat misnamed: even though the man is rich in all three accounts, only in Matthew is he said to be young, and only in Luke is he said to be a ruler! See Matt 19:20 and Luke 18:18).

Matthew 19:16–17

Then someone came to him and said. "Teacher, what good deed must I do to have eternal life?" And he said to him, "Why do you ask me about what is good? There is only one who is good. If you wish to enter into life, keep the commandments. . . ."

Mark 10:17–18

As he was setting out on a journey, a man ran up and knelt before him, and asked him, "Good Teacher, what must I do to inherit eternal life?" Jesus said to him, "Why do you call me good? No one is good but God alone. You know the commandments. . . ."

If you were to compare these accounts with what you find in Luke 18:18–19, you would find that Luke agrees with Mark nearly word for word (apart from the introduction to the scene). This means that Mark, rather than Matthew, must have been Luke's source, since the only reason for thinking that any of them is the source for the others is their verbal agreement. But was Matthew the source for Mark or was Mark the source for Matthew?

One of the interesting things about this passage is that the man who approaches Jesus uses the term "good" in both accounts, but in Matthew he uses it to refer to the "deed" he must do, whereas in Mark he uses it to refer to Jesus. As a result, the ensuing dialogue in Mark makes good sense: Jesus wants to know why the man has called him good when only God is good. But the flow of the dialogue in Matthew seems a bit strange: why would Jesus object to the man asking him about what is good, on the grounds that God alone is good?

One way to understand the redaction of the story is to think that Matthew's dialogue has become convoluted because he has shifted the adjective away from Jesus and onto the deed that must be done. By doing so, he interrupted the flow of the conversation. What would have compelled him to make the change? One possibility is that there was something in Mark's story that proved disturbing to him. On closer examination you may be able to detect what it was. One way to read (or misread) Mark's account is to think that Jesus is claiming not to be good ("Why do you call *me* good? There is no one good but God!") If Matthew realized that Jesus' words might be taken in this way, he may have changed the account slightly simply by moving the adjective "good."

Finally, most scholars are convinced that of the two Gospels that utilized Q, Luke is more likely than Matthew to have preserved its original sequence. This is chiefly because when Matthew used Mark, he often gathered together in one place stories scattered throughout his Markan source. As a much noted example, Matthew assembled miracle stories dispersed throughout Mark chapters 1, 2, 4, and 5 into one large collection of miracles in Matthew 8–9. If this propensity for reorganizing similar kinds of stories was also at work in his treatment of Q, it would make sense that Matthew combines various sayings of Jesus scattered in different portions of

SOME MORE INFORMATION

Box 6.2 The Contents of Q

We cannot know the full contents of Q, but this has rarely stopped scholars from trying. One popular and widespread view, for example, is that Q did not contain a Passion narrative but consisted entirely of sayings of Jesus, and that it was therefore very similar to the *Gospel of Thomas*, a collection of 114 sayings of Jesus without any stories of his deeds or experiences and no references to his death and resurrection. I will be dealing more fully with this view in chapter 12, when I discuss the *Gospel of Thomas*.

Despite the exuberant claims of some scholars, we cannot fully know what Q contained because the document has been lost. We have access to it only through the materials that Matthew and Luke both decided to include in their accounts, and it would be foolish to think that one or both of them included the entire document. Indeed, if only one of them included a passage from Q, then we would have no solid grounds for knowing that it came from Q rather than, say, M or L. It is entirely possible, for example, that Q had a Passion narrative, and that neither Matthew nor Luke chose to use it, or that only one of them chose not to do so (so that some of the verses of Matthew's or Luke's Passion narrative not found in Mark actually derive from Q). At the same time, it is equally possible that Q was almost entirely sayings, without a Passion narrative (or nearly any other narrative). Regrettably, we will never know, unless, of course, Q itself should serendipitously turn up!

Among the materials that we *can* say were found in Q are some of the most memorable passages in the Gospels, including the following (for simplicity, verse references only from Luke are given):

—The preaching of John the Baptist (Luke 3:7–9, 16–17)
—The three temptations in the wilderness (Luke 4:1–13)
—The Beatitudes (Luke 6:20–23)
—The command to love your enemies (Luke 6:27–36)
—The command not to judge others (Luke 6:37–42)
—The healing of the centurion's slave (Luke 7:1–10)
—The question from John the Baptist in prison (Luke 7:18–35)
—The Lord's Prayer (Luke 11:2–4)
—The need for fearless confession in light of the coming judgment (Luke 12:2–12)
—The command not to worry about food and clothing (Luke 12:22–32)
—The parable of the unfaithful slave (Luke 12:39–48)
—Entering the kingdom through the narrow Door (Luke 13:23–30)
—The parable of the great wedding feast (Luke 14:15–24)

Luke. The Beatitudes and the Lord's Prayer, for example, are in different sections of Luke (chaps. 6 and 11) but are joined together as part of the Sermon on the Mount in Matthew (chaps. 5–6). It would make less sense to think that Luke arbitrarily disrupted this kind of unity. Luke's version is therefore probably closer to the original sequence of the stories in Q.

The M and L Sources

We are even less informed about the sources designated M and L. Since these are sources that provide material found in either Matthew or Luke alone, there is nothing to compare them with in order to decide their basic character. We do not know, for instance, whether M (or L) was only one source or a group of sources, whether it was written or oral. It could represent a single document available to the author of Matthew (or Luke), or several documents, or a number of stories that were transmitted orally, or a combination of all of these things. What is clear is that these stories came from somewhere, since it appears unlikely that the Gospel authors simply made them up.

Included in these special sources are some of the most familiar passages of our New Testament Gospels. For example, the stories from M include the visit of the Magi (Matt 2:1–12), the flight to Egypt (Matt 2:13–23), Jesus' instructions on almsgiving and prayer (Matt 6:1–8), and his parables of the treasure hidden in the field (Matt 13:44), the pearl of great price (Matt 13:45–46), the dragnet (Matt 13:47–50), the unmerciful servant (Matt 18:23-35), and the ten virgins (Matt 25:1–12). Among the stories drawn from L are the birth of John the Baptist and the annunciation to Mary (Luke 1:5–80), the shepherds visiting the infant Jesus, the presentation in the Temple, and Jesus as a twelve-year old (Luke 2:1–52), the raising of the widow's son at Nain (Luke 7:11–17), the healing of the ten lepers (Luke 17:11–19), Zachaeus in the sycamore tree (Luke 19:1–10), and the parables of the good Samaritan (Luke 10:29–37), the prodigal son (Luke 15:11–32), Lazarus and the rich man (Luke 16:19–31), and the unjust judge (Luke 18:1–8).

THE METHODOLOGICAL SIGNIFICANCE OF THE FOUR SOURCE HYPOTHESIS

The Synoptic sources problem is an important one because if we have an author's source, we can determine how he has changed it, and knowing how he changed it can give us some clues as to his overriding emphases. If Matthew, for example altered a story he found in Mark, we can assume that his changes tell us something about his own theology or interests. This is not to say that the changes that Matthew and Luke introduced into the stories of Mark are the only things that should concern us when trying to interpret their Gospels. Nor is it to say that redaction criticism, the study of how these authors used their sources, is the only appropriate way to approach them. Quite the contrary. We could just as well study Matthew (or Luke) following the literary-historical method that we used for Mark; and if we undertook the study with sufficient care, we would uncover many of the same points we will find when we apply a redactional approach.

In some sense though, a redactional analysis provides a kind of shortcut to seeing what really matters to an author. It will by no stretch of the imagination indicate everything that is important, but it will help us to discern an author's overarching concerns and emphases.

SUGGESTIONS FOR FURTHER READING

Farmer, William. *The Synoptic Problem: A Critical Analysis*. New York: Macmillan, 1964. One of the best attempts in recent times to argue, against the majority of scholars (and against the perspective found in the present chapter), that Matthew was the first Gospel to be written; for advanced students.

Kloppenborg, J. *The Formation of Q: Trajectories in Ancient Wisdom Collections*. Philadelphia: Fortress, 1987. The fullest available discussion of the development of the tradition that later came to be embodied in the Q document.

Nickle, Keith. *The Synoptic Gospels: Conflict and Consensus*. Atlanta: John Knox, 1980. A brief and clear discussion of the Synoptic Problem, excellent for beginning students.

Sanders, E. P. and Margaret Davies. *Studying the Synoptic Gospels*. Philadelphia: Trinity Press International, 1989. A detailed and thorough discussion of the complexities involved in the Synoptic Problem, including some of the major difficulties with accepting the existence of a Q source; especially suited to advanced students.

Stein, Robert. *The Synoptic Problem: An Introduction*. Grand Rapids, Mich.: Baker Book House, 1987. A good book-length treatment of the range of issues involved in the Synoptic Problem; for beginning students.

Streeter, B. H. *The Four Gospels*. London: Macmillan, 1924. The classic discussion of the Synoptic Problem, which mounts compelling arguments for the priority of Mark and the existence of Q.

CHAPTER 7

Jesus, the Jewish Messiah: The Gospel according to Matthew

The Gospel of Matthew was one of the most highly treasured accounts of Jesus' life among the early Christians. This may explain why it was given pride of place as the first Gospel in the New Testament canon. Its popularity continues unabated today, in no small measure because it preserves many cherished and revered teachings of Jesus such as the memorable sayings of the Sermon on the Mount, including the Beatitudes, the Golden Rule, and the Lord's Prayer—teachings that have inspired Christian readers through the ages and convinced them of Jesus' genius as a teacher of religious principles.

We can begin our discussion of Matthew by reflecting on several of the points that we have already learned. We do not know the name of its author: the title found in our English versions ("The Gospel according to Matthew") was added long after the document's original composition. It is true that according to an old tradition the author was none other than Matthew, the tax collector named in Matthew 9:9. This tradition, however, arose some decades after the Gospel itself had been published, and scholars today have reasons to doubt its accuracy. For one thing, the author never identifies himself as Matthew, either in 9:9 or anywhere else. Also, certain features of this Gospel make it difficult to believe that this Matthew could have been the author. Why, for example, would someone who had been with Jesus rely on another author (Mark) for nearly two-thirds of his stories, often repeating them word for word (including the story of his own call to discipleship; 9:9–13)? And why would he never authenticate his account by indicating that he himself had seen these things take place?

Proto-orthodox Christians of the early centuries, of course, needed to "know" who wrote Matthew before they could include it in their canon of apostolic writings. Even though critical scholars today are not as certain about the author's precise identity, there are a few general things that we can say about him. Since he produced his Gospel in Greek, presumably for a Greek-speaking community, he was probably located somewhere outside of Palestine (since most early Christians in Palestine would have spoken Aramaic as their native tongue). To construct his narrative about Jesus, he made use of a variety of sources available to him, both written documents and oral reports that he had heard, possibly from Christian evangelists and teachers within his own community. Among his written sources were Mark's Gospel and the collection of traditions that scholars designate as Q. If Mark was produced around 65 or 70 C.E., then Matthew was obviously written later, but it is difficult to know how much later. Most scholars are content to date the book sometime during the latter part of the first century, possibly, as a rough guess, around 80 or 85 C.E.

Matthew, as I will continue to call the author for the sake of convenience, chose to follow his predecessor Mark by bringing together stories about Jesus into a connected narrative of Jesus' words and deeds culminating in his death and resurrection. An ancient reader would have recognized the book as a kind of Greco-Roman biography, and so would have entertained certain expectations about what to find in it. Such a reader would have expected the book to describe Jesus' life according to some kind of chronology, highlighting those sayings, actions, and experiences that revealed his essential character. Moreover, he or she would have expected this portrayal to be established by the events described at the very outset.

As was the case with Mark, we will by no stretch of the imagination cover everything of interest and importance in this Gospel. One of the most recent scholarly commentaries on Matthew fills three volumes, the first of which alone runs nearly 800 pages! Here we will discuss the entire book in a fraction of that space and so, merely scratch the surface. But if you scratch a surface in the right places, you can at least get an idea of what lies beneath.

Once again we could take a variety of approaches to our task, including the literary-historical approach we used for Mark. But I have chosen instead to utilize the redactional method discussed in chapter 6. By looking at some of the significant ways Matthew differs from Mark, we can gain some key insights into his understanding of Jesus. Given the importance of beginnings for Greco-Roman biographies, we can start this comparison by examining Matthew's opening chapters.

THE IMPORTANCE OF BEGINNINGS: JESUS THE JEWISH MESSIAH IN FULFILLMENT OF THE JEWISH SCRIPTURES

Matthew follows his predecessor Mark in beginning his Gospel by identifying Jesus as the Christ. He will therefore have a similar task of explaining how Jesus could be the glorious and powerful messiah of the Jews when he was known to have experienced a public humiliation and ignominious death by crucifixion. Far from shrinking from the task, Matthew approaches it head on, in the very opening verse, by emphasizing Jesus' credentials as the messiah: he was "the son of David, the son of Abraham." As Matthew's readers would realize full well, Abraham was thought to be the father of the Jews. And David was their greatest king, whose descendant was to resume his rule, enthroned in Jerusalem and reigning over a sovereign state of Israel as God's anointed. This son of David would be the messiah.

Thus Matthew begins his Gospel by indicating that Jesus was a Jew (from Abraham) in the line of the ancient kings (from David). One is immediately impressed by a distinctive feature of this narrative: Jesus is portrayed as thoroughly and ineluctably Jewish. He was Jewish in Mark's Gospel as well, of course; but here the emphasis is yet stronger. Matthew's narrative will show that Jesus was the ultimate fulfillment of the hopes of the Jews.

The Genealogy of Jesus the Messiah

The Jewish identity of Jesus is confirmed by what follows. Unlike Mark, Matthew provides a genealogy of Jesus, tracing his family line all the way back to the father of the Jews, Abraham himself. Genealogies are not among the most popular reading for students of the Bible today, but this one is remarkable for a number of reasons. It is structured around several key persons in the history of the nation Israel, many of whom are well known from stories preserved in the Jewish Scriptures (e.g., Abraham, Isaac, Jacob; David, Solomon, Rehoboam; Ahaz, Hezekiah, Manasseh). The text consistently, almost monotonously, traces fathers and sons first from Abraham (v. 2) to King David (v. 6), then from David to the deportation to Babylon (v. 12), and then from the deportation to Jacob, the father of Joseph (v. 16). At this point, however, a problem arises: it turns out that the genealogy is of Joseph, the husband of Mary, the woman to whom Jesus is born. According to Matthew, however, Joseph is not Jesus' father, for in this Gospel (unlike Mark, which says not a word about Jesus' birth) Jesus' mother is a virgin. For this reason, Matthew is forced to shift from his description of father and son relationships when he comes to the conclusion of his genealogy at the end of verse 16: "Matthan was the father of Jacob, and Jacob was the father of Joseph, the husband of Mary, of whom was born Jesus, called the Christ."

But what would be the point of tracing Jesus' bloodline back to David and Abraham, when in fact he is not connected to this line? His only link to it is through Joseph, a man who is not his father.

To be sure, the matter is perplexing, even though the basic point that the author is trying to make is relatively clear. He is trying to show that Jesus has Jewish roots and, more specifically, that he can legitimately claim to be of the line of David, as would be necessary for the "son of David," the messiah. Thus, even though the genealogy may appear irrelevant at first glance, in that Jesus doesn't belong to the bloodline that it delineates, it is clearly meant to make a statement about him; because Joseph was in some sense Jesus' "father" (through adoption?), Jesus is related through him to the greats of Israel's past.

Yet more striking is verse 17, which summarizes the genealogy in such a way as to show its real motivation. There were fourteen generations between Abraham and David, fourteen between David and the deportation to Babylon, and fourteen between the deportation to Babylon and the messiah, Jesus. This coincidence is amazing. Between the father of the Jews and the greatest king of the Jews were fourteen generations, as there were between the greatest king of the Jews and the greatest catastrophe of the Jews (the destruction of their nation by the Babylonians) and between the greatest catastrophe of the Jews and the ultimate deliverer of the Jews, the messiah.

The genealogy suggests—indeed, it almost demonstrates—that the entire course of Israel's history has proceeded according to divine providence. Moreover, this history has culminated in Jesus. At every fourteenth generation something cataclysmic happens in Israel's history: their greatest king, their worst disaster, and now their ultimate salvation. Jesus' birth fourteen generations removed from the Babylonian deportation shows that in him God was going to do something significant, something unprecedented for his people Israel.

But is this sequence of fourteen-fourteen-fourteen actually viable? It is not difficult to check since some two-thirds of the names in the genealogy are known to us from the Jewish Scriptures, Matthew's own source for the generations from Abraham to the deportation to Babylon. When the sequence is checked against this source, there do appear to be some problems. The most glaring one comes in verse 8, where Joram is said to be the father of Uzziah, for we know from 1 Chronicles 3:10–12 that Joram was not Uzziah's father; he was his great-great-grandfather. (Read the 1 Chronicles passage for yourself, but bear in mind that Uzziah is called Azariah in this book, as can be seen by comparing 2 Kgs 14:21 with 2 Chron 26:1.) Why, then, would Matthew say that he was his father?

The answer should be obvious. If Matthew were to include all the generations between Joram and Uzziah (his father Amaziah, grandfather Joash, and great-grandfather Ahaziah), he would no longer be able to claim that there were fourteen generations between David and the deportation to Babylon! This would disrupt the entire notion that at every fourteen generations a cataclysmic event happens in the history of the people. And this, in turn, would compromise his implicit claim that because of when he was born, Jesus must be someone special and significant in the divine plan for Israel (see box 7.1).

Thus the genealogy cannot be historically right. But at this stage, we are less interested in pursuing the question of what really happened in the life of the historical Jesus than in seeing how Matthew meant to portray him. Matthew begins right off the bat by informing us, through a genealogy that is not found in his predecessor, Mark, that Jesus was intimately connected with the history of the people of Israel. Indeed, the connection of Jesus with the Jewish people will be a key theme of the Gospel. Jesus will be portrayed in no uncertain terms as the Jewish messiah, come to the Jewish people in fulfillment of their greatest hopes. As the Jewish Savior sent from the Jewish God, he will embrace the Jewish Law and require his followers to do so as well. He will nonetheless come to be rejected by the Jewish leaders, who will mislead most of the Jewish people into rejecting him.

Obviously this portrayal of Jesus is not contradictory to Mark's, since most of Mark's stories have made it into Matthew, but the focus of attention, and therefore the basic portrayal of Jesus, is somewhat different. Here the center of attention is located even more squarely on the nature of Jesus' relationship to Judaism.

SOMETHING TO THINK ABOUT

Box 7.1 Matthew's Scheme of Fourteen

Since Matthew apparently had to manipulate Jesus' genealogy in order to have something of major significance happen every fourteen generations, for example, by leaving out some of the names, we are justified in wondering whether the number fourteen was of particular importance to him. (You'll notice, incidentally, that the final sequence contains only thirteen names, even though Matthew claims that it contains fourteen.) Is there something significant about the number fourteen itself?

Over the years, interpreters of Matthew have puzzled over this question and put forth a variety of theories to account for it. Let me mention two of the more interesting ones.

First, in ancient Israel, as in a number of other ancient societies in which numbers had symbolic significance, the number seven was of supreme importance as a symbol of perfection or divinity (we'll see a lot of sevens when we come to study the Book of Revelation). The ancients divided the week into seven days, probably because they believed that there were seven planets. For some ancient Jews there were seven stages in a person's life and seven parts to the human soul; there were seven heavens, seven compartments of hell, and seven divisions of Paradise; there were seven classes of angels and seven attributes of God; and so on. Consider the words of the famous first-century Jewish philosopher Philo: "I doubt whether anyone could adequately celebrate the properties of the number seven, for they are beyond words" (*On the Creation of the World*, 30).

If seven is a perfect number, a number associated with the divine, what then is fourteen? *Twice* seven! In cultures for which numbers matter, it would have been a doubly perfect number. Did Matthew set up Jesus' genealogy to show the divine perfection of his descent?

A second theory ties the genealogy yet more closely into Matthew's own portrayal of Jesus. As I will discuss further in a later context (see Chapter 24), ancient languages typically used the letters of the alphabet to represent numerals, so that one could add up the letters in a name and come up with a numerical value. As we have already seen, Matthew emphasizes Jesus' messianic character as a descendant of King David. In Hebrew, David's name is spelled with three letters, equivalent to our letters *D*, *V*, and *D* (ancient Hebrew did not use vowels). Interestingly enough, the *D* in Hebrew is worth 4 and the *V* is worth 6, so the numerical value of David's name is fourteen! Has Matthew emphasized the number fourteen in Jesus' genealogy in order to stress his Davidic roots as the messiah of the Jews?

The Birth of the Messiah

This strong focus on Jesus' Jewish roots is confirmed in the birth narrative that follows (chaps. 1 and 2). What is perhaps most striking about Matthew's account is that it all happens according to divine plan. The Holy Spirit is responsible for Mary's pregnancy and an angel from heaven allays Joseph's fears. This conception fulfills a prophecy of the Hebrew Scriptures (1:23). Indeed, so does every other event in the narrative: Jesus' birth in Bethlehem (2:6), the family's flight to Egypt (2:14), Herod's slaughter of the innocent children of Bethlehem (2:18), and the family's decision to relocate in Nazareth (2:23). These stories occur only in Matthew.

Matthew's emphasis on Jesus' fulfilling of the Scripture occurs not only in his birth narrative, but throughout the entire book. On eleven separate occasions (including those just mentioned), Matthew uses a phrase that scholars have sometimes labeled a "fulfillment citation." The formulae of these citations vary somewhat, but they typically run something like this: "this occurred in order to fulfill what was spoken of by the prophet." In each instance, Matthew then cites the passage of Scripture that he has in mind, showing that Jesus is the long expected messiah of the Jews. These fulfillment citations are not drawn from Mark, and among all four New Testament Gospels they occur only in Matthew. Even more than his predecessor, then, Matthew explicitly and emphatically stresses that Jesus is the fulfillment of the Jewish Scriptures.

Jesus fulfills the Scripture in two different ways for Matthew, the first of which is easy to grasp. The Hebrew prophets occasionally made predictions about the future messiah. According to Matthew, Jesus fulfills these predictions. For example, Jesus is born in Bethlehem because this is what was predicted by the prophet Micah (2:6), and his mother is a virgin because this is what was predicted by the prophet Isaiah (1:23).

The second way in which Jesus fulfills the Scripture is a little more complicated. Matthew portrays certain key events in the Jewish Bible as foreshadowings of what would happen when the messiah came. The meaning of these ancient events was not complete until that which was foreshadowed came into existence. When it did, the event was "fullfilled," that is, "filled full of meaning." In the birth narrative, for example, Matthew indicates that Jesus' family flees to Egypt to escape the wrath of Herod "in order to fulfill what was spoken by the Lord through the prophet, saying, 'Out of Egypt I have called my son'" (2:15). The quotation is from Hos 11:1 and originally referred to the exodus of the children of Israel from their bondage in Egypt. For Matthew, Jesus fills this event with meaning. The salvation available to the children of Israel was partial, looking forward to a future time when it would be made complete. With Jesus the messiah, that has now taken place.

Understanding this second way in which Jesus fulfills the Scripture for Matthew helps to explain certain aspects of the opening chapters of Matthew's Gospel (chaps. 1-5) that have long intrigued scholars. Think about the following events in rough outline, and ask yourself how they might have resonated with a first-century Jew who was intimately familiar with the Jewish Scriptures. A male child is miraculously born to Jewish parents, but a fierce tyrant in the land (Herod) is set to destroy him. The child is supernaturally protected from harm in Egypt. Then he leaves Egypt and is said to pass through the waters (of baptism). He goes into the wilderness to be tested for a long period. Afterwards he goes up on a mountain, and delivers God's Law to those who have been following him.

Sound familiar? It would to most of Matthew's Jewish readers. Matthew has shaped these opening stories of Jesus to show that Jesus' life is a fulfillment of the stories of Moses (read Exodus 1–20). The parallels are too obvious to ignore: Herod is like the Egyptian pharoah, Jesus' baptism is like the crossing of the Red Sea, the forty days of testing are like the forty years the children of Israel wandered in the wilderness, and the Sermon on the Mount is like the Law of Moses delivered on Mount Sinai. These parallels tell us something significant about Matthew's portrayal of Jesus. Certainly he agrees with Mark that Jesus is the suffering Son of God, the messiah, but here Jesus is also the new Moses, come to set his people free from their bondage (to sin 1:21) and give them the new Law, his teachings.

We have seen that among first-century Jews there was not just one set of expectations concerning their future deliverer. Many hoped for a future king like David, who would lead his people to military victory over their oppressors and establish Israel as a sovereign state in the Promised Land. Others anticipated the appearance of a cosmic figure on the clouds of heaven, coming in judgment to the earth. Still others looked forward to an authoritative priest who would guide the community through divinely inspired interpretations of the Mosaic Law. One other form that the future deliverer sometimes took is of particular relevance for understanding Matthew's portrayal of Jesus. Some Jews hoped that a prophet like Moses would appear, who not only brought salvation from the hated oppressors of Israel, the Egyptians who had enslaved them for 400 years, but also disclosed the Law of God to his people. Indeed, according to the ancient traditions, Moses himself had said that there would be another prophet like him who would arise among his people (Deut 18:15–19). The hope for a messianic figure like Moses, one chosen by God to bring salvation and new direction, was very much alive among some Jews in the first century.

Unlike later Christians like Marcion (see Chapter 1), who insisted that a person had to choose between Moses and Jesus, Matthew maintains that the choice is instead between Moses without Jesus and Moses with Jesus. For him, false religion involves rejecting Jesus, precisely because Jesus is a new Moses. This new Moses does not replace the old one, however. Quite the contrary, he is the true and final interpreter of what the earlier Moses recorded in his Law. Jesus also gives the divine Law in this Gospel, but for Matthew this law does not stand at odds with the Law of Moses; it is a fulfillment of that Law (5:17). Followers of Jesus must follow the Law of Moses, not abandon it; moreover, they must follow it by understanding it in the way prescribed by the new Moses, Jesus the messiah.

Just as Moses was a prophet who was confronted and rejected by those who refused to recognize his leadership, like all of the prophets in the Jewish Scriptures, according to Matthew, so too Jesus in Matthew is constantly opposed by the leaders of his own people. We have already seen this basic motif of Jesus' rejection in Mark. In many respects, Matthew emphasizes the antagonism even more, and Jesus engages in a far more active counterattack, accusing his opponents of placing a higher value on their own traditions than on the law of God, attacking their wicked motives, and above all charging them with hypocrisy, that is, knowing and teaching the right thing to do but failing to do it.

The Rejected King of the Jews

We do not have to wait long to find Matthew portraying the Jewish leaders as hypocrites, who know the truth but do not follow it. They are presented this way at the outset of the Gospel, while Jesus is still an infant.

The story of the visit of the Magi (2:1–12), found only in Matthew, is one of the most interesting tales of the New Testament. Here we are less interested in the historical problems that the story raises (e.g., how *can* a star stand over a particular house?) than in the point of the story in Matthew's Gospel. Ancient readers would have recognized the Magi as astrologers from the East (perhaps Assyria), who could read the course of human events from the movements of the stars. These wise men are pagans, of course, whose astral observation have led them to recognize that a spectacular event has transpired on earth, the birth of a child who will be king.

The text never explains why Assyrian scholars would be interested in the birth of a foreign king. Perhaps their worship of him indicates that they understand him to be far greater than a mere mortal, king or otherwise. The reader of this account already realizes this, of course, since the child is said to have no human father. What the Magi evidently do not know is where the child is to be born. The star takes them to Jerusalem, the holy city of the Jews, the capital of Judea. There they make their inquiries. Herod, the reigning king of the Jews, hears of their presence and is naturally distraught. Israel has room for only one king, and he himself sits on the throne. He has a reason of his own, then, to locate the child: not to worship him but to destroy him.

Herod calls in the Jewish chief priests and the scholars trained in the Scriptures for counsel, and here we find the key irony of the account. The Jewish leaders know perfectly well where the messiah is to be born: Bethlehem of Judea. They can even quote the Scriptures in support and do so before Herod, who informs the wise men.

Who, then, goes to worship Jesus? Not those who knew where he was to be born, not the Jewish chief priests or the Jewish Scripture scholars or the Jewish king. They stay away. It is the Gentiles, the non-Jews who originally did not have the Scriptures but who learn the truth from those who do, who go to worship the king of the Jews. The Jewish authorities, on the other hand, as represented by Herod their king, plot to kill the child.

This story functions in Matthew's Gospel to set the stage for what will happen subsequently. Jesus fulfills the Scripture and urges his followers to do so as well; he is nonetheless rejected by the leaders of his own people, who plot his death. There are others, however, who will come and worship him. We find this particular Matthean theme played out not only in stories that Matthew has added to his Markan framework but also in the changes that he has made to stories he inherited from Mark. The theme can be seen in the next account of his narrative, where Jesus meets his forerunner, John the Baptist.

JESUS AND HIS FORERUNNER FROM MATTHEW'S PERSPECTIVE

After the birth narrative, Matthew immediately launches into an account of Jesus' baptism. It is at this point that he begins to pick up stories from the Gospel of Mark. As I indicated in Chapter 6, a redactional study of the Gospel examines not only what an author has added to his source (e.g., the entire first two chapters) but even more what he has changed in the stories that he borrowed. This method can be used to examine the first story that Matthew and Mark have in common, Jesus' baptism by John.

The best way to engage in a redactional study is to read the two accounts side by side, making careful and detailed notes on where they differ. These differences may provide a clue into Matthew's overall agenda, since, as we have seen, he presumably would not have changed his source unless he had a reason.

Matthew does change the narrative of Jesus' baptism in a number of ways, many of them reasonably obvious, some of them fairly significant. To begin with, his account is much longer than Mark's. In Matthew's version, John sees a group of Pharisees and Sadducees coming to be baptized, and he lambastes them in harsh terms not found in Mark:

> You brood of vipers! Who warned you to flee from the wrath to come? Bear fruit worthy of repentance. Do not presume to say to yourselves, 'We have Abraham as our ancestor'; for I tell you, God is able from these stones to raise up children to Abraham. Even now the ax is lying at the root of the tree; every tree therefore that does not bear good fruit is cut down and thrown into the fire. (3:7–10)

As the story continues, the reader is struck by the fact that John not only stops the Jewish leaders from being baptized but later tries to stop Jesus as well, although for a totally different reason. The Pharisees and Sadducees are too wicked to be baptized, but Jesus is too good. In fact, Jesus is the one who should baptize John, the superior baptizing the inferior (3:14–15). Jesus, however, convinces John that it is right for him to be baptized, in a dialogue found only in Matthew: "Let it be so now; for it is proper for us in this way to fulfill all righteousness" (3:15).

The baptism scene is similar to Mark's, although a couple of interesting changes occur. The most significant of these is probably the voice from heaven: now rather than addressing Jesus alone ("You are my beloved Son"), it makes an open pronouncement, presumably to the bystanders ("This is my beloved Son"; 3:17).

Having observed these various differences from Mark's account, we are now in a position to ask the redactional question: what do they tell us about Matthew's portrayal of Jesus? For one thing, Matthew's changes highlight the contrasts between Jesus and the Jewish leaders. The latter are sinister vipers, destined for destruction; Jesus on the other hand is superior even to God's chosen prophet, John. This message is obviously not entirely unlike what we found in Mark's Gospel, but here it receives greater emphasis.

In Mark's Gospel we do not find (until the passion narrative) a person who correctly perceives who Jesus is. The same cannot be said of Matthew. We have already seen several people who recognize Jesus' identity: his family (Joseph and Mary), the wise men from the East (who come to worship him), and now, in light of the conversation recorded only in Matthew's account, John the Baptist. This same notion, that Jesus' identity was public, not secret, is also evident in the change of the voice from heaven, which announces to all who can hear that Jesus is the Son of God.

These changes in the baptism narrative coincide with what happens throughout the entire Gospel, for Matthew has strongly curtailed Mark's insistence that Jesus tried to keep his identity secret and that the disciples never recognized who he was. According to Matthew, Jesus was openly proclaimed the messiah during his lifetime and was worshipped as such. Consider, for example, the later episode in which Jesus walks on the water. In Mark the disciples are amazed but totally unable to understand what it all means: "And they were utterly astounded, for they did not understand . . . but their hearts were hardened" (Mark 6:51). In Matthew, on the other hand, they know full well what it means and react by falling down in worship: "And those in the boat worshiped him, saying, 'Truly you are the Son of God'" (14:33).

How can we account for such changes? Why is Jesus acknowledged for who he is in this Gospel?

One possibility is that Matthew has altered Mark's account precisely in order to emphasize the guilt of those who reject Jesus, in particular, the Jewish leaders, who come under more rigorous attack in this narrative. If Jesus' identity is public knowledge, then those who above all others should be in the know, the Jewish authorities, are all the more culpable for rejecting, and even persecuting, him.

One final shift in emphasis in Matthew's account of Jesus' baptism has to do with John's preaching. I have already pointed out that it is much more detailed than in Mark. What is even more striking, though, is the shift in its focus. By adding material drawn from the Q source, Matthew has highlighted the apocalyptic nature of John's proclamation. As we will see in greater detail in Chapter 15, "apocalypticism" was a popular worldview among Jews in the first century. Apocalyptic Jews maintained that the world was controlled by unseen forces of evil but that God was soon going to intervene in history to overthrow these forces and bring his good kingdom to earth. Such Jews believed that they were living at the end of time; the new age was soon to appear. We have already seen elements of this worldview in the Gospel according to Mark, especially in Jesus' lengthy discourse in chapter 13, in which he describes the cosmic upheavals that are going to transpire when the Son of Man arrives in judgment. Moreover, even in Mark, Jesus anticipates that this cataclysmic event is very near: his own generation will not pass away before it takes place (13:30).

Matthew emphasizes the apocalyptic character of Jesus' proclamation even more strongly, as is already evident in the preaching of Jesus' forerunner. John predicts that divine judgment is coming ("Who warned you to flee from the wrath to come?"), that indeed it is almost here ("even now the ax is lying at the root of the trees"). Those who are not prepared will be destroyed ("Every tree therefore that does not bear good fruit is cut down and thrown into the fire"). Moreover, simply being a Jew is no guarantee of salvation ("Do not presume to say among yourselves 'We have Abraham as our ancestor'; for I tell you, God is able from these stones to raise up children to Abraham"). Instead, a person must prepare for the end by living an appropriate life ("Bear fruit worthy of repentance"). These themes proclaimed early on by John will recur on the lips of Jesus throughout this Gospel.

THE PORTRAYAL OF JESUS IN MATTHEW: THE SERMON ON THE MOUNT AS A SPRINGBOARD

Since I am intent on applying a redactional method of analysis to Matthew's Gospel, rather than a literary-historical one, I will not follow the procedure I used with Mark of tracing the development of the narrative and showing how the unfolding of the plot gives an indication of the identity of its main character. Some scholars prefer to use that approach for all narratives, and as we have seen with Mark, the fruit that it bears can be quite satisfying. But there are numerous ways to approach texts, and here we are exploring another.

If we had sufficient time and space, of course, we could proceed through the entire Gospel as we have started, asking how the author has added to, subtracted from, and otherwise changed the one source that we are reasonably certain that he had, the Gospel of Mark. I have opted instead simply to analyze portions of the Sermon on the Mount, one of the most memorable portions of Matthew's narrative, for by examining several of its key passages we can uncover themes that recur throughout the rest of the Gospel.

Jesus: The New Moses and the New Law

The Sermon on the Mount (chaps. 5–7) is the first of five major blocks of Jesus' teaching in Matthew (the others: chap. 10, Jesus' instructions to the apostles; chap. 13, the parables of the kingdom; chap. 18, other teachings on the Kingdom; chaps. 23–27, the "woes" against the scribes and Pharisees and the apocalyptic discourse describing the end of time). We have seen that Matthew appears to portray Jesus as a new Moses. Some scholars have suggested that this collection of his teachings into five major blocks of material is meant to recall the five books of the Law of Moses.

As I have already indicated, a good deal of the material in the Sermon on the Mount comes from Q. Since these Q passages are scattered throughout Luke's Gospel, rather than gathered together in one place, it appears that the Sermon on the Mount may be Matthew's own creation. By taking material dispersed throughout his sources, Matthew has formed them into one finely crafted collection of Jesus' important teachings.

One of the overarching messages of the sermon is the connection between Jesus and Moses. If the Law of Moses was meant to provide divine guidance for Jews as the children of Israel, the teachings of Jesus are meant to provide guidance for his followers as children of the kingdom of heaven (see the summary statement at the end of the sermon, 7:24–28). As I have already intimated, this does not mean that Jesus' followers are to choose between Moses and Jesus; they are to follow Moses by following Jesus. For Matthew, Jesus provides the true understanding of the Jewish Law, and his followers must keep it.

The sermon is thus largely about life in the kingdom of heaven, which according to the statement in 4:17 (immediately before the sermon) was the main emphasis of Jesus' teaching: "Repent, for the kingdom of heaven has come near." This kingdom of heaven does not refer to the place people go when they die. Rather, it refers to God's presence on earth, a kingdom that he will bring at the end of this age by overthrowing the forces of evil. When God does this, the weak and oppressed will be exalted, and the high and mighty will be abased. This appears to be the point of the beginning of the sermon, the Beatitudes (the descriptions of those who are blessed,) found in 5:3–10:

> Blessed are the poor in spirit, for theirs is the kingdom of heaven. Blessed are those who mourn, for they will be comforted. Blessed are the meek, for they will inherit the earth. Blessed are those who hunger and thirst for righteousness, for they will be filled. Blessed are the merciful, for they will receive mercy. Blessed are the pure in heart, for they will see God. Blessed are the peacemakers, for they will be called children of God. Blessed are those who are persecuted for righteousness' sake, for theirs is the kingdom of heaven.

It is not at all clear how we are to interpret these Beatitudes. Given the fact that John the Baptist sets the stage for Jesus' teaching by proclaiming that the end (that is, the kingdom) is near, and that Jesus himself proclaims that "the kingdom of heaven is at hand" (4:17), it seems probable that they refer to the coming kingdom. Even so, scholars have long debated the precise function of these words. Is Jesus setting up the requirements for entrance into the kingdom? Is he saying that people have to become poor in spirit, for example, in order to receive the kingdom? While this is possible, Jesus does not appear to be issuing commands so much as making statements of fact. It would be hard, for example, to think that he was telling people that if they didn't mourn they wouldn't be allowed into the kingdom. Perhaps, then, we should see the Beatitudes as assurances to those who are presently lowly and oppressed, weak and suffering, for when the kingdom of heaven comes, they will receive their reward. Those who now mourn will be comforted, those who now hunger for justice will be granted it, and those who are now persecuted for doing what is right will be vindicated.

Taking Jesus' words in this way, however, creates another problem of interpretation. Do the Beatitudes suggest that everyone experiencing problems will be exalted in the coming kingdom? Or are they instead directed just to those who were following Jesus, the ones to whom Jesus was actually speaking (5:1–2)? This issue cannot be resolved until we examine more fully what it means, for Matthew, to follow Jesus.

Jesus and the Law

Contrary to what many Christians have thought throughout the ages, for Matthew following Jesus does not mean abandoning Judaism and joining a new religion that is opposed to it. Even in Matthew's day some Christians appear to have thought that this is what Jesus had in mind—that he sought to overturn the Law of Moses in his preaching about the way of God. For Matthew, however, nothing could be further from the truth. The keynote of the sermon is struck soon after the Beatitudes in this statement, found only in Matthew's Gospel:

> Do not think that I have come to abolish the law or the prophets; I have come not to abolish but to fulfill. For truly I tell you, until heaven and earth pass away, not one letter, not one stroke of a letter will pass from the law until all is fulfilled. Therefore, whoever breaks one of the least of these commandments, and teaches others to do the same, will be called least in the kingdom of heaven; but whoever does them and teaches them will be called great in the kingdom of heaven. For I tell you, unless your righteousness exceeds that of the scribes and the Pharisees, you will never enter the kingdom of heaven. (5:17–20)

In Matthew, Jesus is not opposed to the Law of Moses. He himself fulfills it, as seen in the important events in his birth, life, and death, events that are said to be fulfillments of the prophecy of Scripture. Moreover, Jesus in Matthew also requires his followers to fulfill the Law, in fact, to fulfill it even better than the Jewish leaders, the scribes and the Pharisees. Matthew indicates what he means the very next passage, the famous "Antitheses" (5:21–48).

Jesus' Followers and the Law

An "antithesis" is a contrary statement. In the six antitheses recorded in the Sermon on the Mount, Jesus states a Jewish law and then sets his interpretation of that law over and against it. I should emphasize that Matthew does not portray Jesus as contradicting the Law; for example, he does not say, "You have heard it said, 'You shall not commit murder,' but I say to you that you should." Instead, Jesus urges his followers to adhere to the law, but, to do so more rigorously than even the religious leaders of Israel. The contrasts of the antitheses, then, are between the way the law is commonly interpreted and the way Jesus interprets it. In all of these antitheses Jesus goes to the heart of the law in question, to its root intention, and insists that his followers adhere to that, rather than the letter of the law as strictly interpreted.

For example, the Law says not to murder (5:21). This law functions to preserve the harmony of the community. The root of disharmony (which leads to murder) is anger against another. Therefore, if one wants to fulfill the Law, by obeying its root intention, he or she must not even become angry with another. The Law also says not to commit adultery (5:27), that is, not to take the wife of another. This law preserves ownership rights, since in ancient Israel, as in many ancient societies, the wife was seen as the property of her husband (e.g., see the Tenth Commandment, where wives are grouped together with houses, slaves, oxen, and donkeys as property of one's neighbor that is not to be coveted; Exod 20:17). The root of adultery, in this view, is a man's passionate desire for another man's wife. Therefore, those who want to keep the law completely should not passionately desire a person who belongs to another.

The Law says to take an eye for an eye, a tooth for a tooth (5:38). This law serves to guarantee justice in the community, so that if a neighbor knocks out your tooth, you cannot lop off his head in exchange. Contrary to the way in which this law is commonly understood today, it was originally meant to be merciful, not vindictive; the penalty should fit and not exceed the crime. Since, however, the root of this law is the principle of mercy, Jesus draws the radical conclusion: instead of inflicting a penalty on another, his followers should prefer to suffer wrong. Therefore, someone who is struck on one cheek should turn the other to be struck as well.

As can be seen from these examples, far from absolving his followers of the responsibility to keep the Law, Matthew's Jesus intensifies the Law, requiring his followers to keep not just its letter but its very spirit. This intensification of the Law, however, raises a number of questions. One that has occurred to many readers over the years is whether Jesus can be serious. Is he really saying that no one who becomes angry, or who lusts, or who returns a blow can enter into the kingdom?

Readers of Matthew have frequently tried to get around this problem by softening Matthew's rigorous statements by importing views not presented in the text itself. For example, it is commonly suggested that Jesus means to set up an ideal standard that no one could possibly achieve to force people to realize that they are utter sinners in need of divine grace for salvation. The point of Jesus' words, then, would be that people cannot keep God's Law even if they want to. The problem with this interpretation is that Jesus in Matthew does not suggest that it is impossible to control your anger or lust, any more than the author of the Torah suggests that it is impossible to control your coveting.

At the same time, Matthew is not simply giving a detailed list of what Jesus' followers must do and not do in order to enter into the kingdom. On the contrary, his point seems to be that overly scrupulous attention to the detail of the Law is not what really matters to God. Even scribes and Pharisees can adhere to laws once they are narrowly enough prescribed, for example, by not murdering and not committing adultery and not eating forbidden foods. God wants more than this kind of strict obedience to the letter of the Law.

The Fulfillment of the Law

What then is the real purpose of the Law? We get a hint of Matthew's answer already in the Sermon on the Mount, in Jesus' famous expression of the golden rule. We know of other ancient teachers who formulated similar guidelines of behavior (see box 7.2), but Jesus' particular formulation is important: "In everything do to others as you would have them do to you; for this is the law and the prophets" (7:12). The final phrase of the saying is the key; the entire Law with all of its commandments can be summarized in this simple principle, that you treat others as you want them to treat you.

For Jesus in Matthew, the true interpretation of the Law does not require nuanced descriptions of how precisely to follow each of its commandments; it involves loving others as much as one's self. This principle can be found in other passages of Matthew's Gospel, most strikingly in 22:35–40, where in response to a question from a "lawyer" (i.e., an expert in the Jewish Law) Jesus summarizes the entire Torah in terms of two of its requirements: that "you shall love the Lord your God

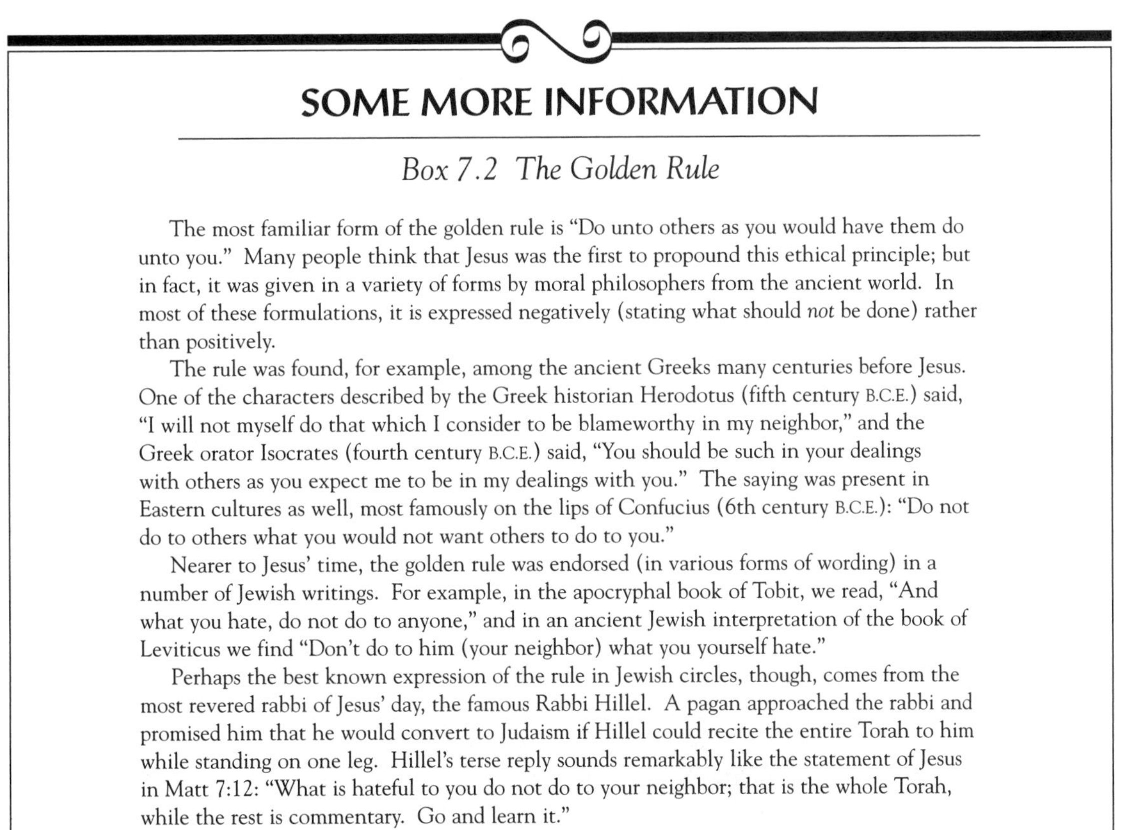

SOME MORE INFORMATION

Box 7.2 The Golden Rule

The most familiar form of the golden rule is "Do unto others as you would have them do unto you." Many people think that Jesus was the first to propound this ethical principle; but in fact, it was given in a variety of forms by moral philosophers from the ancient world. In most of these formulations, it is expressed negatively (stating what should *not* be done) rather than positively.

The rule was found, for example, among the ancient Greeks many centuries before Jesus. One of the characters described by the Greek historian Herodotus (fifth century B.C.E.) said, "I will not myself do that which I consider to be blameworthy in my neighbor," and the Greek orator Isocrates (fourth century B.C.E.) said, "You should be such in your dealings with others as you expect me to be in my dealings with you." The saying was present in Eastern cultures as well, most famously on the lips of Confucius (6th century B.C.E.): "Do not do to others what you would not want others to do to you."

Nearer to Jesus' time, the golden rule was endorsed (in various forms of wording) in a number of Jewish writings. For example, in the apocryphal book of Tobit, we read, "And what you hate, do not do to anyone," and in an ancient Jewish interpretation of the book of Leviticus we find "Don't do to him (your neighbor) what you yourself hate."

Perhaps the best known expression of the rule in Jewish circles, though, comes from the most revered rabbi of Jesus' day, the famous Rabbi Hillel. A pagan approached the rabbi and promised him that he would convert to Judaism if Hillel could recite the entire Torah to him while standing on one leg. Hillel's terse reply sounds remarkably like the statement of Jesus in Matt 7:12: "What is hateful to you do not do to your neighbor; that is the whole Torah, while the rest is commentary. Go and learn it."

Jesus, in short, was not the only teacher of his day who taught the golden rule, or who thought that the essence of the Law of Moses could be summed up in the commandment to love.

with all your heart, and with all your soul, and with all your mind" (Deut 6:5) and that "you love your neighbor as yourself" (Lev 19:18). Mark has this story as well, but Matthew tacks a different ending onto it: "On these two commandments hang all the law and the prophets" (22:40). For Matthew, the entire Law is thus at its very core a commandment to love: to love God with one's entire being and to love one's neighbor as one's self. This is the real intent of the Law, and the followers of Jesus must adhere to it in order to enter into the kingdom of heaven.

Another question naturally emerges from Jesus' insistence that his followers keep the Law. The laws that we have examined so far, for example, in the antitheses and the golden rule, would not have been seen as distinctively Jewish by many people in the ancient world. Most other people in Roman antiquity would have agreed that you should not commit murder or take your neighbor's wife or mete out unfair punishment. What about the laws of Scripture, though, that were widely recognized as making Jews a separate people from non-Jews, for example, the laws that required Jews to circumcise their baby boys, to keep the Sabbath day holy, and to observe certain dietary restrictions? We know from other evidence that by the time Matthew wrote his Gospel, these laws were not being followed by many Gentile Christians. Indeed, as we will see when we come to the letters of Paul (which were written before Matthew and the other Gospels), there were many Christians, including Paul himself, who insisted that Gentile believers should not keep these laws. What, then, about Matthew? Does he think that Jesus radicalized these laws as well as the others? Does Matthew's Jesus expect his followers to keep them?

JESUS AND THE JEWISH CULTIC PRACTICES PRESCRIBED BY THE LAW

Matthew never addresses head on the question of keeping such distinctively Jewish laws. Several points, however, can be raised. The first is that Jesus never disavows any of these Scriptural laws in Matthew or instructs his followers not to keep them. Moreover, in a number of passages not found in Mark, Jesus appears to affirm aspects of traditional Jewish piety. For instance, he castigates the hypocritical ways that the Pharisees give alms, pray, and fast, but he restates the importance of engaging in these practices themselves (6:1–18). He attacks the scribes and Pharisees for tithing "mint and dill and cumin" while neglecting "the weightier matters of the law," such as "justice and mercy and faith," but he goes on to say that both the tithing practices and the weightier matters are to be observed (23:23). He insists that someone who is estranged from another should be reconciled before making an offering in the Temple, but in saying this he implies that it is good and right for the person then to make the offering (5:23–24). He asserts that as the Son of God he is not himself obligated to pay the Temple tax, but he pays it anyway, so as not to give offense (17:24–27).

Similar emphases are found in the changes Matthew made in stories taken from Mark. For example, in Mark's apocalyptic discourse, Jesus speaks of the coming disaster and tells his disciples to "pray that it not be in winter" (because it would then be harder to escape; Mark 13:18). Interestingly, Matthew takes over this verse but adds the words "or during the Sabbath" (Matt 24:20). Why? Apparently because, for Matthew, extensive travel on the Sabbath was forbidden to Jesus' followers, as those who kept the Law. Perhaps more significantly, Matthew has changed Mark's story of Jesus' controversy with the Pharisees over their practice of washing hands prior to a meal (Mark 7:1–23; Matt 15:1–20). In both accounts Jesus argues that what matters is what comes out of people (their behavior), not what goes into them. Mark, however, interprets this to mean that Jesus "declared all foods clean," and thus overturned the Jewish food laws. Matthew, strikingly, omits the line.

All of these examples would make it appear that Jesus in Matthew is not intent on requiring his followers to abandon traditional forms of Jewish piety as rooted in the Torah. He simply assumes, for the most part, that they will practice them as they practice the entire Law (5:17–20).

At the same time, it appears that Jesus in Matthew thinks that his opponents are wrong in placing the highest priority on keeping the cultic requirements of the Law, rather than on emphasizing the commandment to love that lies at its core. This becomes especially clear in stories that Matthew took over from Mark but modified. One example is Mark's account of the call of Levi the tax collector (Mark 2:13–17; in Matthew's account, it is the call of Matthew!). When the Pharisees see Jesus eating in Levi's home with "tax collectors and sinners," they disparage him for mixing with such tainted company. Evidently their own emphasis on ritual purity before God precludes their eating with others who were not equally pure. In Mark, Jesus replies that it is the sick who need a physician, not the well, and that he has come to call sinners, not the righteous. In Matthew, Jesus' reply includes an appeal to the Scriptures: "Go and learn what this means, 'I desire mercy, not sacrifice' [Hos 6:6]. For I have come to call not the righteous but sinners" (Matt 9:13). Thus, according to Matthew, the Pharisees are more concerned with proper observance of the food laws of the Torah than with helping others; Jesus, on the other hand, is principally concerned with reaching out to those in need (for a similar lesson, see Matt 12:1–8).

In sum, it appears that Matthew assumes that Christians in his community (many of them? most of them?) will follow traditional forms of Jewish piety and cultic practices (see box 7.3), but ultimately, for him, these are of secondary importance. The Law is to be obeyed to the fullest extent possible (5:17–20), but in obeying the Law what really matters is human need. For this reason, love is the greatest commandment, and everything else is subservient to it. Even though a similar view was propounded by other rabbis of Jesus day (see box 7.2), Jesus' proclamation stands at odds with the religion advocated by Jewish leaders as portrayed in the Gospel of Matthew.

JESUS REJECTED BY THE JEWISH LEADERS

When Jesus' strong affirmation of the Torah of Moses is set over against his strong opposition to the Jewish leadership, perhaps the most striking aspect of Matthew's Gospel emerges. On the one hand, Jesus is portrayed as altogether Jewish. He is the Jewish messiah sent by the Jewish God to the Jewish people in fulfillment of the Jewish Scriptures. He is also the new Moses who gives the true interpretation of the Mosaic Law. On the other hand, he violently opposes Judaism as it is configured in this Gospel among the Jewish leadership. Somewhat paradoxically, then, in this Gospel Jesus commands his followers to adhere to the Jewish religion as it should be (i.e., as he himself interprets it), while urging them to reject the Jewish authorities, who are portrayed as evil hypocrites, opposed to God and his people.

The hypocrisy of the Jewish leaders was hinted at in the story of the Magi, which we have already considered. It is also found in the Sermon on the Mount, where the "hypocrites" pray, give alms, and fast simply in order to be seen and revered as holy, not out of true devotion to God (6:1–8). These, of course, are stories unique to Matthew. The same emphasis can be seen in stories that Matthew has taken over from Mark. You can see this for yourself by comparing, for instance, the stories of Matthew 12 with those of Mark 2:1–3:6.

A crescendo builds in Jesus' controversies with his opponents, reaching a climax in chapters 21–23, where Jesus himself takes the offensive. As in Mark, he "cleanses the Temple" (21:12–13), rousing the ire of the authorities. But in Matthew they become particularly incensed when they see him heal the blind and the lame and when they hear young children proclaim him the Son of David (21:14–15, only in Matthew). Jesus responds to their indignation by quoting the Psalms: "Out of the mouths of infants and nursing babes you have prepared praise for yourself" (21:16). Despite having witnessed his miracles, the Jewish leaders refuse to believe.

More than that, they attack Jesus by disputing his authority (21:23). In response, Jesus tells a parable (unique to Matthew) of a father with two sons, one of whom said that he would do his father's bidding and yet did not, the other of whom said that he would not but then did (21:28–32). Jesus likens his opponents to those who agree to do what their father (God) requires but fail to do so. He ends by claiming that the most despised of sinners—tax collectors and prostitutes—will enter into the kingdom of heaven ahead of them (21:32).

SOMETHING TO THINK ABOUT

Box 7.3 Gentiles in Matthew's Community

If we had no indication that Christianity spread among non-Jews soon after Jesus' death, we might simply assume that Matthew's community was comprised of Jews who continued to adhere to the law even if they disagreed with the Pharisees over how best to do so. Gentiles, however, were joining the Christian church well before Matthew wrote his Gospel; indeed, at this time there were probably more Gentiles who claimed to be followers of Jesus than Jews. Does Matthew think that these Gentiles Christians are to keep kosher, to observe the Sabbath, and, if male, to be circumcised? It is an intriguing question because, as we will see later, the apostle Paul was adamant that they should not.

Matthew does not address this issue directly. In this Gospel Jesus does give numerous indications that Gentiles will become his followers and inherit the kingdom of heaven; but nowhere does he indicate whether or not any of these converts will be required to be circumcised or to keep sabbath or to keep Jewish food laws. Consider one of the most dramatic statements concerning the heirs of the kingdom to come from Jesus, a statement in response to a Roman (non-Jewish) centurion's trust in his powers:

"Truly I tell you, in no one in Israel have I found such faith. I tell you, many will come from east and west and will eat with Abraham, Isaac, and Jacob in the kingdom of heaven, while the heirs of the kingdom will be thrown into the outer darkness, where there will be weeping and gnashing of teeth" (8:8–10).

The point of Matthew's inclusion of this Q story is clear: many non-Jews will enter into the kingdom, whereas many Jews will be excluded. Whether these Gentiles are expected first to convert to Judaism, however, is something that is not discussed.

The same difficulty occurs in the "Great Commission" at the end of this Gospel. After his resurrection, Jesus appears to his disciples (unlike in Mark) and sends them forth to "make disciples of all nations, baptizing them in the name of the Father and of the Son and of the Holy Spirit, and teaching them to obey everything that I have commanded you" (28:19-20). The disciples are sent to convert not only Jews but Gentiles as well ("nation" and "Gentile" are English translations of the the same Greek word). Moreover, they are not commanded to circumcise these converts but to baptize them; and they are not told to teach them the laws of Torah but the words of Jesus—teachings that sum up these laws in the golden rule and in the love commandment. At the same time, it remains unclear whether or not those who adhere to this teaching have to become Jewish (as was Jesus) and adhere to traditional Jewish piety (as did Jesus).

The ambiguity can also be detected in another story found only in Matthew where Jesus describes the scene of the judgment of the nations (Jews and Gentiles, presumably, or possibly just Gentiles) in 25:31-46. The nations are gathered before the cosmic judge, the Son of Man. Some are sent away to eternal punishment. Why? Not because they failed to observe the distinctive cultic practices of the Jews (circumcision, kosher food laws, Sabbath observance, and the like), but because they did not feed the hungry, give drink to the thirsty, welcome the stranger, clothe the naked, care for the sick, or visit the prisoner. Others are welcomed into the eternal kingdom. Why? Because they did all these things. For Matthew, entry into the kingdom means living for others, loving others as yourself, treating others as you would have them treat you. Those who do so are true followers of Jesus, whether they are Jews or Gentiles. Would the Gentiles who come to believe in him naturally be expected to adopt Jewish ways? Matthew never explicitly indicates one way or the other.

His assault continues in the parables that follow. The Jewish leaders are like those who have been given charge of a vineyard, who, instead of delivering the fruit that is produced to the master, try to keep it all for themselves, beating and killing the messengers that he sends, and finally his own son (21:33–44). The parable is from Mark and its message is clear. The vineyard represents the people of God, the messengers are the prophets, and the son is Jesus. Matthew has changed the ending of the story, however, and in a significant way. Jesus now says that the owner of the vineyard (God) will destroy the resisting farmers (the Jewish leaders) and give the vineyard over to others (the Gentile leaders of the Christian church?), who will deliver the fruit that is required (21:41, 43). As in Mark, the chief priests and Pharisees know that he is speaking against them, and they plot to have him arrested (21:45–46).

But not before Jesus has his full say. He continues to teach by telling a parable that Matthew has drawn from Q, in which the Jewish leaders are likened to those who are invited by a king to a grand wedding feast but make various excuses not to come (22:1–14). In a climactic statement that has no parallel in Luke, Jesus describes the king's wrath against them: "He became enraged, and sending forth his troops he destroyed those murderers and burned their city" (22:7; perhaps a reference to the destruction of Jerusalem in 70 C.E.). Others were then invited to come, and these did so willingly (the coming of the Gentiles into the kingdom; 22:9–10).

The vitriolic castigation of the Jewish leaders reaches its climax in chapter 23, which contains the "Seven Woes" against the Pharisees. Here Jesus condemns his enemies, the "scribes and Pharisees," in no uncertain terms: they are concerned only with praise and admiration, not with doing what is right before God; they are hypocrites, blind guides concerned with minutia instead of with what really matters; they are whitewashed tombs, clean on the outside but full of rot and corruption within; they are a brood of vipers, murderers of the righteous prophets of God, false leaders who shed innocent blood.

Jesus' Passion in Matthew

According to Matthew, the Jewish authorities are fully responsible for the blood of Jesus as well. Many of the stories of Matthew's passion narrative are taken over from Mark, and a detailed study of the ways in which they have been changed can pay rich dividends. Many of the changes work to emphasize both Jesus' innocence and the corresponding guilt of the Jewish leaders who demand his death. As in Mark, for example, Pilate offers to release a prisoner to the Jewish crowds in honor of the Passover feast. In Matthew's account, however, he more clearly prefers to release Jesus rather than the notorious Barabbas (27:15–18). In part, Pilate acts on advice from his wife, who tells him that she has suffered a bad dream about Jesus, whom she knows to be innocent (27:19, found only in Matthew). The "chief priests and elders," however, stir up the crowds to demand Barabbas instead. Pilate insists that Jesus does not deserve punishment, since he has done nothing wrong (27:22), but the people become persistent, and demand his crucifixion (27:23).

Then comes a well-known and ill-fated account, found only in Matthew. Pilate calls for water and washes his hands of the blood of Jesus, proclaiming, "I am innocent of this man's blood;

Figure 7.1 An Ancient Depiction of Pilate Washing His Hands at Jesus' Trial, from an Early Christian Sarcophagus.

see to it yourselves" (27:24). The entire crowd responds in words that have served hateful purposes ever since: "His blood be on us and our children" (27:25). Here the Jews gathered in Jerusalem claim responsibility for Jesus' unjust execution. Over the centuries, this verse has been used for all kinds of malicious acts of anti-Semitism—as if Jews who were not present at the scene could possibly be held responsible for the actions of those who were.

Matthew, however, does not himself portray all Jews as wicked opponents of God, as "Christ-killers" (an anti-Semitic slogan derived largely from this passage). Quite the contrary. As we have seen, Jesus himself is a Jew in this Gospel, as are all of his disciples. He is the Jewish messiah descended from David, the new Moses who urges his followers to fulfill the Jewish law. Nowhere in the Gospel does Jesus condemn Jews for being Jews. Whenever Jesus lambastes specific opponents in Matthew, they are in every instance Jewish *leaders* (Pharisees, scribes, chief priests, and so on). Even in Jesus' trial before Pilate, where Matthew appears to lay the blame of miscarried justice on all the Jewish people who are present, the real culprits are the "chief priests and elders," who stir up the crowds to say what they do (v. 19). Thus, the problem for Matthew is never the Jews or the Jewish religion per se; it is the Jewish authorities. This Gospel consistently affirms Judaism, at least Judaism as it was interpreted by Matthew's Jesus.

MATTHEW AND HIS READERS

On the basis of the portrayal of Jesus in this Gospel, we can hypothesize some things about the context of the author and his audience. Matthew's insistence that Jesus continued to adhere to traditional forms of Jewish piety, and that he advanced the true interpretation of the Law of Moses, suggests that the author himself and some, perhaps most, of his audience were themselves Jewish (see boxes 7.3 and 7.4). Would non-Jews be this interested in seeing Jesus as a thoroughly Jewish teacher intent on keeping the Law who insisted that his disciples followed suit? For Jewish Christians, however, this emphasis seems fairly natural. Moreover, believing in Jesus did not require abandoning the ancestral traditions that stem from Moses. On the contrary, Jesus showed how to understand these traditions and commanded his followers to obey them.

At the same time, there must have also been a good number of Gentiles in Matthew's congregation (see box 7.3). This would explain Jesus' claim that many outsiders would enter into the kingdom ahead of Jews (8:8–10), and also the "Great Commission," which urged missionary work principally among "the Gentiles" (28:19–20). In short, Matthew's congregation appears to be mixed, comprising both Jews and Gentiles. Many scholars have thought that it makes sense to locate it somewhere near Palestine in a major urban area (where Jews and Gentiles might congregate in large numbers), for instance, in Antioch of Syria, where the second-century authors who first quote the book of Matthew happen to have resided.

Perhaps the best way to explain Matthew's extensive criticism of the Jewish authorities is to say that his own community continued to experience opposition from non-Christian Jews, especially influential scribes and rabbis of the local synagogue(s), who accused them of abandoning Moses and the Law, of becoming apostate from the Jewish religion through their ill-advised faith in Jesus.

Matthew, an anonymous Jewish leader of the Christian community (assuming that his strong literary skills, indicative of a higher education, gave him a place of prominence there), penned a Gospel narrative to show that Jesus was in fact the Jewish messiah, who like Moses gave the law of God to his people. More precisely, he was the prophet like Moses who gave the Jewish people the true interpretation of Moses' Law, and beyond that he was a Savior who died for the sins of his people (1:21) and was vindicated by God by being raised from the dead. Moreover, Matthew went out of his way to affirm more strongly than his predecessors Mark and Q that Jesus did not annul the ancient Law of Moses but fulfilled it himself and insisted that all his followers, both Jews and Gentiles, do so as well. This they could do by holding on to Jesus' teachings and by following the principle at the heart of the Torah, given long ago to Jesus' forerunner Moses: to love God with their entire being and their neighbor as much as themselves, "for on these two commandments hang all the law and the prophets."

SOMETHING TO THINK ABOUT

Box 7.4 Was Matthew a Jew?

Some scholars have come to doubt that Matthew was a Jew despite the heavy emphasis on Jesus' own Jewishness in this Gospel. One of the more intriguing pieces of evidence that is sometimes cited involves Matthew's interpretation of passages drawn from the Hebrew Bible, especially Zechariah 9:9, as quoted in Matthew 21:5: "Look your king is coming to you, humble, and mounted on a donkey, and on a colt, the foal of a donkey."

Anyone who has studied the Jewish Scriptures extensively recognizes the literary form of this passage. Throughout the Psalms and other books of poetry, Hebrew authors employed a kind of parallelism in which a second line of a couplet simply repeated the ideas of the first line using different words. Here the parallelism is between the "donkey" of the first line and the "colt, the foal of a donkey" in the second.

Matthew, however, appears to have misunderstood the parallelism, or at least to have understood it in a highly unusual way. For he seems to have thought that the prophet was speaking of two different animals, one of them a donkey and the other a colt. So, when Jesus prepares to ride into Jerusalem, his followers actually acquire two animals for him, which he straddles for the trip into town (21:5–7; contrast Mark 11:7)! Some scholars have argued that no educated Jew would have made this kind of mistake about the Zechariah passage (none of the other Gospel writers, it might be pointed out, does so), so this author could not have been Jewish.

Most other scholars, however, have not been convinced, in part because we know all sorts of educated authors from the ancient world (as well as the modern one) who seem to misread texts that derive from their own contexts. This includes ancient Jewish interpreters of their own Hebrew Scriptures, some of whom produce interpretations that are no more bizarre than Matthew's interpretation of Zechariah. On these grounds, at least, the identity of Matthew has to be left as an open question.

SUGGESTIONS FOR FURTHER READING

Brown, Raymond. *The Birth of the Messiah: A Commentary on the Infancy Narratives in Matthew and Luke*. 2d ed. Garden City, N.Y.: Doubleday, 1993. A massive and exhaustive discussion of the birth narratives of Matthew and Luke, suitable for those who want to know simply everything about every detail.

Carter, Warren. *What Are They Saying about Matthew's Sermon on the Mount?* New York: Paulist, 1994. The best introductory sketch of the scholarly debates concerning the formation and meaning of Matthew's Sermon on the Mount.

Edwards, Richard A. *Matthew's Story of Jesus*. Philadelphia: Fortress, 1985. A nice introductory overview of the major themes of Matthew's Gospel for beginning students.

Kingsbury, Jack D. *Matthew: Structure, Christology, Kingdom*. Minneapolis, Minn.: Fortress, 1975. A solid introduction to the Gospel of Matthew from a literary perspective.

Nickle, Keith. *The Synoptic Gospels: Conflict and Consensus*. Atlanta: John Knox, 1980. A fine introduction to the major themes of Matthew's Gospel.

Overman, J. A. *Matthew's Gospel and Formative Judaism: The Social World of the Matthean Community*. Minneapolis, Minn.: Fortress, 1991. This is the best overall study of the community behind Matthew's Gospel, examined from a socio-historical perspective.

Senior, Donald. *What Are They Saying about Matthew?* New York: Paulist Press, 1983. An overview of scholarly views of Matthew's Gospel, excellent for beginning students.

CHAPTER 8

Jesus, the Savior of the World: The Gospel according to Luke

I have had two overarching goals in our study of the early Christian Gospels to this point. The first has been to explain different methods that scholars have used in their investigation of these texts; the second has been to apply these methods to uncover the distinctive emphases of each Gospel. My underlying assumption has been that the results of our investigation are no more compelling than the methods that we use to attain them. That is to say, while it is important to know what a text means, it is also important to recognize how we know (or think we know) what it means. Moreover, it is useful not only to understand what our methods involve in theory but also to see how they work in practice.

Thus we applied the literary-historical method to discuss the Gospel of Mark and the redactional method to study Matthew. These particular Gospels do not have to be examined in these particular ways. We could just as easily have used a literary-historical method to study Matthew and, at least theoretically, a redactional method to study Mark (although the latter would have proved somewhat difficult since we do not have direct access to any of Mark's sources). My point is that there are a number of approaches that scholars have taken to the Gospels, each with its own benefits and limitations, as they work toward the common goal of explaining the important features of each text.

The methods that we have discussed so far, of course, could also be used for our study of the Gospel according to Luke. Indeed, both of them have been used to this end, with some considerable success. Nonetheless, in staying with my pattern I have opted to introduce yet a third method, one that could just as well have been used with both Mark and Matthew.

This third method has not been discussed as extensively by scholars of the Gospels; it is nonetheless a useful approach and can be explained and justified rather easily. It is most closely aligned with the redactional method that we used with Matthew, but it avoids some of its pitfalls and has a somewhat different theoretical rationale. For the purposes of our study, I will simply call it the "comparative method."

THE COMPARATIVE METHOD AND THE GOSPEL OF LUKE

Perhaps the best way to explain how the comparative method works is to point out two problems that some recent scholars have found with redaction criticism. The first objection is that examining how a redactor has changed a source will not necessarily give a complete account of what he or she considered to be important. This is because the redactor has actually made two kinds of decisions: not only about what to change but also about what to keep. Sometimes it is just as important to know what an author has decided to leave intact as to know what he or she has decided to alter.

This is a valid objection to redaction criticism as it is sometimes practiced; seeing the alterations that authors have made in their sources can only serve as a shortcut to understanding their distinctive emphases. A complete redactional analysis would need to consider, in detail, both the similarities and the differences of the texts in question. As we will see, this is true of the comparative method as well.

The second objection to the redactional method has been raised with even greater vigor. Redaction criticism, opponents say, is necessarily built on assumptions about an author's sources; if these assumptions are found to be false, then the entire method collapses on itself. If, for example, Matthew did not use Mark as a source, that is, if our proposed four-source hypothesis for the Synoptics is wrong, then the study of how Matthew changed Mark is obviously of little use. Since scholars continue to debate the Synoptic Problem, and not everyone is convinced about Markan priority (some scholars continue to think Matthew was written first), are we not compelled to give up redaction criticism as a method? For many scholars the answer is a loud and resounding yes.

This decision, however, may be a bit too hasty, for redaction critics do not simply assume Markan priority, they mount arguments in its favor. Even though the arguments may not be absolutely and universally compelling, they continue to be convincing to the majority of scholars. Furthermore, even if the arguments for Markan priority were somehow proven to be wrong, Matthew's differences from Mark may still be of use in determining Matthew's particular emphases. To see how this is so, we can turn to the comparative method, which establishes the meaning of a text by comparing it to other related texts, without being concerned over whether any of them happened to be among its sources.

I have argued that we can learn something new only in light of what we already know, because there is nothing in our experience as humans that is completely unlike everything else. If there were, we would have no way of sensing, experiencing, understanding, or explaining it. All knowledge—not only of literary texts but of people, the world around us, our experiences and sensations—is necessarily relational. We know what we know only in relation to everything else that we know.

This fundamental principle has been advanced by modern theoreticians of language, who point out that words mean what they do only in relation to other words. We know what one term means because it is not exactly the same as some other term. For instance, we (as English speakers) know what the word "cat" signifies, not because the word has some kind of inherent meaning, but because it is different from other closely related terms, such as "bat," "hat," and "gnat." Moving beyond the term to the thing that it signifies, we know that the thing sitting on our lap is a cat because it is in some ways like, and in other ways unlike, other things in our experience. For example, it is like other things we call animals and unlike things we call plants. As an animal, it is like a mammal rather than, say, a reptile or a bird. And as a mammal, it is both similar to and different from other mammals, such as walruses, dogs, and aardvarks.

This principle of knowing something by its similarities to and differences from other things applies not only to individual terms and the things they signify; it also applies to combinations of words into sensible units such as phrases, clauses, sentences, paragraphs, chapters, and books. We understand the meaning of one book, not in and of itself, but in relationship to everything else that we know, including every other book that we know.

The relevance of this principle for our study of the Gospels should be obvious. We can study any one of the Gospels by comparing it to others, to see its similarities and differences, and thereby come to a more adequate understanding of it. This approach is not unique to the study of early Christian literature, of course, any more than any of our other methods is. In fact, some scholars would argue that since all learning is relational, people necessarily understand everything they read, whether they are cognizant of it or not, by comparing it to everything else they have read.

For our study of Luke, we will try to be cognizant of what we are doing and so, self-consciously, apply the comparative method. The method does not require us to think that Luke used Mark as a source; those who think that he did (as most scholars do) are of course quite free to limit their considerations to seeing how he utilized that source and the others at his disposal (e.g., Q). This is the approach we took in studying Matthew. In this chapter, however, we will overlook the question of sources and focus instead on how Luke compares and contrasts with other texts that are in many ways similar, in particular the two Gospels we are now most thoroughly acquainted with, Matthew and Mark. These similarities and differences will enlighten us concerning several important features of Luke's portrayal of Jesus and thus be useful as an introduction to some of the key themes of his Gospel.

A COMPARATIVE OVERVIEW OF THE GOSPEL

We have already learned several basic points about Luke's Gospel in relation to Matthew and Mark. Like them, Luke is a kind of Greco-Roman biography of Jesus. It too is anonymous and appears to have been written by a Greek-speaking Christian somewhere outside of Palestine. The author evidently penned his account somewhat later than the Gospel of Mark, perhaps at about the same time as the Gospel of Matthew. In the second century, the book came to be attributed to Luke, the traveling companion of the apostle Paul (we will consider the merits of this attribution in the following chapter).

Perhaps the most obvious difference between this Gospel and all others from antiquity (not just Matthew and Mark) is that it is the first of a two-volume set. The unknown author provided a continuation of the story in volume two, the Acts of the Apostles. The Gospel of Luke provides a sketch of the life and death of Jesus, and the book of Acts narrates the birth and life of the Christian church that emerged afterwards. The author appears to have meant these books to be read together. For the purposes of our comparative study, however, we will restrict ourselves in this chapter to an analysis of Luke, reserving an investigation of Acts for the chapter that follows.

THE PREFACE TO LUKE'S GOSPEL

Given the importance that I have attached to the ways each of the other Gospels has begun, we do well to start our comparative study of Luke by considering his introduction. Unlike Mark and Matthew, Luke begins with a formal preface, found in the opening four verses of his account. Readers conversant with a wide range of Greco-Roman literature will have no difficulty understanding the significance of this beginning, for it is quite similar to other prefaces of the period, particularly among works by Greek historians. By beginning his Gospel with a standard "historiographic" preface, written in a much better style of Greek than anything found in Mark or Matthew, Luke alerts his reader both to his own abilities as a writer and to the scope of his work. His book is to be taken as a serious piece of historical writing, at least according to ancient readers' expectations of "history."

Historiographic prefaces in Greco-Roman literature typically indicate that the author has done extensive research of the historical topics under discussion. They commonly refer to the sources that were at his disposal, and they not infrequently suggest that the final product of the author's labors, the volume being read, is far superior to anything previously written on the subject. Sometimes the preface includes the name of the person to whom the work is being dedicated.

All of these features are found in Luke 1:1–4. The author (whom I will continue to call Luke for convenience) indicates that he has had several predecessors in writing a narrative of the life of Jesus (v. 1) and that these narratives are ultimately based on stories that have been passed down by "eyewitnesses and ministers of the word" (v. 2). In other words, the author concedes that his Gospel is based on oral traditions that were circulating among Christian congregations of the first century and that he has made use of other written sources. As we have seen, two of these earlier "narratives of the things which have been accomplished among us" are the Gospel of Mark and the document scholars call Q. Some readers have been struck by the tone of Luke's reference to these predecessors. He claims that his narrative, evidently in contrast with theirs, will be orderly (1:3) and that he is writing so that his reader will now learn the "truth concerning the things about which you have been instructed" (1:4). An intriguing comment this: is Luke making a negative, if implicit, evaluation of Mark's Gospel?

Luke dedicates his work to someone he calls "most excellent Theophilus." Unfortunately, he never tells us who this is. Luke does, however, use the title "most excellent" on three other occasions, each of them in reference to a governor of a Roman province (in the second volume of his work; Acts 23:26; 24:3; 26:25). On these grounds, some scholars have thought that Luke's two volumes were written for a Roman administrative official. If this is correct, one might wonder why a Christian would give a non-Christian governor books on the life of Jesus and the beginnings of the Christian church. According to one point of view, he did so to show someone in power that Jesus and the religion he founded are in no way to be seen as a threat to the social order, and that there is therefore no reason to persecute Christians, since neither they nor their founder have ever opposed the empire or done anything to merit opposition.

Not everyone accepts this view, for reasons that I will explain shortly. If it were true, however, it would help to make sense of several aspects of Luke's portrayal of Jesus. He shows a special concern, for example, to relate the history of Jesus to the broader historical events transpiring within the empire (e.g., 2:1–2; 3:1–2). Moreover, his narrative goes to some lengths to show that Jesus was executed by the state only because Pilate's hand was forced by the leaders of the Jews. In this Gospel, Pilate declares on three different occasions that he finds no guilt in Jesus (23:4, 14–15, 22), and after Jesus dies, the centurion responsible for his execution also proclaims that he was innocent (23:47). Could this Gospel, then, along with its sequel, Acts, have been written as an "apology," that is, an informed defense of Christianity in the face of official opposition of the state (see box 8.1)?

Even though this view can account for some of the features of Luke's narrative, it cannot explain a large number of others, including most of its prominent themes (as we shall see). Moreover, if Luke's overarching purpose was to curry the favor

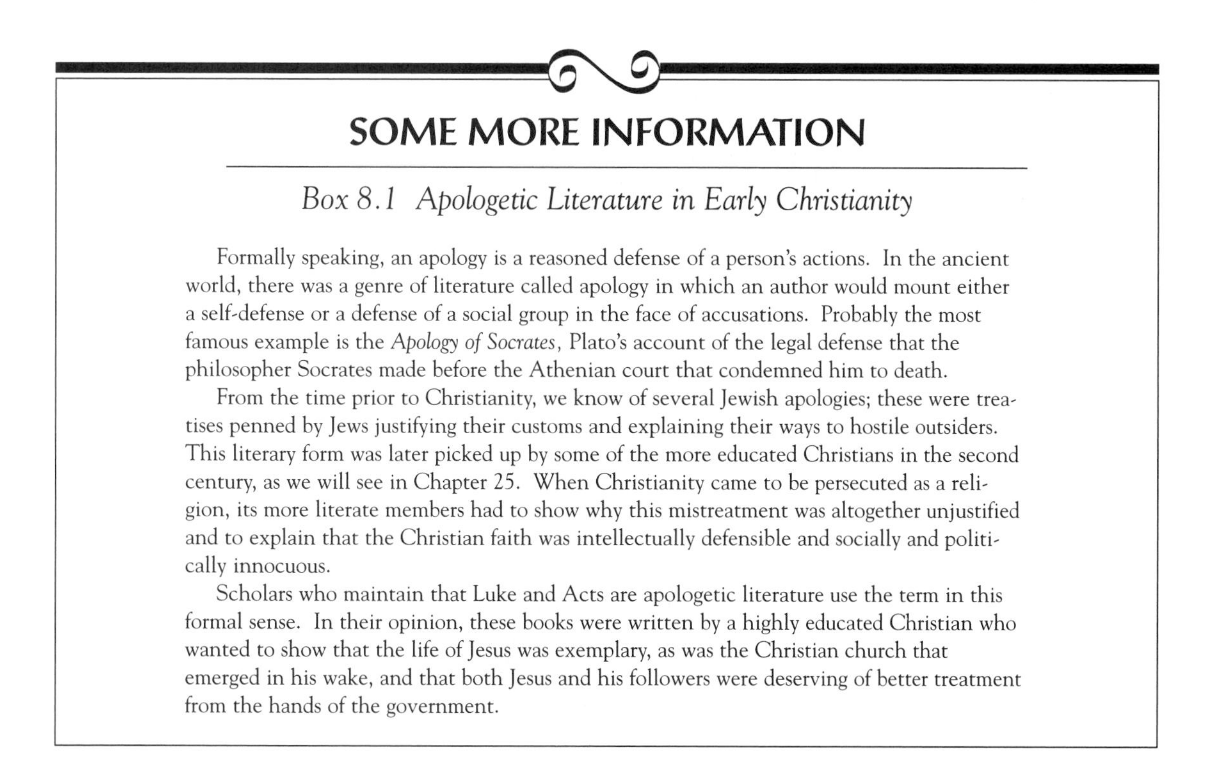

SOME MORE INFORMATION

Box 8.1 *Apologetic Literature in Early Christianity*

Formally speaking, an apology is a reasoned defense of a person's actions. In the ancient world, there was a genre of literature called apology in which an author would mount either a self-defense or a defense of a social group in the face of accusations. Probably the most famous example is the *Apology of Socrates*, Plato's account of the legal defense that the philosopher Socrates made before the Athenian court that condemned him to death.

From the time prior to Christianity, we know of several Jewish apologies; these were treatises penned by Jews justifying their customs and explaining their ways to hostile outsiders. This literary form was later picked up by some of the more educated Christians in the second century, as we will see in Chapter 25. When Christianity came to be persecuted as a religion, its more literate members had to show why this mistreatment was altogether unjustified and to explain that the Christian faith was intellectually defensible and socially and politically innocuous.

Scholars who maintain that Luke and Acts are apologetic literature use the term in this formal sense. In their opinion, these books were written by a highly educated Christian who wanted to show that the life of Jesus was exemplary, as was the Christian church that emerged in his wake, and that both Jesus and his followers were deserving of better treatment from the hands of the government.

of Roman officials, it is odd that he did not portray them in a more favorable light. Pilate, for example, is depicted as a weak administrator who bows to pressure from his own subjects, a portrayal that, in fact, does not square well with the public record of his governorship. Other officials are portrayed in yet less favorable terms in the book of Acts. Most problematic of all, it is nearly impossible to imagine any tangible historical context within which a Christian would write two such large volumes (together, they take up approximately one-fourth of the entire New Testament) and deliver them over to a Roman official with any real expectation that he would read them, let alone be influenced by them.

It is much more likely that these books, along with all of the other Gospels, were "in-house" literature, written by Christians for Christians, rather than evangelistic or propagandistic texts. Who in the outside world would bother to read them? Who on the inside would be foolish enough to think they would? It is worth noting that the first reference to any outsider having any clue as to what was in these books does not come for nearly a hundred years after the production of Luke-Acts (the reference is in an anti-Christian writer named Celsus).

If Luke's Theophilus is not a Roman administrator, who might he be? The name was fairly common in Greek antiquity. Literally translated, it means either "lover of God" or "beloved of God." For this reason, some scholars have plausibly argued that Luke's addressee is a code name for the Christians (the "beloved of God") to whom he writes. Just as other historians prefaced their works by dedicating them to a patron who had provided material support, or to some other person deemed worthy of honor, so Luke may have dedicated his work to his fellow believers, who were worthy of the greatest accolades as those whom God loves, or "Theophilus." If this view of the matter is correct, then the apologetic aspects of the narrative would be directed not to outsiders but to those within the church. Luke's aim may have been to show the Christians themselves that their movement has been nonviolent and socially respectable from the very beginning, thus perhaps providing them with answers they needed when confronted by the objections of outsiders.

There is one final issue to address before leaving the preface and jumping into the narrative itself. Prefaces such as the one in Luke are normally found in historiographic works, but ancient historiography was a different genre of literature from biography, as we will see in greater detail in Chapter 9. This raises the question of whether Luke is essentially comparable to Mark and Matthew as a kind of Greco-Roman biography.

Some scholars have argued that since Luke wrote two volumes, the entire work has to be considered when deciding about genre. According to this view, since Acts is not about the life and death of Jesus but about the church that spread throughout the world after his death, the Gospel itself must be something other than a biography. At the same time, the features of the biographical genre that we found in Mark and Matthew are present in Luke as well. Indeed, in some respects, these features are even stronger. By beginning his Gospel with the miraculous birth of the main character, ending it with his ascent into heaven, and narrating his spectacular deeds and inspired teachings in between, Luke has made his first volume more like biographies of other religious men than either of the other Synoptics. It shares more features, for example, with Philostratus's *Life of Apollonius of Tyana* (see Chapter 2).

What, then, can we conclude about this book's genre? It seems that Luke wrote two closely related works, one a biography of the founder of Christianity and the other a general history of the early Christian movement. In terms of overall conception and significant themes the two volumes are closely related, but their different subject matters required the use of different genres, one a Greco-Roman biography and the other a Greco-Roman history (we will be discussing the genre of Acts in the Chapter 9).

If this is the case, then, the preface to Luke, which would belong more naturally to a history than a biography, can be seen as an introduction to the entire two-volume work. It is structured as a historiographic preface because the work as a whole will comprise not only a biography of the founder of this religion but also a sketch of its early history.

LUKE'S BIRTH NARRATIVE IN COMPARATIVE PERSPECTIVE

The two lengthy chapters that begin Luke's account contain stories relating the births of Jesus and his predecessor, John the Baptist. By beginning with a birth narrative, Luke has an obvious point of contact with Matthew. Mark, you will recall, begins with Jesus as an adult.

There are some very broad and basic similarities between the birth narratives of Matthew and of Luke. In both, for example, Jesus is born in the city of Bethlehem to a virgin named Mary, who is betrothed to a man named Joseph. For most readers, however, what is far more striking are the differences between these accounts. Indeed, none of the specific stories of Luke's narrative occurs in Matthew, just as none of Matthew's appears here. You can see this easily by making a list of everything that happens in Luke and a separate list of everything that happens in Matthew, and comparing the lists. In one of them you will find the shepherds, in the other the Magi; one describes the journey to Bethlehem, the other the flight to Egypt; one records an angel's words to Mary, the other the angelic words to Joseph; and so forth. These are two discrete narratives, and the Christmas story recounted by Christians every December is a conflation of the two.

From a comparative perspective, perhaps the most important feature of these infancy narratives is not simply that they differ from one another but that they do so in ways that are extremely hard to reconcile. These differences give us an excellent opportunity to apply the comparative method of analysis.

An Illustration of the Comparative Method: Joseph and Mary's Hometown

One of the telling differences between the two accounts has to do with the question of Mary and Joseph's hometown. Most people simply assume that the couple lived in Nazareth. In the familiar story of Luke's Gospel, Mary and Joseph leave town for a trip to register for the census in Bethlehem. Mary happens to give birth there (2:1–7), and the couple then returns home just over a month later (2:39; following the law spelled out in Leviticus 12).

Before examining this account in greater detail, we should recall what Matthew says about the same event. Matthew gives no indication at all that Joseph and Mary made a trip from Galilee in order to register for a census. On the contrary, Matthew intimates that Joseph and Mary originally came from Bethlehem. This is suggested, first of all, by the story of the wise men (found only in Matthew), who arrive to worship Jesus after making a long journey in which they followed the star that appeared in the heavens to indicate his birth. They find Jesus in Bethlehem in a "house" (not a stable or a cave; Matt 2:11). Unless one had reason to think otherwise—and Matthew gives readers no reason for doing so—one would assume that the house is where Jesus and his family normally live.

Consider next what Herod does in Matthew's account when he learns from the Magi the time at which they had first seen the star. Based on this information, he sends forth his troops to slaughter every boy in Bethlehem who is two years and under (2:16). In other words, the "slaughter of the innocents" did not occur immediately after Jesus' birth, but some months, or perhaps a year and some months later: otherwise, Herod would have been quite safe to slaughter only the newborns. According to Matthew's account, Joseph and Mary are still in Bethlehem at this time, presumably because that is simply where they live.

Perhaps most telling of all, some time after they had fled to Egypt to escape Herod's wrath, Joseph learns in a dream that he can now return home. But where does he plan to go? The answer is quite clear. He intends to return to the place whence they came, the town of Bethelehem. Only when he learns that the ruler of Judea is Archelaus, a potentate worse than his father Herod, does he realize that they can't return there. For this reason Joseph decides to relocate his family in Galilee, in the town of Nazareth (2:22–23). Thus in Matthew's account, Joseph and Mary appear to have originally lived in Bethlehem, but they relocated to Nazareth when Jesus was a boy and raised him there.

In the Gospel of Luke Jesus is also born in Bethlehem and raised in Nazareth, but the way this comes about is altogether different (see box 8.2). In this account Joseph takes his betrothed Mary from their hometown Nazareth to Bethlehem for a worldwide census ordered by Caesar Augustus, while Quirinius was governor of Syria (2:1–5). Mary goes into labor while in town, so Jesus' birthplace is Bethlehem. After about a month (Luke 2:22–23, 39; see Lev 12:4–6), the family returns to their home in Nazareth, where Jesus is raised (2:39–40). As you might realize, the family's direct return north in Luke does not seem to allow time for Matthew's wise men to visit them in their home in Bethlehem a year or so later, or for their subsequent flight to Egypt.

Of course, it might be possible to reconcile these two narratives if we worked hard enough at it, and certainly Matthew and Luke do not explicitly contradict each other. But the two narratives are quite different from one another, and interestingly, the differences are highlighted by their one overarching similarity. Both authors indicate that Jesus was born in Bethlehem but raised in Nazareth, even though this happens in strikingly different ways in their two narratives (see box 8.3).

The Salvation of the Jews: Luke's Orientation to the Temple

For understanding Luke's overall narrative, perhaps the single most significant feature of these opening chapters is the way they repeatedly emphasize, unlike in Matthew, that the beginning of Jesus' story is closely associated with the Temple in Jerusalem. For Luke, the message of God's salvation comes first to the Jews, to the capital of Judea, to the most sacred location of the most sacred city. Luke's Gospel (and the subsequent narrative in the book of Acts) is oriented toward showing how this salvation comes largely to be rejected in the city of God by the people of God, the Jews themselves. This rejection leads to its dissemination elsewhere, principally among the non-Jews, the Gentiles.

This Lukan orientation is established at the outset of the narrative by the focus on the Temple in passages unique to the Third Gospel. It is here that the birth of Jesus' forerunner John is announced to Zechariah, the priest, faithfully ministering to God in the sanctuary (1:8–23). The parents of John are upright before God as strict observers of traditional Jewish piety. To Jews such as these, God first announces—in the Temple—the coming of his salvation.

Jesus himself comes to be born in nearby Bethlehem during the fortuitous journey of his mother with her betrothed to enroll for the census (2:1–20). He is circumcised on the eighth day, in accordance with the Jewish Law (2:21). Some days later he is brought into the Temple to be consecrated to God (2:22). While in the Temple, he is

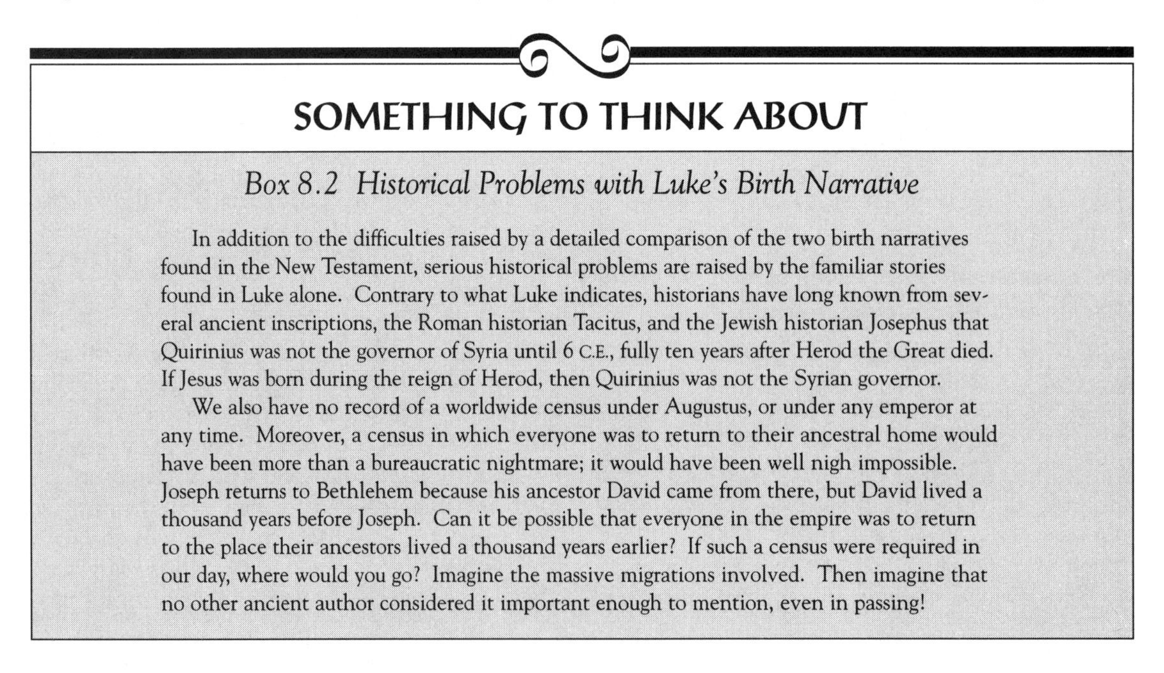

SOMETHING TO THINK ABOUT

Box 8.2 Historical Problems with Luke's Birth Narrative

In addition to the difficulties raised by a detailed comparison of the two birth narratives found in the New Testament, serious historical problems are raised by the familiar stories found in Luke alone. Contrary to what Luke indicates, historians have long known from several ancient inscriptions, the Roman historian Tacitus, and the Jewish historian Josephus that Quirinius was not the governor of Syria until 6 C.E., fully ten years after Herod the Great died. If Jesus was born during the reign of Herod, then Quirinius was not the Syrian governor.

We also have no record of a worldwide census under Augustus, or under any emperor at any time. Moreover, a census in which everyone was to return to their ancestral home would have been more than a bureaucratic nightmare; it would have been well nigh impossible. Joseph returns to Bethlehem because his ancestor David came from there, but David lived a thousand years before Joseph. Can it be possible that everyone in the empire was to return to the place their ancestors lived a thousand years earlier? If such a census were required in our day, where would you go? Imagine the massive migrations involved. Then imagine that no other ancient author considered it important enough to mention, even in passing!

SOMETHING TO THINK ABOUT

Box 8.3 The Virgin Birth in Matthew and Luke

Both Matthew and Luke make it quite clear that Jesus' mother was a virgin, but they appear to understand the significance of Jesus' virgin birth differently. In Matthew, Jesus' birth is said to fulfill the prediction of the Hebrew prophet Isaiah, who foretold that "a virgin shall conceive and bear a son" (1:23). Luke neither quotes this Isaiah passage nor indicates that Jesus' birth fulfills Scripture. What the event means for Luke is suggested in the story of the Annunciation (1:28–38, a passage found only in Luke), where the angel Gabriel assures Mary that her son "will be great, and will be called the Son of the Most High, and the Lord God will give to him the throne of his ancestor David." Mary is disturbed by this pronouncement: how can she bear a son if she has never had sexual relations (1:34)? The angel's reply is striking: "The Holy Spirit will come upon you, and the power of the Most High will overshadow you; therefore the child to be born will be holy; he will be called Son of God" (1:35).

Why, then, is Jesus born of a virgin in Luke? Evidently, because Jesus really is God's son ("therefore . . . he will be called the Son of God"). In other words, his father is not a human but God himself.

As we will see later, Luke is generally thought to have been writing to a Christian community that was largely Gentile. It may be that he has molded his portrayal of Jesus for these converts from other Greco-Roman religions. He presents the story of Jesus' birth in a way that would make sense to a pagan reader who was conversant with tales of other divine beings who walked the face of the earth, other heroes and demigods who were born of the union of a mortal with a god.

recognized as the long awaited messiah by a righteous and devout holy man, Simeon (2:25–36), and an elderly Jewish prophetess, Anna, who spends day and night in the Temple, praying and fasting (2:36–38). In the Temple, his parents offer a sacrifice and do all that the Law commands (2:25, 39).

In the very next account, the only story of Jesus as a youth in the entire New Testament, his parents bring him to Jerusalem as a twelve-year-old boy for the Passover feast. When they leave, he remains behind without telling them. After a three-day search, they finally track him down in the Temple where he is engaged in discussion with Jewish authorities. When his mother upbraids him for causing them distress, Jesus replies, "Did you not know that I had to be in my Father's house?" (i.e., in the Temple; 2:49).

Thus, unlike both Mark and Matthew, Luke stresses Jesus' early association with the Temple in Jerusalem. It is there, in the heart of Judaism, that God's message of salvation comes. This emphasis on Jerusalem and its Temple can be found in other important passages of Luke, as revealed through a comparative analysis. The following paragraphs give just three outstanding examples.

1. In both Matthew and Luke, Jesus experiences three temptations by the Devil in the wilderness (Matt 4:1–11, Luke 4:1–13). The accounts are almost verbally identical. The sequence of temptations, however, differs. In Matthew's account they appear to become increasingly difficult. The first is to turn stones into bread, a temptation difficult to resist, since Jesus has been fasting for forty days. The second is to leap from the top of the Temple, evidently a temptation for Jesus to prove to the crowds below that he is the messiah by being swooped up by the angels before he hits bottom. The third is to worship Satan, a temptation both more subtle and more terrible than the others: Satan promises him the lordship of the earth in exchange, a lordship that will otherwise require his death on the cross.

Figure 8.1 Picture of Saint Luke from a tenth-century manuscript of the Gospels. Notice the five books of Moses resting on his lap and the Old Testament prophets that he lifts up—graphic portrayals of Luke's view that the life and death of Jesus were a fulfillment of the Law and the prophets (see Luke 24:26–27).

The crescendo effect of Matthew's account is muted in Luke, where the second and third temptations are reversed. But the switch has a thematic payoff, for in Luke's sequence the temptations end with Jesus in the holy city Jerusalem, at the holy sanctuary, the Temple. For Luke, this is where God's salvation comes and where the real cosmic battle is waged over God's people, the Jews, many of whom will succumb to Satan and reject the message of Jesus.

2. Whereas in the other Gospels Jesus' final trip to Jerusalem is narrated in rather quick order (e.g., in Mark, it happens only in chap. 10), in Luke it takes up a major portion of the Gospel. Jesus leaves for Jerusalem in chapter 9 and does not arrive until chapter 19, spending the interim period, on the way, healing and teaching. Why such an extensive account of Jesus going to Jerusalem? Perhaps to highlight the significance of the event: God's salvation comes to the heart of Judaism, only to be rejected there.
3. The Gospel not only begins in the Temple of Jerusalem, it also ends there. Unlike in Mark, where the women are instructed to tell the disciples to go to Galilee to see Jesus, and unlike in Matthew, where they actually do go and encounter him there, in Luke they are told not to go outside Jerusalem; they remain there for some weeks after seeing Jesus on the day of his resurrection (24:49). Finally, after their last encounter with the resurrected Lord, they watch him take his leave from just outside the city and return, not to their homeland, Galilee, but to the Temple, where they spend their days worshipping God (24:50–52).

For Luke, the message of God comes to his people in their most sacred city, Jerusalem, in the most sacred of all sites, the Temple, but this message is not meant only for the Jews. In Luke's view, it is a message of salvation for all people. This can be seen by applying a comparative analysis to another passage of Luke's early chapters, his genealogy of Jesus.

The Salvation of the Gentiles: Luke's Orientation to the Whole World

We spent some time examining Matthew's genealogy of Jesus (actually, his genealogy of Joseph, the husband of Jesus' mother). Luke too has a genealogy (3:23–38). One of the most obvious differences between them is that they are, in fact, different genealogies! Both of them do trace Jesus' lineage through Joseph, even though in neither Gospel is Joseph Jesus' father, and in both of them Joseph is a descendant of King David. What is striking, though, is that Joseph's ties to David are traced through different lines in the two accounts. In Matthew, Joseph is a direct descendant (from father to son) of David's son Solomon; in Luke he is descended through a different line, from David's other son Nathan. The discrepancy, when looked at from the other direction by moving from Joseph backwards, leads to some questions. Who was Joseph's father? Was it Jacob (as in Matthew) or Heli (as in Luke?) Was his paternal grandfather Matthan or Matthat? Was his paternal great-grandfather Eleazar or Levi? His great-great-grandfather Eliud or Melchi? And so forth. One of the fascinating aspects of scholarship is to see how readers have attempted to explain these differences over the years. Some have claimed, for instance, that one of the genealogies is Joseph's and the other is Mary's. The problem, of course, is that both of them explicitly trace the ancestry of Joseph (Matt 1:16; Luke 3:23).

A second difference is perhaps even more obvious to a first-time reader of Luke. Unlike Matthew's, Luke's genealogy does not occur where you might expect, in the narrative of Jesus' birth, but after his baptism (3:23–38). Why would Luke wait until Jesus is a grown man of "about thirty" to describe his genealogy (3:23)? Possibly the best way to answer this question is to consider an important connection between Jesus' baptism and his genealogy in Luke. Both passages conclude by showing that Jesus is the Son of God. The baptism ends with the declaration from heaven that Jesus is God's own son (3:22). The genealogy ends by implicitly making the same declaration but in a radically different way. Here Jesus' lineage is traced not just to David or to Abraham or even to Adam, the first human being. The genealogy goes all the way back to God, the "father" of Adam—making Jesus the Son of God by direct descent!

The third significant difference between these two genealogies is closely related. Luke's genealogy does not so much stress Jesus' Jewishness, as one descended from the father of the Jews, or his messiahship, as the Son of David. Jesus' human lineage goes far beyond both of these figures who are so important for the history of Judaism, back to the man responsible for the human race itself, Adam. Thus, if Matthew's genealogy was important in showing that Jesus belonged to the Jews, Luke's is important in showing that he belongs to all people, both Jews and Gentiles.

Here we have an important indication that for Luke the message of salvation that begins in the heart of Judaism is a message for all the nations of earth. In fact, as we will see, Luke devotes virtually his entire second volume to showing how this message came to be rejected by Jews, and so went forth to the Gentiles. Indeed, a careful reader of Luke's work does not need to wait for volume two to get this message. It is embodied here in the Gospel itself, as the comparative method of analysis can clearly demonstrate.

FROM JEW TO GENTILE: LUKE'S PORTRAYAL OF JESUS THE REJECTED PROPHET

We have already seen that both Mark and Matthew establish essential aspects of their portrayals of Jesus by the way they describe the beginning of his public ministry. Mark, for example, uses his early narratives to show that Jesus was an authoritative leader, teacher, and healer; Matthew uses his to portray Jesus as the new Moses bringing the authoritative interpretation of God's Law. In Luke, Jesus' ministry begins with a sermon in the synagogue that infuriates his fellow Jews, who then make an attempt on his life. It is not an auspicious beginning.

In order to begin Jesus' ministry in this way, Luke narrates a story that does not occur until nearly halfway through both Mark's and Matthew's account of the ministry (Mark 6:1–6; Matt 13:53–58; Luke 4:16–30). This is the famous narrative of Jesus' sermon in his hometown of Nazareth, a story that is much longer and more detailed in Luke than in the other Gospels and that, as the opening account, sets the stage for Luke's overall portrayal of Jesus. As a visitor to the synagogue, in Luke, Jesus is given the opportunity to read and comment on the Scripture. He reads from the book of Isaiah, in which the prophet claims to be anointed with the spirit of God in order "to bring good news to the poor . . . to proclaim release to the captives and recovery of sight to the blind, to let the oppressed go free, to proclaim the year of the Lord's favor" (4:18–19).

After reading the Scripture, Jesus sits and begins to proclaim that the predictions of the prophet have now come to fulfillment—by implication, in him. Those in the synagogue are incredulous; they know, after all, who Jesus is (or

Figure 8.2 The remains of the synagogue in Capernaum. The surviving building represents a structure that was built on the spot where Jesus himself would have visited several centuries earlier.

think they do; they call him "Joseph's son" in v. 22). Jesus understands their reaction: they want him to prove himself by doing miracles for them like he has done in Capernaum. This may strike the reader as a somewhat peculiar request, since in this Gospel, unlike Mark and Matthew, Jesus has not yet gone to Capernaum or done any miracles.

In any event, Jesus responds by launching into an extended sermon, not found in the other Gospels, in which he recounts two familiar stories from the Jewish Scriptures about prophets who were sent by God, not to Jews but to Gentiles. He tells how Elijah was sent to assist a widow in the city of Zarephath during an extended drought and how Elisha was sent to heal not the lepers of Israel but Naaman, the leper king of Syria (4:25–27). In both instances God sent his prophet, not to help his people the Israelites, but to pronounce judgment against them for having turned against him. These prophets ministered to Gentiles outside of the people of God.

These are the stories that Jesus uses to explain how he fulfills the prophecy of Isaiah. His message is clear: he too is a prophet of God who will not receive a warm welcome among his own people in Israel, who like their ancestors have rejected God along with his prophets. Because of this rejection, Jesus' message will be taken to the Gentiles.

Jesus' sermon is not a smashing success; in fact, it is very nearly a smashing failure. The Jews in the synagogue rise up in anger and try to throw him off a cliff. Jesus escapes, leaves town, and takes his message elsewhere (4:28–30). For Luke, this reaction marks the beginning of the fulfillment of the sermon that Jesus has just preached. The prophet of God is opposed by his own people, and they will eventually call for his death. As a prophet, he knows that this is to happen. Indeed, it has all been predicted in the Jewish Scriptures.

Rejecting him, the people have rejected the God that he represents. This compels the prophet to take his message elsewhere. Eventually, the message will go not simply to another city of Israel, but to another people, indeed to all other peoples, the nations of the earth.

LUKE'S DISTINCTIVE EMPHASES THROUGHOUT HIS GOSPEL

The passages that we have examined from the outset of Luke's narrative intimate many of the key themes that you will find throughout the rest of the Gospel, themes relating to Luke's understanding of Jesus and to the way his salvation affects the entire world. As we will see, many of these themes continue to play a significant role in the second volume of Luke's work, the Acts of the Apostles.

Jesus the Prophet

Our comparative analysis has begun to show that Luke understood Jesus to be a prophet sent by God to his people. For ancient Jews, a prophet was not a crystal ball gazer, a person who made inspired predictions about events far in the future. He was a spokesperson for God, a messenger sent from God to his people. Often the message was quite straightforward, involving a call to the people of God to mend their ways and return to God by living in accordance with his will. Throughout the Hebrew Scriptures, of course, prophets make predictions; usually (but not always) these are dire. If the people of God do not repent and begin to live in accordance with God's Law, he will punish them through plague, famine, or military disaster. Prophets tend to see into the future only insofar as it affects those who reject or accept their message.

Jesus as a Prophet in Life. Mark and Matthew, of course, also understand Jesus to be a prophet. In both Gospels he speaks God's word and predicts the coming destruction of Jerusalem and his own death at the hands of his enemies. But Luke places an even greater emphasis on Jesus' prophetic role as the spokesperson for God who comes to be rejected by his own people. This emphasis can be seen not only in the inaugural story of Jesus' ministry, the sermon in Nazareth, but also in a number of other stories that occur in Luke but in neither of the other Gospels.

In fact, the prophetic character of Jesus is seen even before the rejection scene in Nazareth, for in this Gospel Jesus is born as a prophet. Scholars have long noted that the birth narrative of Luke 2 appears to be closely modeled on the account of the birth of the prophet Samuel, as narrated in the Jewish Scriptures (1 Sam 1–2). In both instances, a devout Jewish woman miraculously conceives, to the joy and amazement of her family, and she responds in song, praising the God of Israel, who exalts those who are humble and humbles those who are exalted (compare the song of Hannah in 1 Sam 2:1–10 with the "Magnificat" of Mary in Luke 1:46–55). Anyone conversant with the Jewish Scriptures would recognize these allusions and conclude that Jesus is born like a prophet.

Moreover, when Jesus begins his public ministry, he explicitly claims to be anointed as a prophet who will proclaim God's message to his people. Recall his opening sermon in Nazareth, the fullest text of which is found in Luke. And not only does Jesus preach as a prophet in this Gospel, he also does miracles as a prophet. Among our surviving Gospels, Luke alone relates the story in which Jesus raises the only son of a widow from Nain from the dead (7:11–17). The story is clearly reminiscent of a miracle of the prophet Elijah, who in the Jewish Scriptures raises the only son of the widow from Zerephath from the dead (1 Kings 17:17–24). The similarity of the events is not lost on Jesus' companions. When they see what he has done, they proclaim "A great prophet has arisen among us" (7:16).

Jesus as a Prophet in Death. Not only is Luke's Jesus born as a prophet, and not only does he preach as a prophet and heal as a prophet, he also is said to die as a prophet. There was a long-standing tradition among Jews that their greatest prophets, both those about whom stories were told in the Scriptures (e.g., Elijah and Elisha) and those who penned scriptural books themselves (e.g., Jeremiah, Ezekiel, and Amos), were violently opposed and sometimes even martyred by their own people. In Luke's account, Jesus places himself in this prophet-

SOMETHING TO THINK ABOUT

Box 8.4 Jesus' Bloody Sweat in Luke

One of the most striking things about Luke's account of Jesus' Passion is that Jesus does not appear to experience any deep anguish over his coming fate. This becomes clear in a comparative study of what Jesus does prior to his betrayal and arrest (Luke 22:39–46; Mark 14:32–42). In Mark's account, Jesus is said to become "distressed and agitated" (14:33). Luke's version says nothing of the sort. In Mark, Jesus tells his disciples that his soul is sorrowful unto death (14:34), words not found in Luke. In Mark, Jesus leaves his disciples and falls to his face on the ground to pray (14:35). In Luke, he simply takes to his knees. In Mark, Jesus prays fervently three times for God to "remove this cup from me" (14:36, 39, 41). In Luke, he asks only once, and prefaces his prayer with "if you are willing." Thus in comparison with Mark, Luke's Jesus does not appear to be in gut-wrenching distress over his coming fate. But consider the famous verses found in the middle of the scene, Luke 22:43–44, where an angel from heaven comes to give Jesus much needed support and where his sweat is said to have become "like great drops of blood falling to the ground"? Don't these verses show Luke's Jesus in profound agony?

They do indeed. But the question is whether these verses were originally penned by Luke or were added by later scribes who felt somewhat uneasy over the fact that Jesus in this version does not seem distraught by his coming fate. If you are using the New Revised Standard Version (or any of a number of other modern translations as well), you will notice that the verses are placed in double brackets. These show that the translators are fairly confident that the verses did not originally form part of Luke's Gospel but were added by well-meaning scribes at a later time. One reason for thinking so is the fact that these verses about Jesus' bloody sweat are absent from our oldest and many of our best manuscripts of the New Testament.

In Chapter 28 we will be considering in greater detail the ways early Christian scribes changed their texts. At that time, I will say a few things about how we are able to decide what the original words of the New Testament were, given the fact that we no longer have the originals, but only copies made centuries after the originals had been lost. For now, I need simply point out that this famous passage describing Jesus' bloody sweat may not have originally been part Luke's Gospel, so that without exception, Jesus remains calm and in control of his destiny, assured of God's ongoing concern and able to face his fate with confidence and equanimity.

ic line. In a passage that is again unique to Luke, Jesus laments for Jerusalem, anticipating that he will suffer there the fate of a prophet:

> Listen, I am casting out demons and performing cures today and tomorrow, and on the third day I finish my work. Yet today, tomorrow, and the next day I must be on my way, because it is impossible for a prophet to be killed outside of Jerusalem. Jerusalem, Jerusalem, the city that kills the prophets and stones those who are sent to it! How often have I desired to gather your children together as a hen gathers her brood under her wings, but you were not willing. (13:32–34)

Jesus' knowledge that he must die as a prophet may explain some of the unique features of Luke's Passion narrative. These features can be highlighted by comparing Luke's account with the one we have studied so far in the greatest depth, Mark's.

In Mark's Passion narrative, as we have seen, Jesus appears somewhat uncertain of the need for his own death up until the very end. He does, of course, predict that he is soon to die and at one point he even explains why it is necessary ("as a ransom for many"; 10:45), but when the moment arrives, he appears torn with uncertainty (see Chapter 5).

There is no trace of uncertainty, however, in Luke's account. Here Jesus the prophet knows full well that he has to die, and shows no misgivings or doubts, as can be seen by making a detailed comparison of the two accounts of what Jesus does prior to his arrest in the Garden of Gethsemane (Mark 14:32–42; Luke 22:39–46; see box 8.4).

The same contrast appears in the accounts of Jesus' crucifixion. We have seen that in Mark's Gospel Jesus is silent throughout the entire proceeding. (Is he in total shock?) His only words come at the very end, after everyone (his disciples, the Jewish leaders, the crowds, the Roman authorities, the passersby, and even the two other criminals on their crosses) has either betrayed, denied, condemned, mocked, or forsaken him. Then he cries out, "My God, my God, why have you forsaken me?" and dies.

Luke paints a very different portrayal of Jesus in the throes of death. For one thing, Jesus is not silent on the way to crucifixion. Instead, when he sees a group of women weeping for him, he turns and says to them, "Daughters of Jerusalem, do not weep for me, but weep for yourselves and for your children" (23:28). Jesus does not appear to be distraught about what is happening to him; he is more concerned for the fate of these women. This note of confidence and concern for others is played out in the rest of the narrative. While being nailed to the cross, rather than being silent, Jesus asks forgiveness for those who are wrongfully treating him: "Father forgive them, for they do not know what they are doing" (23:34). While on the cross, Jesus engages in an intelligent conversation with one of the criminals crucified beside him. Here (unlike in Mark) only one of the criminals mocks Jesus; the other tells his companion to hold his tongue, since Jesus has done nothing to deserve his fate. He then turns to Jesus and asks, "Jesus, remember me when you come into your kingdom" (23:42). Jesus' reply is stunningly confident: "Truly I tell you, today you will be with me in Paradise."

Jesus is soon to die, but as a prophet he knows that he has to die, and he knows what will happen to him once he does: he will awaken in paradise. And this criminal who has professed faith in him will awaken beside him. Most striking of

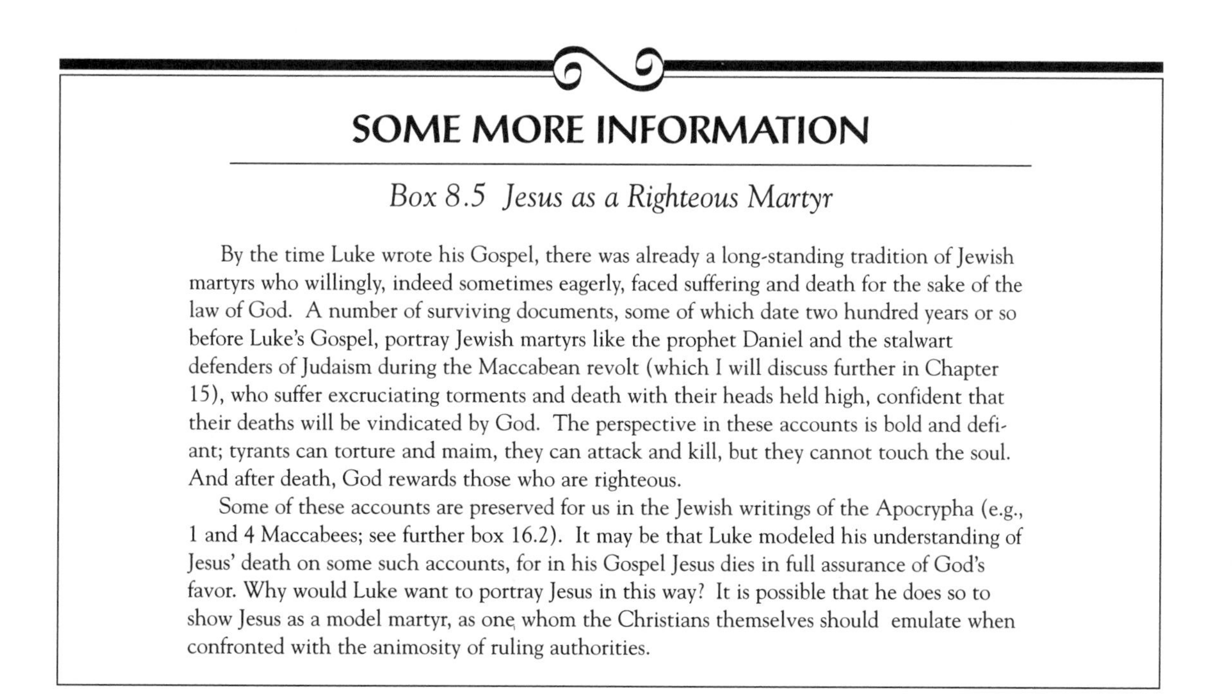

SOME MORE INFORMATION

Box 8.5 Jesus as a Righteous Martyr

By the time Luke wrote his Gospel, there was already a long-standing tradition of Jewish martyrs who willingly, indeed sometimes eagerly, faced suffering and death for the sake of the law of God. A number of surviving documents, some of which date two hundred years or so before Luke's Gospel, portray Jewish martyrs like the prophet Daniel and the stalwart defenders of Judaism during the Maccabean revolt (which I will discuss further in Chapter 15), who suffer excruciating torments and death with their heads held high, confident that their deaths will be vindicated by God. The perspective in these accounts is bold and defiant; tyrants can torture and maim, they can attack and kill, but they cannot touch the soul. And after death, God rewards those who are righteous.

Some of these accounts are preserved for us in the Jewish writings of the Apocrypha (e.g., 1 and 4 Maccabees; see further box 16.2). It may be that Luke modeled his understanding of Jesus' death on some such accounts, for in his Gospel Jesus dies in full assurance of God's favor. Why would Luke want to portray Jesus in this way? It is possible that he does so to show Jesus as a model martyr, as one whom the Christians themselves should emulate when confronted with the animosity of ruling authorities.

all is the way in which the scene ends. Whereas in Mark Jesus appears to die in despair, forsaken not only by friends, companions, and fellow Jews, but even by God himself—in Luke's Gospel he dies in full assurance of God's special care and favor. Here he does not cry out in anguish, "My God, my God, why have you forsaken me?" Instead, he offers up a final prayer, indicative of his full confidence in God's love and providential care: "Father, into your hands I commend my spirit" (23:46).

These differences are significant and should not be downplayed, as if Mark and Luke were portraying Jesus in precisely the same way. When modern readers act as if they were, for example, by thinking that Jesus said all of these things on the cross, some of them recorded by Mark and others by Luke, they take neither account seriously, but rather create their own account, in which Jesus is portrayed as all things at one and the same time. But Mark has one way of portraying Jesus and Luke another, and readers who combine their two portraits form a different Gospel, one that is neither Mark nor Luke.

In Mark Jesus is in real agony at the end. In Luke he dies in calm assurance. Each author wanted to emphasize something significant about Jesus' death. We have already seen Mark's emphasis. Luke's is somewhat different. Luke emphasizes that Jesus died as a righteous, blameless martyr of God. As a prophet he knew that this had to happen (see box 8.5).

Jesus' Death in Luke

One other important aspect of Luke's portrayal of Jesus in his death emerges when we consider the events that transpire at the close of the scene. As we saw in Mark's Gospel, the view that Jesus' death was an atoning sacrifice was suggested by the tearing of the curtain in the Temple immediately after he expired and the confession of the centurion that "this man was the Son of God." Oddly, Luke includes both events but narrates them in ways that differ significantly from the accounts in Mark (and in Matthew).

In Luke's Gospel the curtain is torn in half, not after Jesus breaths his last, but earlier, when darkness comes upon the land as the light of the sun fails (due to an eclipse? 22:45). Scholars have long debated the significance of this difference, but most think that for Luke the tearing of the curtain does not show that Jesus' death brings access to God, since here, it is torn before he dies, but rather that God has entered into judgment with his people as symbolized by this destruction within the Temple. In this Gospel, Jesus himself proclaims to his enemies among the Jewish authorities that "this is your hour and the power of darkness" (22:53). The torn curtain accompanies the eerie darkness over the land as a sign of God's judgment upon his people, who have rejected his gift of "light to those who sit in darkness and in the shadow of death" (1:79).

Moreover, in Luke the centurion does not make a profession of faith in the Son of God who had to die ("Truly this man was God's Son," Mark 15:39; Matt 27:54); here his words coincide with Luke's own understanding of Jesus' death: "Certainly this man was innocent" (Luke 23:47). For Luke, Jesus dies the death of a righteous martyr who has suffered from miscarried justice; his death will be vindicated by God at the resurrection. What both of these differences suggest is that Luke does not share Mark's view that Jesus' death brought about atonement for sin. An earlier statement in Mark corroborates his perspective; Jesus' own comment that "the Son of Man came not to be served, but to serve, and to give his life a ransom for many" (10:45; Matt 20:28). It is striking and significant that this saying is not found in Luke.

Jesus, then, must die because he is a prophet who comes to be rejected by God's people. His death does not appear to bring salvation in and of itself, and yet the death of Jesus must relate to salvation for Luke. But how? This is a puzzle we will take up further when we study the second volume of his work, the Acts of the Apostles. For now I can point out that the salvation that Jesus preaches in Luke is similar to the salvation preached by the prophets of the Hebrew Scriptures. The people of God need to repent of their sins and return to God. When they do so, he will forgive them, and grant them salvation. For Luke, the biggest sin of all was killing God's prophet. As we will see in our study of Acts, when people realize what they have done in this grotesque miscarriage of justice, they are driven to their knees in repentance. And when they turn to

Figure 8.3 The Last Supper, as portrayed in a sixth-century manuscript called the "Rossano Gospels." The scene at the right is Jesus washing the disciples' feet, from John 13.

God in recognition of their guilt, he responds by forgiving their sins. Thus, what brings a right relationship with God for Luke is not Jesus' death per se but the repentance that his death prompts.

The Gentile Mission

We have already seen that Luke places considerable emphasis on Jesus' significance for the Gentile as well as the Jew. This emphasis is not unique, of course. Mark himself may have been a Gentile, and almost certainly a large portion of his audience was. Matthew also appears to have written to a mixed congregation of Jews and Gentiles, even though he was himself probably Jewish. For both authors, salvation in Jesus comes to all people. Even more than in Matthew and Mark, however, this is a special emphasis in Luke, as we have seen already in his genealogy. For Luke, salvation comes to the Jewish people in fulfillment of the Jewish Scriptures, but since they reject it, the message goes to the Gentiles. This too, as we will see in our study of Acts, happens in fulfillment of the Scriptures.

One of the unmistakable indications that Luke is especially concerned for the Gentile mission is the fact that he is the only Gospel writer who includes a sequel recounting the spread of the religion throughout the empire, particularly among non-Jews. This concern is also found elsewhere in the Gospel. As we have seen, after Jesus' death the disciples are not told to go to Galilee (contrast 24:6, 49 with the instructions to the women in Mark 16:7). They remain in Jerusalem, where they encounter the resurrected Jesus (contrast chap. 24 with Matt 27:10, 16–20). On this occasion, Jesus explains that everything that happened to him was in fulfillment of the Scriptures; indeed, so is the Gentile mission that is yet to take place, for "repentance and forgiveness of sins is to be proclaimed in his name to all nations [same word as "Gentiles"], beginning from Jerusalem" (24:47).

The Divine Plan

Thus the Gentile mission was all part of God's plan, in place, according to Luke, since time immemorial. As we will see, the spread of the Christian church in the book of Acts occurs under the powerful direction of the Holy Spirit. This is the reason it proves so successful: since God is behind it, it cannot be stopped. The divine plan is at work in the Gospel as well, where Luke places a careful emphasis on terms like the "will" and the "plan" of God (e.g., see, 4:43; 13:33; 22:37; 24:7, 26, 44).

SOMETHING TO THINK ABOUT

Box 8.6 The Institution of the Lord's Supper in Luke

We have already seen that some of the ancient manuscripts of the New Testament differ from one another in significant ways (see box 8.4, and the more complete discussion in Chapter 28). One of the places that this matters is in Luke's account of the Last Supper (22:14–23). One peculiarity of this passage is that in some manuscripts, including those on which most of our English translations are based, Jesus does more than give his disciples the bread and the cup of wine, as he does in Mark. In these manuscripts, and most translations, he gives his disciples the cup, and then the bread, and then the cup again.

Of still greater interest is what Jesus actually says in these verses. In verse 19, he speaks of his body "which is given for you," and in verse 20 he calls the (second) cup the "the new covenant in my blood." Nowhere else in Luke's Gospel does Jesus claim that his death is a sacrifice that brings salvation. In fact, Luke is missing all such claims that are present in both Mark and Matthew (e.g., Mark 10:45; Matt 20:28). What, though, are we to make of these particular verses, which do make such a claim?

Some of our ancient manuscripts do not include this portion of the passage. Indeed, the early Christian writers who quote Luke's account of the Last Supper did not know that the verses exist. Thus, they may well have been added to this Gospel later by well-meaning scribes. This finding is significant, for apart from these verses, Luke nowhere expresses Mark's view that Jesus' death was a sacrifice that brought an atonement for sin.

The Delay of the End of Time

Luke's idea of the divine plan relates to one other distinctive aspect of his Gospel. In Mark and Matthew, as we saw, Jesus predicts the imminent end of the world. In Luke all of these predictions about the end are worded differently. In Luke, Jesus does not envisage the end of the age happening immediately. How could he? First the Christian church had to be spread among the Gentiles, and this would take time.

Consider the differences between the apocalyptic predictions of Mark and those of Luke. In Mark 9:1, Jesus claims that some of his disciples will not taste death "until they see that the kingdom of God has come with power." Luke has the same story, but here the disciples are told simply that some of them will not taste death until "they see the kingdom of God" (9:27; note that they are not promised to see its "coming in power," i.e., with the coming of the Son of Man). For Luke, the disciples already see the kingdom of God, because for him the kingdom of God is already present in Jesus' ministry. This becomes clear in several stories found only in Luke: the kingdom of God is said to have "come near" in the ministry of Jesus' disciples (10:9, 11), it is said to have already "come to you" in Jesus' own ministry (11:20), and it is said already to be "among you" in the person of Jesus himself (17:21). To be sure, even in Luke there is to be a final cataclysmic end to history at the end of this age (21:7–32), but this will not come during the disciples' lifetime.

Luke's emphasis on the delay of the end also explains the difference in Jesus' reply when interrogated by the high priest. Whereas in Mark Jesus stated that the high priest would "see the Son of Man seated at the right hand of the Power, and coming with the clouds of heaven" (14:62), in Luke his response is simply that "from now on the Son of Man will be seated at the right hand of the power of God" (22:69). Luke appears to know full well that this high priest would not live to see the Son of Man coming in his glory to bring the end of the age; in his version of the story, Jesus never predicts that he will.

Other differences in Luke's account point in the same direction. For example, only in Luke is Jesus said to have delivered the parable of the pounds, precisely in order to disabuse those who thought that "the kingdom of God was to appear immediately" (19:11–27; contrast the parable of the talents in Matt 25:14–30). One final Lukan emphasis also relates closely to the delay of the end: Jesus' social concerns.

The Social Implications of the Gospel

Throughout the history of religion, people committed to the belief that the end is near have occasionally withdrawn from society and shown little concern for its ongoing problems. Why commit oneself to fighting poverty and oppression if the world is going to end next week? In Luke's Gospel, Jesus knows that the end is *not* imminent, and this may explain one other way in which his Gospel stands out as unique. More than either of the other Synoptics, Luke emphasizes Jesus' concern for the social ills of his day.

Luke contains many of the beatitudes found in Matthew, but they are worded differently, and the differences clearly illustrate Luke's social agenda. Whereas Jesus in Matthew says, "Blessed are the poor in spirit" (5:3), in Luke he says, "Blessed are you who are poor" (6:20). Luke's concern here is for literal, material poverty. Whereas Matthew's Jesus says, "Blessed are those who hunger and thirst for righteousness" (5:6), in Luke he says, "Blessed are you who are hungry now" (6:21) Moreover, in Luke Jesus not only blesses the poor and oppressed; he also castigates the rich and the oppressor: "Woe to you who are rich. . . . Woe to you who are full now. . . . Woe to you who are laughing now" (6:24–26).

Luke's social agenda is also evident in the attention that Jesus pays to women among his followers here (see further the discussion in Chapter 23). As we will later see, the negative attitudes toward women that exist today were rooted early in Western culture. From a feminist perspective, things were much worse at the beginning of the Christian era than they are now. In Luke's Gospel, on the other hand, Jesus associates with women, has women among his followers, and urges his women followers to abandon their traditional roles as caretakers so they can heed his words as his disciples (e.g., see 8:1–3 and 10:38–42, stories unique to Luke).

CONCLUSION: LUKE IN COMPARATIVE PERSPECTIVE

We are in a position now to wrap up our reflections on the Gospel according to Luke. Here as in Matthew and Mark we have a kind of Greco-Roman biography, in which the things Jesus says, does, and experiences reveal who he is to the attentive reader. Had we chosen, we could have examined this Gospel without recourse to these other biographies of Jesus, following the literary-historical method that we used to study Mark. Alternatively, we could have analyzed it strictly in light of how the author modified his sources, as we did for Matthew. Instead we explored this text in light of similar biographies of Jesus, irrespective of whether Luke used any of them as sources. Has this approach proved useful?

Our comparative analysis has shown that Luke has a number of distinctive emphases. Luke stresses that the salvation that came in Jesus was first directed to the heart of Judaism, but Jesus as a Jewish prophet was rejected by his own people. The message was then to be sent into the whole world for the salvation of all people, Jew and Gentile, a message of forgiveness of sins to all who would repent. The worldwide mission envisioned by Jesus was planned from time immemorial by God himself and would be completed before the end of the age could come. Since the end was not to be imminent in Jesus' own day, the mission involved not only preaching the news of God's salvation but also working to right the ills of society in a world beset by poverty and oppression.

We might ask what these distinctive emphases can tell us about the author of this book and his audience. The question might be premature, however, for the Gospel of Luke is the first volume of a two-volume work, which ultimately must be read as a unit if we are to understand the full message of its author.

SUGGESTIONS FOR FURTHER READING

Brown, Raymond. *The Birth of the Messiah: A Commentary on the Infancy Narratives in Matthew and Luke*. Garden City, N.Y.: Doubleday, 1977. A massive and exhaustive discussion of the birth narratives of Matthew and Luke, suitable for those who want to know simply everything about every detail.

Brown, Raymond. *The Death of the Messiah: From Gethsemane to the Grave*. 2 vols. London: Doubleday, 1994. A detailed and thorough discussion of the passion narratives of the four Gospels, in all of their aspects and for all of their verses.

Cadbury, H. J. *The Making of Luke-Acts*. 2d ed. London: SPCK, 1968. A classic study that shows how the author of Luke and Acts used the traditions and sources at his disposal to produce a unified narrative; for advanced students.

Conzelmann, Hans. *The Theology of St. Luke*. New York: Harper & Row, 1960. This classic study of Luke from a redaction-critical perspective argues that Luke modified the traditions he received particularly in light of the delay of the end of time; for advanced students only.

Esler, Philip F. *Community and Gospel in Luke-Acts: The Social and Political Motivations of Lucan Theology*. Cambridge: Cambridge University Press, 1987. A study of Luke-Acts from the perspective of the social sciences, which shows the relationship between the theological perspectives of these books and the sociopolitical context of the author and his community; for advanced students.

Juel, Donald. *Luke-Acts: The Promise of History*. Atlanta: John Knox, 1983. A clearly written discussion of the background of Luke-Acts and its overarching themes; ideal for beginning students.

Maddox, R. *The Purpose of Luke-Acts*. Edinburgh: T & T Clark, 1982. An intelligent and incisive discussion of the prominent themes and major emphases of Luke and Acts.

Powell, M. A. *What Are They Saying about Luke?* New York: Paulist, 1989. An excellent survey of modern scholarly views of Luke's Gospel, for beginning students.

CHAPTER 9

Luke's Second Volume: The Acts of the Apostles

For people interested in knowing what happened to Jesus' followers after his death and resurrection, the Acts of the Apostles has always been the first place to turn. This is our earliest account of the Christian church, an account that speaks of massive conversions to the faith, of miraculous deeds performed by the apostles, of opposition and persecution by nonbelievers, of the inner workings of the apostolic band and their interactions with significant newcomers like Paul, and above all of the dramatic spread of the Christian church from its inauspicious beginnings among the few followers of Jesus in Jerusalem to the heart of the empire, the city of Rome.

Although the book of Acts is the second volume of a two-volume work, it is not the same kind of book as the first volume. The Gospel of Luke portrays the life of Jesus, the rejected Jewish prophet, from his miraculous conception to his miraculous resurrection. The portrayal is comparable in many ways to the Gospels of Mark and Matthew and is best classified, in terms of genre, as a Greco-Roman biography. The book of Acts, however, is quite different. Here there is no solitary figure as a main character; instead, the book sketches the history of Christianity from the time of Jesus' resurrection to the Roman house arrest of the apostle Paul.

THE GENRE OF ACTS AND ITS SIGNIFICANCE

Some scholars have argued that since the two volumes were written as a set, they must be classified together, in the same genre. The books are structured quite differently, however. The book of Acts is concerned with the historical development of the Christian church. Moreover, the narrative is set within a chronological framework that begins with the origin of the movement. In these respects the Acts of the Apostles is closely related to other histories produced in antiquity.

Historians in the Greco-Roman world produced a number of different kinds of literature. Some ancient histories focused on important leaders or episodes in the history of a particular city or region. Others were broader in scope, covering significant events in the history of a nation. Sometimes these histories were arranged according to topic. More commonly, they were set forth in a chronological sequence. Chronological narratives could be limited to a single, but complicated, event (as in Thucydides's account of the Peloponnesian War) or to a series of interrelated events (as in Polybius's account of the rise of Rome to dominance over the Mediterranean). Sometimes they covered the most ancient events in the memory of a nation, as a way of showing how the people became who they were.

The book of Acts is most like this final kind of history, one that traces the key events of a people from the point of their origin down to near the present time, to show how their character as a people was established. Scholars sometimes call this genre general history. One well-known example, produced at approximately the same time as Acts, was written by the Jewish historian Josephus. His twenty-volume work, *The Antiquities of the Jews*, sketches the significant events of Judaism all the way from Adam and Eve down to his own day.

Unlike biographies, ancient histories have a number of leading characters, sometimes, as in Josephus, a large number of them. Like biographies, though, they tend to utilize a wide range of subgenres, such as travel narratives, anecdotes, private letters, dialogues, and public speeches. On the whole, histories from Greco-Roman antiquity were creative literary exercises rather than simple regurgitations of names and dates; historians were necessarily inventive in the ways they collected and conveyed the information that they set forth.

All histories, however, whether from the ancient world or the modern, cannot be seen, ultimately, as objective accounts of what happened in the past. Because so many things happen in the course of history (actually, billions of things, every minute of every day), historians are compelled to pick and choose what to mention and what to describe as significant. They do so according to their own values, beliefs, and priorities. Thus, we can almost always assume that a historian has narrated events in a way that encapsulates his or her understanding of the meaning of those events.

This aspect of limited objectivity is particularly obvious in the case of historians living in antiquity. Theirs was a world of few written records but abundant oral tradition. Indeed, many ancient historians expressed a preference for hearing an account from an oral source rather than finding it in a written record. This approach stands somewhat at odds with the modern distrust of "mere hearsay," but there is some logic behind it: unlike written documents, oral sources can be interrogated to clarify ambiguities. Still, one can imagine the difficulties of determining what really happened on the basis of oral accounts. Moreover, when it came to the written record, ancient historians obviously had no access to modern techniques of data retrieval. For these reasons, they generally had little concern for, and less chance of, getting everything "right," at least in terms of the high level of historical accuracy expected by modern readers.

Nowhere can this be seen more clearly than in the case of sayings and speeches that are recorded in the ancient histories. On average, speeches take up nearly a quarter of the entire narrative in a typical Greco-Roman history. What is striking is that ancient historians who reflected on the art of their craft, like the Greek historian Thucydides (fifth century B.C.E.), admitted that speeches could never be reconstructed as they were really given: no one took notes or memorized long oratories on the spot. Historians quite consciously made the speeches up themselves, composing discourses that seemed to fit both the character of the speaker and the occasion.

Not only in presenting the speeches of their protagonists but also in relating the events of the narrative itself, ancient historians were somewhat less ambitious than their modern counterparts. They strove not for absolute objectivity but for verisimilitude. They worked to produce a narrative account that rang true, that made sense in light of what they had uncovered through their interrogation of oral sources and their perusal of written records.

We will see that many of these aspects of ancient histories apply to the book of Acts as a general history. Before pursuing our study of the book, however, we should return to the issue of genre. Is it plausible that the two volumes of Luke's work represent two distinct genres?

To understand why the author would have chosen two different literary genres for these closely related books, we need to recognize the constraints under which he was operating. The design of Luke's work was different from anything that had yet been produced, so far as we know, by the burgeoning Christian church. In it Luke set out to sketch the history of the early Christian movement. This movement, though, could not be explained apart from the history of its founder, Jesus.

Since the first part of the history of this movement was concerned with the life and teachings of Jesus, the subject matter itself, not to mention the models available to the author (e.g., the Gospel of Mark), more or less determined the genre of the first volume. It was to be a biography. The second volume was to sketch the development of the movement after the death of Jesus. The biographical genre was much less applicable here, as there were more characters and events to consider. Thus, Luke wrote a general history of the movement for his second volume, providing a chronologically arranged account of the spread of the Christian movement after the death of its founder, Jesus.

SOMETHING TO THINK ABOUT

Box 9.1 The Book of Acts: An Ancient Novel?

Some recent studies of the genre of Acts have concluded that it is more like an ancient novel than a general history. Novels in the Greco-Roman world were fictionalized narratives written almost exclusively for entertainment. They normally told the tale of lovers who were separated by misfortune and experienced numerous trials in their attempt to become reunited. One of the themes that permeates these books is the persecution and oppression of the main characters, who are (usually) innocent of any wrongdoing. Among the subgenres typically employed in the novels are travel narratives, shipwreck scenes, dialogues, speeches, and private letters—all of which are found in the book of Acts.

Other scholars are not persuaded by this thesis. Acts is not about estranged lovers; indeed, there is no romance of any kind here (in contrast to every surviving Greek or Roman novel). Moreover, this book does not focus from beginning to end on the exploits of the chief protagonist(s) in the ways the novels do: the main character (Paul) does not come on the scene until a third of the way through the narrative. Finally, and perhaps most importantly, Luke does not appear to have written this book as a fictionalized narrative chiefly for entertainment. There may indeed be fictional elements in the account, as we will see; but judging from the preface to volume one, from the subject matter of the narrative (the spread of the Christian church), and from the main characters themselves (who are, after all, historical persons), we can more plausibly conclude that Luke meant to write a history of early Christianity, not a novel. Moreover, all of the ancient Christian authors who refer to the book appear to understand it in this way.

There are, however, a number of novelistic touches in the book, and we would be remiss not to recognize them. The narrative is entertaining in places, and it does indeed embody a number of the story-telling techniques common among ancient writers of fiction, including the various subgenres and themes mentioned above.

What can we say now, in general terms, about the significance of the genre of Acts and the relationship of the book to the Gospel of Luke? When we read the book of Acts as a general history, we should expect to find a narration of events that the author considers significant for understanding the early Christian movement. Furthermore, if we are interested in reading his book as an ancient reader would, we should not evaluate it strictly in terms of factual accuracy. In addition, we should be looking for themes and points of view that parallel those found in volume one, the Gospel of Luke. Finally, since this book is also a chronologically arranged narrative, even though of a different kind from the Gospel, we might expect our ancient author to set the tone for the rest of the account at the very outset.

THE THEMATIC APPROACH TO ACTS

For each of the Gospels examined so far I have explained and used a different method of analysis: a literary-historical method with Mark, a redactional method with Matthew, and a comparative method with Luke. Theoretically, each of these methods could be used with the book of Acts as well, even though it is the only general history preserved within the New Testament. A literary-historical approach would explore the development of the characters and plot of the story in light of the expectations of its audience, based on their knowledge of the genre and the background information that the author appears to presuppose.

A redactional method would determine the sources available to the author and ascertain how he has modified them—a somewhat complicated business with Acts since none of its sources survives (although this has not stopped scholars from trying). A comparative method would consider the message of Acts in light of the writings of other early Christian authors, such as the apostle Paul, one of the main characters of Acts' narrative. Here, however, we will explore the possibilities of yet a fourth approach, one that might be labeled the "thematic method."

Every author has major ideas that he or she tries to communicate in writing. A thematic approach attempts to isolate these ideas, or themes, to come to some understanding of the author's overarching emphases. Themes can be isolated in a number of ways—as we will see, the other methods can be useful in this regard—but the focus of attention is not on how the narrative plot unfolds (as in the literary-historical method) or on how the work compares and contrasts with another (as in the redactional and comparative). The focus is on the themes themselves and the ways they are developed throughout the work.

As with all methods, the thematic approach is best explained by showing it at work. In this introduction to Acts, we will focus on two portions of the narrative that provide special promise for understanding Luke's main emphases: the opening scene, which relates the work back to what has already transpired in the Gospel of Luke and anticipates what will take place in the narrative to follow, and the speeches of the main characters, which are scattered throughout the text and appear to represent compositions of the author himself.

FROM GOSPEL TO ACTS: THE OPENING TRANSITION

The first and most obvious thing to notice in the opening verses of Acts is that this book is also dedicated to "Theophilus," who is reminded of the basic content of the first volume of the work, namely, "all that Jesus did and taught from the beginning until the day when he was taken up to heaven" (1:1–2). This kind of opening summary statement was common in multivolume works of history in antiquity, as a transition from what had already been discussed. The dedication to Theophilus and the accurate summary of the first volume, as well as the similar themes and consistent writing style of Luke and Acts, have convinced virtually all scholars that the same author produced these two books.

The story of Acts begins with Jesus' appearances over a course of forty days after his resurrection. During this time, he convinces his former disciples that he has come back to life and he teaches them about the kingdom (v. 3). In keeping with Luke's emphasis in volume one on Jerusalem as the place to which salvation came, the disciples are told to remain in Jerusalem until they receive the power of the Holy Spirit (v. 4; contrast the resurrection narratives of Mark and Matthew). In Acts the message of God's redemption goes forth from the holy city because it is rejected there. Just as Jesus the prophet was rejected by his own people in Jerusalem, so too his apostles will be rejected in Jerusalem. The spreading of the message was anticipated in the sermon of Jesus in Luke 4: because Jews will reject the message, it will be taken outside, to the Gentiles. The book of Acts is largely about this movement of the gospel from Jew to Gentile, from Jerusalem to the ends of the earth.

This theme is announced in these opening verses. The disciples inquire whether this is the time that the Kingdom will be brought to Israel (v. 6). They expect that now is the time in which their apocalyptic hopes will be realized, when God will intervene in history and establish his glorious kingdom for his people. We saw in the Gospel that Luke rejected the idea that the end was to come during the lifetime of Jesus' disciples. Here as well Jesus tells his disciples not to be concerned about when the end will come. Instead, they are to work in the present to spread the gospel through the power of the Holy Spirit:

> It is not for you to know the times or periods that the Father has set by his own authority. But you will receive power when the Holy Spirit has come upon you; and you will be my witnesses in Jerusalem, and in all Judea and Samaria, and to the ends of the earth. (1:7–8)

This injunction to engage in the Christian mission foreshadows what is to take place throughout the rest of the book; indeed, the spread of the church provides the organizing motif for the entire narrative. In broad terms, it happens as follows. As anticipated, the Holy Spirit comes upon the apostles in the next chapter, on the day of Pentecost. The Spirit works miracles on their behalf and empowers them to proclaim the gospel of Christ. Thousands upon thousands of Jews convert as a result (chaps. 3–7), but opposition arises among the Jewish leadership, as it did in the case of Jesus himself in the gospel. Christians scatter from the city, taking the gospel with them, first to Judea and Samaria (chap. 8). The most significant convert in these early years is a former opponent of the church, Saul, also known as Paul (chap. 9). Largely, though not exclusively, through Paul's work, the gospel is taken outside of Palestine and spreads throughout several of the provinces of the Empire. Over the course of three missionary journeys (see fig. 9.1), Paul establishes churches in major cities in Cilicia, Asia Minor, Macedonia, and Achaia (which roughly correspond to modern-day Turkey and Greece; chaps. 13–20).

Eventually, he makes a fateful journey to Jerusalem (chap. 21), where he is arrested by Jewish leaders, put on trial, allowed to make several speeches in his own defense, and finally sent to stand before Caesar in Rome (chaps. 22–27). The book ends with Paul under house arrest in Rome, preaching the good news to all who will hear (chap. 28). This appears to fulfill Luke's anticipation that the gospel would go to the "ends of the earth," for the message of Christ has now spread far and wide, and is proclaimed in the very heart of the empire, in the capital city itself.

The geographical spread of the Christian church is not Luke's only concern in Acts. In some ways, he is even more dedicated to showing how the gospel came to cross ethnic boundaries. Indeed, he goes to great lengths to explain, and justify, how the Christian gospel ceased being a message only to Jews. To be sure, the earliest con-

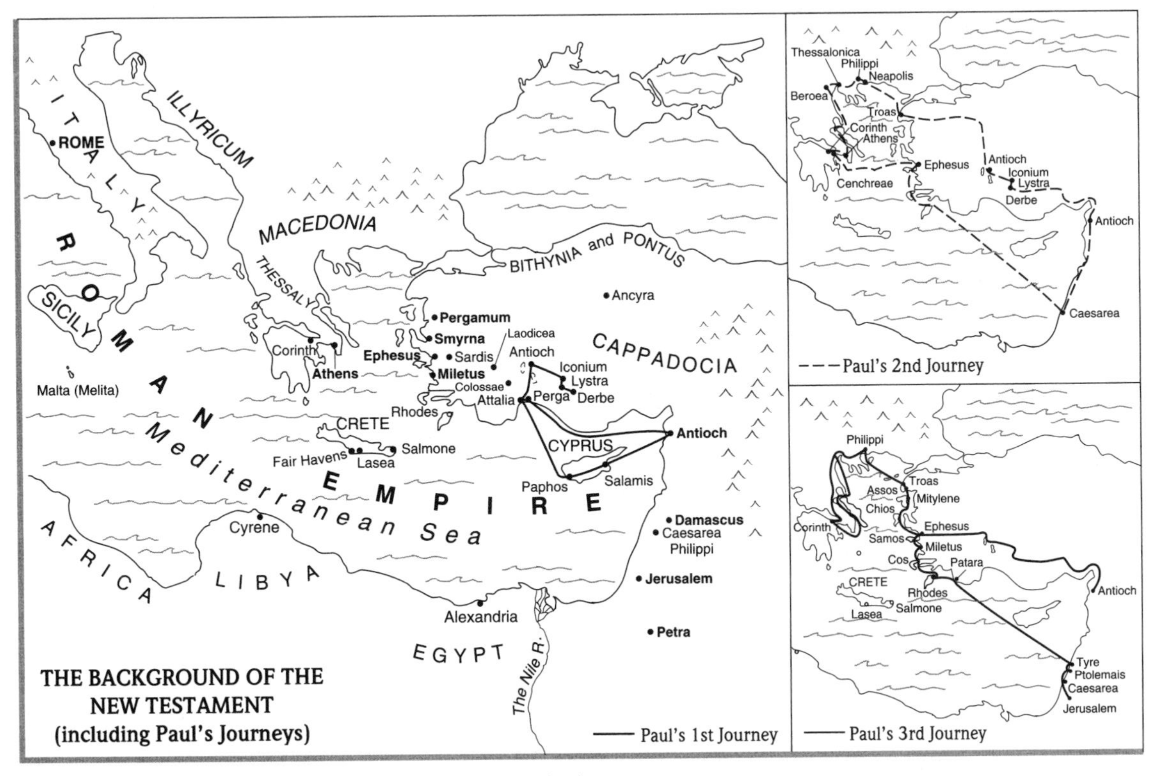

Figure 9.1 Paul's Missionary Journeys According to the Book of Acts.

SOMETHING TO THINK ABOUT

Box 9.2 Luke's Mysterious Two Men

Who are the mysterious "two men in white robes" who appear to the disciples in Acts 1:10–11 to tell them that Jesus will return from heaven in the way that he has ascended? A careful reader will recall having seen two such persons before, at the conclusion of Luke's Gospel where "two men in dazzling clothes" appear to the women in Jesus' empty tomb and tell them that he has risen from the dead (24:4; contrast Mark 16:5 and Matt 28:5). Are they also the "two men" who appear yet earlier still, "in glory" on the Mount of Transfiguration (Luke 9:30–31)? It is striking that Luke uses similar terms to describe these figures in all three passages. What is more, he tells us who they are in their first appearance (9:30). They are Moses, the greatest lawgiver of the Jews, and Elijah, the greatest Hebrew prophet (so great that he was taken directly into heaven without dying; see 2 Kgs 2:9–12).

A number of interpreters have recognized the symbolic significance of these two figures in Luke-Acts as embodiments of the Law and the prophets (i.e., the Hebrew Scriptures). Thus, for Luke, the Scriptures themselves, as personified in Moses and Elijah, provide testimony to the climactic moments of Jesus' existence: his mission on earth that leads to his death (Luke 9:31), his resurrection from the dead (Luke 24:4), and his ascension into heaven and return in glory (Acts 1:11). In other words, Luke uses these mysterious two men to show that every aspect of Christ's work of salvation occurs in fulfillment of God's plan, as set forth in the Jewish Scriptures.

verts were Jews, as were Jesus himself and his closest disciples, but many Jews rejected this gospel. According to Luke, God therefore opened up the faith to the non-Jew. This first happens in chapter 8 with the conversion of a number of Samaritans, people who lived in Samaria who were considered to be "half-Jews" by many who lived in Judea. Soon thereafter, the apostle Peter learns through a vision that God means for the Gentiles also to hear and accept the message of salvation in Christ (chaps. 10–11). Much of the rest of the book shows how the gospel meets continual opposition among Jews in every province to which it goes but finds ready acceptance among Gentiles, especially those associated with the Jewish synagogues. The main character involved in spreading this gospel is Paul.

This emphasis on the Gentile mission of the church naturally raises some pressing questions. If the message of salvation that came to the Jews goes to the Gentiles, do these Gentiles first have to become Jews? To put the matter somewhat differently, if (as Luke's Gospel itself indicates) Jesus was a Jew, sent from the Jewish God as a Jewish prophet to the Jewish people, in fulfillment of the Jewish Scriptures, then isn't this religion Jewish? Surely for a person to become a follower of Jesus he or she must first adopt Judaism. The author of Acts does not think so. As we will see, he devotes a good portion of his history to explaining why.

But if Gentiles coming into the church do not need to become Jewish, then hasn't the religion itself ceased being Jewish? And haven't its representatives, such as Paul, made an irreparable break with their Jewish past? Again, the author of Acts does not think so. And again, he devotes a good portion of his account to explaining why.

Before examining these explanations in the themes set forth in the speeches in Acts, we should complete our investigation of the opening passage. It ends with Jesus physically ascending into heaven. Two men in white robes suddenly appear to the apostles as they watch him depart (see box 9.2). They tell the apostles not to stand by gaping into heaven; for just as Jesus departed from them, so he will return (vv. 10–11).

These words of comfort to the apostles may suggest that for Luke, even though the end of the age was not to come in the lifetime of Jesus' disciples, it was still destined to come soon. Indeed, Luke may have anticipated that it would come in his own lifetime; Jesus had yet to return on the clouds of heaven in judgment to set up his kingdom on earth. For Luke himself the end still is at hand, and the gospel needs to be proclaimed with yet greater urgency, as Jew and Gentile join together in their faith in the Christ of God.

Thus we can see many of the major themes of Luke's Gospel repeated at the outset of Acts, and we can anticipate their recurrence throughout the narrative. These themes include the focus on Jerusalem, the proclamation of the gospel beginning with the Jews but moving to the Gentiles, the necessary delay of the end while this worldwide proclamation takes place, and, perhaps most importantly, the divine guidance of the Christian mission by the Holy Spirit. For Luke, it is God who directs the movement of the Christian church from start to end.

Before turning to an examination of some of the speeches in which Luke's main themes recur, we should look at a few of the other ways that Luke develops them in some of the more artistic aspects of his narrative.

LUKE'S ARTISTRY AS A STORYTELLER

Readers of the New Testament have long noticed many clear similarities between what happens to Jesus in the Gospel of Luke and to Christian believers in the book of Acts. These parallels show that Luke was no mere chronicler of events, set on providing an objective account of the early years of the Christian movement. He compiled this history with a clear purpose, part of which was to show that the hand of God was behind the mission of the church as much as it was behind the mission of Jesus. Thus, for example, at the beginning of Jesus' ministry in Luke, he is baptized and receives the Holy Spirit; when new believers are baptized in the book of Acts, they also receive the Spirit. The Spirit empowers Jesus to do miracles and to preach in Luke; so too it empowers the apostles to do miracles and to preach in Acts. In Luke, Jesus heals the sick, casts out demons, and raises the dead; in Acts, the apostles heal the sick, cast out demons, and raise the dead. The Jewish authorities in Jerusalem confront Jesus in Luke; the same authorities confront the apostles in Acts. Jesus is imprisoned, condemned, and executed in Luke; some of his followers are imprisoned, condemned, and executed in Acts.

These parallels are not simply interesting coincidences. One author has produced both books, and he uses the parallel accounts to make a major point: the apostles continue to do Jesus' work and thereby prolong his mission through the power of the same Spirit. Thus they engage in similar activities, experience similar receptions, and suffer similar fates. The author's literary artistry is not limited, however, to creating parallels between Luke and Acts. Just as interesting are the parallels between the main characters in the narrative of Acts itself, particularly between Peter, the main character of chapters 1–12, and Paul, the main character of chapters 13–28.

Several examples of these parallels stand out. Both Peter and Paul preach sermons to Jewish crowds, and what they have to say is in many respects remarkably similar (e.g., see the speeches in chaps. 3 and 13). Both perform amazing miracles; both, for example, cure the sick without having any direct contact with them. Thus Peter's shadow can bring healing (5:15), as can Paul's handkerchiefs (19:12). Both are violently opposed by leaders among the Jews but vindicated by God; they are imprisoned for their proclamation yet delivered from their chains by divine intervention (12:1–11; 16:19–34). Perhaps most importantly of all, both become absolutely convinced, on the grounds of divine revelation and the success of their proclamation, that God has decided to admit Gentiles into the church without their first becoming Jews (chaps. 10–11, 15).

These parallels reinforce our earlier impression that throughout this narrative Luke is intent on showing that God is at work in the Christian mission. Those who are faithful to God give similar speeches with similar results; they perform similar miracles, receive similar revelations, and experience similar fates. Luke's artistry, then, serves a clear thematic purpose. Nowhere can this be seen more clearly than in the speeches that Luke has devised for his leading characters.

Figure 9.2 Peter, Jesus, and Paul, the three most important characters of Luke-Acts, from a catacomb painting in Rome.

Ancient historians typically wrote the speeches of their main characters themselves, choosing words that they judged to be appropriate to their character and suitable for the occasion. Since Luke appears to stand in line with the Greek historians, we can assume that he too wrote the speeches of his main characters. This would help explain why the same themes recur in these speeches no matter who delivers them.

THEMES IN THE SPEECHES IN ACTS

As with most general histories, speeches figure prominently in the book of Acts. Indeed, they take up nearly one-quarter of the entire narrative, about average for histories of the period. To isolate some of the important Lukan themes in the book, we will examine several examples of different kinds of speeches.

One of the ways to classify the speeches in Acts is to consider the different kinds of audiences to which they are delivered, on the assumption that speakers will stress different things in different contexts. Some of the speeches are delivered by Christian leaders to other Christians as a means of instruction or exhortation, others are addressed by Christians to potential converts in the context of evangelism, and yet others are given by Christians to legal or religious authorities, as apologies (see box 8.1).

Speeches to Christian Believers

Peter's Opening Speech. The first speech in the book is delivered by Peter, right at the outset of the narrative. After seeing Jesus ascend into heaven, the eleven disciples return to Jerusalem and devote themselves to prayer with Jesus' female followers and family. The first concrete action that the group takes is to elect a new member of "the Twelve" to replace Judas Iscariot, who after betraying Jesus suffered an ignominious death, (see box 9.3). Peter arises and delivers a speech on how they ought to proceed in their new circumstances (1:15–22). The speech anticipates many of the central themes of the book, including the important issue of how this new religious movement relates to its Jewish roots. Before delving into Peter's view of this relationship (at least, as Luke portrays it), we should consider the broader context.

To most Jews in the book of Acts, the Christian claims about Jesus are altogether unacceptable, and throughout the narrative the principal instigators of persecution against the Christians are Jews. From a historical perspective, this opposition is understandable. Christians claimed that Jesus was the messiah; but the messiah, in the expectations of most Jews, was to be a figure of power and grandeur. Jesus, on the other hand, was a crucified criminal. In the opinion of most Jews (both historically and in the narrative of Acts), those who proclaim Jesus as the messiah have not only lost touch with their Jewish roots, they have also violated the clear teaching of Scripture.

Luke has a different perspective. We have already seen that some of Luke's predecessors and contemporaries (e.g., Mark and Matthew) claimed that Jesus was the fulfillment of the Jewish Scriptures. Luke takes this view a step further.

SOMETHING TO THINK ABOUT

Box 9.3 The Death of Judas

Only two passages of the New Testament describe the death of Judas: Acts 1:18–19 and Matt 27:3–10. It is interesting to compare their similarities and differences. In Matthew, Judas tries to return the thirty pieces of silver that he has been paid to betray Jesus. When the priests refuse to take it, he throws it down in the Temple and goes out and hangs himself. The priests are unable to use the silver for the Temple treasury, since it is "blood money," that is, money tainted with the blood of Jesus' execution, so they decide to use it to purchase a piece of property as a place to bury strangers. From that time on, since the place was bought with blood money, it was appropriately called the "Field of Blood."

In Acts Judas's death is again associated with the Field of Blood but for an entirely different reason. In Peter's speech we learn that Judas himself purchased the field, after which he died a bloody death. He does not appear to have hanged himself, however; Peter says that he fell headlong and burst in the middle so that his intestines "gushed out." It is hard to know what Luke, the author of Peter's speech, has in mind as the cause of death (was it a suicide? Did Judas fall on a sword? did he jump off a cliff? did he spontaneously swell up and burst?). In any event, Luke clearly thinks the Field of Blood obtained its name from Judas's blood being spilled on it.

These two accounts are difficult to reconcile, but in some respects it is their similarities that are of greatest interest. Why do they both connect the name of this Field of Blood with the death of Judas? Is it possible that there actually was a field in Jerusalem made up of red clay and called the Field of Blood because of its color? A slight piece of evidence for this conclusion derives from Matthew, who indicates that it was a "potter's field" (27:10), that is, a field from which clay was extracted for pottery. It is difficult to decide whether Judas actually killed himself there, whether he was at some point its owner, or whether his blood money was used to purchase it. At the least, we can say that later Christians came to associate this clay lot with the disciple who had betrayed his master and then experienced an ignominious death.

The entire Christian movement *after* Jesus is a fulfillment of Scripture as well. This theme is played out in the narrative of Acts and is anticipated already by the opening words of Peter's first speech. Peter argues that the death of Judas, and the need to replace him with someone else, was predicted by David in the Psalms.

Peter cites two Psalm texts to support his view (1:20). Since he is addressing a friendly audience, he evidently does not need to provide a rationale for the way he interprets these passages. But if you read these quotations in their original contexts (Psalm 69 and Psalm 109), you will probably find it hard to understand how anyone could think that they predict what was going to happen hundreds of years later to one of the messiah's followers. In Luke's account, though, Peter interprets them in precisely this way. This in itself can tell us something about what was happening at the time of Luke, the author of Peter's speech. In Luke's own day, Christians were evidently combing the Jewish Scriptures to find indications of what had been fulfilled in their midst, not only in the life of Jesus but also in the life of their own communities (see Chapter 16). From Luke's perspective, the history of Christianity fulfills the Scriptures.

This basic approach to the Jewish Scriptures was not unique to early Christianity (see Chapter 15, especially on the Dead Sea Scrolls). In any event, since Luke understood not only Jesus but also the entire Christian movement to be a fulfillment of the Jewish Scriptures, he did not see it as standing in opposition to Judaism. Rather, it was in direct continuity with it. Why, then, would Christianity be rejected by leaders among the Jews? Luke's reader is left to infer that those who opposed Jesus' followers were necessarily opposed to their own religion and, as a consequence, to their own God. This is a strong statement, even if made only by implication.

Perhaps more obviously, as a corollary, Luke's view that Christianity is a fulfillment of Scripture indicates that God himself was behind the Christian movement. This indeed is perhaps the overarching theme of the entire narrative. This movement comes from God (see especially 5:33–39). God's involvement is clearly seen in one other way in this early scene, although not directly in Peter's speech itself. How do the disciples elect a new member to join the Twelve? They pray and cast lots. This was an ancient method of determining the divine will for an action. A jar containing two or more stones or bones was shaken until one of them fell out. Since the process could not be rigged, the lot that fell was understood to be God's choice. Evidently, the procedure worked to Luke's satisfaction: Matthias becomes the twelfth apostle.

This takes us back to Peter's speech and the final theme for us to consider. The speech is meant to persuade the believers to engage in a particular course of action. This is a typical feature of speeches to believers in this book, but what is curious in the context of the wider narrative is the particular course of action that Peter urges. Peter persuades the assembly to elect a new member of the Twelve to be a witness of Jesus' resurrection. This is to be a person who had accompanied the other disciples throughout the whole of Jesus' ministry, from his baptism by John to his ascension (1:21–22). The first requirement is itself somewhat odd, in that Jesus does not call his disciples until well after his baptism in Luke (see Luke 5); in any event, the new member of the apostolic band was to have been with Jesus from the outset of his ministry.

What is more perplexing is that the speech intimates that the election of this new apostle is crucial for the propagation of the Christian gospel that is to take place in the subsequent narrative. In fact, this is not the case at all. After Matthias is elected to be an apostle, he is never mentioned again in the book of Acts. Why, then, does Luke compose a speech urging his election? To put the question into a broader context I should point out that Matthias is not the only apostle who fails to appear in the rest of the narrative. Most of the Twelve do not. Why would a book entitled "The Acts of the Apostles" not discuss the acts of the apostles?

As already seen, the titles of our New Testament books were not original but were added by later Christian scribes. In this case, at least, the title is not at all apt. For the book is not about the deeds of the apostles per se but about the spread of the Christian religion through the labors of only a few of them, and about the reactions that it provoked among those who refused to accept it. Indeed, there are only two main characters in the book (along with numerous minor characters), one of whom, the chief protagonist for most of the narrative, is Paul, who was not one of the Twelve.

Why is it so important for Luke to begin his account with the election of a twelfth apostle, if neither he nor most of his companions figure prominently in the narrative, whereas one who is not among their number ends up being the central character? Perhaps the answer relates to another of Luke's prominent themes: the notion of continuity in early Christianity. We have already seen one form of continuity in our discussion of Luke's Gospel, namely, the continuity between Jesus and Judaism; we have uncovered a second form in our study of Acts to this point, the continuity between Judaism and Christianity. But there is yet a third form at work in Luke's narrative—a continuity between Jesus and his church. This continuity is assured by the Twelve, who start out as Jesus' original disciples and provide the transition as the leaders of the Christian community in Jerusalem. The thematic point for Luke is that Christianity is not something that begins, strictly speaking, after Jesus' death. It is something that is intimately connected with his life. Those who were closest to Jesus in his lifetime were responsible for the original dissemination of this religion after his death.

Indeed, even though the twelve apostles rarely appear individually (with the chief exception of Peter and the partial exception of John), they play a prominent role in the founding of the church at the outset of the narrative. They are present en masse when Peter preaches his first evangelistic sermon, converting several thousand Jews (2:14); they are the teachers of the newfound community of faith, a community unified around their instruction (2:42); they perform miracles, convincing others to believe (2:43; 5:12); they edify believers by testifying to the resurrection of Jesus (4:33); and they organize and run that early community, distributing funds that are raised and taking care of those in need (4:35–36). Moreover, they make all of the key decisions affecting the church throughout the world. This final theme comes to prominence in the series of speeches delivered to believers assembled for the famous Jerusalem Council in chapter 15, another critical juncture of the narrative.

The Speeches at the Jerusalem Council. The narrative backdrop of these speeches is as follows. After Paul is converted by a vision of Jesus on the road to Damascus (chap. 9), the apostle Peter learns through a vision and an encounter with a group of believing Gentiles that God makes no distinction between Jew and Gentile, that Gentiles can belong to the people of God without first becoming Jews (chaps. 10–11). Soon thereafter, Paul and his companion Barnabas are set apart by the church in Antioch as missionaries to other lands; they engage in an evangelistic campaign (Paul's "first missionary journey"), chiefly in Asia Minor. Some Jews convert, but many others resist; Paul comes to be opposed, sometimes with violence, by leaders of the Jewish synagogues (chaps. 13–14). This Jewish opposition in turn forces Paul and Barnabas to proclaim their faith to the Gentiles, many of whom come to faith.

When they arrive back in Antioch, Paul and Barnabas are confronted by Christians from Judea who insist that Gentile men must be circumcised to experience salvation. This leads to a major controversy. Paul, Barnabas, and several others are appointed to go to Jerusalem to discuss the matter with the apostles. At this conference, Peter and James, the brother of Jesus, give their opinions in speeches delivered to the assembled body of believers.

Many of the themes that we have already isolated in Acts can be found here as well (15:7–21): God has been totally in charge of the Christian mission, including the conversion of the Gentiles (vv. 7–8); he makes no distinction between Jew and Gentile in that all are saved on equal grounds (vv. 9–11); and the salvation of the Gentiles represents a fulfillment of Scripture (vv. 15–19). Once the apostles, along with the other leaders of the Jerusalem church, have heard these speeches, they are unified in their judgment and write a letter to the Gentile churches explaining their decision. The net result is that not just the Jerusalem church but all of the churches in the empire (e.g., those founded by Paul and Barnabas) stand under the leadership of the apostles, the original eyewitnesses of Jesus, who are themselves totally unified in their teaching.

In Sum: Speeches to Believers in Acts. What can we say in conclusion about the important themes found in the speeches of Christians to other believers in Acts? Above all, they tell us something about Luke's view of the nature of the early Christian church. Strictly speaking, the church for Luke is not a new thing. On the one hand, it represents the fulfillment of the Jewish Scriptures; on the other hand, it stands in direct continuity with Jesus through the twelve apostles. These apostles may not have been directly involved in the spread of this religion after the opening scenes of the narrative—it is chiefly Paul, who is not one of their number, who takes the Gospel abroad—but they are the ones who bear ultimate responsibility for this mission. They began the process in Jerusalem and continue to guide and direct the church along the paths ordained by God. Moreover, these apostles are in complete agreement on every important issue confronting the church. The church begins with a golden age of peace and unity under the leadership of the apostles.

Evangelistic Speeches: Peter's Speech on the Day of Pentecost

We can now turn to consider several of the speeches delivered by Christians to potential converts. Each speech, of course, has unique elements of its own, but certain basic themes recur throughout them all. Our thematic approach will isolate these recurring motifs in the first evangelistic speech of the narrative, the one delivered by Peter on the Day of Pentecost in chapter 2 (see also the speeches in 3:12–26; 4:8–12, 23–30; 7:1–53; and 13:16–41).

The Pentecost speech immediately follows the coming of the Holy Spirit, an event predicted by Jesus in both Luke and Acts. After the election of Matthias, the followers of Jesus are gathered together in one place when they suddenly hear a sound like strong wind and see tongues like fire alighting on one another's heads. They begin to speak in foreign languages that none of them has previously learned. A large number of Jews from around the world has gathered in Jerusalem for the feast of Pentecost. Crowds descend upon the spirit-filled apostles and their colleagues; the foreigners are shocked to hear "Galileans" speaking to them in their own native languages. Some of the bystanders begin to mock the apostolic band as a group of drunk and rowdy revelers.

This development provides Peter with an occasion to make a speech and an audience to hear it. He declares that what has happened is nothing less than a fulfillment of the plan of God as foretold by the prophet Joel:

> In the last days it will be, God declares, that I will pour out my Spirit upon all flesh, and your sons and your daughters shall prophesy, and your young men shall see visions, and your old men shall dream dreams. (2:17, quoting Joel 2:28)

Peter is particularly emphatic that the Spirit that has come among the believers has been sent by none other than Jesus. The sermon quickly shifts to who Jesus is and to the way in which he can affect a person's standing before God (2:22–36). Here we come to one of the most interesting aspects of the theology of Luke, the author of the speech. Jesus is portrayed here as a mighty man who did fantastic miracles, who was lawlessly executed by evil people but vindicated by God, who raised him from the dead in fulfillment of the Scriptures. After this brief narration of Jesus' story, Peter moves quickly to the climax of his speech: "God has made him both Lord and Messiah, this Jesus whom you crucified" (v. 36).

The point is quite clear. Jesus was the innocent victim of miscarried justice and the people who hear the sermon are themselves to blame, but God reversed their evil action by raising Jesus. The message has its desired effect: cut to the quick, the crowds ask what they should do, that is, how they might make amends for their evil deed. Peter's gives an immediate response: they must repent of their sins and be baptized in the name of Jesus. Those who do so will find forgiveness (vv. 38–39).

As you can see, the way Peter describes Jesus and the salvation he brings corresponds with the views that we found in the Gospel according to Luke. Jesus' death does not bring an atonement (contrast Mark's Gospel). It is a miscarriage of justice. Nor does Jesus' resurrection, in itself, bring salvation. It instead demonstrates Jesus' vindication by God. How then do Jesus' death and resurrection affect a person's standing before God, according to this evangelistic speech in Acts? When people recognize how maliciously Jesus was treated, they realize their own guilt before God—even if they were not present at Jesus' trial. They have committed sins, and the death of Jesus is a symbol of the worst sin imaginable, the execution of the prophet chosen by God. The news of Jesus' death and vindication drives people to their knees in repentance. When they turn from their sins and join the community of Christian believers (through baptism), they are forgiven and granted salvation.

Thus salvation for Luke does not come through the death of Jesus per se; it comes through repentance and the forgiveness of sins. This theme is played out in all of the missionary sermons of Acts. As the Christian preachers emphasize time and again, Jews have a history of disobedience to God, a history that has climaxed in their execution of God's Son Jesus. They must realize how wrong they have been and turn to God to make it right.

Most of the Jews in the book continue to manifest an attitude of disobedience, from Luke's perspective. They not only resist the message of salvation, they actively reject it, by opposing the Christian mission and persecuting the Christian missionaries. The persecution begins in Jerusalem but continues everywhere the message is proclaimed. It leads to the first martyrdom in early Christianity, that of Stephen, following a lengthy missionary speech (chaps. 7–8). Before long, the opposition is headed by Saul of Tarsus (Paul), who participates in Stephen's death but, as we have seen, soon converts to Christianity and becomes its leading missionary.

Paul's conversion does nothing to abate the Jewish opposition to the faith. If anything, it intensifies it. In virtually every city and town that he enters, after experiencing some initial success among Jews in the synagogues, he is violently opposed by Jewish authorities, who drive him out. After making three missionary journeys to Asia Minor, Macedonia, and Achaia, he makes a final fateful trip to Jerusalem (compare Jesus in the Gospel). There he is arrested by the authorities at the instigation of the unbelieving Jews and forced to stand trial, on a number of occasions, for his faith.

Paul's arrest and trials take up a substantial portion of the narrative in Acts (chaps. 21–28; comparable to the space devoted to Jesus' last days in Luke). Much of this final third of the book is devoted to speeches in which Paul defends himself against accusations by Jewish leaders that he has violated the Torah and is a menace to the Empire. By considering some of the themes of these "apologetic" speeches we will see yet further aspects of Luke's overall conception of the early Christian church.

Apologetic Speeches: Paul's Final Appeal to Jews in Rome

Before we examine the themes of the apologetic speeches we should review the basic narrative. Paul is arrested in Jerusalem while making an offering in the Temple, which was meant to show that he was in no way opposed to the Law of Moses (chap. 21). He is taken into Roman custody and allowed to make a defense to the Jewish crowds (chap. 22). He is then made to stand trial before the Jewish Sanhedrin (chap. 23). When the Roman tribune learns of a plot to assassinate him, he has him removed to Caesarea to await trial before the governor Felix (chap. 23). He there makes his defense, but Felix, hoping for a bribe, leaves him in prison for two years (chap. 24). Felix is replaced as governor by Porcius Festus, who also puts Paul on trial. Rather than heeding Paul's plea of innocence Festus chooses to ingratiate himself with the Jewish leaders by offering to let Paul stand trial before them in Jerusalem. Realizing the slim odds of a fair hearing there, Paul demands his rights as a Roman citizen to stand before the emperor himself (chap. 25). Before departing for Rome, Paul has opportunity to speak before the visiting king of the Jews, Herod Agrippa II (chap. 26).

Every time Paul defends himself in these chapters, the ruling authorities have ample opportunity to recognize his innocence. But either because of a desire for a bribe (Felix), or as a favor to the Jewish leaders (Festus), or because of Paul's appeal to Caesar (Festus and Agrippa), nothing is done to release him. He is instead sent to Rome to stand trial before Caesar. On the way, he experiences a number of harrowing adventures at sea, including shipwreck (chap. 27). He miraculously survives, however, and makes it to Rome, where the book ends with him under house arrest for two years. As he awaits trial, he preaches to all who would hear and defends himself against all charges (chap. 28).

As was the case with the speeches to believers and to potential converts, each of the apologetic speeches in Acts has its own orientation and emphasis. Here again, a number of themes recur throughout. One of the shortest speeches of the entire book is delivered to the local Jewish leaders in Rome, who appear before Paul in the final chapter:

> Brothers, though I had done nothing against our people or the customs of our ancestors, yet I was arrested in Jerusalem and handed over to the Romans. When they had examined me, the Romans wanted to release me, because there was no reason for the death penalty in my case. But when the Jews objected, I was compelled to appeal to the emperor—even though I had no charge to bring against my nation. For this reason therefore I have asked to see you and speak with you, since it is for the sake of the hope of Israel that I am bound with this chain. (28:17–20)

Here are sounded the characteristic themes of Paul's apology: *(a)* he has done nothing against the Jewish people or Jewish customs, but on the contrary continues to subscribe in every way to the religion of Judaism; *(b)* he was found to be innocent by the Roman authorities; and *(c)* his current problems are entirely the fault of recalcitrant Jewish leaders. The final theme we have already seen throughout the book of Acts. What might we say about the other two?

Just as Jesus is portrayed as fully Jewish in the Gospel of Luke (see, for example, the early emphasis on the Temple and Jerusalem), and the earliest Christian movement is portrayed as fully Jewish in the opening chapters of Acts (where Christians spend their days in the Temple), so Paul is shown to be devoted to his ancestral traditions even after his conversion. He is a Jewish Christian who does nothing at any time contrary to the Law of Moses. To be sure, he is accused of violating the Law—when he is arrested in chapter 21, he is charged with bringing Gentiles into an area of the Temple reserved for Jews—but Luke goes out of his way to show that the charge is categorically false. Paul's companions in the Temple were Jews. They were fulfilling their sacred vows as prescribed in the Torah. Paul himself was there to pay for these vows and to perform a sacrifice of cleansing. Thus Paul is portrayed here as incontrovertibly Jewish.

This portrayal of Paul is consistent throughout the entire narrative of Acts. Never does Paul renounce his faith in the God of Israel, never does he violate any of the dictates of Torah, never does he spurn Jewish customs or practices. His sole "faults" are his decisions to embrace faith in Jesus and to take his message to the Gentiles. For Paul himself, however, neither his newfound faith nor his Gentile mission compromise his Jewish religion; quite the contrary, these represent fulfillments of Judaism.

Throughout his speeches in Acts, Paul stresses that his new faith is rooted in Jesus' resurrection from the dead (the "hope of Israel," 28:20). Moreover, he insists that belief in the resurrection is the cornerstone of the Jewish religion. For him, failure to believe in Jesus' resurrection results from a failure to believe that God raises the dead. And failure to believe that God raises the dead is to doubt the Scripture, to deny the central affirmation of Judaism. For this reason, according to Paul's speeches, faith in Jesus' resurrection is an affirmation of Judaism, not a rejection of it.

This does not mean that Paul (as portrayed by Luke) maintained that Gentiles have to become Jews in order to belong to the people of God. In fact, Gentiles are allowed to remain Gentiles, and are not compelled to practice circumcision or to keep kosher food laws. For Luke this is far from a rejection of Judaism; throughout this book, Jews like Paul remain Jewish, even after coming to faith in Christ.

Thus, part of Paul's defense in Acts is to show that he has not compromised his Judaism one iota by becoming a believer in Jesus. The other part relates to his standing before the Roman Empire. His opponents claim that he is a dangerous person who must be destroyed. As you might expect, Luke has a different opinion. Indeed, his narrative shows that Paul was innocent of any wrongdoing, just as Jesus was in the Gospel. As Paul himself proclaims in his apologetic speeches, he has violated no laws and caused no problems for the ruling authorities. Problems erupt only because those who hear Paul's proclamation oppose him and create disturbances. As we have seen, in most instances it is Jews who are at fault (interestingly, Luke never portrays these rabble rousers as being punished; for him, it is only the innocent who suffer!) On occasion there are pagans to blame (e.g., see the riot in Ephesus in chap. 17). In no case is Paul himself responsible for any wrongdoing, as even the governors before whom he appears attest. Nonetheless, just as Pilate condemned Jesus to death after declaring him innocent, so the Roman administrators of Acts treat Paul as if he is guilty, knowing full well that he is not.

In one sense, as a prominent spokesperson of the emerging Christian church, Paul represents the entire Christian movement for Luke. Here is one who remained faithful to his Jewish roots and in full compliance with the laws of the state. The narrative of his trials and defenses shows that the disturbances that erupted during the early years of the Christian movement could not be laid on the Christians themselves. They are innocent of all wrongdoing, whether judged by the Torah or by rulers of the empire.

IN SUM: PROMINENT THEMES OF LUKE-ACTS

We have now spent considerable time examining the principal emphases of the two-volume work that scholars call Luke-Acts, exploring the Gospel through the comparative method and the book of Acts through the thematic approach. In doing so, we have isolated a number of important motifs that run through the two works:

- an emphasis on the Jewish origins of Christianity, its fulfillment of the Jewish Scriptures, and its continuity with Judaism
- the portrayal of Jesus as a Jewish prophet, rejected by his own people
- the consequent movement of the religion from the Jews to the Gentiles, with its concomitant geographical shift from the holy city of Jerusalem to the ends of the earth
- the proclamation to Jew and Gentile alike of salvation through the repentance of sins and the forgiveness of God
- the stress that Gentiles who accept this offer of salvation need not adopt all the ways of Judaism
- the delay of the time of the end to make this Christian mission a possibility
- the rightness of this religion in both the divine sense (it came from God in fulfillment of the Scriptures) and the human one (it did nothing to violate Jewish custom or Imperial law)
- the complete unity and harmony of the church as guided by the apostles, who agree on every issue and resolve every problem through the direction of the Spirit
- ultimately, the hand of God directing the course of Christian history behind the scenes, from Jesus' own life and death to the life and ministry of the apostles that he left behind

THE AUTHOR OF LUKE-ACTS AND HIS AUDIENCE

Luke-Acts was written anonymously, but the question of authorship is more complicated here than with Matthew and Mark, for those narratives give no concrete clues concerning the identity of their authors. With Luke-Acts there may be clues. To evaluate them we must address three interrelated questions: What is the evidence that Luke-Acts was written by someone named Luke? Is this evidence convincing? Why does the author's identity matter?

The Identity of the Author

Whereas the other authors that we have studied utilize the third person throughout their entire narratives, the author of Luke-Acts occasionally speaks in the first person. This does not happen in the Gospel of Luke, but it does occur in four passages that describe Paul's journeys in Acts (16:10–17, 20:5–16, 21:1–18, 27:1–28:16). In these accounts, the author speaks not of what "they" (Paul and his companions) were doing but of what "we" were doing.

The natural implication of these passages, at least in the judgment of many readers, is that the author is describing events in which he himself participated. One reason that this might matter has to do with the historical value of Acts as an account of the life and teachings of the apostle Paul. If one of Paul's own companions wrote the book, then surely, according to some scholars at least, it preserves an accurate description of the things that Paul said and did. At the same time, as always happens in the seesaw of scholarly debate, there are other scholars who take a different position. These argue that despite these "we" passages the author of Acts was not one of Paul's companions and that, even if he were, his account would not necessarily be accurate.

Before setting out the pros and cons of each view, we need to look a bit further into the evidence itself. Specifically, how does one get from the presence of these "we" passages in Acts to the conclusion that the author of these books was a companion of Paul named Luke? Most scholars agree that the stress on the Gentile mission in Acts, in which Gentiles don't have to become Jews in order to be Christians, suggests that the author was himself a Gentile (although Paul himself had a similar view, and he was certainly a Jew). The question then arises, do we know of any Gentile companions of the apostle Paul from his own writings? In fact, three such persons are mentioned in the letter to the Colossians, which is attributed to Paul: Epaphras, Demas, and Luke the beloved physician. We know that they are Gentiles because the author names them in Col. 4:14 *after* he has mentioned other companions who were "of the circumcision" in 4:11. The same three are mentioned by name, along with Mark and Aristarchus, in Paul's letter to Philemon (vv. 23–24). Of these three, Demas is mentioned elsewhere as having abandoned Paul at some point (2 Tim 2:10); he would not, therefore, be a likely candidate as the author of Acts. Epaphras is described as the founder of the church of Colossae, a community that is never mentioned in Acts, as one might expect it to have been, had its founder been the author. That leaves Luke. As a medical doctor he would have been literate, and he is mentioned as a close companion of Paul again in 2 Tim 4:11. Could it be that this Gentile physician penned the lengthiest corpus of the New Testament?

For a long time, scholars were convinced that corroborating evidence could be found in the vocabulary used throughout Luke-Acts. It appeared at first glance that the two books used an inordinate number of medical terms (compared to the other writings of the New Testament), indicating, perhaps, that the author was a physician. As it turns out, this impression is altogether false. When scholars actually went to the trouble of comparing the medical terminology with that found in works by other Greek authors of the period, they discovered that "Luke" uses such terms no more frequently than other educated writers of his day.

Now, then, what concrete arguments can be made from the other direction? Is there any evidence against identifying the author of these books as Luke, Paul's Gentile traveling companion? The first thing to point out is that of the three Pauline passages that mention "Luke," two of them occur in books that are widely thought by scholars not to have been written by Paul himself. As we will see in Chapter 22, the vast majority of scholars do not think that Paul himself actually wrote 2 Timothy, and the authorship of Colossians is hotly debated. This means that there is only one certain reference to Luke in Paul's writings, Philemon 24, which neither calls him a Gentile nor identifies him as a physician. There would be no more reason for thinking that this person wrote Luke-Acts than anyone else Paul mentions in any of his letters.

Were the books written by one of Paul's companions, even if we don't know the name of this person? The most important thing to say is that even if they were, this would provide no guarantee of their historical accuracy. We have no way of knowing how long this alleged companion of Paul was with him, whether he knew him well, or, if he did know him well, whether he presented him accurately and fairly. Actually, this final statement is not altogether true—for there is one way of determining whether the portrayal of Paul in Acts is accurate and fair: we can compare what Acts says about Paul with what Paul says about Paul. Unfortunately, when we do so (as we will see in Chapter 17), a number of significant differences emerge—both discrepancies of detail, such as where Paul was at certain times and with whom, and broader discrepancies in the actual teachings of Paul.

Even if one of Paul's companions did write the book, then, there is no guarantee that what he says about Paul is what Paul would have said about himself. For this we need to turn to Paul's own letters. What, though, can we say about the so-called "we" passages of Acts? One curious feature of these accounts is how abruptly they begin and end. The author never says, "Then I joined up with Paul in Philippi, and from there we set out for Thessalonica," or anything like it. Instead, he begins using the first-person pronoun without advance warning, in midstride as it were, and ends using it similiarly. Look for yourself at the first occurrence of its use by reading 16:10–17 carefully. Someone might make sense of the abrupt way the author begins to speak of what "we" did by assuming that he joined Paul immediately before his journey over to Philippi. But how could one explain that the author left Paul's company between the time the possessed slave girl began following them around (v. 17) and the time Paul cast out the evil spirit (v. 18, or perhaps v. 19)?

If it is hard to explain these "we" passages as personal reminiscences by the author of Acts, is there some other explanation for their presence in the book? There are, in fact, plenty of other explanations, but here I will mention only one that is commonly proposed, as a way to help you to begin thinking about the problem on your own. We know that Luke used sources for his Gospel narrative; he tells us so explicitly. Did he not use sources for the book of Acts as well? Yes, certainly he must have. Is it possible that one of his sources was a fragmentary travel diary of some sort, a travelogue that one of Paul's companions had kept, and that Luke simply incorporated it into his document without changing it, just as, in places, he incorporated Mark and Q in his Gospel? This must at least be a possibility and would explain the abruptness with which he begins and ends his use of the first-person pronoun. Perhaps he used the source that he had and wrote a story "around" it.

The Author and His Themes in Context

In some ways, the entire discussion of authorship is irrelevant to the task that we started out to accomplish. Knowing the name of the author of this book, or even knowing that he was a companion of one of its main characters, does not help us very much in trying to understand what he wanted to emphasize about the history of the early Christian church. Conversely, though, discerning the distinctive emphases of the narrative can tell us something about the author and about his audience.

A good place to begin is with some of the observations we made in our discussion of the first volume, the Gospel of Luke. We might ask, for example, why the author of Luke modified Mark's account of Jesus' demeanor in the face of death. Jesus in Luke is portrayed as a kind of ideal martyr for the faith. Throughout the book of Acts as well, Christian leaders face opposition boldly, refusing to bow to the unreasonable demands of those who oppose them. It is possible that these narratives were meant to bolster the confidence and courage of Luke's readers, who themselves confronted hostility in the world around them.

Why does Luke emphasize that the end was not supposed to have come in the lifetime of Jesus' disciples? Obviously because it had not come, and perhaps many or most of Jesus' disciples were already dead. For Luke, though, this clearly was according to plan. The divine purpose of the Christian church was to spread the gospel through the lands of the Gentiles. This, of course, requires time; time itself, therefore, could not come to a screeching halt. By the time Luke was writing, however, the gospel had already been preached to the "ends of the earth," for the book of Acts concludes in Rome, in the heart of the empire, where the gospel was brought by Paul. What more needed be done before the end? Perhaps nothing, for Luke. He and his congregation may have expected to be the last generation before the end.

Luke could provide no absolute assurance of this, however, so he emphasizes to his readers that their ultimate concern should not be with the future but with the present. Thus they should act on the social implications of Jesus' message in the Gospel (by helping the poor and the oppressed) and continue spreading the good news in Acts. This author wants to stress that the delay of the end cannot be used to nullify the truth of the Christian message. It is likely that some nonbelievers in the author's locality were using the delay precisely to this end, by pointing out that Jesus' failure to return in judgment was a sure sign that the Christians had been wrong all along. In opposition to such a view, Luke stresses that God did not mean for the end to come right away. More importantly, he indicates that despite the delay of the end there is good reason to believe that God was and still is behind the Christian mission. Otherwise, from Luke's perspective, it would be impossible to explain the miraculous success of the Christian mission throughout the world. The hand of God was behind this mission, and there was nothing that any human could ever do to stop it.

Finally, we should look at two of Luke's themes that might appear at first glance to be at odds with one another: his emphasis on the Jewish roots of Christianity and his concern for the Gentile mission. Why would Luke focus on Jesus' fulfillment of the Jewish scriptures if he was writing for Gentiles who did not have to become Jews to be Jesus' followers? Why would he stress that Christianity itself was predicted in Jewish texts, if most of the converts to the religion were not Jewish? Why, in short, would Luke situate this increasingly Gentile movement so squarely in the context of Judaism?

One possible answer to these questions lies outside of our investigation of the books per se, in the world in which they were written and read. Even into the late second century of the Common Era, when Christianity had become a distinct religion, separated from Judaism, the intellectual defenders of Christianity—the "apologists," as they were known (see box 8.1)—continued to stress the claims made by Luke, that Christianity was not something new but something old, older even than the Jewish prophets, as old as the author of the Torah and Moses himself. They stressed this claim because of a common notion shared by most persons of the ancient world (whether pagan, Jewish, or Christian) that anything new—an idea, a philosophy, a religion—was automatically suspect. Unlike in the modern age, where creative ideas and new technologies are widely recognized as good (the newer the better!), in the ancient world older was better. There was a strong regard for antiquity in antiquity. This was particularly the case when it came to religion. If a religion was new, it could scarcely be true.

Christians in the Roman world were confronted with a basic problem. Everyone knew that Jesus was crucified under Pontius Pilate when Tiberius was emperor. Even by the second century, Jesus was considered "recent." If something recent is automatically suspect, then a religion based on Christ is in peril. To deal with this problem, the second century apologists appealed to the Jewish roots of the religion, as already stressed, for example, by the Gospel of Luke and (perhaps for a different reason) by the Gospel of Matthew. According to these later authors, Christianity was not a new thing but an old thing. It was predicted by the prophets and anticipated by Moses. As the apologists pointed out, Moses wrote 800 years before the greatest Greek philosopher, Plato, and 400 years before the oldest Greek poets, Homer and Hesiod. If Jesus was predicted by the Jewish prophets and Moses, then the religion he established is very old indeed.

It is at least possible that Luke, a Gentile living in a largely pagan environment, wanted to stress the Jewish roots of Christianity for precisely such reasons. The religion founded on Jesus is ancient; it is a fulfillment of the Jewish Scriptures. It is, in fact the true expression of faith in the God of Israel, whose people the Jews have long disobeyed him and have now done so once too often. Now they have rejected the great prophet of God, God's own Son, whose message of salvation has as a consequence gone forth to the Gentiles.

SUGGESTIONS FOR FURTHER READING

In addition to the works listed at the end of chapter 8, see the following studies.

Hengel, Martin. *Acts and the History of Earliest Christianity*. Trans. J. Bowden. Philadelphia: Fortress, 1980. A detailed study, for advanced students, that argues (in contrast to the present chapter) that the book of Acts for the most part presents a historically reliable account.

Keck, Leander E. and J. Louis Martyn, eds. *Studies in Luke-Acts*. Nashville: Abingdon, 1966. A superb collection of significant essays on Luke and Acts. Especially important is P. Vielhauer, "Paulinisms of Acts," pp. 35–50, a classic study that mounts convincing arguments that the portrayal of Paul in Acts does not coincide with Paul's portrayal of himself.

Powell, M. A. *What Are They Saying about Acts?* New York: Paulist, 1991. An overview of modern scholarship on the book of Acts, for beginning students.

Reardon, B. P., ed. *Collected Ancient Greek Novels*. Berkeley: University of California Press, 1989. A very nice collection of all the ancient Greek novels, useful for comparison with the book of Acts for those who think that Acts contains novelistic features.

CHAPTER 10

Jesus, The Man Sent from Heaven: The Gospel according to John

The Gospel of John has always been one of the most popular and beloved books of the New Testament. It is here that Jesus makes some of his most familiar and yet extraordinary declarations about himself, where he says that he is "the bread of life," "the light of the world," "the good shepherd who lays down his life for his sheep," and "the way, the truth, and the life." This is the Gospel that identifies Jesus as the Word of God "through whom all things were made." It is here that he makes the astonishing claim that "before Abraham was, I am," where he confesses that "I and the Father are one," and where he tells Nicodemus that "you must be born again." And it is in this Gospel that Jesus performs many of his most memorable acts: turning the water into wine, raising his friend Lazarus from the dead, and washing his disciples' feet.

These sayings and deeds, and indeed many more, are found only in the Fourth Gospel, making it a source of perpetual fascination for scholars of the New Testament. Why are such stories found in John but nowhere else? Why is Jesus portrayed so differently here than in the other Gospels? Why, for example, does he talk so much about his own identity in John but scarcely at all in the Synoptic Gospels? And why does this Gospel identify Jesus as God's equal, when none of the earlier Gospels does?

These questions will be at the forefront of our investigation in this chapter. Before beginning our study, however, I should say a word about how we will proceed. Historians are responsible not only for interpreting their ancient sources but also for justifying these interpretations. This is why I have deliberately introduced and utilized different methods for analyzing each of the books we have studied: the literary-historical method for Mark, the redactional method for Matthew, the comparative method for Luke, and the thematic method for Acts. As I have indicated, there is no reason for historians to restrict themselves to any one of these approaches: each could be applied to any one of these books.

To illustrate this point, we will apply all four methods to the Gospel of John. This exercise will show how a variety of approaches can enrich the process of interpretation, but it will also provide us with the data we need to understand yet a fifth method that scholars have used in their study of the early Christian literature, one that might be called the "socio-historical method." In a nutshell, the socio-historical method seeks to understand how a literary text reflects the social world and historical circumstances of the author who produced it. We have already explored this issue with each of the other Gospels, but only in passing. In this chapter we will learn how to pursue the matter with greater rigor and in fuller detail. Since one of the prerequisites for applying this method is a detailed knowledge of the text itself, we can begin by examining the Fourth Gospel from the literary-historical, thematic, comparative, and redactional perspectives.

THE GOSPEL OF JOHN FROM A LITERARY-HISTORICAL PERSPECTIVE

Despite its wide-ranging differences from the Synoptics, the Gospel of John clearly belongs in the same Greco-Roman genre. It too would be perceived by an ancient reader as a biography of a religious leader: it is a prose narrative that portrays an individual's life within a chronological framework, focusing on his inspired teachings and miraculous deeds and leading up to his death and divine vindication.

As was the case with the other Gospels, the portrayal of Jesus is established at the very outset of the narrative, by the introductory passage known as the Johannine Prologue (1:1–18). This prologue, however, is quite unlike anything we have seen in our study of the Gospels to this stage. Rather than introducing the main character of the book by name, it provides a kind of mystical reflection on the "Word" of God, a being from eternity past who was with God and yet was God (v. 1), who created the universe (v. 3), who provided life and light to all humans (vv. 4–5), and who entered into the world that he had made, only to be rejected by his own people (vv. 9–11). John the Baptist testified to this Word (vv. 6–8), but only a few received it; those who did so became children of God, having received a gift far greater even than that bestowed by the servant of God, Moses himself (vv. 12–14; 16–18).

It is not until the end of the prologue that we learn who this "Word" of God was. When the Word became a human being, his name was Jesus Christ (v. 17). Up to this point, that is, through the first eighteen verses of the book, the ancient reader may not have realized that he or she was reading an introduction to a biography. Rather, the prologue appears to be a philosophical or mystical meditation. Beginning with 1:19, however, the book takes on a biographical tone that continues to the very end.

What can we make of the prologue, then, from the literary-historical perspective? Since ancient biographies typically established the character traits of the protagonist at the outset of the narrative, it is perhaps best to assume that an ancient reader, once he or she realized that this book is a biography of Jesus, would be inclined to read the rest of the story in light of what is stated about him in the mystical reflection at the outset. This is no biography of a mere mortal. Its subject is one who was with God in eternity past, who was himself divine, who created the universe, who was God's self-revelation to the world, who came to earth to bring light out of darkness and truth out of error. He is a divine being who became human to dwell here and reveal the truth about God. This Gospel will present a view of Jesus that is far and away the most exalted among our New Testament narratives.

A more complete literary-historical analysis would examine some of the critical incidents that occur early on in the narrative, and perhaps focus on key events that transpire throughout. Here I would like simply to introduce the possibilities of this method for the Fourth Gospel, rather than utilize it at length, and so will summarize the major developments of the plot and indicate something about how the narrative itself is structured.

After the prologue, the Gospel readily divides itself into two major blocks of material. The first twelve chapters narrate events in Jesus' public ministry, which appears to extend over a two- or three-year period (since there are three different Passover feasts mentioned). This section begins with John the Baptist and several of his disciples, who recognize Jesus as one who was specially sent from God. Most of this first section (chaps. 1–12) is devoted to recording Jesus' own declarations of who he is (the one sent from heaven to reveal God) and the miraculous "signs" that he does to demonstrate that what he says about himself is true. Altogether, Jesus performs seven such signs (in chaps. 2, 4, 5, 6, 9, and 11), most of them directly tied to his proclamations (see box 10.1). Thus, for example, he multiplies the loaves of bread and claims that he is the "bread of life" (6:22–40), he gives sight to the blind and says that he is "the light of the world" (9:1–12), and he raises the dead and calls himself "the resurrection and the life" (11:17–44).

Also included in these stories of Jesus' public ministry are several discourses not directly tied to the signs. In these speeches Jesus explains his

SOME MORE INFORMATION

Box 10.1 Jesus' Signs in the Fourth Gospel

The following are the seven miraculous signs that Jesus performs in the Fourth Gospel:

- Turning water into wine (2:1–11)
- Healing the Capernaum official's son (4:46–54)
- Healing the paralytic by the pool of Bethzatha (5:2–9)
- Feeding the 5,000 (6:1–14)
- Walking on water (6:16–21)
- Healing the man born blind (9:1–12)
- Raising Lazarus from the dead (11:1–44)

Jesus performs no other public miracles in John; but notice the statement near the end of the book: "Now Jesus did many other signs in the presence of his disciples, which are not written in this book. but these are written so that you may come to believe that Jesus is the Messiah, the son of God, and that through believing you may have life in his name." (20:30–31)

identity at greater length, for instance to Nicodemus in chapter 3 and to the Samaritan woman at the well in chapter 4. Closely connected to these self-revelations are stories of Jesus' rejection by his enemies, "the Jews" (see box 10.2), and his denunciatory responses in which he castigates those who fail to recognize him as the one sent from God (see chaps. 5, 8, and 10).

The plot of the Fourth Gospel unfolds, then, like this. Jesus proclaims that he is the one sent from heaven to reveal the truth about God, and he does signs to demonstrate that he is who he says he is. Some people accept his message, but most, especially the Jewish leaders, reject it. He condemns their failure to believe, and at the end of the first section, in chapter 12, decides to do no more work among them. From this point on, Jesus removes himself from the public eye, delivering no more self-proclamations to Jewish outsiders and performing no more signs to establish his identity.

Indeed, starting with chapter 13, there is not much time left before Jesus is to return to his heavenly home. Whereas the first twelve chapters stretch over two or three years, chapters 13–19 take place within a single twenty-four hour period. These chapters begin by recounting the events and discussions at Jesus' final meal with his disciples. After he washes his disciples' feet (13:1–20), and announces that he will soon be betrayed (13:21–30), he launches into his longest discourse of the Gospel, commonly known to scholars as the "Farewell Discourse." Here Jesus states that he is soon to leave the disciples to return to the Father; they are not to be dismayed, however, for he will send them another comforter, the Holy Spirit, who will assist and instruct them. When Jesus leaves, his disciples will be hated by nonbelievers in the world, but they are to continue doing his commandments, confident of his presence among them in the Spirit.

This speech consumes more than three chapters. In chapter 17, Jesus offers a final prayer to his Father for his disciples, that they may remain faithful even after he has gone. The rest of the book, chapters 18–21, presents Jesus' Passion and resurrection in stories more or less similar to those

SOME MORE INFORMATION

Box 10.2 "The Jews" in the Fourth Gospel

You will notice in reading through the Fourth Gospel that the phrase "the Jews" is almost always used as a negative term of abuse. The Jews are portrayed as the enemies of Jesus who are consequently opposed to God and aligned with the Devil and the forces of evil (see especially 8:31–59). Vitriolic statements of this kind may sound anti-Semitic to our ears—as indeed, they should. As we will see in Chapter 24, hateful acts of violence have been perpetrated over the years by those who have taken such charges as divine sanctions for oppression and persecution. But we will also see that our modern notion of anti-Semitism may not be appropriate for understanding the meaning of such comments in the early Christian literature.

Despite these harsh statements about Jews in the Gospel of John, even here Jesus and his followers are portrayed as Jews who subscribe to the authority of Moses and participate in the Jewish cult and the Jewish festivals. If Jesus and his followers are Jews, how can all Jews be lumped together and branded as the enemies of God? I will try to answer this question later in the chapter when we consider the Gospel from a socio-historical perspective. At this stage, we need only note "the Jews" is a technical term of disapprobation throughout this narrative; thus, when I refer to John's own comments, I will place the term in quotation marks.

found in the Synoptics. As he predicted, Jesus is betrayed by his own disciple Judas; he is interrogated by the high priest, denied by his disciple Peter, and put on trial before the Roman governor Pilate. At the instigation of his enemies among the Jews, he is condemned to crucifixion. He dies and is buried by Joseph of Arimathea, but on the first day of the week, he is raised from the dead. Chapters 20–21 narrate various appearances to his followers, whom he convinces that he is both alive and divine.

THE GOSPEL OF JOHN FROM A THEMATIC PERSPECTIVE

Whereas the literary-historical approach to the Gospels focuses on the conventions of the biographical genre, and so determines how a book portrays its main character through the unfolding of the plot as he interacts with those around him, the thematic approach isolates prominent themes at key points of the narrative and traces their presence throughout, more or less overlooking questions of plot and character interaction. If we were to examine John from a strictly thematic point of view, we might follow the pattern we established for the book of Acts and look at some of the salient motifs established at the outset in the prologue, and in some of the speeches of the main character.

From a thematic point of view, it is interesting to note that although the prologue identifies Jesus as the Word of God who has become human, he is never explicitly called this anywhere else in the Gospel. Nonetheless, certain other aspects of the prologue's description recur throughout the narrative. For example, just as the Word is said to be "in the beginning" with God, so Jesus later speaks of possessing the glory of the Father "before the world was made" (17:5); just as the Word is said to be "God," so Jesus says "I and the Father are one" (10:30); just as in the Word "was life," so Jesus claims to be "the resurrection and the life" (11:25); just as this life is said to be the "light that enlightens all people," so Jesus says that he is "the light of the world" (9:5); just as the Word is said to have come from heaven into this world, so Jesus maintains that he has been "sent" from God (e.g., 17:21, 25); and just as the Word is said to be rejected by his own people, so Jesus is rejected by "the Jews" (chap. 12), and later unjustly executed (chap. 19).

A full analysis, of course, would look at each of these themes at length. It would also consider other ideas found elsewhere, for example, in Jesus' discourses. These include (a) his first public speech in chapter 3, where he indicates to Nicodemus that only through a birth from "above" can one enter into the kingdom of God and that only the one who comes from above (i.e., he himself, Jesus) can reveal what is necessary for this heavenly birth; (b) his final public speech in chapter 12, where he proclaims that those who have seen him have seen the Father who sent him, whereas those who reject him have rejected God; and (c) his prayer in chapter 17, which more or less functions as his final speech in the presence of his disciples, where he affirms that he has come from God and is now soon to return to him.

If we were to follow this line of inquiry further, one of the interesting observations that we could make is that, contrary to what you might expect, some of the themes of the Fourth Gospel are not developed consistently. Instead, they appear to be understood differently at different points of the narrative. Rather than pursue this issue here, though, we will save it for our discussion of the socio-historical method, which uses divergent thematic emphases of a text to understand the social history lying behind it.

THE GOSPEL OF JOHN FROM A COMPARATIVE PERSPECTIVE

One of the most striking features of the Fourth Gospel is the way in which some of the distinctively Johannine themes stand in such stark contrast to those in the other early Christian writings that we have examined so far. Even to the casual reader, the Fourth Gospel may seem somewhat different from the other three within the canon. Nowhere in the other Gospels is Jesus said to be the Word of God, the creator of the universe, the equal of God, or the one sent from heaven and soon to return. Nowhere else does Jesus claim that to see him is to see the Father, that to hear him is to hear the Father, and that to reject him is to reject the Father. Exactly how different is the Fourth Gospel from the others? The comparative approach seeks to answer this question.

Comparison of Contents

Despite the important and significant differences among the Synoptic Gospels, they are much more similar to one another than any one of them is to John. Suppose we were to list the most significant accounts of the Synoptics. In two of them Jesus is said to be born in Bethlehem, to a virgin named Mary. In all three, his public ministry begins with his baptism by John, followed by a period of temptation in the wilderness by the Devil. When he returns, he begins to proclaim the coming kingdom of God. This proclamation is typically made through parables; in fact, according to Mark's Gospel (4:33–34), this is the only way that Jesus taught the crowds. In addition to teaching, of course, Jesus also performs miracles. In Mark, his first miracle involves the exorcism of a demon. Throughout the first part of his ministry, then, Jesus engages in exorcisms (and other miracles) and teaching, principally in parables. Halfway through these Gospels, he goes up onto a high mountain and is transfigured before his disciples; it is there that he reveals to them his glory. Otherwise, it remains hidden. Indeed, he does not speak openly of his identity in these books (even in Matthew, where it is occasionally recognized), and he commands the demons and others who know of it to keep silent. At the end, he has a last meal with his disciples, in which he institutes the Lord's Supper, distributing the bread ("This is my body . . . ") and then the cup ("This is the cup of the new covenant in my blood . . . "). He afterwards goes out to pray in the Garden of Gethsemane, where he asks God to allow him to forgo his coming Passion. He is then arrested by the authorities and made to stand trial before the Jewish authorities of the Sanhedrin, who find him guilty of blasphemy before delivering him over to the Romans for trial and execution.

These stories make up the backbone of the Synoptic accounts of Jesus. What most casual readers of the New Testament do not realize is that none of them is found in John.

Figure 10.1 Portrayal of Jesus washing the disciples' feet, one of the stories of the Fourth Gospel that does not occur in the Synoptics, from a mosaic in the basilica of Saint Mark in Venice.

Read the text carefully for yourself. There is no word about Jesus' birth in Bethlehem here or about his mother being a virgin (in John, as in Mark, Jesus appears for the first time as an adult). Jesus is not explicitly said to be baptized by John. He does not go into the wilderness to be tempted by the Devil. He does not proclaim the kingdom of God that is coming and he never tells a parable. Jesus never casts out a demon in this Gospel. He does not go up onto the Mount of Transfiguration to reveal his glory to his disciples in a private setting, nor does he make any effort to keep his identity secret or command others to silence. Jesus does not institute the Lord's Supper in this Gospel, nor does he go to the Garden of Gethsemane to pray to be released from his fate. In this Gospel, he is not put on trial before the Sanhedrin or found guilty of committing blasphemy.

If John does not have these stories about Jesus, what stories does he have? The majority of John's stories are unique to John; they are found nowhere else. To be sure, many of the same characters appear in this Gospel: Jesus, some of his family, his male disciples, several female followers, John the Baptist, the Jewish leaders, Caiaphas, Pontius Pilate, and Barabbas. Moreover, some of the same (or similar) stories are found in John and the Synoptics, including, for example, the feeding of the 5,000, the walking on the water, and many of the events of the Passion narrative: Jesus' anointing, his entry into Jerusalem, his betrayal and arrest, the denial by Peter, the Roman trial, and the crucifixion. But most of the events of the Synoptics, except for the Passion narrative, are not found in John, just as, by and large, the words and deeds recorded in John occur only in John. Only here, for example, do we hear of some of Jesus' most impressive miracles: the turning of water into wine (chap. 2), the healing of the lame man by the pool of Bethzatha (chap. 5), the restoration of sight to the man born blind (chap. 9), and the raising of Lazarus from the dead (chap. 11). Only here do we get the long discourses, including the dialogues with Nicodemus in chapter 3, with the Samaritan woman in chapter 4, with his opponents among the Jews in chapters 5 and 8, with his disciples in chapters 13–17. Just in terms of content, then, John is quite different from the Synoptics.

Comparison of Emphases

The differences between John and the Synoptics are perhaps even more striking in stories that they have in common. You can see the differences yourself simply by taking any story of the Synoptics that is also told in John and comparing the two accounts carefully (as we did for the trial of Jesus in Chapter 3). A thorough and detailed study of this phenomenon throughout the entire Gospel would reveal several fundamental differences. Here we will look at two differences that affect a large number of the stories of Jesus' deeds and words.

First, the deeds. Jesus does not do as many miracles in John as he does in the Synoptics, but the ones he does are, for the most part, far more spectacular. Indeed, unlike in the Synoptics, Jesus does nothing to hide his abilities; he performs miracles openly in order to demonstrate who he is. To illustrate the point, we can compare two stories that have several striking resemblances: the Synoptic account of the raising of Jairus's daughter (Mark 5:21–43) and John's account of the raising of Lazarus (John 11:1–44). Read them for yourself. In both, a person is ill and a relative goes to Jesus for help. Jesus is delayed from coming right away, so that by the time he arrives the person has already died and is being mourned. Jesus speaks of the person as "sleeping" (a euphemism for death). Those present think that he has come too late and that now he can do nothing, but Jesus approaches the one who has died, speaks some words, and raises the person from the dead. Both accounts end with Jesus' instructions to care for the person's well-being.

Although the two stories are similar in kind, they differ in the details of how the miracle is portrayed. First of all, in the story in Mark, Jesus is delayed inadvertently; he has an encounter with someone in the crowd, and in the meantime, the young girl dies. In John's Gospel, on the other hand, Jesus intentionally stays away until Lazarus dies (v. 6). Why would he want Lazarus to die? The text of Jesus' words tells us in no uncertain terms: "Lazarus is dead; and for your sake I am glad that I was not there, so that you may believe" (v. 15). In John's Gospel, Lazarus has to die so that Jesus can raise him from the dead and convince others of who he is. As Jesus himself puts it: "This illness . . . is for the glory of God, so that the Son of God may be glorified by means of it" (v. 4).

There is another significant difference between the accounts. In Mark, Jesus heals the girl in private, taking only her parents and three of his disciples with him. In John, Jesus makes the healing a public spectacle, with crowds looking on. We have already discussed why Mark may have wanted to portray Jesus as performing his miracles in secret, but why the publicity in John? A complete study of John would show why: unlike the Synoptics, the Fourth Gospel uses Jesus' miracles to convince people of who he is. Indeed, as Jesus states in this Gospel, "Unless you see signs and wonders, you will not believe" (4:48; see box 10.3).

It is striking that in the Synoptic Gospels Jesus refuses to do miracles in order to prove his identity. When the scribes and Pharisees approach him and ask him to do a "sign" (Matt 11:38), he bluntly refuses, maligning them as sinful and adulterous for wanting a sign when his own preaching, superior to that of Jonah and Solomon (both of whom converted the disbelieving by their proclamations), should suffice. A similar lesson is conveyed through the Synoptic story of Jesus' temptation in the wilderness (drawn from Q; Matt 4:1–11, Luke 4:1–13). As you will recall, at one point Jesus is tempted to jump off the pinnacle of the Temple. A thoughtful reader may wonder why this would be alluring. One can understand why fasting for forty days might make Jesus tempted to turn stones into bread, but why would anyone be tempted to jump off a ten-story precipice? The text itself provides an explanation: if Jesus jumps, the angels of God will swoop down and catch him before he hits bottom. One must assume that the crowds of faithful Jews down below would see this supernatural intervention on Jesus' behalf—this is in the Jerusalem Temple—and so become convinced of who he was. Thus, in the Synoptic temptation narrative, when Jesus is tempted to prove his identity by doing a miracle, he resists the temptation as Satanic.

Neither of these stories—the request for a sign or the account of the temptation—is found in the Fourth Gospel. For in this Gospel, far from spurning the use of miracles to reveal his identity, Jesus

Figure 10.2 The Raising of Lazarus (see John 11), as Portrayed on an Ancient Piece of Painted Glass.

SOMETHING TO THINK ABOUT

Box 10.3 Signs and Faith in the Fourth Gospel

A number of scholars think that Jesus' statement to the Capernaum official in John 4:48, "Unless you see signs and wonders, you will not believe," is meant as a reproach and shows that Jesus was put off because this man needed proof before he would believe, whereas true faith requires no proof. Supporting evidence might be found in John 20:28, where the resurrected Jesus appears to rebuke doubting Thomas on similar grounds: "Have you believed because you have seen me? Blessed are those who have not seen and yet have come to believe."

Some scholars take this way of reading the Gospel a step further. They maintain that the author of the "signs source" from which the author derived his stories of Jesus' miraculous deeds (a source we will be discussing later in this chapter; see also box 10.1) had a rudimentary understanding of the relationship between Jesus' miraculous deeds and faith: Jesus' deeds prove that he is the Son of God. The author of the Fourth Gospel, however, had a more nuanced view of the matter (according to this view). For him, the miracles were not unambiguous proofs, they were literally significant only for those who were open to the truth about Jesus as the one who reveals God. This is why some people could benefit from Jesus' miracles and yet still not understand what they signified (e.g., see, 2:23–25; 3:2–10; 6:26; and 11:45–48).

Other scholars take a different position altogether. For them, Jesus' miraculous deeds in the Fourth Gospel are not irrefutable proofs of his identity but are nonetheless clear and necessary indicators ("signs") of who he is. In other words, the author of the Fourth Gospel believed that no one could come to understand Jesus as the one sent from God without first seeing what he had done. In this way of reading the Gospel, Jesus' statement in 4:48 is not a reproach but a statement of fact: no one will believe without seeing Jesus' signs, for these are the deeds that reveal who he is. This does not mean that everyone who sees these deeds necessarily comes to faith, but everyone who comes to faith has necessarily seen these deeds.

What though about those who were not there to see them? Evidently, for the author of John, such people can come to faith by hearing or reading about these signs. This is why he penned his account in the first place, as seen in the conclusion that he took over from his signs source (and thereby affirmed): "Jesus did many other signs in the presence of his disciples, which are not written in this book. But these are written so that you may come to believe." (20:30–31)

performs them for precisely this purpose. Thus, the Fourth Gospel does not actually call Jesus' spectacular deeds "miracles," which is a Greek word that means something like "demonstration of power" (and is related to our English word "dynamite"); instead it calls them "signs," for they are signs of who Jesus is.

What, then, is the function of the miraculous deeds in the Fourth Gospel? Unlike in the Synoptics, they are done publicly in order to convince people of Jesus' identity so that they may come to believe in him. This purpose is made plain by the words of the Fourth Evangelist himself, in his concluding comment on the significance of Jesus' great deeds: "Jesus did many other signs in the presence of his disciples, which are not written in this book. But these are written so that you may come to believe that Jesus is the Messiah, the Son of God, and that through believing you may have life in his name" (20:30–31).

John's unique understanding of Jesus' miracles is matched by his distinctive portrayal of Jesus' teachings. In the Synoptic Gospels, you will have noticed that Jesus scarcely ever speaks about himself. There his message is about the coming kingdom of God and about what people must do to prepare for it. His regular mode of instruction is the parable. In John, however, Jesus does not speak in parables. nor does he proclaim the imminent appearance of the kingdom. He instead focuses his words on identifying himself as the one sent from God (see box 10.4).

In the Fourth Gospel, Jesus has come down from the Father and is soon to return to him. His message alone can bring eternal life. He himself is equal with God. He existed before he came into the world. He reveals God's glory. Only those who receive his message can partake of the world that is above, only they are in the light, and only they can enter into the truth. Jesus himself is the only way to God: "I am the way, and the truth, and the life. No one comes to the Father except through me" (14:6).

Whereas Jesus scarcely ever talks about himself in the Synoptics, that is virtually all he talks about in John, and there is a close relationship here between what he says and what he does. He says that he is the one sent from God to bring life to the world, and he does signs to show that what he says is true.

In short, John is markedly different from the Synoptics in both content and emphasis and with respect to both Jesus' words and his deeds. As I indicated at the outset, historians must try to explain these different portrayals of Jesus. One of the ways they have done so is to use the socio-historical method. Before looking at how this method works, however, we should see what important features of the Fourth Gospel can be uncovered through a redactional approach.

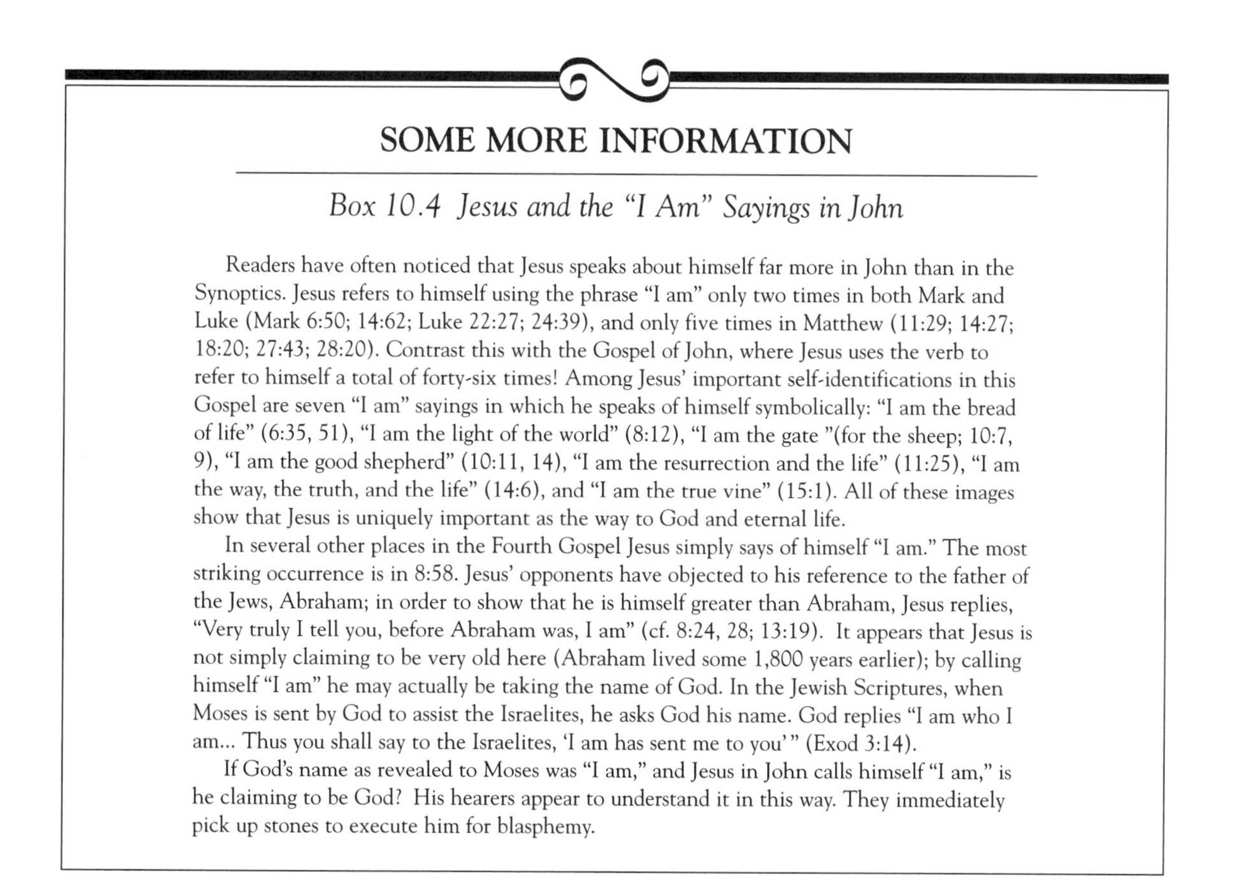

SOME MORE INFORMATION

Box 10.4 Jesus and the "I Am" Sayings in John

Readers have often noticed that Jesus speaks about himself far more in John than in the Synoptics. Jesus refers to himself using the phrase "I am" only two times in both Mark and Luke (Mark 6:50; 14:62; Luke 22:27; 24:39), and only five times in Matthew (11:29; 14:27; 18:20; 27:43; 28:20). Contrast this with the Gospel of John, where Jesus uses the verb to refer to himself a total of forty-six times! Among Jesus' important self-identifications in this Gospel are seven "I am" sayings in which he speaks of himself symbolically: "I am the bread of life" (6:35, 51), "I am the light of the world" (8:12), "I am the gate "(for the sheep; 10:7, 9), "I am the good shepherd" (10:11, 14), "I am the resurrection and the life" (11:25), "I am the way, the truth, and the life" (14:6), and "I am the true vine" (15:1). All of these images show that Jesus is uniquely important as the way to God and eternal life.

In several other places in the Fourth Gospel Jesus simply says of himself "I am." The most striking occurrence is in 8:58. Jesus' opponents have objected to his reference to the father of the Jews, Abraham; in order to show that he is himself greater than Abraham, Jesus replies, "Very truly I tell you, before Abraham was, I am" (cf. 8:24, 28; 13:19). It appears that Jesus is not simply claiming to be very old here (Abraham lived some 1,800 years earlier); by calling himself "I am" he may actually be taking the name of God. In the Jewish Scriptures, when Moses is sent by God to assist the Israelites, he asks God his name. God replies "I am who I am... Thus you shall say to the Israelites, 'I am has sent me to you'" (Exod 3:14).

If God's name as revealed to Moses was "I am," and Jesus in John calls himself "I am," is he claiming to be God? His hearers appear to understand it in this way. They immediately pick up stones to execute him for blasphemy.

THE GOSPEL OF JOHN FROM A REDACTIONAL PERSPECTIVE

As we have seen in our earlier discussions, redaction criticism works to understand how an author has utilized his or her sources. Scholars have successfully used this method with the Gospels of Matthew and Luke, where they have posited two sources with reasonable certainty (Mark and Q). The rationale for using this method is somewhat more tenuous in the case of the Fourth Gospel, since this author's sources are more difficult to reconstruct. Still, John must have derived his stories about Jesus from somewhere (since he evidently didn't make them all up).

One perennial question is whether John had access to and made use of the Synoptic Gospels. The question is somewhat thorny, and we cannot delve into all of its complexities here. Instead, I will simply indicate why many scholars continue to be persuaded that he did not utilize the Synoptics. As we have seen, the principal grounds for assuming that one document served as a source for another is their wide-ranging similarities; when they tell the same stories and do so in the same way, they must be literarily related to one another. Thus Matthew, Mark, and Luke must have sources in common because they agree with one another on a number of occasions, often word for word. This is not the case for the Fourth Gospel. Most of John's stories outside of the Passion narrative are found only in John, whereas most of the stories in the Synoptics are not found in John. If this author had used the Synoptics as sources, why would he have omitted so many of their stories? Or—to put the burden of proof in its proper place—why should someone think that John used the Synoptics as sources when they do not have extensive verbatim agreements, even in the stories that they happen to share?

When thinking about the relationship of the New Testament writings to one another, we must constantly bear in mind that in the ancient world books were not published as they are today. In the modern world, books are mass-produced and sold all over the world, with the distribution of copies taking weeks at the most. In the ancient world, books were copied one at a time and distribution was haphazard at best. In-house literature was not advertised, and circulation was random and uncontrolled. Suppose, for, example that the Gospel of Luke was produced in Asia Minor; Christians in Alexandria may not have heard about it until years later. Or if Matthew was produced in Syria, the Christians of Corinth may not have known of it for decades. Thus there is no guarantee that simply because John was penned some ten or fifteen years after the Synoptics, its author would have known them. On the contrary, given the sizable differences between them, it appears unlikely that he did.

How then can we account for the similar stories that John and the Synoptics tell on occasion? The simplest explanation is that they would have been independently drawn from the oral traditions circulating about Jesus. In different regions of the world, both where there were written accounts about Jesus and where there were not, some of the same stories would naturally have been told. The story of Jesus' Passion is one example. It appears that Christians in many places told of how Jesus was betrayed by one of his own disciples, denied by another, and abandoned by all the rest, and of how he was confronted by the Jewish religious leaders, turned over to Pontius Pilate, and crucified for claiming to be king of the Jews. The similarities between John and the Synoptics in such stories may simply derive from related oral traditions in circulation in their respective communities.

Evidence of Sources in John

Just because John does not appear to have used the Synoptic Gospels as sources, however, does not mean that he did not use other written documents. Indeed, scholars have typically pointed to three pieces of evidence to suggest that he did.

Differences in Writing Style. Every author has a distinctive style of writing. When you are familiar enough with the way someone writes, you are able to recognize his or her work when you see it. For example, if someone were to insert a page of James Joyce into a story by Mark Twain, a careful reader would immediately recognize the difference. Apart from the change of subject matter, the style itself would be a dead giveaway.

Nothing quite so radical occurs with the changes of style in the Fourth Gospel, but there are passages that appear to come from different writers. We have already looked briefly, for example, at the prologue. Scholars have long recognized the poetic character of this passage, which makes it quite unlike the rest of the narrative. Indeed, it appears to be almost hymnic in quality, as if it were composed to be sung in praise of Christ. Notice, for instance, how the various statements about "the Word" are linked together by key terms, so that the end of one statement corresponds to the beginning of the next. This pattern is even easier to see when the passage is read in the original Greek, as a literal translation can show: "In the beginning was the Word, and the Word was with God, and God was the Word . . . in him was life, and the life was the light of humans, and the light shines in the darkness, and the darkness has not extinguished it."

Interestingly, this careful poetic pattern is broken up in the two places where the subject matter shifts away from the Word to a discussion of John the Baptist (vv. 6–8, 15). It may be that the original hymn did not include these verses. You will notice that when they are taken out, the passage flows quite smoothly without a break.

Is it possible that this hymn was written by someone other than the author of the Fourth Gospel, who borrowed it for the beginning of his biographical account of Jesus? Most scholars find this view entirely plausible. Recall that the central theme of the prologue, that Jesus is the Word made flesh, occurs nowhere else in the entire Gospel. This may indicate that whoever composed these opening verses did not produce the rest of the narrative. Thus we may be dealing with different authors.

Repetitions. There are several passages in this Gospel that appear redundant, where similar accounts are repeated in slightly different words. These passages may derive from different sources. For example, chapters 14 and 16 (parts of the Farewell Discourse) are remarkably alike in their key themes. In both chapters Jesus says that he is leaving the world but that the disciples should not grieve because the Holy Spirit will come in his stead; the disciples will be hated by the world, but they will be instructed and encouraged by the Spirit present among them. Why would this message be given twice in the same speech? It may have been repeated for emphasis, but the repetition seems less emphatic than simply redundant. Another explanation might be that the author had access to two different accounts of Jesus' last words to his disciples, which were similar in their general themes but somewhat different in their wording. When he composed his Gospel, he included them both.

The Presence of Literary Seams. The two preceding arguments for sources in John may not seem all that persuasive by themselves. The third kind of evidence, however, should give us pause. Inconsistencies in John's narrative, sometimes called literary seams, provide the strongest evidence that the author of John used several written sources when producing his account.

Authors who compose their books by splicing several sources together don't always neatly cover up their handiwork but sometimes leave literary seams. The Fourth Evangelist was not a sloppy literary seamster, but he did leave a few traces of his work, which become evident as you study his final product with care. Here are several illustrations.

1. In chapter 2, Jesus performs his "first sign" (2:11) in Cana of Galilee by changing the water into wine. In chapter 4, he does his "second sign" (4:54) after returning to Galilee from Judea, healing the Capernaum official's son. The problem emerges when you read what happens between the first and second signs, for John 2:23 indicates that while Jesus was in Jerusalem many people believed in him "because they saw the signs that he was doing." How can this be? How can he do the first sign, and then other signs, and then the second sign? This is an example of a literary seam; in a moment I will explain how it indicates that the author used sources.
2. In John 2:23, Jesus is in Jerusalem, the capital of Judea. While there, he engages in a discussion with Nicodemus that lasts until 3:21. Then the text says, "After this Jesus and his

disciples went into the land of Judea" (3:22). But they are already *in* the land of Judea, in fact, in its capital. Here, then another literary seam. (Some modern translations have gotten around this problem by mistranslating verse 22 to say that they went into the "countryside of Judea," but this is not the meaning of the Greek word for "land.")

3. In John 5:1, Jesus goes to Jerusalem, where he spends the entire chapter healing and teaching. The author's comment after his discourse, however, is somewhat puzzling: "After this, Jesus went to the other side of the Sea of Galilee" (6:1). How could he go to the other side of the sea if he is not already on one of its sides? In fact, he is nowhere near the Sea of Galilee; he is in Jerusalem of Judea.
4. At Jesus' last meal with his disciples, Peter asks, "Lord, where are you going?" (13:36). A few verses later, Thomas says to Jesus, "Lord, we do not know where you are going" (14:5). Oddly enough, several minutes later, Jesus states, "But now I am going to him who sent me; yet none of you asks me, 'Where are you going?' " (16:5)!
5. At the end of chapter 14, after delivering a speech of nearly a chapter and a half, Jesus says to his disciples, "Rise, let us be on our way" (14:31). The reader might expect them to get up and go, but instead Jesus launches into another discourse: "I am the true vine, and my Father is the vinegrower . . . " (15:1). This discourse is not just a few words spoken on the way out the door. The speech goes on for all of chapter 15, all of chapter 16, and leads into the prayer that takes up all of chapter 17. Jesus and the disciples do not leave until 18:1. Why would Jesus say, "Rise, let us go," and then not leave for three chapters?

Readers have devised various ways of explaining these kinds of literary problems over the years, but the simplest explanation is probably that the author decided to weave different written sources into his narrative. To show how this theory works, we can consider the Farewell Discourse. Recall the various problems in this portion of the Gospel: there appears to be a repetition of material between chapters 14 and 16, and there are at least two literary "seams" here, one involving the question of where Jesus is going (13:36; 14:5; 16:5) and the other involving Jesus' injunction for them all to get up and leave (14:31; 18:1).

The theory of sources can solve these problems. Suppose for the sake of argument that the author had two different accounts (A and B) of what happened at Jesus' last meal with his disciples. Suppose further that account A told the stories that are now located in chapters 13, 14, and 18, and account B told the stories found in chapters 15, 16, and 17 (see fig. 10.3). If the author of the Fourth Gospel had taken the two accounts and spliced them together, inserting account B into account A, between what is now the end of chapter 14 and the beginning of chapter 18, this would explain all the problems we have discussed. There is a repetition between chapters 14 and 16 because the author used two accounts of the same event and joined them together. Moreover, Jesus states that "no one asks me, 'Where are you going?'" because in account B, (chaps. 15–17) no one *had* asked him where he was going; the questions of Peter and Thomas were originally found in the other account, (A). Finally, in account A Jesus had said, "Arise, let us go," and he and his disciples immediately got up and went. In the final version of John they do not get up and go for three chapters because account B was interposed between two verses (14:31 and 18:1) that stood together in account A.

Character of the Sources in John

Thus the theory of written sources behind the Fourth Gospel explains many of the literary problems of the narrative. These sources obviously no longer survive, but we can make some inferences about them.

The Signs Source. Some of the seams that we have observed appear to suggest that the author incorporated a source that described the signs of Jesus, written to persuade people that he was the messiah, the Son of God. There are seven signs in the Gospel; it is possible that these were all original to the source. You may recall that seven is the perfect number, the number of God; is it an accident that there were seven signs?

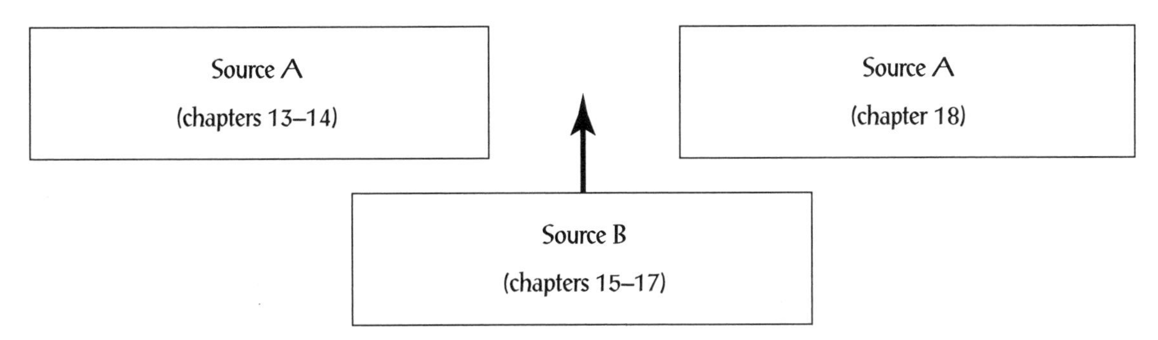

Figure 10.3 Sources in the Farewell Discourse.

The source may have described the signs that Jesus did in sequence and enumerated each one ("This is the first sign that Jesus did," "This is the second sign," and so on). If so, the evangelist kept the first two enumerations (2:11 and 4:54) but for some unknown reason eliminated the others. Keeping the first two, however, left a seam in his narrative, since Jesus does other signs between them (2:23).

The signs source may well have concluded after its most impressive sign, the raising of Lazarus, with the words that are now found in 20:30–31: "Now Jesus did many other signs in the presence of his disciples, which are not written in this book. But these are written so that you may come to believe that Jesus is the Messiah, the Son of God, and that through believing you may have life in his name." The book of signs, then, would have been some kind of missionary tractate designed to convince Jews of Jesus' identity through his miraculous deeds. At some point the events it describes would have been combined with sayings of Jesus that related closely to the things he did. Thus, in John, Jesus not only feeds the 5,000 but also claims to be the bread of life, he not only heals the blind but also claims to be the light of the world, he not only raises the dead but also claims to be the resurrection and the life.

Discourse Sources. Jesus' lengthy speeches in this Gospel appear to have come from a source; indeed, as we have seen, there must have been more than one of them. This, at least, is the best explanation for the literary problems in the Farewell Discourse (chaps. 13–17). The other sayings may derive from the same or similar sources.

Passion Source. Most scholars are persuaded that John's Passion narrative (chaps. 18–20) derives from a source that was similar in many ways to the narrative that is found in Mark. It is difficult to know, however, whether the source was written or oral.

Other Sources. We have already seen that the prologue to the Gospel appears to have been derived from a source, possibly an early Christian hymn to Christ. Something similar can be said of the last chapter, in which Jesus makes a final appearance to several of his disciples after his resurrection (he had already appeared to them in chapter 20). An earlier edition of the Gospel appears to have ended with the words I have just quoted from 20:30–31, which certainly sound like the ending of a book. The final chapter was added later to record one other incident of significance to the author (see box 10.5). It is here that Jesus indicates that Peter will be martyred for his faith and where he is mistakenly understood to say that the unnamed "beloved disciple" will not die prior to his own return.

THE SOCIO-HISTORICAL METHOD

Now that we have examined the Fourth Gospel in light of all of the other methods of analysis that we have learned, we are in a position to explore yet another approach that scholars have taken in studying the New Testament narratives. The socio-

SOMETHING TO THINK ABOUT

Box 10.5 The Death of the Beloved Disciple in the Johannine Community

John 21:21–3 preserves an interesting conversation between the resurrected Jesus and Peter. When Peter asks him about the unnamed "beloved disciple," Jesus responds, "If it is my will that he remain until I come, what is that to you? Follow me." The author goes on to explain that some people misunderstood Jesus' words as a promise that this disciple would not die before Jesus returned from heaven at the end of the age, but that, in fact, Jesus had not explicitly said this. Why would the author of this story want to correct this misunderstanding?

In the opinion of some scholars, it was because some members of the Johannine community had expected that their beloved leader, this unnamed disciple, would not die before the coming of the end. When he did, they were thrown into confusion. Had the Lord gone back on his promise? This author constructs the story to explain that Jesus never had said "that he would not die" (21:23). If this interpretation is correct, then the Gospel would have been published in its final form, with the addition of chapter 21, only after the death of the beloved disciple, and probably after the martyrdom of Peter as well (see 21:18–19).

historical method asks an entirely different set of questions from those we have already addressed, but it bases these questions, and their answers, on the kinds of information that we have just uncovered in our study. We have seen that the author of the Fourth Gospel created a Greco-Roman biography of Jesus based on a number of written and oral sources that were available to him. We have examined some of the important themes in his final product and have seen how these themes differ from those found in other early Gospels.

I have hinted, though, that the themes found in the Fourth Gospel are not always internally consistent, that is to say, there appear to be several different perspectives embodied here, rather than only one. This should come as no surprise given what we have seen about the sources of this book. The author utilized earlier accounts written by other authors, and no doubt each author had his or her own perspective on Jesus and the meaning of what he said and did. By adopting a variety of sources, the author necessarily incorporated a range of views about Jesus.

Different people have different ways of looking at the world and of interpreting important events, and not only because they have different personalities and different brains. People also look at the world differently because they have experienced it differently. The average New Yorker and the average Muscovite had very different perceptions of the cold war, in no small measure because their experiences of it were so different. Accounts of World War II written by American, German, and Russian soldiers might contain similar information, but each would be slanted differently, depending on the perspective of the author, as derived from their personal experiences.

Investigators using the socio-historical approach to a text are interested in knowing how the historical experiences of an author and his or her social group (e.g., a family, a church, army, nation, or any other group of persons who are united together under some conditions) affected the presentation of the material. They focus on the relationship between a literary text and the social history of its author and his or her community.

The theory behind the method can be stated simply: the social history of a community will affect the way it preserves its traditions. Let me illustrate the theory with a modern example, before applying it to the traditions about Jesus preserved in the Fourth Gospel. On any given Sunday, thanks to the use of a standardized lectionary in many Christian denominations, church congregations around the globe read the same passage of Scripture and hear sermons based on these passages. Even within the same city, different churches hear different kinds of sermons, despite the fact that the Scriptural passages are the same. These differences relate not only to the personality and training of the preachers but also to the life experiences of the audiences that they are addressing. To take an obvious example, someone in a black church in Soweto, South Africa, in the 1980s, when apartheid was official policy, would have heard a very different kind of sermon from someone in a white upper-class church in suburban America. For preachers attempt to relate a biblical text to the experiences of their communities, to show how it continues to speak to them in their struggles, whatever these might be.

Theoretically, it would be possible to listen to a set of sermons from an unknown church and reconstruct aspects of the congregation's social context on the basis of what was heard. For instance, if a sermon offers divine solace to those who suffer under the oppressive policies of a powerful minority, one might reasonably assume that the congregation has experienced such policies and requires such solace. If a sermon on the same text challenges the complacency of those who feel secure and who have no care for the downtrodden, one might conclude, depending on what else is said, that it was delivered to a relatively affluent congregation as a call for them to heed their Christian duties. Thus there is a close interrelationship between an author's social experiences and the text (in this case, the sermon) that he or she produces.

What if we do not have direct access to these social experiences, but only to the text? Then if we want to learn something about the underlying social history we have no recourse but to use the text itself, reasoning backwards from what it says to the social experiences that it appears to presuppose. This is obviously a tricky business, but it can yield some interesting results if done carefully. As with all of the other methods we have examined, it is much easier to show how the method works in practice than to explain it in the abstract. When applied to the Fourth Gospel, the method works like this: We have reason to think that there were several sources lying behind this author's account. These sources must have come from different periods in the community's history, since all the authors would presumably not be writing at precisely the same moment. Moreover, in some important aspects these sources have different ways of understanding their subject matter. It is at least possible that the social experiences of the authors who produced these sources contributed to their distinctive understandings. If so, then it is also possible, in theory, for us to analyze the sources of the Fourth Gospel in order to trace the social history of the community of the authors who produced them.

THE GOSPEL OF JOHN FROM A SOCIO-HISTORICAL PERSPECTIVE

To begin we should examine the different thematic emphases in the stories of John, which ultimately may derive from different sources. We know that one of the distinctive features of this Gospel is the exalted view of Jesus, which is emphasized in so many of its narratives. But you may have noticed in your own reading of the Gospel that not every story shares this exalted perspective. In fact, a number of John's stories portray Jesus, not as an elevated divine being come from heaven, but as a very human character. To use the jargon employed by historians of Christian doctrine, portions of this narrative evidence a "high" christology, in which Jesus is portrayed as fully divine, and others show a "low" christology, in which he is portrayed as human, and nothing more.

In the modern world, many Christians subscribe to both a high and a low christology, in which Jesus is thought to be both fully divine and fully human. Did both of these perspectives develop simultaneously, so that the earliest Christians already thought of Jesus as God and man? In the Synoptic Gospels, even though Jesus is portrayed

Figure 10.4 Two portrayals of Jesus as the Good Shepherd (see John 10), one from an ancient Christian sarcophagus and the other from a sixth-or seventh-century mosaic in the Basilica of Saint Apollinaris on the outskirts of Ravenna, Italy.

somewhat as a Hellenistic divine man, like Apollonius of Tyana, for example, there was no sense there that he had existed in eternity past, that he was the creator of the universe, or that he was equal to the one true God. Scholars have long recognized that the notion of Jesus' divinity may have developed over a period of time; as Christians began to reflect more and more on who Jesus was, they began to ascribe greater and greater honors to him. Indeed, in the Fourth Gospel we are able to trace the development of christology within one particular community, from its early reflections of Jesus as a human chosen by God to fulfill the task of salvation to its later conclusion that Jesus was himself divine and the full equal with God. This development appears to have been intimately related to the social experiences of the community that told the stories.

Divergent Christologies in the Johannine Community

An interesting example of an account that embodies a low christology comes in the story of the first disciples in 1:35–42. We are probably justified in supposing that the story was in circulation prior to the writing of the Fourth Gospel, and that the author of this Gospel heard it (or read it) and incorporated it into his narrative after the prologue, which he derived from a different source. In what social context would the story have been told originally?

You will notice that Jesus is called three different things in this account: John the Baptist calls him "the lamb of God" (v. 36), the disciples who follow him call him "rabbi" (v. 38), and one of them, Andrew, calls him the "messiah" (v. 41). Each of these terms makes sense as an identification of Jesus within a Jewish context. As we have seen, the "lamb of God" refers to the Passover lamb that was sacrificed in commemoration of the exodus from Egypt; for John, Jesus is the lamb because his death brings about the salvation celebrated in the Passover meal (see Chapter 3). The term "rabbi" was a common designation for a Jewish teacher, and the term "messiah" referred to the future deliverer of the people of Israel.

None of these terms suggests that the author of

this story understood Jesus to be divine in any way. Neither passover lambs nor rabbis were divine, and the messiah was a human chosen by God, not God himself. Moreover, these are terms that would make sense to a Jewish, rather than to a Gentile, audience. What might this tell us about the social context within which a story like this was told? Here is an account of two Jews who come to Jesus and discover that he is the one they have been waiting for, the messiah. It appears to be the kind of story that would have originally been told by Jews to other Jews, to show them that Jesus is to be recognized as the Jewish messiah (and a rabbi, and the lamb of God).

One other feature of this story should be noted. On three occasions the author explains the terms that he uses; he interprets "rabbi" as "teacher" (v. 38), "messiah" as "Christ" (v. 41), and "Cephas" as "Peter" (v. 42). These interpretations are necessary because the three terms are not Greek, the language of the Fourth Gospel, but Aramaic. Why would some of the key terms of the story be in Aramaic, and why would the author have to translate them? Perhaps the most likely explanation is that the story was originally told in Aramaic; when it was eventually translated into Greek, several of its important terms were left in the original language, as sometimes happens, for example, with a punch line when an anecdote is told to a bilingual audience. The author of the Fourth Gospel, who incorporated the story into his account, realized that his readers (or at least some of them) did not know Aramaic, and so he translated the terms for them.

If this reconstruction of events is correct, then the story would be very old by the time it came to the author of the Fourth Gospel. It would have originally been told among Aramaic-speaking Christians converted from Judaism, presumably those living in Palestine, perhaps not too distant in time from Jesus himself. The story is about how Jesus fulfills the expectations of Jews, and it is designed to show how Jews might come to believe in him as the messiah. There is nothing in this story, however, to suggest that he is divine.

There are other stories, however, in which Jesus *is* portrayed as divine, and in which this is the single most important thing to know about him. His divinity, for example, is one of the leading points of the prologue. In addition, the prologue, along with many other stories in the Gospel, gives no indication of being originally composed in Aramaic. Thus, the prologue might not be as old as the story of the call of the first disciples. Moreover, the prologue, and other stories like it, do not have the kind of friendly disposition toward the Jews that we find here in the account of the call of the disciples (e.g., see 1:11).

How does one explain these thematic differences among the stories of John? Social historians would argue that the history of the community affected the ways that people told the stories about Jesus and that critical events in this history led to changes in the community's understanding of Jesus and his relationship to the people to whom he came. Scholars who have developed this idea have traced the community's history through three stages.

The History of the Johannine Community

Stage One: In the Synagogue. The oldest stories of the Fourth Gospel appear to indicate that the Johannine community originated as a group of Jews who came to believe that Jesus was the messiah and who nonetheless continued to maintain their Jewish identity and to worship in their Jewish synagogue. We do not know where exactly this community was originally located, it may have been someplace in Palestine where Aramaic was spoken.

The evidence for these historical conclusions comes from our only source of information, the Gospel of John itself. Some of John's stories emphasize Jesus' Jewishness and narrate how some Jews came to identify him as the Jewish messiah. Since this identification of the messiah would have been of no interest to pagans (it's a reference to the deliverer of Israel), it makes sense that the stories would have been told within Jewish communities. Since the stories presuppose knowledge of Jesus' own mother tongue, Aramaic, they appear to be among the most ancient accounts of the Gospel.

The Johannine community of Jewish believers may have owed its existence to a follower of Jesus whom they later called "the beloved disciple." This enigmatic figure appears several times in the course of the Gospel and appears to have enjoyed a position of prominence among those who told the stories (e.g., see John 13:23; 19:26–27; 20:2–8).

It appears that these Jewish converts attempted to proselytize other members of their Jewish synagogue. Evidence for this hypothesis is found not only in such stories as the call of the disciples, which presumably would have been told in order to show how some Jews had recognized Jesus as their messiah, but also, perhaps, in the signs source. You may recall the theory that this source ended with the words now found in 20:30–31: "Jesus did many other signs in the presence of his disciples, which are not written in this book. But these are written so that you may come to believe that Jesus is the Messiah, the Son of God, and that through believing you may have life in his name." The purpose of the signs source, in other words, was missionary. It recorded the miraculous deeds of Jesus to convince Jews that Jesus was the messiah. Originally, then, the signs were not designed to show that Jesus was God; they indicated that he was empowered by God as his representative. Jesus was still understood to be a special human being at the stage of the community's history in which the stories were first told, but he was not yet thought of as divine.

Stage Two: Excluded from the Synagogue. It is impossible to say how long the Jews of this community remained in their synagogue without causing a major disturbance. What does become clear from several of the stories of the Fourth Gospel is that a significant disruption eventually took place in which the Jews who believed in Jesus were excluded from the synagogue. There is no indication of exactly what led to this exclusion, but it is not difficult to paint a plausible scenario. First-century Jews by and large rejected any idea that Jesus could be the messiah. For most of them, the messiah was to be a figure of grandeur and power, for example, a heavenly being sent to rule the earth or a great warrior-king who would overthrow the oppressive forces of Rome and renew David's kingdom in Jerusalem. Jesus was clearly nothing of the sort. On the contrary, he was an itinerant preacher who was executed for treason against the state.

So long as the Jews who believed in Jesus kept a low profile, keeping their notions to themselves, there was probably no problem with their worshipping in the synagogue. From its earliest days, however, Christianity was a missionary religion, dedicated to converting others to faith in Jesus. In the Johannine community, as in most other Jewish communities, the Christians were no doubt rejected by the majority of the Jews and probably mocked and marginalized. This may have led, on the one hand, to increased antagonism from non-Christian Jews and, on the other, to heightened efforts at evangelism on the part of the Christian Jews. Eventually, these believers in Jesus became something more than a headache. Perhaps because of their persistent badgering of the skeptical and their refusal to keep their views to themselves, or perhaps for some other unknown reason, this group of believers in Jesus was forced to leave the Jewish community.

There is some evidence within the Gospel of John itself that the Jewish Christians within the synagogue were at some point forced to leave. Several scholars have found the most compelling piece of evidence in the healing story of John 9. In this account, Jesus heals a man who had been born blind. The Jewish authorities take umbrage at this action because it occurred on the Sabbath. They interrogate the man who has been healed, trying to learn how he gained his sight. When he identifies Jesus as the one who healed him, they refuse to believe it and call in his parents to uncover the truth. His parents, however, refuse to answer their questions, insisting that since he is of age they should ask the man himself. Then the author explains why the man's parents refuse to cooperate, in one of the most intriguing verses of the entire Gospel: "His parents said this because they were afraid of the Jews; for the Jews had already agreed that anyone who confessed Jesus to be the Messiah would be put out of the synagogue" (9:22).

This verse is significant from a socio-historical perspective because we know that there was no official policy against accepting Jesus (or anyone else) as messiah during his lifetime. On the other hand, some Jewish synagogues evidently did begin to exclude members who believed in Jesus' messiahship towards the end of the first century. So the story of Jesus healing the blind man reflects the experience of the later community that stood behind the Fourth Gospel. These believers in Jesus had been expelled from the Jewish community, the community, presumably, of their families and friends and neighbors, in which they had worshipped God and had fellowship with one another.

This expulsion from their synagogue had serious implications for the Christian community's social life and for the way it began to understand its world and its stories about its messiah, Jesus.

Stage Three: Against the Synagogue. Sociologists have studied a number of religious communities that have been excluded from larger social groups and forced to carry on their communal activities on their own. The findings of these various studies are of some interest for understanding how the views of the Johannine community appear to have developed with the passing of time.

Religious groups (sometimes called "sects") that split off from larger communities often feel persecuted, many times with considerable justification, and build ideological walls around themselves for protection. A kind of fortress mentality develops, in which the small splinter group begins to think that it has been excluded because those of the larger society are willfully ignorant of the truth, or evil, or demonically possessed. There can arise a kind of "us versus them" mentality, in which only those on the inside are "in the know" and stand "in the light." On the outside, in the large community that has excluded them, there is only falsehood and error; to dwell there is to dwell in the darkness.

The later traditions embodied in the Gospel of John appear to be rooted in such dualities of truth versus error, light versus darkness, the children of God versus the children of the devil, the followers of Jesus versus "the Jews." This latter phrase has puzzled readers of the Gospel over the years. How can the enemies of Jesus so consistently be called "the Jews"? Weren't Jesus and his own followers Jews? How then can all of the Jews be bad?

The answer appears to lie in the experiences of the Christian community at the time. Even though its members had originally been drawn from the Jewish community, most Jews in the local synagogue had by and large rejected their message. The synagogue therefore became the enemy and took on a demonic hue in their eyes. Why had its members so thoroughly and vigorously rejected the message of Jesus? In the view of the Johannine Christians, it must have been because they were alienated from the truth and could not understand it even if they heard it. Jesus was the representative of God, and the enemies of God could not possibly accept his representative. Indeed, the message of Jesus was so thoroughly divine, so completely focused on things of heaven, that those whose minds were set on things of this world could not perceive it. Jesus was from above, and those who recognized only the things of this earth could not perceive him (3:31–36).

Thus, it appears that the christological focus of this community shifted radically after its exclusion from the synagogue. Jesus, to be sure, was still thought of as a rabbi, as the lamb of God, and as the messiah, but he was much more than that. For these excluded Christians, Jesus was unique in knowing about God; he was the one who brought the truth of God to his people. How did he know this truth? The community came to think that Jesus knew God because he had himself come from God. He was the man sent from heaven, come to deliver the message of God to his people before returning to his Father. Only those who ultimately belonged to God could receive this truth; only those who were born "from above" could enter into God's kingdom (3:3).

The social context of exclusion from the synagogue thus led these Johannine Christians to see Jesus as something more than a man representing God or as one sent to deliver God's message. He came to be understood as the embodiment of that message itself. Jesus was himself God's Word. As his Word, he had existed with God from the beginning and was himself God, in a sense. He was God's equal, existent from eternity past, who became human to communicate God's truth to his own. Those who saw him saw the Father, those who heard him heard the Father, and those who rejected him rejected the Father.

In the later stages of the Johannine community a number of memorable stories, and redactions of earlier stories, came to be told, such as the stories in which Jesus claims, "Before Abraham was, I am" (8:58) and "I and the Father are one" (10:30). Also, at some point in its later history, someone within this Christian community composed a hymn to Christ as the Word of God become flesh: "In the

SOMETHING TO THINK ABOUT

Box 10.6 John's De-Apocalypticized Gospel

We have already seen that Luke's Gospel tones down the apocalyptic character of Jesus' proclamation, as it is found, for example, in the Gospel of Mark. In John's Gospel, the apocalyptic message is toned down even more. For John, eternal life is not a future event. As the author puts it early on in the narrative, using the present tense: "Whoever believes in the Son has eternal life" (3:36). Eternal life in this Gospel does not come at the end of time, when the Son of Man arrives on the clouds of heaven and brings in the kingdom. Eternal life is here and now, for all who believe in Jesus. That is why Jesus does not deliver an "apocalyptic discourse" in this Gospel (cf. Mark 13) or speak about the coming Son of Man or the imminent kingdom of God. The kingdom of God is entered by those who have faith in Jesus, in the present (cf. 3:3).

That a person's standing before God is determined not by the future resurrection, but the present relationship with Jesus is illustrated by John's account of the dialogue between Jesus and Martha in the story of Lazarus. Jesus informs Martha that her brother will rise again (11:23). She thinks he is referring to the resurrection at the end of time, and agrees with him (11:24), but he corrects her. He is referring to possibilities in the present, not the future. "I am the resurrection and the life. Those who believe in me, even though they die, will live, and everyone who lives and believes in me will never die" (11:25–26).

In Chapter 15 we will see that Jewish apocalyptists maintained a dualistic view of the world, in which this age belonged to the forces of evil whereas the age to come belonged to God. In John's Gospel this dualism does not have a temporal dimension (this age and the future age) but a spatial one (this world and the world that is above). Those who are from the world that is above belong to God, those from below belong to the Devil. How does one belong to the world that is above? By believing in the one who has come from that world, Jesus (3:31). Thus, in this Gospel Jesus' proclamation is no longer an apocalyptic appeal to repent in the face of a coming judgment; it is an appeal to believe in the one sent from heaven so as to have eternal life in the here and now. John, in short, presents a de-apocalypticized version of Jesus' teaching. (For a remnant of the older apocalyptic view, found even here, see 5:28–29.)

beginning was the Word, and the Word was with God, and the Word was God. All things came into being through him, and without him not even one thing came into being. In him was life, and the life was the light of all people. And the Word became flesh and lived among us, and we have seen his glory" (1:1–14). The author of the Fourth Gospel eventually attached this moving hymn to his narrative, providing a prologue that explained his understanding of Jesus, as narrated in the various stories that he had inherited from his tradition.

THE AUTHOR OF THE FOURTH GOSPEL

Like Mark, Matthew, Luke, and Acts, the Gospel of John was written anonymously. Since the second century, however, it has been customarily attributed to John the son of Zebedee, commonly thought to be the mysterious "beloved disciple."

The idea that one of Jesus' own followers authored the book has traditionally been based on a couple of comments made in the text itself: (a) the reference to an eyewitness who beheld the water and blood coming from Jesus' side at his crucifixion (19:35) and (b) the allusion to the beloved disciple as the one who bore witness and wrote about these things (21:24).

Most modern scholars question, however, whether these verses should be taken to indicate that the beloved disciple authored the Gospel. For example, 19:35 says nothing about who actually wrote down the traditions, but only indicates that the disciple who witnessed Jesus' death spoke the truth ("He who saw this has testified so that you also may believe. His testimony is true, and he knows that he tells the truth"). Furthermore, 21:24 indicates that, whoever this disciple may have been, it was someone other than the author of the final form of the book. Notice how the verse differentiates between the "disciple who is testifying to these things and has written them" and the author who is describing them: "we [i.e., someone other than the disciple himself] know that his testimony is true."

Some of the traditions of this Gospel, then, may ultimately go back to the preaching of one of the original followers of Jesus, but that is not the same thing as saying, that he himself wrote the Gospel. Could this unnamed disciple have been John the son of Zebedee? One of the puzzling features of this Gospel is that John is never mentioned by name here. Those who think that he wrote the Gospel claim that he made no explicit reference to himself out of modesty. Not surprisingly, those who think he did not write it argue just the opposite, that he is not named because he was an insignificant figure in Jesus' story for the members of this community. Indeed, the evidence could probably be read either way. For what it is worth, the book of Acts suggests that John, the son of Zebedee, was uneducated and unable to read and write (the literal meaning of the Greek phrase "uneducated and ordinary"; Acts 4:13).

In any event, it should be clear from our analysis that the Fourth Gospel was probably not the literary product of a single author. Obviously, one person was responsible for the final product, but that person, whoever he or she was, constructed the Gospel out of a number of preexisting sources that had circulated within the community over a period of years. The author appears to have been a native speaker of Greek living outside of Palestine. Since some of the traditions presuppose a Palestinian origin (given the Aramaic words), it may be that the community relocated to a Greek-speaking area and acquired a large number of converts there at some point of its history. Whether the author accompanied the community from the beginning or was a relative latecomer is an issue that can probably never be resolved.

SUGGESTIONS FOR FURTHER READING

Brown, Raymond. *The Community of the Beloved Disciple*. New York: Paulist, 1979. A superb and influential study that uses a socio-historical method to uncover the history of the community behind the Fourth Gospel.

Culpepper, Alan. *The Anatomy of the Fourth Gospel: A Study in Literary Design*. Philadelphia: Fortress, 1983. A study of John from a literary perspective, in which the history of the tradition is bypassed in order to understand the Gospel as a whole as it has come down to us; for somewhat more advanced students.

Kysar, Robert. *John the Maverick Gospel*. Atlanta: John Knox, 1976. All in all, probably the best introduction to the unique features of John's Gospel for beginning students.

Martyn, J. Louis. *History and Theology in the Fourth Gospel*. 2d ed. Nashville, Tenn.: Abingdon, 1979. A fascinating and highly influential study of the social history that led to the development of John's traditions of Jesus.

Sloyan, Gerard S. *What Are They Saying about John?* New York: Paulist, 1991. A very nice introductory sketch of the modern scholarly debates concerning major aspects of John's Gospel.

Smith, D. Moody. *John among the Gospels: The Relationship in Twentieth-Century Research*. Minneapolis, Minn.: Fortress, 1992. A very clear discussion of the relationship of John and the Synoptics, as seen by scholars of the twentieth century.

Smith, D. Moody. *The Theology of John*. Cambridge: Cambridge University Press, 1994. A clearly written and incisive discussion of the major themes of the Fourth Gospel, for beginning students.

CHAPTER 11

From John's Jesus to the Gnostic Christ: The Johannine Epistles and Beyond

Three other books in the New Testament stem from the same community as the Fourth Gospel. The Johannine epistles are located near the end of the New Testament among the other "general epistles." These epistles are called "general" or "catholic," from a Greek word that means "universal," in part because they were traditionally thought to address general problems experienced by Christians everywhere, as opposed to the Pauline epistles, which were directed to particular situations. As will become increasingly clear, however, this classification is not particularly apt: each of the general epistles also deals with specific problems of specific communities.

Nowhere is this more evident than with 1, 2, and 3 John. These books are particularly important for our study of early Christianity because they address members of the Johannine community some time after the Gospel was produced. Just as we can use that earlier writing to reconstruct the history of the community from its early days down to the penning of the Gospel, we can also use the epistles to determine some of the key events that transpired subsequently

THE QUESTIONS OF GENRE AND AUTHOR

As we have already seen, well over half of the New Testament writings (seventeen out of twenty-seven books) are epistles. An epistle is a letter, that is, a piece of private or public correspondence sent through the ancient equivalent of the mail. Usually this involved having someone agree to hand deliver the letter, either a person sent specially for the task or someone known to be traveling in the right direction.

Letters were a common form of written communication in the ancient world, and people wrote a number of different kinds, as can be seen in the thousands of samples that have survived from antiquity. Some letters were collected and published by famous authors like Cicero, Seneca, and Pliny the Younger. Others were written by private and otherwise unknown individuals and discarded by their recipients, only to be discovered in modern times by archaeologists who make a living out of digging through ancient trash heaps buried in the sands of Egypt.

In the modern world, different kinds of letters require different kinds of writing conventions. A cover letter that you send with your resume to a prospective employer will look very different from a letter that you send home from school or a note that you dash off to a boyfriend or girlfriend. Likewise, in the ancient world, private letters to friends differed from open letters to be read by everyone; letters of recommendation differed from literary letters discussing important topics for educated audiences, and public letters persuading a community to engage in a certain course of action differed from private letters to governmental officials that petitioned for a particular cause.

Private letters in the ancient world, unlike modern ones, generally began by identifying the person writing the letter, either by name or, in rare cases, by some other descriptive term (see 3 John 1). This identification was followed by an indication, usually by name, of the person being addressed. Normally the author included some form of greeting and well-wishing at the outset, perhaps a prayer on the recipient's behalf and an expression of thankfulness to the gods for them. In interpreting ancient letters, these introductory conventions cannot be taken too literally as expressing the author's real feelings, any more than modern conventions (for example, we use "Dear" in formal business letters).

After these introductory items would come the body of the letter, in which the subject of the letter and the author's concerns were expressed. Finally, the letter might conclude with some words of encouragement or consolation or admonition, an expression of hope of being able to see one another face to face, greetings to others in the family or community, a farewell, and sometimes a final prayer and well-wishing (see box 11.1).

The letters of the Johannine community that have made it into the New Testament are not nearly so difficult to read as the Gospels. The epistles of 2 John and 3 John take up only a page each, about average for most letters from the ancient world. These, in fact, are the two shortest books of the entire New Testament. One of the first things that may strike you as you read these two letters is that they make full use of the standard conventions of letters that I have just mentioned. There is therefore little doubt that these two books are actually letters, that is, hand-delivered pieces of correspondence. The letter of 2 John is written by someone who identifies himself as "the elder" to a mysterious person called "the elect lady." In the course of his letter, however, the author stops speaking to this "lady" and begins to address a group of people ("you" plural, starting in v. 6). This shift has led most scholars to assume that the term "elect lady" refers to a Christian community, a group of people who are considered to be the

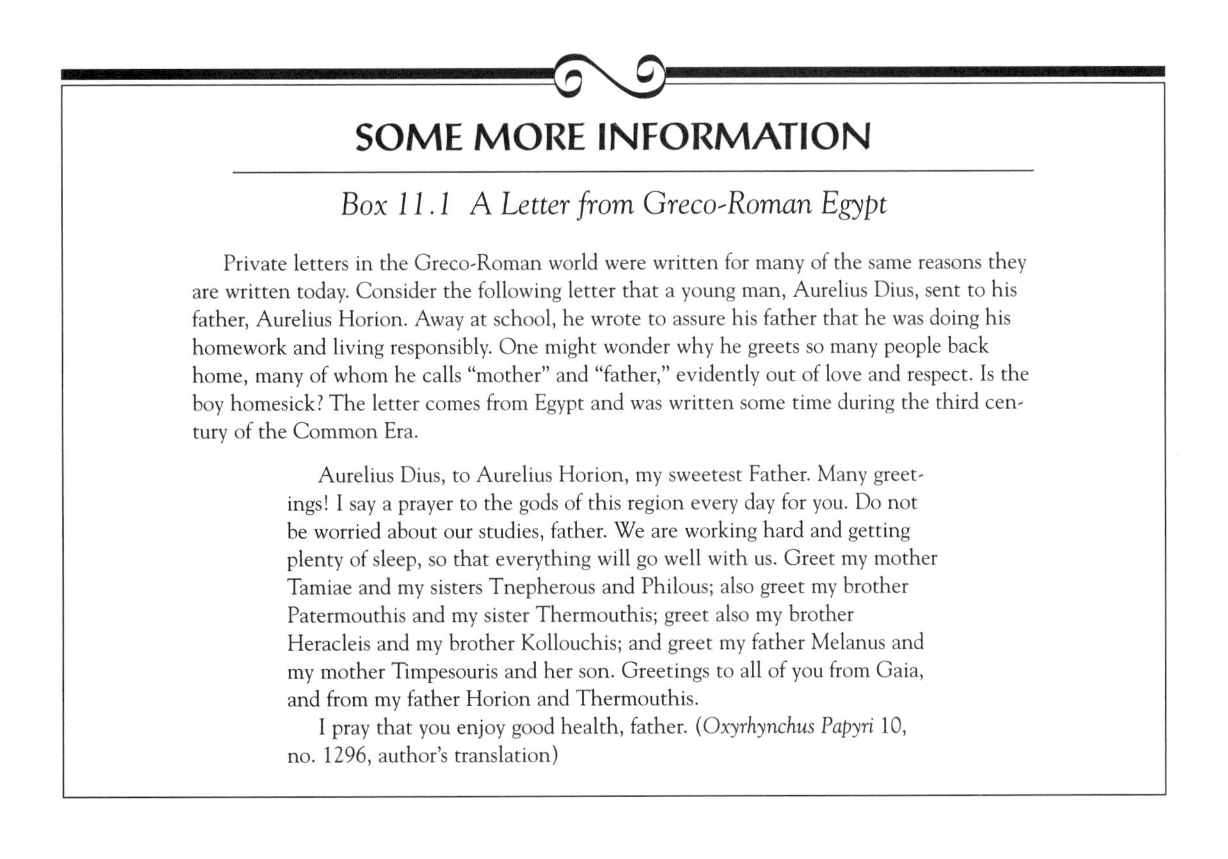

SOME MORE INFORMATION

Box 11.1 A Letter from Greco-Roman Egypt

Private letters in the Greco-Roman world were written for many of the same reasons they are written today. Consider the following letter that a young man, Aurelius Dius, sent to his father, Aurelius Horion. Away at school, he wrote to assure his father that he was doing his homework and living responsibly. One might wonder why he greets so many people back home, many of whom he calls "mother" and "father," evidently out of love and respect. Is the boy homesick? The letter comes from Egypt and was written some time during the third century of the Common Era.

> Aurelius Dius, to Aurelius Horion, my sweetest Father. Many greetings! I say a prayer to the gods of this region every day for you. Do not be worried about our studies, father. We are working hard and getting plenty of sleep, so that everything will go well with us. Greet my mother Tamiae and my sisters Tnepherous and Philous; also greet my brother Patermouthis and my sister Thermouthis; greet also my brother Heracleis and my brother Kollouchis; and greet my father Melanus and my mother Timpesouris and her son. Greetings to all of you from Gaia, and from my father Horion and Thermouthis.
>
> I pray that you enjoy good health, father. (*Oxyrhynchus Papyri* 10, no. 1296, author's translation)

chosen of God. If this assumption is correct, then 2 John is a letter in which a Christian leader (the elder) is addressing problems in a local church of a different community.

The letter of 3 John appears to have the same author. The writing style and many of the themes are the same, and again the author identifies himself as "the elder." In this instance, however, he addresses not an entire community but an individual named Gaius, lending his support to Gaius's side in a dispute that has arisen in the church.

Whereas both of these writings appear to be genuine letters, 1 John does not share their literary conventions. Notice that the author does not introduce himself or address his recipients directly at the outset, nor does he offer a greeting, prayer, or thanksgiving on their behalf. Moreover, at the end there are no closing greetings, well-wishings, final prayers, or even a farewell. On the other hand, the author does speak to his audience as those to whom he is "writing" (1:4; 2:12–14). 1 John is therefore less like an actual letter and more like a persuasive essay written to a community, a treatise intended to convince its recipients to engage in a certain course of action. There are other actual letters from antiquity that served as persuasive essays; this particular one appears to have been sent without the conventions typically found in epistles. Possibly it was sent with a separate cover letter that no longer survives. For the sake of convenience, I will continue to call this book of 1 John an epistle, even though technically speaking it is not.

It is reasonably clear that the author who wrote the letters of 2 and 3 John also produced this essay. Much of the vocabulary and many of the themes are the same, as is the writing style and the historical situation that the book appears to presuppose. Was this author also the one who produced the final version of the Gospel of John near the end of the first century? Scholars have debated the issue extensively. Today, the majority of scholars believe that this writer was not the author of the Gospel; rather, he was someone living in the same community at a somewhat later time, a person who knew the teachings found in the Gospel and who addressed problems that had arisen in the community after the Gospel had been circulated.

On the one hand, the author of 1, 2, and 3 John seems to understand the Christian faith in terms quite similar to those found in the Fourth Gospel, for a number of themes that are important in the Gospel appear here in the epistles as well (see box 11.2). Yet the writing styles are not the same, and the problems in the community appear to be quite different. As one salient example, the problem of the community's relationship to the Jewish synagogue, one of the primary concerns of the Fourth Gospel, is completely missing from these epistles. Perhaps with the passing of time, the pain of this earlier crisis faded and new problems arose; then a new author, intimately familiar with his community's Gospel and influenced by the ways it understood the faith, wrote to address these problems. This would explain both the similarities of the epistles to the Gospel and the differences.

THE NEW TESTAMENT EPISTOLARY LITERATURE AND THE CONTEXTUAL METHOD

With the Johannine epistles we come to the first New Testament writings of our study that are not, strictly speaking, narratives. The Gospels each narrate accounts of Jesus' words, deeds, and experiences, and the book of Acts recounts the words, deeds, and experiences of several of his apostles. The epistles, on the other hand, are writings of Christian leaders to individuals or churches to address problems that have arisen in their communities. Indeed, it is safe to say that all of the New Testament epistles are written in response to situations that the authors felt a need to address. Given the "occasional" nature of these letter (meaning they were written for certain occasions), how should we go about studying them? The question, of course, relates not only to the Johannine epistles but to all the others as well, including those appearing under the name of the apostle Paul.

It would be possible to apply some of the methods we have already examined in relationship to the Gospels and Acts as we study of the epistles, but here we will explore an approach that scholars

SOME MORE INFORMATION

Box 11.2 The Gospel and Epistles of John: Some Thematic Similarities

The Johannine epistles share a number of their distinctive themes with the Fourth Gospel, often expressing them in exactly the same words. It seems reasonable to assume, therefore, that all four books derive from the same community, which had developed characteristic ways of understanding its religious traditions.

Among the shared themes are the following:

- The images of light and darkness (1 John 1:5–7; 2:9–11; cf. John 8:12; 12:46)
- The new and old commandments (1 John 2:7; cf. John 13:34)
- Abiding in Christ (1 John 2:27–28; cf. John 15:4, 6)
- The command to love one another within the community (1 John 3:11; cf. John 13:34–35)
- Being hated by the world (1 John 3:13; cf. John 15:18–19; 17:13–16)
- Christ "laying down his life" for others (1 John 3:16; cf. John 10:11, 15, 17–18; 15:12–13)
- Christ as the one sent by God into the world out of love (1 John 4:9; cf. John 3:16)

have used extensively with this kind of occasional literature: contextual analysis. The method is particularly useful to historians who are interested in knowing not only what this literature says or teaches but also the specific historical circumstances that led to its production. As you will see, this approach is closely related to the socio-historical method described in Chapter 10. That method focuses on the social history of the community as it can be traced over a period of time, and the text is used to provide evidence for reconstructing that history. In the contextual method, the principal concern is the literary text itself; the social history of the community that is presupposed by the text is used to explain some of its important features.

The concern to understand the socio-historical context within which an occasional writing was produced is rooted in a theoretical view of language shared by many scholars, that knowing a document's historical context is absolutely vital for its interpretation. According to this view, words convey meaning only within a context; thus, when you change the context of words, you change what they mean.

This is because, as we have seen, words and phrases do not have any inherent meaning but mean what they do only in relationship to other words and phrases, so that words and phrases can be made to mean a wide variety of things (practically anything, according to some theorists). Let me illustrate the point through a brief example. Suppose you were to hear the phrase "I love this course." It would obviously mean something altogether different on the lips of your roommate when he is about to break 80 on the eighteenth hole at the country club than it would coming from your precocious younger sister at her first posh restaurant in the midst of the second course of a five-course meal. And it would mean something quite different still if spoken by your buddy at his favorite race-car track or by the woman sitting behind you in your New Testament class after hearing yet another scintillating lecture.

You might think, though, that the phrase in all of these cases means basically the same thing. Somebody appreciates something called a "course." But suppose you are in the middle of the most boring lecture by your most boring professor, wondering why you are there instead of outside catching some rays, when you hear a guy in the back row whisper the same words, "I *love* this course," and then snicker? You know full well what the words mean; they mean just the opposite of what they meant for the woman in the scintillating lecture. Thus, words mean what they do only in light of their context. If you change the context, you change what the words mean. This is true of all words in every language.

One practical implication of this insight is that if we are to understand a person's words we have to understand the context within which they are spoken. This principle applies not only to oral communication but written communication as well. With ancient literature, however, only in rare instances do we have solid evidence for the historical context within which words were spoken or written. Thus, we have to work hard at reconstructing the situation that lies behind a text if we want to understand the context within which it was produced. Only then can we use these contexts to help us interpret the texts.

Unfortunately, in many instances the only way we can know about the precise historical context of a writing is through clues provided by the writing itself. Doesn't this procedure, then, involve a kind of circular reasoning: to interpret a writing we have to understand the context, but we cannot understand the context until we interpret the writing?

The procedure probably is circular on some level, but it does not have to be completely so, for some ways of understanding the context within which a writing is produced will make better sense of the writing than others. Consider this analogy. Have you ever listened to someone talk on the phone and, on the basis of what he or she said, figured out what the other person was saying as well? What you did was to reconstruct the words you did not hear on the basis of the words that you did, and understood the words you heard in light of those you did not. To put it differently, you reconstructed the context of what you heard and made sense of the words in light of it. Now in some cases you may be wrong, but if you listen carefully enough, and if the speaker gives you enough to go on (and does not simply grunt in agreement every now and then), you can in many instances understand the full conversation based on your reconstruction of the words coming from the other end of the line.

Something like this happens when we apply the contextual method to a New Testament writing. On the basis of the conversation that we do hear, we try to reconstruct the conversation that we do not, and thereby come to a better understanding of what the author is trying to say. For some of the books of the New Testament, including the Johannine epistles, this method can prove to be quite enlightening. To be sure, there are some serious limitations to this approach, some of which have been overlooked by scholars for whom it is the method of choice. But the nature of these limitations cannot be fully appreciated in the abstract; they will make sense as we apply the method to specific texts, such as the Johannine epistles.

THE JOHANNINE EPISTLES FROM A CONTEXTUAL PERSPECTIVE

I will be treating these letters as a group of works produced by the same author at roughly the same time. The first is an open letter or persuasive treatise written to a community (1 John), the second a personal letter to the same community (2 John), and the third a personal letter to an individual within it (3 John). There are clues within the letters themselves concerning the historical context that prompted the author to produce them. The first step in the contextual method of interpretation is to examine these clues and use them to reconstruct the situation.

The most important event in the recent history of this community is that it has experienced a serious rift. The author of 1 John indicates that a faction from within the community has split off from the rest of the group and left in a huff:

> They went out from us, but they did not belong to us; for if they had belonged to us, they would have remained with us. But by going out they made it plain that none of them belongs to us. (1 John 2:19)

Why did this Christian community split, with some members leaving, presumably to start their own congregation? In the next few verses the author designates those who left as "liars" and "antichrists," a word which literally means, "those who are opposed to Christ." He then contrasts them with those who have remained, who "know the truth." What do these antichrists believe that makes them so heinous to this author? He indicates that they have "denied that Jesus is the Christ" (2:22). The author's language may appear to suggest that those who have seceded from the community, a group that some scholars have labeled the "secessionists," are Jews who failed to acknowledge that Jesus is the messiah. But they used to belong to the community, that is, they were Christians. In what sense, then, could they deny that Jesus is the Christ?

There are two other places where the author discusses these "antichrists." In 1 John 4:2–3 the author claims that unlike those who belong to God, the antichrists refuse to confess that "Jesus Christ has come in the flesh." A similar statement occurs in 2 John 7, where the antichrists are called "deceivers who have gone out into the world" and are said to deny that "Jesus Christ has come in the flesh." These descriptions suggest the secessionists may have held a point of view that we know about from other sources from about the same period, such as the writings of Ignatius (which we will be discussing at greater length in Chapter 25). Ignatius opposed a group of Christians who, like Marcion a few years later (see Chapter 1), maintained that Jesus was not himself a flesh-and-blood human being but was completely and only divine. For these persons, God could not have a real bodily existence; God is God—invisible, immortal, all-knowing, all-powerful, and unchanging. If Jesus was God, he could not have experienced the limitations of human flesh. For these people, Jesus only seemed to experience these limitations. Jesus was not really a human; he merely appeared to be.

These Christians came to be known by their opponents as "docetists," a term that derives from the Greek verb for "appear" or "seem." They were opposed by Christian leaders like Ignatius who took umbrage at the idea that Jesus and the things he did, including his death on the cross, were all a show. For Ignatius, Jesus was a real man, with a real body, who shed real blood, and died a real death.

It may be that the secessionists from the Johannine community had developed a docetic kind of christology. In the words of the author, they "denied that Jesus Christ had come in the flesh." If they were, in fact, early docetists, then a number of other things that the author says in these letters make considerable sense. Take, for instance, the opening words of 1 John. Readers who do not realize that the essay is being written because a group of docetic Christians have seceded from the community may not understand why the author begins his work the way he does, with a prologue that in many ways is reminiscent of the Prologue to the Fourth Gospel (with which he was probably familiar):

> We declare to you what was from the beginning, what we have heard, what we have seen with our eyes, what we have looked at and touched with our hands, concerning the word of life—this life was revealed, and we have seen it and testify to it, and declare to you the eternal life that was with the Father, and was revealed to us. (1:1–2)

Once a reader knows the historical context of the epistle, however, this opening statement makes considerable sense. The author is opposing Christians who maintain that Jesus is a phantasmal being without flesh and blood by reminding his audience of their own traditions about this Word of God made manifest: he could be seen, touched, and handled; that is, he had a real human body. And he shed real blood. Thus, the author stresses the importance of Jesus' blood for the forgiveness of sins (1:7) and of the (real) sacrifice for sins that he made (2:2; 4:10).

What led a group of Johannine Christians to split from the community because of their belief that Jesus was not a real flesh-and-blood human being? We have seen that after the community was excluded from the synagogue, it developed a kind of fortress mentality that had a profound effect on its christology. Christ came to be seen less and less as a human rabbi or messiah and more and more as a divine being of equal standing with God, who came to reveal the truth of God to his people only to be rejected by those who dwelt in darkness. Those who believed in him claimed to understand his divine teachings and considered themselves to be children of God. By the time the Fourth Gospel was completed, some members of the Johannine community had come to believe that Jesus was on a par with God.

It appears that Christians in this community did not stop developing their understandings of Jesus with the completion of the writing of the Gospel. Some of them took their christology a step further. Not only was Jesus equal with God, he was God himself, totally and completely. If he was God, he could not be flesh because God was not composed of flesh; Jesus therefore merely appeared to be a human.

This view proved to be too much for some of the other members of the community; battle lines were drawn and a split resulted. The Johannine epistles were written by an author who thought that the secessionists had gone too far. For this author, Christ was indeed a flesh-and-blood human being; he was the savior "come in the flesh," whose blood brought about salvation from sin. Those who rejected this view, for him, had rejected the community's confession that the man Jesus was the Christ; in his view, they were antichrists.

The charges that the author levels against the secessionists do not pertain exclusively to their ideas about Christ. He also makes moral accusations. He insinuates that his opponents do not practice the commandments of God (2:4), that they fail to love the brothers and sisters in the community (2:9–11; 4:20), and that they practice sin while claiming to have no contact with it (1:6–10). It is possible that, in the mind of the author at least, these moral charges related closely to the doctrinal one. If the secessionists undervalued the fleshly existence of Jesus, perhaps they undervalued the importance of their own fleshly existence as well. In other words, if what really mattered to them was the spirit rather than the flesh, then perhaps they were unconcerned not only about Jesus' real body but also about their own. Thus, they may well have appeared totally uninterested in keeping the commandments that God had given and in manifesting love among the brothers and sisters of the community. This would explain why the author stresses in his letters the need to continue to practice God's commandments and to love one another, unlike those who have left the community.

REFLECTIONS ON THE CONTEXTUAL METHOD

At this stage you may have recognized one of the difficulties in this kind of contextual analysis. It is very hard for the historian to know for a fact that the Johannine secessionists actually taught that it was unimportant to love one another and to keep God's commandments. The problem is that the only source we have for the secessionists' views is the author of the Johannine epistles, and he was their enemy.

As we know from other kinds of literature, ancient and modern, it is a very tricky business to learn what people say and do on the basis of what their enemies say about them. Imagine trying to reconstruct the beliefs and practices of a modern politician on the basis of what the opposing campaign says! Sometimes enemies misunderstand their opponents' views, or distort them, or misrepresent them, or draw implications from them that the other party does not.

What, then, do we actually know about the Johannine secessionists? Do we know for a fact that they were docetists who taught others to disobey the commandments and live in sin? No, what we know is that this is how the author of 1 John portrayed them. Some scholars are inclined to accept this portrayal as accurate; others are more cautious and say that we only know how the author himself perceived the secessionists. Others are even more cautious and say that we do not really know how the author perceived them, only how he described them. The issue is not easily resolved, and it is one you need to be alert to as you yourself engage in contextual studies of the New Testament writings.

With these caveats in mind, let me summarize what we can probably say about the historical context of the Johannine letters and then show how these letters can be seen as a response to the situation at hand. There is little to suggest that the author of these letters was intentionally duplicitous in his assessment of his opponents, even though we can never know this for certain. Whether or not his perceptions were correct, then, we can at least say how he perceived the situation. From his point of view, a group of former members of the community had split to form their own group; they taught that Jesus was not a real human being, but only divine; they saw no need to keep the commandments and did not manifest love to other members of the community, and were therefore antichrists and liars; and they continued to be a threat to the community's well-being by deceiving others.

If this is the context, as seen through the eyes of the "elder," what more can we say about the historical occasion of 1, 2, and 3 John? The author was a leader of a community at some distance from the one he addresses in these letters. That he was not in the immediate vicinity is demonstrated by his closing remarks in 2 and 3 John, where he indicates that he will visit soon so as not to be forced to rely on the written word to communicate his views (2 John 12; 3 John 13–14). He appears to have seen himself as having authority over the Christians to whom he writes; that is why he can exhort them to believe and act in the ways that he commands.

1 John would have been a treatise to those in this neighboring church who have not joined the secessionists, written as a kind of open letter to persuade them to remain faithful to the author's position and to see it as standing in true conformity with the tradition that they inherited when they joined the community. 2 John would have been a personal letter to the church urging, in shorter fashion, much the same advice; and 3 John would have been a private letter to an individual in this community giving instruction about a particular aspect of the problem that has arisen.

Scholars have expressed different opinions concerning what had happened to create the need for this final letter, the one most closely related to private letters in antiquity. It appears clear, in any event, that Gaius, the recipient of the letter, is in conflict with another leader in the congregation, Diotrephes, and that this conflict has to do with whether the author of these letters and the representatives he sends to the church should be received as authorities. The author sees Diotrephes as an opponent and Gaius and Demetrius (perhaps the carrier of the letter? v. 12) as allies. It could be that Diotrephes has supported the views of the secessionists and is trying to convert the rest of the church, or it could be that he simply does not like the "elder" who writes this letter, or appreciate his barging in to force his opinions upon the church that meets in his home (see box 11.3). Other options are possible, some of which may occur to you as you yourself engage in the contextual analysis of these letters.

BEYOND THE JOHANNINE COMMUNITY: THE RISE OF CHRISTIAN GNOSTICISM

Many of the charges leveled against the secessionists of the Johannine community have reminded scholars of the ways groups of Christian Gnostics are portrayed in sources that have survived from the second century. These groups are called Gnostic because of a fundamental notion that they all appear to have held in common, that "gnosis" (the Greek word for "knowledge") was necessary for salvation. These groups were not unified among themselves: there were lots of different gnostic

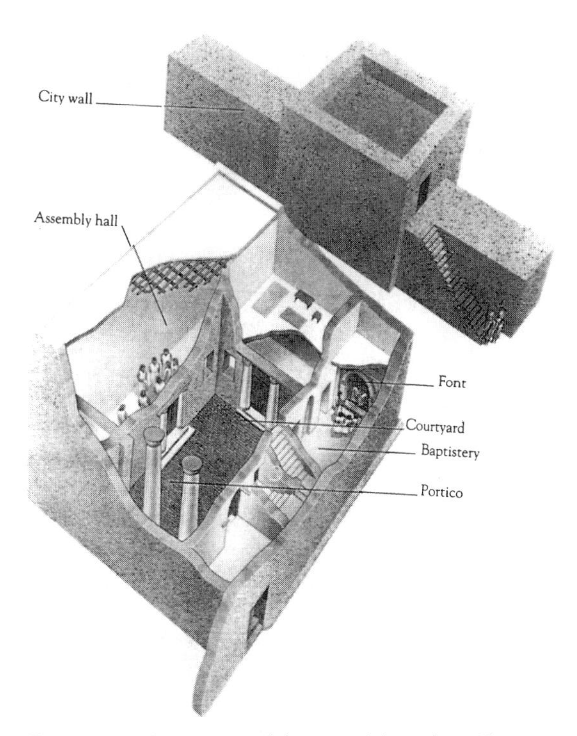

Figure 11.1 Cross-sectional drawing of the earliest Christian church building discovered, a converted house in the eastern Syrian city of Dura.

SOME MORE INFORMATION

Box 11.3 House Churches in Early Christianity

Most people do not realize that for many, many years Christians did not construct church buildings for their services of worship and fellowship. The earliest Christian church building to be uncovered by archaeologists (actually a house that was converted to serve as a church in the city of Dura in Eastern Syria) dates from around the year 250 C.E., well over two centuries after the death of Jesus. The lack of a specially designated sacred space for Christian worship during its first two hundred years made this religion different from almost all others in its world. Pagan cults were centered in temples and shrines, and Jews, of course, worshipped in synagogues (which were themselves sometimes converted homes).

If Christians did not meet in buildings specifically designed for the purpose, where did they meet? References in Paul's letters, the book of Acts, and other early Christian literature show that the early Christian communities were "house churches." Christians gathered together in the private homes of their wealthier members, who alone would have had room to accommodate more than a few persons. One consequence was that the membership and attendance at any given church would have been limited by the size of the house in which they met. Also, within a given city there could have been a number of Christian churches, each possibly with its own leader, who in many cases, presumably, was the person who provided the house.

This situation may shed light on the problem of interchurch conflict addressed in 3 John. It is possible, for example, that Diotrephes owns a relatively large house and meets weekly with a group of Christians, among whom he has assumed the role of leader and patron. Could it be that he sees the "elder" as an interloper who is out of bounds in trying to control what happens within the confines of his own home, among the Christians whom he entertains weekly for a service of worship and fellowship?

groups and they taught lots of different things. In their basic view of the world and their central theological convictions, however, these groups appear to have stood in some continuity. For all of them, this world was an evil place in which to live, and those who wanted to escape needed to acquire the knowledge (gnosis) necessary for liberation. Christ was a divine being who brought this knowledge from on high. Those who received this knowledge were thereby given the means of salvation.

The Fourth Gospel enjoyed particular success among many of the communities of Christian Gnostics, and, in the opinion of some scholars, the views of the secessionists from the Johannine community were closely related to those held by Christian Gnostics of a slightly later period. Moreover, as we will see in the next chapter, some of these Christian Gnostics produced Gospels of their own, books that never became part of the New Testament. If we want to understand the early Christian literature from a historical point of view, we have to learn about these other writings and see how they portray Jesus as a divine redeemer.

The Problems of Definitions, Sources, and Dating

Scholars in the present century have engaged in heated debates over how to define Gnosticism. These debates are intimately related to the problems that we have with the ancient sources that describe Gnostics or were written by Gnostics. Until about a hundred years ago, our only sources for understanding Gnosticism were the writings of its most vocal opponents, the proto-orthodox church fathers of the second, third, and fourth centuries. In our discussion of the Johannine epistles, we have already seen some of the problems with reconstructing a group's beliefs and activities on the basis of an attack by its enemies. With regard to Gnosticism the problems are even more severe. Proto-orthodox church fathers like Justin, Irenaeus, and Tertullian saw Gnosticism as a major threat to the success and unity of Christianity and pulled out all the stops in their assaults on it. Many of their charges—for example, their claim that certain groups of Gnostics engaged in wild sexual orgies and bizarre nocturnal rituals that involved eating babies—must be scrutinized with care (see Chapter 25).

One of the most significant archaeological discoveries of the twentieth century provided us with an entirely new source of information about Gnostics, a source not penned by its opponents but by Gnostic believers themselves. In 1945, just over a year before the discovery of the Dead Sea Scrolls, an Egyptian bedouin stumbled upon a jar containing thirteen ancient books. These books contained some fifty-two literary works, most of them previously unknown. When they finally made their way through antiquities dealers into the hands of competent scholars, it became clear what they were. This bedouin had accidentally unearthed a collection of ancient Gnostic texts written in Coptic, an ancient Egyptian language.

The books themselves were manufactured in the fourth century (we can tell because the scrap paper used to strengthen the bindings includes receipts that are dated), but they contain copies of documents that were produced much earlier, many of them during the second century at the latest. Linguists have established beyond any doubt that the books were originally written in Greek. The newly discovered documents represent translations of these earlier compositions, made perhaps in the third or fourth centuries. In some respects these documents have revolutionized our understanding of early Christian history. For here is a library of texts evidently of some importance to a community of Gnostic believers, texts in some ways like and in other ways unlike those that later became known as the New Testament. They are similar in that they also contain Gospels and other writings allegedly penned by apostles. They are different in that their perspectives on Jesus and God and the created universe are quite at odds with those that made it into the canon. Among the most interesting of these texts are the other Gospels about Jesus, including one allegedly written by his disciple Philip, another by his female companion Mary, and a third by his twin brother Thomas (see box 12.2). Some of these writings report hitherto unknown revelations that Jesus allegedly imparted to his closest apostles after his resurrection; others contain mystical reflections on how the universe came into being and how humans came to occupy a place in it.

Since these writings were discovered near the village of Nag Hammadi, Egypt, they have become known as the "Nag Hammadi library." For historians of Christian Gnosticism they are of unparalleled significance, in no small measure because they allow us to speak more confidently about what Gnostics believed without having to rely completely on the claims and charges of their opponents.

Figure 11.2 The Gnostic books discovered in 1945 near Nag Hammadi, Egypt, and the place where they were found.

This is not to say that scholars have now reached a consensus on every (or any!) important aspect of the study of Gnosticism. Far from it. Just as it is difficult to use the writings of the church fathers to learn exactly what Gnostics really thought, so too it is difficult to use these Gnostic writings themselves. For one thing, the writings found at Nag Hammadi do not share a consistent point of view, and we have no assurance that all of these texts were ever seen as authoritative by any one community, in the way the texts of the New Testament later came to be for orthodox Christians. Moreover, since these texts appear to have been written for the internal consumption of the communities that produced them, they assume a good deal about what their authors and readers already know to be the case. They do not spell out the Gnostic system (or the Gnostic systems) but appear to presuppose it. Thus, to understand these writings we have to reconstruct their underlying world(s) of thought.

Finally, it is difficult to know exactly how to go about interpreting these writings. Some of the church fathers, for example, evidently had access to writings very similar to some of the works discovered at Nag Hammadi, but they misunderstood (or misrepresented) how they were to be read. The anti-Gnostic author Irenaeus, for instance, appears to have read Gnostic poetry that celebrated the mysteries of creation. Instead of allowing for poetic license, however, he interpreted the texts literally as straightforward descriptions of how the universe came into being. As modern interpreters, we do not want to fall into the same trap: imagine how well you would do in an English poetry class if you failed to recognize metaphor when you saw it! But with the Nag Hammadi documents, it is often hard to know whether we are reading historical narrative or metaphysical poetry, propositional truths or mystical reflections.

In what follows I will try to lay out some of the basic assumptions that appear to underlie most of the Gnostic systems that we know about. Before doing so, I need to say a word about the dates of these systems and their relationship to non-Gnostic Christianity, also matters of intense and heated debates among scholars.

The anti-Gnostic church fathers maintained that Gnosticism was a Christian heresy invented by evil persons who corrupted the Christian faith to their own ends. A good deal of modern scholarship has been committed to showing that this perspective cannot be right, that, in fact, Gnosticism originated apart from Christianity but was later merged with it in some religious groups, forming a kind of synthesis, a Gnostic Christianity.

It is difficult to know what cultural forces would have produced Gnosticism, but it appears to represent a creative combination of diverse religious and philosophical perspectives, melded together in an age in which numerous religions and philosophies were widely known and often linked. If this is right, then Gnosticism and Christianity may have started out at about the same time, and because of many of their similarities, which we will see momentarily, came to influence each other in significant ways. It is interesting to note that some of the Gnostic tractates discovered at Nag Hammadi appear to be non-Christian, which would be hard to explain if Gnosticism originated as a Christian heresy.

So, in the way I will be using the term here, "Gnosticism" refers to a diverse set of views, many of them influenced by Christianity, that may have been in existence by the end of the first century but certainly were by the middle of the second. Our best evidence for specific Gnostic groups comes from the second century, the period in which the proto-orthodox opponents of the Gnostics were penning their vitriolic attacks and many of the documents preserved at Nag Hammadi were originally produced.

The Tenets of Gnosticism

Certain basic tenets appear to underlie the various Christian Gnostic religions. These are not explicitly set forth in any of the Gnostic writings that have survived; they do, however, appear to be presupposed by many of them as an underlying worldview.

The World: Metaphysical Dualism. Gnostics understood the world in radically dualistic terms. All of existence could be divided into two fundamental components of reality: matter and spirit. Some aspects of this worldview have struck scholars as similar to certain Eastern religions, such as

SOMETHING TO THINK ABOUT

Box 11.4 How Do You Know a Gnostic When You See One?

One of the major problems for proto-orthodox church fathers who attacked Christian Gnostics was knowing what constituted Gnosticism and, therefore, how to recognize a Gnostic when they met one. Part of the problem was that many different religious ideas could be called Gnostic, and those who might be considered Gnostic were far from agreeing with one another on a number of important issues. Frustration over this predicament is evident in the writings of one of the best-known authors of the second century, the anti-Gnostic church father Irenaeus. In his words: "Since they [the Gnostics] differ so widely among themselves both as respects doctrine and tradition, and since those of them who are recognized as being most modern make it their effort daily to invent some new opinion, and to bring out what no one ever before thought of, it is a difficult matter to describe all their opinions" (*Against the Heresies*, 1. 21. 5).

One thing Irenaeus and his colleagues were convinced of, however, was that even though Gnostics were difficult to recognize they had thoroughly infiltrated many of the churches: "Such persons are to outward appearance sheep; for they appear to be like us, by what they say in public, repeating the same words as we do; but inwardly they are wolves" (*Against the Heresies*, 3. 16. 8). In other words, the Gnostic Christians could agree with everything the proto-orthodox Christians said—they could affirm everything in the proto-orthodox creeds and participate in all the proto-orthodox rituals—but inwardly they understood these things as having deeper, symbolic meanings that the proto-orthodox Christians rejected. No wonder it was so difficult for the anti-Gnostic opponents to drive them out of the churches. It was not easy to recognize a Gnostic when you saw one.

the Zoroastrianism of ancient Persia, with which some Gnostics may have come in contact. Other aspects resemble philosophical views propagated in the West, such as the teachings of Plato and his followers. Wherever Gnostics derived their notions, they appear to have believed that the material and spiritual worlds were at odds with one another and that ultimately the material world was evil and the spiritual world was good.

Unlike representatives of certain Eastern religions, Gnostics did not believe that the struggle between matter (evil) and spirit (good) was eternal. For them, the material world had not always existed but came into being at some point in time. Nor did they subscribe to the view, held by most Jews and proto-orthodox Christians, that the one true God had created this world. For them, the material world was inherently evil. The true God, author of all good, could not have created something that was evil.

According to the Gnostics, the creation of the world was the result of a cosmic catastrophe. The myths that the Gnostics told largely functioned to explain how this catastrophe came about. As I have indicated, it is not a simple matter to determine when these myths are to be taken literally and when they represent mystical reflections on the nature of being. In either case, they appear to reflect the Gnostics' sense of alienation from the material world and to explain how this state of alienation came into being. The myths generally begin before the creation of the world, when there was no material existence at all but only the good realm of the spirit, inhabited by the true God.

The Divine Realm: The Unknowable God and his Aeons. In the beginning, according to some of the Gnostic myths (which I will necessarily simplify here in my brief summary), there was one true God, an all-powerful divine being who was totally spirit. This God was unlike everything we can imagine. He continues to exist even now, of course, but he is so great and so unlike anything human that he is far beyond anyone's capacity to comprehend. He is unknown and unknowable.

At some point in eternity past, this divine spirit produced offspring, other divine beings who were also spirit. These offspring were produced as couples and were sometimes called "aeons." Some of these couples themselves produced offspring, eventually creating a large divine realm, inhabited by spiritual beings at greater or lesser remove from the true God depending on when they came to be generated. According to some of the myths, one of these aeons, sometimes named Sophia (the Greek word for "wisdom"), exceeded her bounds by trying to comprehend the whole of the divine realm. In overreaching herself, she fell from the world of the divine, becoming separated from the other divinities and her own consort. In her fall, she became terrified, angry, and upset. These emotions somehow became personified and took on a life of their own. In a sense, they were the offspring that resulted from her fall, but they were imperfectly formed in that they were generated apart from the union of Sophia with her divine consort.

These, malformed divine beings are responsible for the creation of the world. One of them in particular, named Ialdabaoth in some of the Gnostic texts (a name closely related to the Hebrew name of God in the Jewish Scriptures) is portrayed as the Demiurge, the one who brought the material world into being (see box 11.5).

He did so because he and the other fallen offspring of Sophia wanted to capture her and rob her of her divine power. To prevent her from recovering her strength and returning to the divine realm, they divided her into innumerable pieces and entrapped her in matter. The material world was thus created by these evil deities as a prison where Sophia, or rather her parts, are confined. Specifically, this element of the divine is entrapped in human bodies.

The Human Race: The Divine Spark. The reason Gnostics feel so alienated in this world is because they *are* alienated. Within them is a spark of the divine, entrapped by alien beings who are

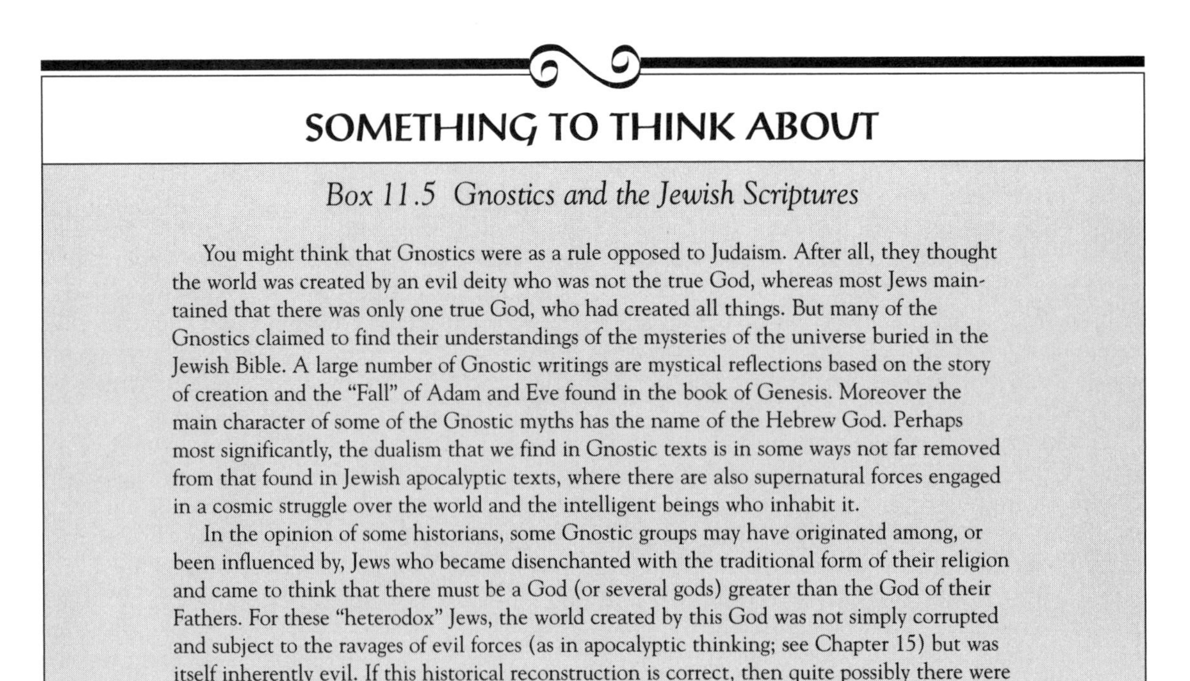

SOMETHING TO THINK ABOUT

Box 11.5 Gnostics and the Jewish Scriptures

You might think that Gnostics were as a rule opposed to Judaism. After all, they thought the world was created by an evil deity who was not the true God, whereas most Jews maintained that there was only one true God, who had created all things. But many of the Gnostics claimed to find their understandings of the mysteries of the universe buried in the Jewish Bible. A large number of Gnostic writings are mystical reflections based on the story of creation and the "Fall" of Adam and Eve found in the book of Genesis. Moreover the main character of some of the Gnostic myths has the name of the Hebrew God. Perhaps most significantly, the dualism that we find in Gnostic texts is in some ways not far removed from that found in Jewish apocalyptic texts, where there are also supernatural forces engaged in a cosmic struggle over the world and the intelligent beings who inhabit it.

In the opinion of some historians, some Gnostic groups may have originated among, or been influenced by, Jews who became disenchanted with the traditional form of their religion and came to think that there must be a God (or several gods) greater than the God of their Fathers. For these "heterodox" Jews, the world created by this God was not simply corrupted and subject to the ravages of evil forces (as in apocalyptic thinking; see Chapter 15) but was itself inherently evil. If this historical reconstruction is correct, then quite possibly there were Gnostic groups made up largely of people who continued to consider themselves Jews.

committed to keeping it imprisoned for their own purposes. True Gnostics know that they do not belong in this material world; heaven is their home.

To be sure, not all humans experience this sensation, for the divine spark does not reside in everyone, only in the elect few. Other humans are simply part of the material world, with nothing divine within them. They are like other animals, created by the Demiurge and destined, eventually, to be destroyed along with all other works of his creation. The Gnostics, however, are destined for better things, for within them is the spark of the divine which can be liberated from this miserable existence. How can the spirit within be set free to return to its heavenly home? Only by acquiring the knowledge necessary for salvation.

Salvation: The Knowledge (Gnosis) That Liberates. Gnostics claimed that a person could be saved from this material world only by acquiring the proper knowledge. Wisdom had become fragmented and ignorance reigned supreme. Salvation meant acquiring the true knowledge of how this state of affairs had come to be and what it would take to change it. Since Gnostics were the imprisoned spirits, this knowledge involved an intricate self-understanding of who they were and how they came to be here, of where they came from and how they could return.

This knowledge, of course, was only for those destined for salvation. Those who were "in the know" (i.e., those who were "gnostic") were capable of receiving and understanding this secret knowledge, which was hidden from the common folk.

We are not certain, exactly, what this knowledge entailed, since the Gnostics who attained it kept it secret. It appears that different Gnostic groups propagated different forms of instruction, probably corresponding to the different myths that they told. Included in many circles would be knowledge of how the spiritual bondage had occurred and what it would take to escape it. In some systems, the soul could be liberated at death only by knowing the passwords that the evil creator gods required for passage through their respective heavenly realms on the journey to the highest world of the true God. What were these passwords? I would tell you, if I were certain that you too were Gnostic.

The knowledge that is necessary for salvation obviously cannot come from within this world. The world is material and therefore evil; there are no material means for discovering the truth of our entrapment or the secrets for our liberation. Saving knowledge must therefore come from outside this world. It must come from the world of God.

In most Gnostic systems, the only way to escape this world is for an aeon to come down from the divine realm to communicate the knowledge necessary for salvation to the sparks that have been entrapped.

Christ: The Divine Redeemer. This emissary from the divine realm obviously could not be human, for to be human means to be entrapped in the realm of matter. Thus, the divine emissary could not have a real flesh-and-blood body, he could not actually be born, could not actually bleed, and could not actually die. For Gnostic Christians the divine being who came into the world was Christ himself. There were two different ways that Gnostics could understand Christ as the one who came from above without being human. One is the docetic view that we examined earlier in relation to the secessionists from the Johannine community and the opponents of Ignatius from roughly the same time period (and Marcion some years later). In this view, Christ was not a real human being but only appeared to be. In the words of the apostle Paul, which some of these Gnostics could quote, Christ came "in the likeness of sinful flesh" (Rom 8:3). For Gnostics, Christ looked like a human but was not.

According to this view, Jesus was a totally spiritual being who communicated to his disciples the gnosis required for their liberation; they in turn passed it along to their own followers by word of mouth. When they committed this knowledge to writing, it was only in veiled texts that were hard for all but insiders to understand (otherwise, the divine gnosis would be available to everybody). For the sake of the public eye, Jesus kept up his human appearance throughout his ministry, seeming to become hungry and thirsty, seeming to bleed and die. But it was all an appearance.

The other Gnostic option was to claim that whereas Jesus was a real flesh-and-blood human being, he was not the same person as the heavenly Christ, a separate being who temporarily inhabited Jesus' body. This view appears to have been more common among Gnostics, so far as we can tell from our surviving sources. In this understanding, Jesus was a righteous man who was chosen by a divine being, the heavenly Christ, as a dwelling place. When Jesus was baptized, the Christ descended from heaven in the form of a dove and entered into him, empowering him to do miracles and to teach the gnosis necessary for salvation. Then, prior to Jesus' death, the Christ departed from him, leaving him alone to suffer and die. As a spiritual being, the Christ himself obviously could not come in contact with pain and death.

According to some of these Gnostics, the Christ returned to Jesus after his crucifixion and raised him from the dead, empowering him to appear to his disciples over a long period of time and to convey the gnosis that they too would need to survive death and return to the heavenly realm.

We are told by the early church writer Irenaeus that some Gnostics with this point of view had a particular attachment to the Gospel of Mark, in which Jesus' public life begins with the baptism, when he receives the Spirit, and ends on the cross, when he cries out, "My God, my God, why have you left me behind?" (as these Gnostics interpreted Jesus' final words).

The Church: The Body of the Elect. As we have seen, not everyone could acquire the knowledge of salvation; not even other Christians were entitled to learn the true mysteries of the faith. It was only for the elite few who had the divine spark within them. The Gnostics therefore kept their knowledge a secret.

According to some Christian Gnostics, the human race could thus be divided up into three classes: (a) the Gnostics themselves, possessors of ultimate knowledge, destined for a glorious salvation when they returned to the heavenly realm whence they came; (b) other Christians, who mistakenly believed that they had the truth, when they had nothing but a superficial knowledge of it through a literal understanding of the sacred writings of the apostles and the doctrines transmitted in the church (see box 11.4); these persons would receive some form of salvation if they had faith and did good works, but their afterlife would not be nearly as glorious as that of the true Gnostics; and (c) all other persons, who had no part of the divine within them and were destined for destruction along with the rest of the material world when the salvation of the divine sparks had been complete.

A large number of Gnostics were active members of Christian churches and would not have been recognized as unique simply on the basis of outward appearances (see box 11.4). They were able to read the sacred books and profess the sacred truths of the Christian religion along with everyone else. But in their hearts they professed to understand the deeper truths of these things and to possess ultimate knowledge and real understanding. Thus, they saw themselves as an elite body within the churches.

The intellectuals among the non-Gnostic Christians (such as Irenaeus and Tertullian) looked upon this Gnostic deeper knowledge as a rejection of the basic truths of Christianity. For these proto-orthodox writers, anyone who claimed that the creator God was not the true God, that the material world he made was evil, that Jesus was not his true son, and that he did not really shed blood and die on the cross—anyone who believed such things could not claim to be Christian and could have no part in the salvation of God.

Most Christians today would probably agree, for they are the spiritual descendants of the group that won these debates in the second and third centuries. As a consequence, they have inherited these proto-orthodox positions. Gnostics, on the other hand, would say that such Christians have simply failed to see the truth, as revealed in the gnosis conveyed by Christ himself.

Ethics: The Ascetic Ideal. One of the most interesting aspects of ancient Gnosticism is that Gnostics were routinely charged by their opponents with engaging in flagrant acts of indecency and immorality. For example, they were accused of engaging in scandalous and offensive sex rituals, of murder, and of cannibalism. In hindsight we might say that Gnostics were charged with these things, not because they did them, but because they were thought to have done them, since in the eyes of their opponents they were secretive purveyors false doctrine.

Throughout the ancient world, a wide range of groups, even groups of Jews and proto-orthodox Christians, were accused of engaging in precisely the same kinds of activities (see Chapter 25). It appears that one way to cast aspersions on one's opponents was to implicate them in such ways, much as today politicians have standard epithets (e.g., "tax-and-spend liberal," "opponent of family values") that they sling at their opponents, hoping they stick, whether or not they happen to conform to reality. In the case of the Gnostics, the logic of the charges would have been convincing to outsiders. Here was a secretive group of people who totally devalued the body. If the body doesn't matter, then surely it doesn't matter what you do with your body.

The Gnostics themselves, however, appear to have employed a different kind of logic in their ethics, one that led to just the opposite results. For the Gnostics, since the body was evil, along with all other material things, no one should become attached to it and submit to its evil physical desires. Thus the Gnostic writings embrace an ascetic lifestyle which condemned gluttony, drunkenness, and sexual activity of every kind in an attempt to contribute to the liberation of the soul.

GNOSTICS AND THE JOHANNINE COMMUNITY

You may have already been struck by certain similarities between of these Gnostic views and those of some members of the Johannine community. While we cannot assume that the secessionists, let alone the author of the Fourth Gospel, considered themselves to be Gnostics, the similarities in their views are nonetheless quite interesting, particularly with respect to christology. As we have seen, the Gospel of John portrays Jesus, not merely as a human being chosen by God to be his messiah, but as a divine being come down from heaven to dwell among humans. In some sense, he is God himself, the Word of God come to speak to the world. His discourses reveal who he is as the one who has been sent from above; his miracles are performed to show that he is right. His ultimate goal is to convey the liberating knowledge that is necessary for salvation: "You will know the truth, and the truth will make you free" (8:32).

Such notions proved quite palatable to Gnostic Christians of the second century, many of whom revered the Fourth Gospel as a sacred text that revealed the mysteries of their faith. Indeed, so far as we know, the first commentary on any Christian text of any kind was the commentary on John written by Heracleon, a Gnostic Christian living around the year 170 C.E.

Unfortunately, we may never know what historical relationship may have existed between this Gnostic commentator of the late second century and the secessionists who withdrew from the Johannine community some three-quarters of a century earlier. It is possible that the secessionists came into contact with a sect of non-Christian Gnostics and adopted many of their perspectives so as to create a kind of hybrid faith, a Christian Gnosticism of the sort described in this chapter. It is equally possible that the sect disappeared from the face of the earth by being integrated into a larger society of gnostically-minded individuals. What we do know with some degree of probability, based on the historical reconstruction sketched earlier, is that prior to leaving the Johannine community the secessionists had already developed perspectives that would have proved compatible with views embraced by various groups of Gnostics, and when they seceded from the community, they took their Gospel with them. From their point of view, of course, their interpretation of the Gospel was the correct one. It was also an interpretation that made sense to various Christian Gnostics of the second century. It did not make sense, however, to the Johannine Christians they left behind or to the proto-orthodox Christians of later years, who condemned the Gnostics and their interpretations and succeeded in advancing their own.

SUGGESTIONS FOR FURTHER READING

Brown, Raymond. *The Community of the Beloved Disciple*. New York: Paulist, 1979. A superb and influential study that uses a socio-historical method to trace the history of the community from the time prior to the Fourth Gospel through the writing of the Johannine epistles, and beyond.

Layton, Bentley. *The Gnostic Scriptures: A New Translation with Annotations*. Garden City, N.Y.: Doubleday, 1987. An invaluable translation of important Gnostic documents, including those discovered at Nag Hammadi and those quoted by the church fathers, with a very useful introductory sketch of Gnosticism.

Lieu, J. M. *The Theology of the Johannine Epistles*. Cambridge: Cambridge University Press, 1991. A good recent discussion of major themes in 1, 2, and 3 John.

Robinson, James, ed. *The Nag Hammadi Library in English*. 3d ed. New York: Harper & Row, 1988. A very convenient English translation of the documents discovered at Nag Hammadi, with brief introductions.

Rudolph, Kurt. *Gnosis: The Nature and History of Gnosticism*. Trans. R. McL. Wilson. San Francisco: Harper & Row, 1987. The best book-length introduction to ancient Gnosticism.

CHAPTER 12

Jesus from Different Perspectives: Other Gospels in Early Christianity

We have already seen that Matthew, Mark, Luke, and John were not the only Gospels produced by the early Christians. They were simply the four that came to be included in the New Testament. Indeed, it is striking that the author of one of them, Luke, indicates that he had "many" predecessors in producing a narrative of the things Jesus said and did. It is unfortunate that apart from the Gospel of Mark, all of these earlier accounts have been lost. Still, by studying the canonical Gospels, we have been able to learn something about their sources, including Q, the collection of Jesus' sayings (and several deeds) that both Matthew and Luke used for their narratives, the signs source used by John for his accounts of Jesus' miracles, several sources for the discourses of Jesus in John, and passion narratives (possibly written) underlying the accounts in Mark and John.

Some scholars have detected additional sources behind .the canonical Gospels. And we know for a fact that Christian communities read and revered yet other Gospel texts. Indeed, thanks to manuscript discoveries over the past century, including the Nag Hammadi library, some two dozen accounts of Jesus' life and teachings now survive from the early centuries of Christianity. We know that others were written as well (which have not survived) because they are discussed, and sometimes quoted, in the writings of the early church fathers.

Only a few of the noncanonical Gospels will be of concern to us here, since most of them were not produced during the earliest period of Christian history (roughly through the first half of the second century) but in the later second, third, and fourth centuries, and on into the Middle Ages. It is important to recognize the existence of these later Gospels, however, because they show that Christians did not stop reflecting on the significance of Jesus or refrain from writing accounts of his life once the books of the New Testament were produced. Stories about him continued to be told, and invented, for centuries. They continue to be invented even today, as you can see by watching any of the versions produced in Hollywood.

In the last chapter we examined the beliefs of the Christian Gnostics and saw that in addition to using the Gospels of Mark and, especially, John they produced Gospels of their own. So too did some of the opponents of Gnostics, for example, the Jewish Christian group known as the Ebionites, who had their own Gospel, also allegedly written by an apostle (see Chapter 1). So did the Marcionites, who opposed the Jewish Christians and the Jewish religion that they embraced. And so did certain groups of proto-orthodox Christians, whose love for the Gospels that became part of the New Testament did not prevent them from penning still other accounts of the words and deeds of Jesus. Among these various noncanonical Gospels, several are of real interest to the historian of earliest Christianity, including the *Gospel of Peter*, which provides an intriguing account of Jesus' death and resurrection, and the *Gospel of Thomas*, touted by some scholars as the "Fifth Gospel" since it appears to preserve actual teachings of the historical Jesus not found in the New Testament.

For the purposes of our study, I will categorize the earliest Gospels into three groups: (a) "narrative" Gospels, which are written accounts of Jesus' sayings, deeds, and experiences; (b) "sayings" Gospels, which are comprised almost exclusively of Jesus' words to his disciples, whether during his ministry or after his resurrection; and (c) "infancy" Gospels, which are narratives of Jesus' birth and youth.

NARRATIVE GOSPELS

Matthew, Mark, Luke, and John can all be considered narrative Gospels. So can some of the written sources underlying these Gospels, for example, the signs source of the Fourth Gospel, and possibly the special sources for Matthew and Luke, called M and L (if these were actual written sources). We know that still other narrative Gospels existed in the early church, for Luke labels the works of his predecessors "narratives." With the exception of Mark, however, none of these earlier accounts has survived intact. What have survived are numerous references to Gospels of this kind in the writings of the church fathers, sometimes with discussions of their contents and quotations from their texts. In addition, we have a fragmentary manuscript of one of the most important of these works, a Gospel that claims to have been written by Jesus' disciple Peter.

The Jewish-Christian Gospels

We have seen that Christianity started out as a movement within Judaism. Jesus, his disciples, and the people they originally converted were Jewish; they read the Jewish Scriptures, observed the Jewish Law, and adhered to Jewish customs. Each of the Gospels we have examined, however, strives to show, in its own way, how Jesus was rejected by his own people, leading to the establishment of a community of believers outside of Judaism. This can be most clearly observed in the case of John, where the Christian community appears to have been excluded from the local synagogue at some point prior to the writing of the Gospel.

We will see throughout our study that most other Christian authors of the first century also attempted to distinguish themselves from the non-Christian Jews in their environment. But not all of them did. We know of Christian communities throughout the second century that were made up of Jews who had converted to belief in Jesus as the messiah but who nonetheless continued to maintain their Jewish identity, keeping kosher food laws, observing the sabbath, circumcising their baby boys, praying in the direction of Jerusalem, and engaging in a number of other Jewish practices. Various "Jewish-Christian" communities were scattered throughout portions of the Mediterranean. We know of some, for example, in the Transjordan region of Palestine (east of the Jordan River) and of others in Alexandria, Egypt. Each of these groups, no doubt, differed from others in specific matters of doctrine and practice.

Some of these Jewish-Christian groups had their own Gospels, accounts of the life of Jesus that portrayed him in ways amenable to the communities' own views, just as the canonical Gospels were amenable to the views of the communities that produced them. We know of three of these Jewish-Christian Gospels from the writings of church fathers who discuss them.

The Gospel of the Nazareans. This Gospel was evidently written in Aramaic, the native language of Jesus and his earliest followers. It may have been produced in Palestine near the end of the first century, that is, at about the time of the Gospel of John. The church fathers who refer to it sometimes claim that it was an Aramaic translation of the Gospel according to Matthew, minus the first two chapters. This claim makes sense, since the Gospel of Matthew is in many respects the most Jewish of our Gospels. It is there, for example, that Jesus instructs his followers to keep the entire law even better than the scribes and the Pharisees (5:17–20). At the same time, Matthew's story of Jesus' miraculous conception would have been unacceptable to Jewish Christians who believed that Jesus was a righteous man chosen to be God's messiah but not himself divine or born of a virgin.

The church fathers who refer to the *Gospel of the Nazareans* intimate that some of its stories differed from the accounts found in Matthew. These differences make it difficult to judge whether the anonymous author of this Gospel (a) had access to a version of Matthew that was somewhat different from the one that later became part of the Christian canon, for example, in lacking a birth narrative, (b) modified the Matthew that we know, for example by deleting the opening chapters, or (c) did not actually use a version of the Gospel of Matthew at all. In the last case, he may have used traditions similar to those found in Matthew, which circulated in the same or a neighboring community, and produced his own version from them.

The Gospel of the Ebionites. This Gospel appears to have been a combination of the Synoptic Gospels, a kind of "Gospel harmony" in which the three accounts were merged to form one longer and fuller version of Jesus' life. It was evidently written in Greek and was possibly used among Jewish Christians living in the Transjordan. One of its striking features is that it recorded words of Jesus to the effect that Jews no longer needed to participate in animal sacrifices in the Temple. Connected with this abolition of sacrifice was an insistence that Jesus' followers be vegetarian. This insistence led to some interesting alterations of stories found in the Synoptics. Simply by changing one letter, for example, the author modified the diet of John the Baptist; rather than eating "locusts" (Mark 1:6; the Greek word is *akrides*) he is said to have eaten "pancakes" (*egkrides*).

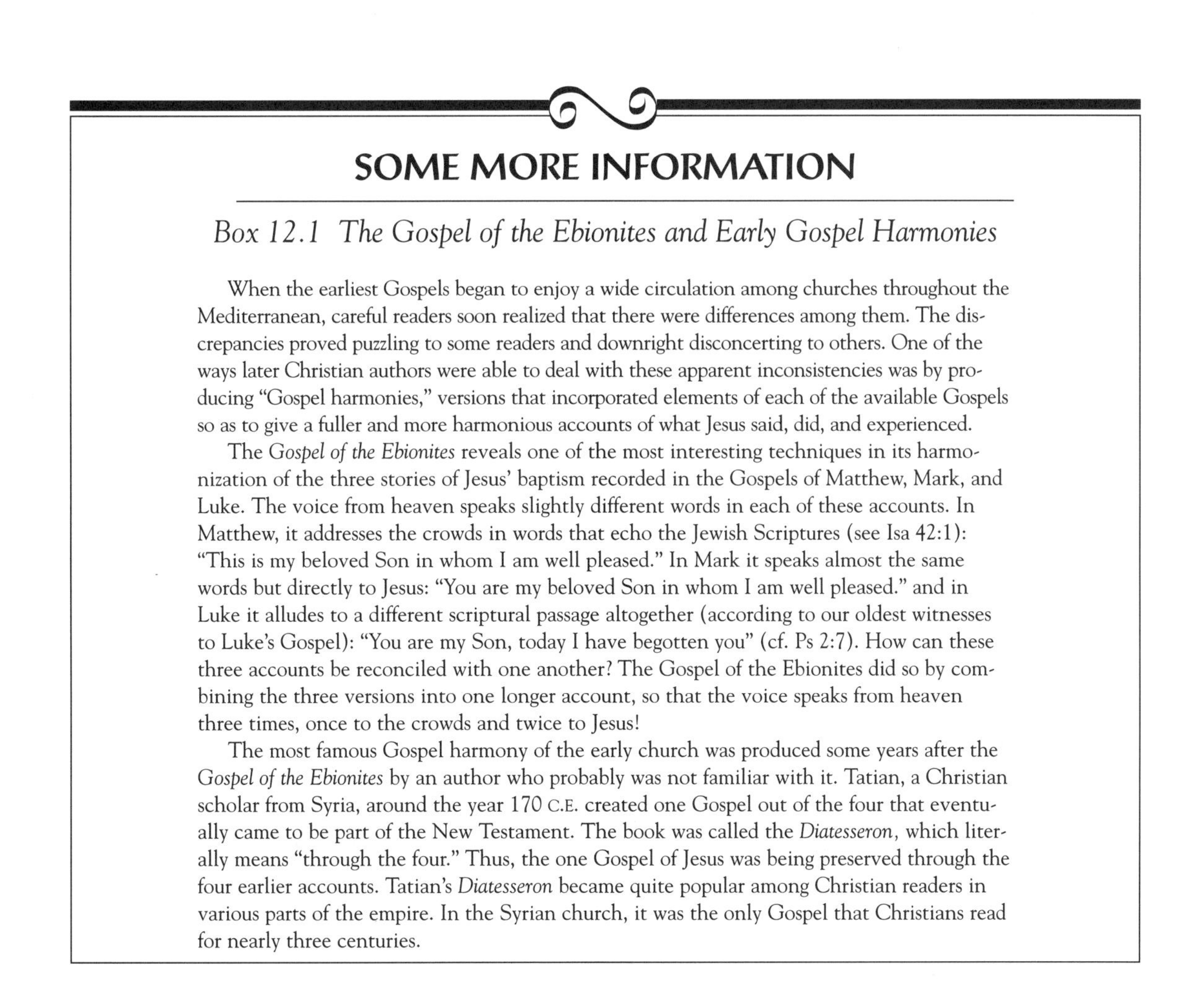

SOME MORE INFORMATION

Box 12.1 The Gospel of the Ebionites and Early Gospel Harmonies

When the earliest Gospels began to enjoy a wide circulation among churches throughout the Mediterranean, careful readers soon realized that there were differences among them. The discrepancies proved puzzling to some readers and downright disconcerting to others. One of the ways later Christian authors were able to deal with these apparent inconsistencies was by producing "Gospel harmonies," versions that incorporated elements of each of the available Gospels so as to give a fuller and more harmonious accounts of what Jesus said, did, and experienced.

The *Gospel of the Ebionites* reveals one of the most interesting techniques in its harmonization of the three stories of Jesus' baptism recorded in the Gospels of Matthew, Mark, and Luke. The voice from heaven speaks slightly different words in each of these accounts. In Matthew, it addresses the crowds in words that echo the Jewish Scriptures (see Isa 42:1): "This is my beloved Son in whom I am well pleased." In Mark it speaks almost the same words but directly to Jesus: "You are my beloved Son in whom I am well pleased." and in Luke it alludes to a different scriptural passage altogether (according to our oldest witnesses to Luke's Gospel): "You are my Son, today I have begotten you" (cf. Ps 2:7). How can these three accounts be reconciled with one another? The Gospel of the Ebionites did so by combining the three versions into one longer account, so that the voice speaks from heaven three times, once to the crowds and twice to Jesus!

The most famous Gospel harmony of the early church was produced some years after the *Gospel of the Ebionites* by an author who probably was not familiar with it. Tatian, a Christian scholar from Syria, around the year 170 C.E. created one Gospel out of the four that eventually came to be part of the New Testament. The book was called the *Diatesseron*, which literally means "through the four." Thus, the one Gospel of Jesus was being preserved through the four earlier accounts. Tatian's *Diatesseron* became quite popular among Christian readers in various parts of the empire. In the Syrian church, it was the only Gospel that Christians read for nearly three centuries.

The Gospel of the Hebrews. This Gospel was also written in Greek and was in use among Jewish Christians in Alexandria, Egypt. Its title was evidently given to it by outsiders to differentiate it from the one used by the Gentile Egyptian nationals, who called the work the *Gospel of the Egyptians*. We know that the *Gospel of the Hebrews* narrated important events in the life of Jesus, including his baptism, temptation, and resurrection, but the brief quotations of it found in the writings of the church fathers show that these stories were not simply borrowed from the other Gospels that we know. It appears that the author collected stories, possibly from the oral tradition, and compiled a narrative of his own much as Mark and John had done. Several of the church fathers' references to this Jewish-Christian Gospel imply that it had a Gnostic slant; this would not be surprising given the use of this Gospel in Alexandria, a major center of early Christian Gnosticism.

Marcion's Gospel

As we saw in Chapter 1, the second-century theologian Marcion stood at the opposite end of the Christian spectrum from the Jewish Christians. Whereas they embraced the Jewish Scriptures and maintained Jewish ways, he rejected Judaism as the religion of a false God. Indeed, for him the true God had sent Jesus to counteract the works of the creator. It was the creator who had chosen Israel and given them his law. His righteous demands, however, were harsh, and the punishment for disobedience was severe. The true God, the God of love, had sent Jesus in the appearance of human flesh to redeem people from this God of the Jews. Jesus himself had no dealings with the creator or his creation.

Marcion claimed the apostle Paul as his authority for these views. Throughout his letters, Paul speaks of his "Gospel," but which Gospel did he mean? Marcion decided that Paul's Gospel differed from the one(s) used in the Christian churches, which had been corrupted by copyists who did not realize that Jesus had nothing to do with the Jewish God or the religion that he established. In their ignorance, they altered the stories they copied by inserting positive references to the Jewish Scriptures and the creation of the world and ascribing them to Jesus. Marcion decided to correct the work of these scribes, and so produced a revised version of the Gospel, which for him represented the original version, one without these references.

Evidently, he used the Gospel of Luke as his starting point. From this Gospel he excised passages that referred positively to the Old Testament and to the Jewish God and his creation. He apparently removed the entire first two chapters as well, which contain the birth narrative, since in his docetic christology Jesus could not have been born. He may also have added several passages to get his point across more firmly; in his Gospel, Jesus allegedly claims to have come "not to fulfill the Law, but to abolish it" (contrast Matt 5:17).

Even though Marcion's Gospel does not survive intact, it is quoted at length by his chief adversary, the proto-orthodox church father Tertullian. It would be wrong to overlook the significance of this text simply because Marcion created it by modifying other Gospels that we already have. Matthew and Luke did so as well: recall how they handled Mark. Moreover, Marcion's Gospel proved to be particularly important in the second century. In certain communities throughout the Mediterranean around the year 200 C.E., Marcionites reading this version of the Gospel outnumbered every other kind of Christian.

The Gospel of Peter

Another narrative Gospel that was popular in some circles of the second century was one allegedly written by Jesus' close disciple Peter. We have known about this book for centuries, thanks to the writings of the fourth-century church father Eusebius, but we have come to know parts of its actual text only over the past hundred years, since a fragment of its final pages was discovered in 1886 in the grave of a Christian monk in Egypt.

Eusebius indicates that the Gospel was popular in parts of Syria during the second half of the second century. According to his account, Serapion, the bishop of Antioch, approved the *Gospel of Peter* for use in the church of Rhossus, even though he had not read it himself. When Serapion was told, though, that the book contained passages that could be used to support a docetic christology, he perused a copy and quickly dashed off a letter forbidding its use and detailing the offensive passages. Eusebius quotes from this letter but does not cite the passages Serapion had in mind. This is particularly to be regretted because without them we cannot be certain that the Greek manuscript discovered at the end of the nineteenth century is from the same *Gospel of Peter* as the one Serapion had read. In any event, the manuscript is of considerable interest in and of itself.

The document consists of only a few pages near the end of the narrative. It is impossible to know how long the entire account was or whether, for example, it included stories of Jesus' entire ministry or only of his Passion. The text begins in the middle of a passage with the statement that "None of the Jews washed his hands, neither did Herod nor any of his judges. As they did not wish to wash, Pilate got up." Evidently the preceding passage narrated the story, otherwise known only from Matthew, of Pilate's washing his hands at Jesus' trial (Matt 27:24). In Peter's account, however, the emphasis is not on Pilate, who is portrayed throughout as innocent of Jesus' death, but on Herod, the King of the Jews, and on the Jewish leaders who collaborated with him. In the next verse, it is Herod who orders Jesus to be taken out and crucified.

The narrative continues with the request of Joseph (of Arimathea) for Jesus' body, the mockery of Jesus, and his crucifixion. These accounts are both like and unlike what we read in the canonical Gospels. For example, in verse 10, Jesus is said to be crucified between two criminals, as in the other Gospels, but then we find the unusual statement that "he was silent as if he had no pain." This last statement could well be taken in a docetic way; perhaps Jesus appeared to have no pain because he in fact did not have any. Some scholars have seen this verse as providing evidence that the document is the "heretical" Gospel known to Serapion. Further confirmation may come several verses later. When Jesus is about to die, he utters his "cry of dereliction" in words similar to, but not identical with, those found in Mark's account: "My power, O power, you have left me" (v. 19). He is then said to be "taken up," even though his body remains on the cross. Is Jesus here bemoaning the departure of the divine Christ from him prior to his death, in keeping with the view of many Gnostics?

The account continues by describing Jesus' burial and then, in the first person, the distress of the disciples: "We fasted and sat mourning and crying night and day until the Sabbath" (v. 27). As in Matthew's Gospel, the Jewish leaders ask Pilate for soldiers to guard the tomb. This Gospel, however, provides more elaborate detail. The centurion in charge is named Petronius, who along with a number of soldiers rolls a huge stone in front of the tomb and seals it with seven seals. They then pitch their tent and stand guard.

Then comes perhaps the most striking passage of the narrative, an actual account of Jesus' resurrection and emergence from the tomb, an account found in none of the other early Gospels. A crowd has come from Jerusalem and its surrounding neighborhoods to see the tomb. During the night hours, they hear a great noise and observe the heavens open up; two men descend in great splendor. The stone before the tomb rolls away of its own accord, and the two men enter. The soldiers standing guard awaken the centurion, who comes out to see the incredible spectacle. From the tomb there emerge three men; the heads of two of them reach into heaven. They are supporting the third, whose head reaches up beyond the heavens. Behind them emerges a cross. A voice then speaks from heaven: "Have you preached to those who are sleeping?" The cross replies, "Yes" (vv. 41–42).

The soldiers run to Pilate and tell him all that has happened. The Jewish leaders beg him to keep the story quiet, for fear that they will be stoned once the Jewish people realize what they have done in putting Jesus to death. Pilate commands the soldiers to silence, but only after reminding the Jewish leaders that Jesus' crucifixion was indeed their fault, not his. The next day at dawn, not knowing what has happened, Mary Magdalene goes with several women companions to the tomb to provide a more adequate burial for Jesus' body, but the tomb is empty, save for a heavenly visitor who tells her that the Lord has risen and gone. The manuscript then ends in the middle of a story that apparently described Jesus' appearance to some of his disciples (perhaps similar to that found in John 21): "But I, Simon Peter, and Andrew, my brother, took our nets and went to the sea; and with us was Levi, the son of Alphaeus, whom the Lord . . . " (v. 60). Here the manuscript breaks off.

Scholars continue to debate certain aspects of this fascinating account. Did this Gospel contain a narrative of Jesus' ministry or only of his Passion? Was it written by a Gnostic? When was it written? Did its author use any of the canonical Gospels as sources? If not, where did he acquire his accounts? Are some of the traditions that are preserved here earlier than those found in the Passion narratives of Matthew, Mark, Luke, and John?

Rather than go into all of the details of these debates, let me simply indicate the view that strikes me as the most reasonable and explain why. This Gospel appears to have been written after the canonical Gospels but not in reliance upon them. It was based on popular stories about Jesus' Passion, which were in circulation in a number of Christian circles. Its author may have had Gnostic leanings and certainly felt considerable antipathy towards non-Christian Jews.

That the *Gospel of Peter* represents a later stage of development in the traditions about Jesus than what we find in the first-century Gospels is suggested first of all by the heightened legendary elements, especially, the (literally) heightened Jesus and the cross that walks behind him and speaks to the heavens. The treatment of "the Jews" in this account is also significant for dating its traditions, for here they are made even more culpable for Jesus' death than in the canonical Gospels. Indeed, Pilate, representing the Roman authorities, is altogether blameless; it is the king of the Jews, Herod, along with the other Jewish leaders, who are totally at fault for Jesus' unjust condemnation. This portrayal coincides with views that were developing in Christian circles in the second century, a period in which Christian anti-Judaism began to assert itself with particular vigor (as we will see in Chapter 24). One byproduct of this increased animosity is that Christians began to exonerate Pilate for Jesus' death and to blame Jews (indeed, all Jews) more and more. In the *Gospel of Peter*, it is Jews who actually do the dirty work of crucifying Jesus; later they regret it, and explicitly express their fears that Jerusalem will now be destroyed as a result of their actions. The interpretation of the destruction of Jerusalem as God's vengeance upon the Jewish people for the execution of Jesus became a common theme in Christian writers of the second century. Further support for the late date of the account comes in the hints of a Gnostic understanding of Jesus' Passion, which we noted earlier. It appears, therefore, that the account as we now have it was written after the Gospels that eventually became part of the New Testament.

Is the *Gospel of Peter* based on any of these earlier narratives? It does have a number of close parallels to the canonical Gospels, particularly to Matthew, where we also read of Pilate washing his hands and the posting of a guard at the tomb. At the same time, we would be hard-pressed to explain why this author left out so many canonical passages that would have suited his purposes so admirably, had he known them, including the cry from the Jewish crowds, in which they assume full responsibility for Jesus' death after Pilate washes his hands ("His blood be upon us and our children"; Matt 27:25), the account of Jesus' carrying his cross, and the mocking of Jesus during his crucifixion. Recall that the only solid grounds for thinking that one document was the source for another is when they have extensive verbal agreements. There are no full sentences that the *Gospel of Peter* shares word for word with the other Gospels; indeed, there are virtually no verbatim agreements of any kind that extend for more than two or three words.

Perhaps it is best, then, to see the accounts of this narrative as having been drawn from stories about Jesus' Passion and resurrection that were in wide circulation among Christians. Some of these stories would have been known in similar forms in different communities; none of them would have been told in exactly the same way, since they were passed along by word of mouth. As Christians told the stories, they modified them, adding legendary details here and there, eliminating parts that appeared irrelevant, and incorporating their own views into the narrative. The author of the *Gospel of Peter*, living perhaps at the beginning of the second century, did what others had done before him and as others would do afterwards; he collected the stories he had heard, or possibly read, and created out of them a narrative of the words, deeds, and experiences of Jesus.

SAYINGS GOSPELS

We have seen that some of the sources lying behind the canonical Gospels may have contained principally, or exclusively, sayings of Jesus. Most, but not all, of the Q material consists of sayings, and at least two sources recounting Jesus' discourses were used by the author of the Fourth Gospel. Unfortunately, we are not able to determine whether these Johannine sources included other traditions as well, for example, stories of what Jesus did and experienced.

For many years, scholars denied that a pure "sayings" Gospel, that is, one filled with Jesus' teachings and nothing else, could have existed in the early church, especially if these sayings made no reference to Jesus' own death and resurrection. This view was based on the prevailing notion that for all early Christians the real significance of Jesus was that he died for the sins of the world and was raised from the dead. To be sure, Jesus' teachings were important to the early church, but according to this view all of the early Christians believed that his death and resurrection alone had brought salvation. The discovery of "sayings" Gospels, especially the *Gospel of Thomas*, has forced scholars to reconsider this view.

The Gospel of Thomas

The *Gospel of Thomas* is without question the most significant book discovered in the Nag Hammadi library. Unlike the *Gospel of Peter*, discovered sixty years earlier, this book is completely preserved. It has no narrative at all, no stories about anything that Jesus did, no references to his death and resurrection. *The Gospel of Thomas* is a collection of 114 sayings of Jesus.

The Sayings of the Collection. The sayings are not arranged in any recognizable order. Nor are they set within any context, except in a few instances in which Jesus is said to reply to a direct question of his disciples. Most of the sayings begin simply with the words "Jesus said." In terms of genre, the book looks less like the New Testament Gospels and more like the Book of Proverbs in the Hebrew Bible. Like Proverbs, it is a collection of sayings that are meant to bring wisdom to the one who can understand. In fact, the opening statement indicates that the correct understanding of these sayings will provide more than wisdom; it will bring eternal life: "These are the secret words which the living Jesus spoke, and Didymus Judas Thomas wrote them down. And he said, 'He who finds the meaning of these words will not taste death'" (*Gosp. Thom.*, 1).

The Jesus of this Gospel is not the Jewish messiah that we have seen in other Gospels, not the miracle-working son of God, not the crucified and resurrected Lord, and not the Son of Man who will return on the clouds of heaven. He is the eternal Jesus, whose words bring salvation.

The Reputed Author. Who is Didymus Judas Thomas, who allegedly penned these words? We know this name from other ancient Christian sources, such as the *Acts of Thomas*. Both "Didymus" and "Thomas" are words that mean "twin" (the first is Greek, the second Semitic); Judas is his proper name. According to the *Acts of Thomas*, he was a blood relation of Jesus, the same one mentioned in the New Testament (Mark 6:3). Thus, Didymus Judas Thomas was Jesus' twin brother (see box 12.2). Who better to relate the secret words of Jesus that can bring eternal life than his own twin brother?

The Character of the Sayings. Many of the sayings of Jesus in this Gospel will be familiar to those who have read the Synoptic Gospels: "If a blind man leads a blind man, the two of them fall into a pit" (*Gosp. Thom.*, 34); "Blessed are the poor, for yours is the Kingdom of Heaven" (54); " The harvest is great, but the workers are few; but beseech the Lord to send workers to the harvest" (73). Other sayings sound vaguely familiar, yet somewhat peculiar: "Let him who seeks not cease seeking until he finds, and when he finds, he will be troubled, and when he is troubled, he will marvel, and he will rule over the All" (2).

Still other sayings of Jesus in the *Gospel of Thomas* sound quite unlike anything known from the New Testament: " . . . On the day when you were one, you became two. But when you have become two, what will you do?" (11); "If the flesh exists because of spirit, it is a miracle, but if spirit exists because of the body, it is a miracle of mira-

SOME MORE INFORMATION

Box 12.2 Judas Thomas as Jesus' Twin Brother

Some of the Christians in Syria thought that Jesus' brother Judas (or Jude), mentioned in Mark 6:3, was actually his twin. Hence the name, Judas Thomas, "Jude, the Twin." This idea is puzzling for most modern readers. If these ancient Syrian Christians believed that Jesus was unique in being born of a virgin, how could they also think that he had a twin brother?

Unfortunately, none of the ancient Syrian texts that allude to this belief answers the question. But we may be able to gain some insight by considering other places in ancient literature in which twins are born, one the son of a mortal and the other the son of a god. The most famous account comes from Greek mythology in the tale of the birth of Heracles (Hercules) and his twin brother, the mortal Iphicles. The story was retold many times, perhaps most memorably in a humorous play titled *Amphitryon*, by the Roman playwright Plautus, in the second-century B.C.E.

The plot goes like this. Amphitryon is a general in the Greek army who leaves his pregnant wife Alcmena in order to go off to war. The night before he returns, Zeus looks down upon Alcmena and becomes awestruck by her ravishing beauty. Assuming the shape of Amphitryon, Zeus comes to her, claiming to have returned from battle. They spend the night in passionate embrace; so much does Zeus enjoy the tryst that he commands the constellations to stop their motion so as to prolong the night. When he finally departs, many hours later, Amphitryon returns home, dismayed and distraught that Alcmena isn't overjoyed at seeing him after his long absence—not understanding, of course, that she thinks she has just spent a wild night frolicking in his arms.

Her divine encounter has left Alcmena doubly pregnant. She eventually gives birth to two sons: Iphicles, the human son of Amphitryon, and Heracles, the divine son of Zeus. Did the ancient Syrian Christians know tales such as this and think that it might be possible for Jesus and Judas to be twins, born at the same time of the same mother, one being the son of God and the other the son of Joseph?

cles. But I marvel at how this great wealth established itself in this poverty" (29); "I stood in the midst of the World, and I appeared to them in the flesh. I found all of them drunk; I did not find any of them thirsting. And my soul was pained for the sons of men because they are blind in their hearts, and they do not see that they came empty into the world. . . . When they have shaken off their wine, then they shall repent" (28); "His disciples said, 'On what day will you be revealed to us and on what day shall we see you?' Jesus said, 'When you undress without being ashamed, and you take your clothes and put them under your feet as little children and tramp on them, then you shall see the Son of the Living One, and you shall not fear'" (37).

The Overarching Message of the Book. The meanings of these sayings are in no way obvious. If they were, they would not be called secret! They will seem far less obscure, however, if you try to understand them in light of the basic gnostic myth explained in the preceding chapter. Many of the most puzzling sayings in this collection appear to reflect the notion that within the hearer is a spark of the divine that had a heavenly origin. This spark has tragically fallen into the material world, where it has become entrapped in a body (sunk into "poverty"), and in that condition it has become forgetful of its origin (or "drunk"). It needs to be reawakened by learning the truth about this material world and the impoverished material body that it inhabits. Jesus is the one who conveys this truth; once the spirit learns the meaning of his words, it will be able to strip off this body of death, symbolized sometimes as garments of clothing, and escape this material world. It will then have salvation, life eternal; it will rejoin the divine realm and rule over all.

There is not a word in the *Gospel of Thomas* about Jesus' crucifixion and resurrection. Indeed, for this author none of Jesus' earthly activities appears to matter; there is no word here of his miracles or encounters or experiences. What matters are Jesus' secret teachings. He brings salvation not through his Passion but by conveying the message necessary for deliverance from this impoverished material existence.

Not only are Jesus' bodily experiences of no importance in the *Gospel of Thomas*, but the physical existence of the believer is irrelevant as well. For this reason, neither human events on the personal level nor history itself is of any consequence. The kingdom of God is not something to be expected in the future: "His disciples said to him, 'On what day will the kingdom come?'" Jesus answers: "It will not come by expectation. They will not say, 'Look here,' or, 'Look there,' but the Kingdom of the Father is spread out on the earth and people do not see it" (*Gosp. Thom.*, 113). The kingdom is here, now, for those who know who they are and whence they have come; it is not a physical place, but a salvation from within. Jesus says "If the ones who lead you say, 'There is the kingdom, in heaven,' then the birds of heaven shall go before you. If they say to you, 'It is in the sea,' then the fish shall go before you. Rather the kingdom is within you and outside you. If you know yourselves, then you will be known, and you will know that you are sons of the living Father. But if you do not know yourselves, then you are in poverty and you are poverty" (*Gosp. Thom.*, 3).

Thus this material world and the body that we inhabit are poor excuses for existence. Only through knowledge—knowledge of who one really is, as revealed by the living Jesus—can we escape and enjoy the riches of the kingdom of the Father.

This is a powerful message, and one that stands in stark contrast to the Gospel proclaimed by other Christians of the early church, who maintained that the material world was good because it was created by God, who taught that the kingdom of God would be a physical presence on earth to come in the near future, and who proclaimed that salvation came not by understanding the secret message of Jesus but by believing in his death and resurrection.

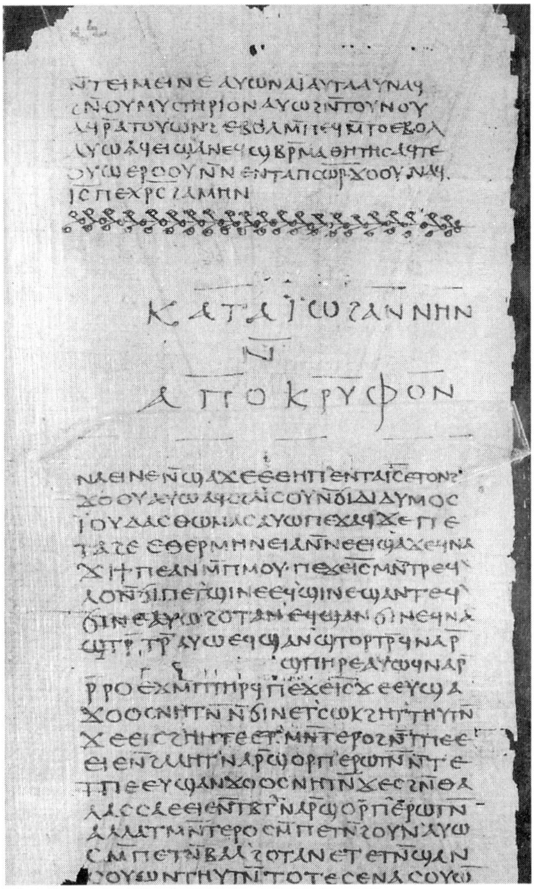

Figure 12.1 The opening of the Coptic *Gospel of Thomas*, which begins (in the middle of the page) with the words "These are the secret words which the living Jesus spoke, and Didymus Judas Thomas wrote them down."

Thomas and the Synoptics. Scholars have naturally raised the question of whether the *Gospel of Thomas* represents a form of Christianity that is early and independent of that preserved, say, in the Synoptic Gospels or whether it represents a later development of Christianity, based in part on the teachings of Jesus found in the Synoptics but modified in light of Gnostic beliefs. As we have seen, some of the sayings in Thomas are like those found in the Synoptics, with slight differences.

Could some of these be closer to the way Jesus actually expressed himself? Other sayings cannot be found in the Synoptics. Could some of these be authentic? Is the entire collection early, from the first century itself, or was it compiled only later?

These are intriguing questions but ones that are not easily answered. Scholars have argued about them intensely since the discovery of the Gospel, and even now, fifty years later, the heat of the debate has not subsided. Let me explain the position that strikes me as the most plausible.

It does not appear that the *Gospel of Thomas* actually used the Synoptic Gospels to formulate its own sayings of Jesus. As we have seen, the burden of proof in such matters is on the one who claims that an author used another document as a source. The surest indicators of reliance upon a source are detailed and extensive verbal parallels, but this is precisely what we do not find with the *Gospel of Thomas* in relation to the Synoptics. There are many similar sayings but few extensive verbal correspondences.

The fact that the *Gospel of Thomas* is written in Coptic rather than Greek, the language of the Synoptics, does not work against this position. Several Greek fragments of Thomas have also survived from antiquity, discovered not at Nag Hammadi but in an ancient trash heap elsewhere in Egypt, in a town called Oxyrhynchus. These small fragments date to some point in the second century, much earlier than the Coptic translation. They show us that the Gospel was originally written in Greek, and they indicate something about the care with which the translator did his work. When studied closely, they confirm our suspicion that extensive verbal similarities did not exist between the original *Gospel of Thomas* and the Synoptics.

Finally, if Thomas did use the Synoptics, it would be especially hard to explain why he left out of his account most of their sayings of Jesus, many of them relevant to his agenda. It is probably better, therefore, to assume that the author who calls himself Thomas knew a number of the sayings of Jesus and understood these sayings in a particular way, based on his knowledge of what I have called the Gnostic myth. He collected these sayings, some of them old, some of them new, and put them into a Gospel designed for his community, where beliefs were rooted not in the death and resurrection of Jesus but in his secret message.

Thomas and the Q Source. The final product reminds many scholars of the Q source. Some have maintained that Q was also composed entirely of the sayings of Jesus and that the community for whom it was written was not concerned about Jesus' activities and experiences, including his death on the cross. If they are right, then something like Thomas's community, even if not quite so Gnostic in its orientation, was already in existence prior to the writing of the New Testament Gospels.

Many other scholars, on the other hand, have their doubts. For one thing, it is not true that Q contained no narratives. As we have seen, two of them survive: the temptation and the healing of the centurion's son. How many others did Q narrate? Unfortunately, despite the extravagant claims of some scholars, we simply cannot know. Even more unfortunately, we cannot know whether the Q source contained a Passion narrative, even though scholars commonly claim that it did not. The reality is that our only access to Q is through the agreements of Matthew and Luke in stories not found in Mark. True, Matthew and Luke do not agree in their Passion narratives against Mark. Does this mean that Q did not have a Passion narrative? Not necessarily. It could mean that when either Matthew or Luke differs from Mark in the Passion narrative, one account was taken from Q and the other was drawn from Mark. Or it could mean that Matthew or Luke, or both, occasionally utilized their other traditions (M and L) for Jesus' Passion, rather than Q.

There is at least one stark difference between Q and Thomas, which relates directly to the beliefs of the communities that preserved them. We have seen that Thomas denies the future coming of the Son of Man in judgment upon the earth; this futuristic hope, however, is an important theme in Q. Some scholars have argued that Q sayings like Luke 12:8–9 (Matt 10:32–33), which speaks of the day of judgment when the Son of Man arrives, were not in the original version of Q but were only added later. Their reason for thinking so, however, is that they believe that the original version of Q was not apocalyptic in its orientation: any apocalyptic ideas would therefore not have been original to it. As you might surmise, this leads to a kind of circular reasoning, no less curious for being so common: if Q was like Thomas, it cannot have had apocalyptic sayings; if we remove the apocalyptic sayings from Q, it is like Thomas; therefore, Q was originally like Thomas.

Conclusion: The Date of Thomas and Its Traditions. Although we cannot know whether a source like Thomas existed during the first century, there are good reasons for thinking that Thomas itself did not. The most obvious is that the full-blown Christian-gnostic myth that many of Thomas's sayings presuppose cannot be documented as existing prior to the second century.

This is not to deny, however, that individual sayings found in Thomas may go back to Jesus himself. Indeed, as we will see later, all of the sayings in Thomas, and in every other source, canonical and noncanonical, must be judged as theoretically going back to Jesus. Moreover, there are grounds for thinking that some of the 114 sayings of this particular Gospel, especially some of the parables, are preserved in an older form than in the canonical Gospels, that is, they may be more like what Jesus actually said (see box 12.3).

Revelation Discourses

The other kind of sayings Gospel is an account in which Jesus appears to one or more of his disciples after his resurrection and conveys the secret revelation that is necessary for their salvation, a revelation which they then dutifully record for those who are chosen. Often these secret revelations have to do with the mysteries of how the universe came into existence, how souls came to be present here, and how they can escape. In other words, the vast majority of these Gospels are Gnostic in their orientation.

One example is the widely circulated *Apocryphon of John* (an apocryphon is a secret book), in which the resurrected Jesus appears to John the son of Zebedee to reveal to him the secrets of the universe and the divine realm, the origin of the evil creator Ialdabaoth, the creation of the human race, and its salvation through the appearance of a divine aeon from on high, who reveals the secret knowledge necessary for deliverance from this material world. The form of the Gnostic myth revealed here is very similar to the account narrated by the church father Irenaeus around the year 180 C.E., so the book appears to have been known in Christian churches by the middle of the second century.

Belonging to the same basic genre and coming from about the same time is the *Apocryphon of James*, discovered near Nag Hammadi. This Gospel is a dialogue between Jesus and his two disciples Peter and James 550 days after his resurrection. In the dialogue Jesus responds to the questions of his followers and urges them to attain salvation by knowing themselves and living in ways appropriate to the children of God.

Not all of the revelation discourses were Gnostic, however. In fact, one of the most interesting is a proto-orthodox writing from the early or mid second century produced in large measure to counter gnostic ideas about the nature of Christ's body. This work does not come from Nag Hammadi but was uncovered in a Coptic translation in Cairo at the end of the nineteenth century. It is called the *Epistle of the Apostles* because it is allegedly a letter written to Christians around the world by the eleven apostles after Jesus' resurrection (Judas having hanged himself). In this letter, the "apostles" claim to have received a special revelation from Jesus warning them to avoid the teachings of the false apostles Simon Magus and Cerinthus, two of the most infamous Gnostics in the eyes of second-century proto-orthodox writers. In particular, the document affirms the idea that Jesus was a real flesh-and-blood human being and emphasizes that those who believe in him are destined to be raised, bodily, from the dead.

INFANCY GOSPELS

As can be seen from the revelation discourses, Christians appear to have been intrigued with the activities of Jesus after his resurrection, perhaps because the earliest traditions said so little about what he did between his resurrection and ascension into heaven. One other period about which the earliest traditions were largely silent was Jesus' infancy and youth. The New Testament Gospels present only a few of stories relating to Jesus' young life, for example, Matthew's account of the worship of the Magi and the flight to Egypt and Luke's story of Jesus' visit to the Temple as a twelve-year-old. After the New Testament Gospels were written—

SOME MORE INFORMATION

Box 12.3 The Older Sayings of the Gospel of Thomas

If the *Gospel of Thomas* was written independently of the Synoptics, what does one make of the sayings of Jesus that they have in common but in slightly different forms? Is it possible that Thomas may preserve an older form of some of these sayings, closer to the way in which Jesus delivered them? Most scholars think that this is at least theoretically possible.

How do we know when a saying is older? We will consider this issue at greater length in Chapter 13. Here let me point out one controversial criterion that some researchers have used. If there are two different forms of a saying, these scholars claim, then the one that is simpler and more direct is more likely to be older. The logic behind this criterion is that sayings were generally embellished and expanded in the retelling.

Not everyone agrees with this criterion, but it at least deserves some consideration. What happens when it is applied to the sayings found in both Thomas and the Synoptics? Sometimes the form found in Thomas can lay claim to being older. Consider the following examples:

Thomas	*The Synoptics*
The disciples said to Jesus, "Tell us, what is the Kingdom of Heaven like?" He said to them, "It is like a mustard seed, smaller than all seeds. But when it falls on plowed ground, it puts forth a large shrub and becomes a shelter for the birds of heaven." (*Gosp. Thom.*, 20)	He also said, "With what can we compare the kingdom of God, or what parable will we use for it? It is like a mustard seed, which, when sown upon the ground, is the smallest of all the seeds on earth; yet when it is sown it grows up and becomes the greatest of all shrubs, and puts forth large branches, so that the birds of the air can make nests in its shade." (Mark 4:30–32)
And he said, "The man is like a wise fisherman who threw his net into the sea. He drew it up from the sea; it was full of small fish. The fisherman found among them a large, good fish. He threw all the small fish back into the sea; with no trouble he chose the large fish. He who has ears to hear, let him hear." (*Gosp. Thom.*, 8)	"Again, the kingdom of heaven is like a net that was thrown into the sea and caught fish of every kind; when it was full, they drew it ashore, sat down, and put the good into baskets but threw out the bad. So it will be at the end of the age. The angels will come out and separate the evil from the righteous and throw them into the furnace of fire, where there will be weeping and gnashing of teeth." (Matt 13:47–50)
Jesus said, "If a blind man leads a blind man, the two of them fall into a pit." (*Gosp. Thom.*, 34)	He also told them a parable: "Can a blind person guide a blind person? Will not both fall into a pit?" (Luke 6:39; the version in Matt 15:14 is somewhat longer)

and possibly earlier, although we have no hard evidence one way or the other—Christians began to tell stories about Jesus as a young boy. For the most part, the legendary character of these creative fictions is easily detected. We are fortunate that later authors collected some of them into written texts, the so-called infancy Gospels, which began to be produced by the first part of the second century at the latest.

One of the earliest is the *Infancy Gospel of Thomas* (not to be confused with the Coptic *Gospel of Thomas* discovered near Nag Hammadi), a document dated by some scholars to around the year 125 C.E. Here is a fascinating account of Jesus' youth beginning at the tender age of five. Behind the narrative lies a question that intrigues some Christians even today: if Jesus was a miracle-working Son of God as an adult, what was he like as a child? In this account, as it turns out, he is more than a little mischievous. When he first appears, he is making clay sparrows by a stream on the Sabbath. A Jewish man passing by sees what he has done and upbraids him for violating the Law by not keeping the Sabbath day holy. Instead of apologizing, the child Jesus claps his hands and tells the sparrows to be gone. They come to life and fly off, thereby destroying any evidence of wrongdoing!

One might have expected that with his supernatural powers Jesus would have been a useful and entertaining playmate for the other children in town. It turns out, however, that the boy has a temper and is not to be crossed. When a child accidentally runs into him on the street, Jesus turns in anger and declares, "You shall go no further on your way." The child falls down dead. (Jesus later raises him from the dead, along with others that he has cursed on one occasion or another.) And Jesus' wrath is not reserved for children. When Joseph sends him to school to learn to read, Jesus refuses to recite the alphabet. His teacher pleads with him to cooperate. Jesus replies with a scornful challenge: "If you really are a teacher and know the letters well, tell me the power of Alpha and I'll tell you the power of Beta." More than a little perturbed, the teacher cuffs the boy on the head, the single largest mistake of an illustrious teaching career. Jesus withers him on the spot. Joseph is stricken with grief and gives an urgent order to his mother: "Do not let him go outside: anyone who makes him angry dies."

As time goes on, however, Jesus begins to use his powers for good. He saves his friends from deadly snake bites, heals the sick, and proves remarkably handy around the house: when Joseph miscuts a board, Jesus corrects his mistake miraculously. The account concludes with Jesus as a twelve-year old teaching in the Temple; he is surrounded by scribes and Pharisees who listen to him and who bless Mary for the wonderful child she has brought into the world.

The blessing of Mary is a theme that is played out in some of the other infancy Gospels, although most of these are dated after the second century. One that may have been written early, however, is the *Gospel of James*. The James of the title is the brother of Jesus, known from other sources. His Gospel, or "proto-Gospel," as it is sometimes called since it narrates events prior to Jesus' birth, describes the miraculous character of their mother, Mary. Jesus obviously did not come into the world in a normal way, in this author's view, since his mother was a virgin. Why, though, was she chosen to bear the Son of God? The accounts of this Gospel provide some pious reflections that give an answer: Mary herself was born miraculously and was set apart for the service of God at a young age. The account describes Mary's birth, early life, and activities prior to and immediately after bearing Jesus through the power of the Holy Spirit, including a more extended description of her relations with Joseph, a narrative of their journey to Bethlehem, and an account of her postpartum examination. Narratives such as this became increasingly important in the early Middle Ages as Christians began to venerate the Blessed Virgin Mary, or "mother of God," as she came to be called.

CONCLUSION: THE OTHER GOSPELS

What can we say in conclusion about the other Gospels, those that did not make it into the New Testament? Most of them are later than the canonical four. This does not mean, however, that

Matthew, Mark, Luke, and John were the earliest accounts to be written. On the contrary, these books were themselves based on earlier sources that have since been lost. Moreover, some of the traditions preserved in the noncanonical Gospels, especially in the Gospels of Thomas and Peter, may be much older than the books themselves, at least as old as some of the traditions in the canonical books. On the whole, though, the noncanonical Gospels are of greater importance for understanding the diversity of Christianity in the second and third and later centuries than for knowing about the writings of the earliest Christians. When the Christians of the second century began to collect apostolic writings into a canon of Scripture, they considered the age of a document to be an important criterion for deciding whether or not it belonged. Those that had been around for a long time, and that were widely known as a result, were more likely to be included in the canon than those that had been penned only recently.

The noncanonical Gospels are important for the study of the New Testament, however, for they show that Christians continued to reflect on the significance of Jesus and to incorporate their views into the stories told about his words and deeds. This process began at the very outset of Christianity itself, when the earliest believers told others about the man in whom they believed. This widespread modification of the tradition explains why we have to approach the surviving Christian Gospels not only from the literary perspective, to see how each Gospel portrays Jesus—a task that we have now completed—but also from the historical perspective, to determine which of the traditions preserved in these Gospels, both canonical and noncanonical, are historically accurate.

SUGGESTIONS FOR FURTHER READING

Cameron, Ron, ed. *The Other Gospels: Non-Canonical Gospel Texts*. Philadelphia: Westminster, 1982. A nice collection of the noncanonical Gospels of the second and third centuries, in English translation with brief introductions.

Cartlidge, David R., and David L. Dungan, eds. *Documents for the Study of the Gospels*. 2d ed. Philadelphia: Fortress, 1994. A valuable selection of ancient literary texts that relate to the New Testament Gospels, including English translations of the Gospels of Thomas and Peter and selections from several other noncanonical Gospels.

Crossan, John Dominic. *Four Other Gospels: Shadows on the Contours of the Canon*. Minneapolis, Minn.: Winston Press, 1987. Presenting an alternative perspective from that sketched in the present chapter, Crossan maintains that some of the noncanonical Gospels were earlier and more reliable than those within the New Testament.

Elliott, J. K. *The Apocryphal New Testament: A Collection of Apocryphal Christian Literature in an English Translation*. Oxford: Clarendon, 1993. An excellent one-volume collection of noncanonical Gospels, acts, epistles, and apocalypses, in a readable English translation with nice, brief introductions.

Hennecke, Edgar, and Wilhelm Schneemelcher, eds. *New Testament Apocrypha*, 2 vols. Trans. by A. J. B. Higgins, et al. Ed. R. McL. Wilson. Philadelphia: Westminster Press, 1991. English translations of all the early noncanonical writings preserved from Christian antiquity, with detailed scholarly introductions; an indispensable resource for advanced students.

Klijn, A. F. J. *Jewish-Christian Gospel Tradition*. Leiden: E. J. Brill, 1992. The most thorough examination of the noncanonical Jewish-Christian Gospels, as attested principally in the writings of the early Christian church fathers.

Pagels, Elaine. *The Gnostic Gospels*. New York: Random House, 1976. An enormously popular and provocative account of the views presented in the Gnostic Gospels discovered near Nag Hammadi, especially in relation to emerging Christian orthodoxy.

CHAPTER 13

The Historical Jesus: Sources, Problems, and Methods

Up to this point in our study we have examined the early Christian Gospels as discrete pieces of literature, uncovering their unique portrayals of Jesus through a variety of methods: literary-historical, redactional, comparative, thematic, and socio-historical. At every stage, we have been interested in learning how an author, and the sources he used, understood and portrayed the life of Jesus. At no point have we moved beyond these literary concerns to ask about what actually happened during the life of Jesus, to find out what he really said, did, and experienced. We are now in a position to explore these other, purely historical issues. Apart from what certain Christian authors said about Jesus long after the fact, what can we know about the man himself, about the actual life of the historical Jesus?

This is a difficult question to answer (even though Christian scholars, preachers, and laypeople seem to answer it easily all the time) because, as we have seen, the earliest accounts of the historical Jesus, the Christian Gospels, vary so widely among themselves. The differences are not restricted to conflicting details scattered here and there among the records, even though differences of this kind do indeed abound, as anyone who does a methodical comparison of the early Gospels can see. The differences go much deeper, to the very heart and soul of how Jesus is understood and portrayed. Think of how differently Jesus appears, for example, in the Gospels of Mark and John and Thomas. Given the variety of portrayals of Jesus and the different accounts of what he said and did, some of them difficult to reconcile with one another, how can the historian decide what really happened during his life? Before addressing this question directly, let me say a word about the grounds of our knowledge about Jesus, or about any other person from the past.

PROBLEMS WITH SOURCES

The only way that we can know what a person from the past said and did is by examining sources from the period that provide us with information. Most of our sources for the past are literary, that is, they are texts written by authors who refer to the person's words and deeds. But sources of this kind are not always reliable. Even eyewitness accounts are often contradictory, and contemporary observers not infrequently get the facts wrong. Moreover, most historical sources, for the distant past at least, do not derive from eyewitnesses but from later authors reporting the rumors and traditions they have heard.

For these reasons, historians have to devise criteria for determining which sources can be trusted and which ones cannot. Most historians would agree that for reconstructing a past event the ideal situation would be to have sources that (a) are numerous, so they can be compared to one another, (b) derive from a time near the event itself, so that they are less likely to have been based on hearsay or legend, (c) were produced independently of one another, so we know the authors could not have been in collusion, (d) do not contradict one another, so some of them are not necessarily in error, (e) are internally consistent, suggesting a basic concern for reliability, and (f) are not biased toward the subject matter, so we know the authors have not skewed their accounts to serve their own purposes.

Before pursuing the question of whether the literary texts that provide us with information about Jesus have these characteristics, we should remember that we are not asking whether the New Testament Gospels are important as religious or theological documents. Instead of passing judgment on their worth or trying to undermine their authority for those who believe in them, we are asking the question of the historian: are these books reliable for reconstructing what Jesus actually said and did?

As a first step toward an answer, we can ask whether any of the Gospel accounts can be corroborated by other ancient sources that describe the life and teachings of the historical Jesus. It is surprising to many students, and unsettling to many historians, that there are very few sources from the ancient world, outside of the Gospels, that even mention the man Jesus. And those few that do exist give us very little information. For organizational purposes we can categorize these other sources as non-Christian (whether Jewish or pagan) or Christian (whether within the canon or outside of it). For fairly obvious reasons, our investigation will be restricted to sources that can be plausibly dated to within a hundred years of Jesus' death, that is, to those written before the year 130 C.E. This is about the length of time that separates us today from William McKinley, the twenty-fifth president of the United States. Sources produced much later than this are almost certainly based on hearsay and legend rather than reliable historical memory.

NON-CHRISTIAN SOURCES

In popular imagination Jesus is generally thought to have had a stupendous impact on the society of his day. Surely he was the talk of the empire, with masses besieging him, Jewish leaders hating him, and Roman authorities fearing him. The political appointees of Rome must have communicated news of this man from Galilee to their superiors; word of his power and presence must have made its way to Rome. Possibly the decision to have him eliminated came down from on high, with the divine emperor himself realizing that he had met his match in this Son of God become man. In this view, Jesus' impact on society was immediate and mind-boggling, like a comet colliding with the earth.

The historian, on the other hand, realizes that this view is pure fiction. If our surviving sources are any indication at all, Jesus' impact on Roman society was slight, less like a comet striking the earth than a stone being tossed into the ocean. This may be hard to believe given the enormous impact his life has made on the history of civilization in the centuries that followed. Historians, though, are compelled to take the evidence seriously, and in this case the evidence is persuasive.

Pagan Sources

Among the hundreds of documents by pagan writers (i.e., those who were neither Jewish nor Christian) that survive from the first century of the Common Era—writings by historians, poets, philosophers, religious thinkers, public officials, and private persons, including literary texts, public inscriptions, private letters, and notes scribbled on scratch paper—Jesus is never mentioned. There are no birth records, official correspondence, philosophical rebuttals, literary discussions, or personal reflections. Nothing written by any pagan author of the first century so much as mentions his name.

The first reference to Jesus in a pagan source comes some eighty years after his death, in a letter written in 112 C.E. by the Roman governor of Bithynia-Pontus, Pliny the Younger, who asks his emperor, Trajan, what he should do about prosecuting Christians in his province. Pliny's letter tells us some interesting things about the followers of Jesus, for example, that they covered a range of ages and socioeconomic classes, but it says nothing about Jesus himself, except that he was worshipped by these people as a god. Thus, while the letter is important in showing how far Christianity had spread and what it was like in the early years of the second century, it is of practically no use in determining what the historical Jesus actually said and did.

A few years later, the Roman historian Suetonius mentions riots that had occurred among the Jews in Rome during the reign of the emperor Claudius (41–54 C.E.). He says they were instigated by a person named "Chrestus." Is this a misspelling of "Christ"? Some scholars think so. Unfortunately, Suetonius tells us nothing about the man. If he does have Jesus in mind, he must be referring only to Jesus' followers, since Jesus himself had been executed some twenty years before these riots swept through the capital.

At about the same time (115 C.E.), another Roman historian, Tacitus, mentions Christians in his *Annals*. He reports that when Nero torched the city of Rome, he placed the blame on the Christians and used them as scapegoats. In his discussion, Tacitus gives us the first bit of historical information about Jesus from a pagan author: "Christus, from whom their [the Christians'] name is derived, was executed at the hands of the procurator Pontius Pilate in the reign of Tiberius" (*Annals* 15.44). Tacitus goes on to indicate that the "superstition" that emerged in Jesus' wake first appeared in Judea (see box 13.1).

It is a pity that Tacitus does not tell us more. One must assume either that he did not consider information about Jesus to be of real historical importance or that this was all that he knew. Some scholars have noted that even this bit of knowledge is not altogether reliable: Pilate was not, in fact, a procurator but a prefect. In any event, Tacitus's report confirms what we know from other sources, that Jesus was executed by order of the Roman governor of Judea, Pontius Pilate, sometime during Tiberius's reign. We learn nothing, however, about the reason for this execution, or about Jesus' life and teachings.

These are the only references to Jesus in pagan sources during the hundred-year period after his death. On the whole, they provide scarcely any information concerning the things Jesus said, did, and experienced. For this kind of information, we are therefore obliged to turn elsewhere.

SOME MORE INFORMATION

Box 13.1 Christianity as a Superstition in the Roman World

Tacitus called Christianity a "superstition," as did a number of our later Roman sources. Authors in the Greco-Roman world used this term to describe any set of religious beliefs and practices that were antisocial, irrational, and motivated by raw fear of divine vengeance. Such beliefs and practices were antisocial in that they involved religious acts that were not sanctioned by the recognized cults and so were out of bounds from the point of view of society at large (see the discussion of magic in Chapter 2) They were irrational in that they could not be justified in terms of the prevailing modes of logic. They were motivated by fear, rather than the more noble virtues of love, truth, and honor, in that they maintained that the gods were bent on punishing those who did not perform their prescribed religious acts regularly and scrupulously.

For many of the highly educated members of Roman society in the second century, Christianity fit this description perfectly. As we will see in Chapter 25, this religion was not sanctioned by the state and was perceived as a secret and mildly dangerous society; its beliefs struck outsiders as irrational, especially its central claim that an executed criminal was the Lord of the universe; and it 's members often preached "fire and brimstone" against all who rejected its message, showing fear of divine retribution. Small wonder that the upper echelons of Roman society were not immediately drawn to this new religion.

Jewish Sources

In contrast to pagan sources, we have very few Jewish texts of any kind that can be reliably dated to the first century of the Common Era. There are references to Jesus in later documents, such as those that make up that great collection of Jewish lore and learning, the Talmud. This compilation of traditions was preserved by rabbis living in the first several centuries of the Common Era, and some of the traditions found in the Talmud may possibly date back to the period of our concern, but scholars have increasingly realized that it is difficult to establish accurate dates for these traditions. The collection itself was made long after the period of Jesus' life; the core of the Talmud is the Mishnah, a collection of Rabbinic opinions about the Law that was not written until nearly two centuries after his death. Moreover, Jesus is never mentioned in this part of the Talmud; he appears only in commentaries on the Mishnah that were produced much later. Scholars are therefore skeptical of the usefulness of these references in reconstructing the life of the historical Jesus.

There is one Jewish author, however, who wrote during our time period (before 130 C.E.) and mentioned Jesus. The Jewish historian Josephus produced several important works, the two best known of which are his insider's perspective on the Jewish War against Rome in 66–73 C.E. (he had been a general in the Jewish army but was captured and then made into a kind of court historian by the Roman emperor Vespasian) and his twenty-volume history of the Jewish people from Adam and Eve up to the time of the Jewish War, a book that he titled *The Antiquities of the Jews*. Scores of important, and less important, Jews, especially Jews in and around Josephus's own time, are discussed in these historical works. Jesus is not mentioned at all in Josephus's treatment of the Jewish War, which comes as no surprise since his crucifixion took place some three decades before the war started, but he does make two tantalizingly brief appearances in the *Antiquities*.

One reference to Jesus occurs in a story about the Jewish high priest Ananus, who abused his power before Rome in the year 62 C.E. by unlawfully putting to death James, whom Josephus identifies as "the brother of Jesus who is called the messiah" (*Ant.* 20.9.1). From this reference we can learn that Jesus was known to have a brother named James, which we already knew from the New Testament (see Mark 6:3 and Gal 1:19), and that he was thought by some people to be the messiah, although obviously not by Josephus himself, who remained a non-Christian Jew.

Josephus's religious perspective has made the other reference to Jesus a source of considerable puzzlement over the years, for he not only mentions Jesus as a historical figure but also appears to profess faith in him as the messiah—somewhat peculiar for a person who never converted to Christianity (see box 13.2). This second passage indicates that Jesus was a wise man and a teacher who performed startling deeds and as a consequence found a following among both Jews and Greeks; it states that he was accused by Jewish leaders before Pilate, who condemned him to be crucified; and it points out that his followers remained devoted to him even afterwards (*Ant.* 18.3.3).

It is useful to know that Josephus had this much information about Jesus. Unfortunately, there is not much here to help us understand specifically what Jesus said and did. We might conclude that he was considered important enough for Josephus to mention, though not as important as, say, John the Baptist or many other Palestinian Jews who were considered to be prophets at the time, about whom Josephus says a good deal more. We will never know if Josephus actually had more information about Jesus at his disposal or if he told us all that he knew. No other non-Christian Jewish source written before 130 C.E. mentions Jesus.

Clearly, we cannot learn much about Jesus from non-Christian sources, whether pagan or Jewish. If we want to know what Jesus actually said and did during his life, we are therefore compelled to turn to sources produced by his followers.

CHRISTIAN SOURCES

Outside the New Testament Gospels

Most of the noncanonical Gospels are legendary and late, dating from the second to eighth centuries. In many cases they are dependent on information gleaned from our earlier sources, especially

SOMETHING TO THINK ABOUT

Box 13.2 The Testimony of Flavius Josephus

Probably the most controversial passage in all of Josephus's writings is his description of Jesus in book 18 of *The Antiquities of the Jews*.

> At this time there appeared Jesus, a wise man, if indeed one should call him a man. For he was a doer of startling deeds, a teacher of people who receive the truth with pleasure. And he gained a following both among many Jews and among many of Greek origin. He was the Messiah. And when Pilate, because of an accusation made by the leading men among us, condemned him to the cross, those who had loved him previously did not cease to do so. For he appeared to them on the third day, living again, just as the divine prophets had spoken of these and countless other wondrous things about him. And up until this very day the tribe of Christians, named after him, has not died out. (*Ant.* 18. 3. 3)

This testimony to Jesus has long puzzled scholars. Why would Josephus, a devout Jew who never became a Christian, profess faith in Jesus by suggesting that he was something more than a man, calling him the messiah (rather than merely saying that others thought he was), and claiming that he was raised from the dead in fulfillment of prophecy?

Many scholars have recognized that the problem can be solved by looking at how, and by whom, Josephus's writings were transmitted over the centuries. They were not preserved by Jews, many of whom considered him to be a traitor because of his conduct during and after the war with Rome. Rather, it was Christians who copied Josephus's writings through the ages. Is it possible that this reference to Jesus has been beefed up a bit by a Christian scribe who wanted to make Josephus appear more appreciative of the "true faith"?

If we take out the Christianized portions of the passage, what we are left with, according to one of the most convincing recent studies, is the following:

> At this time there appeared Jesus, a wise man. For he was a doer of startling deeds, a teacher of people who receive the truth with pleasure. And he gained a following both among many Jews and among many of Greek origin. And when Pilate, because of an accusation made by the leading men among us, condemned him to the cross, those who had loved him previously did not cease to do so. And up until this very day the tribe of Christians, named after him, has not died out. (Meier 1991, 61)

the New Testament Gospels. As we have seen, the *Gospel of Thomas* may provide some independent knowledge of aspects of Jesus' teaching, but we must be alert to the gnostic inclination of its more unusual sayings; and the *Gospel of Peter* may provide information concerning events at Jesus' trial, although the anti-Jewish slant of the report makes even this doubtful. For the historian interested in knowing what really happened, the other noncanonical Gospels do not inspire confidence.

Students are sometimes surprised to learn how little information about the historical Jesus can be gleaned from the New Testament writings that fall outside of the four Gospels. The apostle Paul, who was not personally acquainted with Jesus but who may have known some of his disciples, provides us with the most detail. Regrettably, it is not much. As we will see in somewhat greater length in Chapter 21, Paul informs us that Jesus was born of a woman (Gal 4:4), that he was born as a Jew (Gal

4:4), that he had brothers (1 Cor 9:5), one of whom was named James (Gal 1:19), that he ministered among the Jews (Rom 15:7), that he had twelve disciples (1 Cor 15:5), that he instituted the Lord's Supper (1 Cor 11:23–25), possibly that he was betrayed (1 Cor 11:23, assuming that the Greek term here means "betrayed" rather than "handed over" to death by God), and that he was crucified (1 Cor 2:2). In terms of Jesus' teachings, in addition to the words at the Last Supper (1 Cor 11:23–25), Paul may refer to two other sayings of Jesus, to the effect that believers should not get divorced (1 Cor 7:10–11) and that they should pay their preachers (1 Cor 9:14).

Apart from these few references, Paul says almost nothing about the life and teachings of Jesus, even though he has a lot to say about the significance of Jesus' death and resurrection and his expected return in glory. The other New Testament authors tell us even less. This means that if historians want to know what Jesus said and did they are more or less constrained to use the New Testament Gospels as their principal sources. Let me emphasize that this is not for religious or theological reasons, for instance, that these and these alone can be trusted. It is for historical reasons, pure and simple. Jesus is scarcely mentioned by non-Christian sources for over a century after his death, and the other authors of the New Testament are more concerned with other matters. Moreover, the Gospel accounts outside the New Testament tend to be late and legendary, of considerable interest in and of themselves, but of little use to the historian bent on determining what actually happened during Jesus' lifetime. With the partial exceptions of the Gospels of Thomas and Peter, which even by the most generous interpretations cannot provide us with substantial amounts of new information, the only real sources available to the historian interested in knowing what Jesus said and did are therefore the New Testament Gospels.

The New Testament Gospels

To what extent are these New Testament documents reliable for the historian, and how can they be used to answer historical questions about Jesus? The answers to these questions can be inferred from our earlier analysis of these documents as literary texts. We have seen, for example, that the New Testament Gospels were not written at the time of Jesus' life or immediately thereafter. It appears that Mark, the earliest Gospel, was penned around the year 65 C.E. or so, and that John, the latest, was written perhaps around the year 95 C.E. These are only approximate dates, of course, but they are accepted by virtually all scholars. Thus, the earliest surviving Gospels were produced thirty-five to sixty-five years after the events they narrate. In modern terms, this would be like having written records of John F. Kennedy or Albert Einstein or Babe Ruth appear for the *first time* this year.

We have also seen that the authors of these Gospels were likely not among the earliest followers of Jesus. They themselves do not claim to be disciples; the books are all anonymous, and they give no solid information as to their authors' identity. And there are few reasons for thinking that the later traditions about who they were can be accepted as anything but hearsay.

These circumstances do not in themselves make the Gospels unreliable as historical documents. A book written fifty years after the fact by someone who was not an eyewitness is not necessarily historically inaccurate. What is more telling is the lack of consistency among these earliest accounts of the life of Jesus. For as we have repeatedly seen, both in their details and in their overall portrayals of who Jesus was, what he taught, and what he did, the four Gospels do not stand in perfect harmony with one another. They differ both in the factual information they provide, such as where Jesus' family was from, what he did during his life, when he died, and what his disciples experienced afterwards, and in the ways they understand who he was and what he did, for instance, whether he taught about his own identity and whether he performed miracles in order to demonstrate who he was.

All of the early Christian authors had perspectives on who Jesus was and on how he was significant. These perspectives affected the ways they told stories about him. Moreover, each author inherited a number of his stories from earlier written sources. and each of these sources had its own perspective. And before anyone bothered to write down stories about Jesus, such stories had circulated by word of mouth for years and years among Christians who recounted them for a variety of reasons: to magnify the importance of Jesus, to convince others to believe in him, to instruct them concerning his relationship with God, to show how he understood the Hebrew Scriptures, to encourage his followers with the hope that his words could bring, and so forth. As the stories circulated orally, they were changed to suit the purposes at hand. They were modified further when they were written down and further still when they were later redacted. Recall that this view is not based simply on scholarly imagination; we have evidence for it, some of which I have laid out in earlier chapters.

Figure 13.1 Jesus as the Good Shepherd. This is one of the earliest paintings of Jesus to survive from antiquity (from about two centuries after Jesus' death), from the catacomb of San Callisto in Rome.

Because these documents were of such importance to people who believed in Jesus as the Son of God, their concerns, to put it somewhat simplistically, were less historical than religious. Those who passed along the traditions and those who wrote them down were not interested in providing the brute facts of history for impartial observers; they were interested in proclaiming their faith in Jesus as the Son of God. This was "good news" for the believer, but it is not necessarily good news for historians, who are interested in getting behind the perspectives of the authors of the Gospels, and those of their sources, to reconstruct what Jesus really said, did, and experienced. How can "faith documents" such as the Gospels, produced by believers for believers to promote belief, be used as historical sources?

Over the course of the past century, historians have worked hard to develop methods for uncovering historically reliable information about the life of Jesus. In this hotly debated area of research, reputable and intelligent scholars have expressed divergent views concerning both the methods to be applied to the task and the conclusions to be drawn, even when there is a general agreement about method. I will sketch several of the methodological principles that have emerged from these debates in the pages that follow. As you will see, there is a logic behind each of them that is driven by the character of the sources. All of these principles can be applied to any tradition about Jesus, early or late, Christian or non-Christian, preserved in the New Testament Gospels or elsewhere. Anyone who does not find these principles satisfactory must come up with others that are better; in no case, however, can we simply ignore the problems of our sources and accept everything they say about Jesus' words and deeds as historically accurate. Once it is acknowledged that these Gospels are historically problematic, then the problems must be dealt with in a clear and systematic fashion. My sketch of the historical Jesus in Chapter 15 will be based on the application of these various principles.

THE HISTORICAL APPROACH TO THE GOSPELS: CRITERIA FOR RECONSTRUCTING THE LIFE OF JESUS

In many respects, the historian is like a prosecuting attorney. He or she is trying to make a case and is expected to bear the burden of proof. As in a court of law, certain kinds of evidence are acknowledged as admissible, and witnesses must be carefully scrutinized. Given the circumstances that our "witnesses" are the documents from antiquity that speak about Jesus, we can use three criteria to make a case for what actually happened during his life.

Independent Attestation

In any court trial, it is better to have a number of witnesses who can provide consistent testimony than to have only one, especially if we can show the witnesses did not confer with one another to get their story straight. A strong case will be supported by several witnesses who independently agree on a point at issue. So too with history. An event mentioned in several independent documents is more likely to be historical than an event mentioned in only one. This principle does not deny that individual documents can provide reliable historical information. but without corroborating evidence it is often impossible to know whether or not an individual source has made up an account, or perhaps provided a skewed version of it.

For the life of Jesus, we do in fact have a number of independent sources. Mark, the apostle Paul, and the authors of Q, M, L, and the signs source probably all wrote independently of one another; that is, it appears that Mark had not read the signs source, that Paul had not read Q, and so on. Moreover, we have seen that the *Gospel of Thomas*, possibly the *Gospel of Peter*, probably the Johannine discourse sources, and certainly Josephus were all produced independently of the other surviving accounts. Therefore, if there is a tradition about Jesus that is preserved in more than one of these documents, no one of them could have made it up, since the others knew of it as well, independently. If a tradition is found in several of these sources, then the likelihood of its going back to the very beginning of the tradition from which they all ultimately derive, back to the historical Jesus himself, is significantly improved.

This criterion does not work for sources that are not independent. For example, the story of Jesus and the so-called rich young ruler is found in three of the Gospels (Matt 19:16–22; Mark 10:17–22; and Luke 18:18–23; see box 6.1), but since Matthew and Luke took the story over from Mark—assuming the view of Markan priority that we discussed in Chapter 6—it is not independently attested. Thus, the criterion of independent attestation does not work for stories found in all three Synoptic Gospels, since the source for such stories is Mark, or among any two of them, since these are either from Mark or Q.

In other circumstances, however, the criterion does work. Some simple examples can help to clarify the matter. First, stories in which John the Baptist encounters Jesus at the beginning of his ministry can be found in Mark, in Q (where John's preaching is expounded), and in John. Why did all three sources, independently of one another, begin Jesus' ministry with his association with John the Baptist? Possibly because it really did start this way. Second, Jesus is said to have brothers in Mark (6:3), John (7:3), and Paul's first letter to the Corinthians (9:5). Moreover, Mark, Paul (Gal 1:19), and Josephus all identify one of his brothers as James. Conclusion: Jesus probably did have brothers and one of them was probably named James. Finally, Jesus tells parables in which he likens the kingdom of God to seeds in Mark, Q, and the *Gospel of Thomas*. Conclusion: Jesus probably did tell such parables. All of these examples involve independent sources.

Obviously there are limitations to the criterion of independent attestation. Merely because a tradition is found in only one source—for example, Jesus' visit to the Temple as a twelve-year-old or the parable of the Good Samaritan—it is not automatically discounted as historically inaccurate. The criterion shows which traditions are more likely to be authentic, but it does not show which ones are necessarily inauthentic—a critical difference!

At the same time, multiply attested traditions are not necessarily authentic either; they are simply more likely to be authentic. If a tradition is attested independently, then at the very least it must be older than all of the sources that record it, but it does not necessarily go all the way back to Jesus. It could well be, for example, that a multiply attested tradition derives from the years immediately after Jesus' death, with different forms of the story being told in a variety of communities thereafter. For this reason, our first criterion has to be supplemented with others.

Dissimilarity

The most controversial criterion that historians use, and sometimes misuse, to establish authentic tradition from the life of Jesus is commonly called the "criterion of dissimilarity." Any witness in a court of law will naturally tell things the way he or she sees them. Thus, the perspective of the witness has to be taken into account when trying to evaluate the merits of a case. Moreover, sometimes a witness has a vested interested in the outcome of the trial. A question that perennially comes up, then, involves the testimony of interested parties: are they distorting, or even fabricating, testimony for reasons of their own? The analogy does not completely work, of course, for ancient literary sources (or for modern ones either, for that matter). Authors from the ancient world were not under oath to tell the historical facts, and nothing

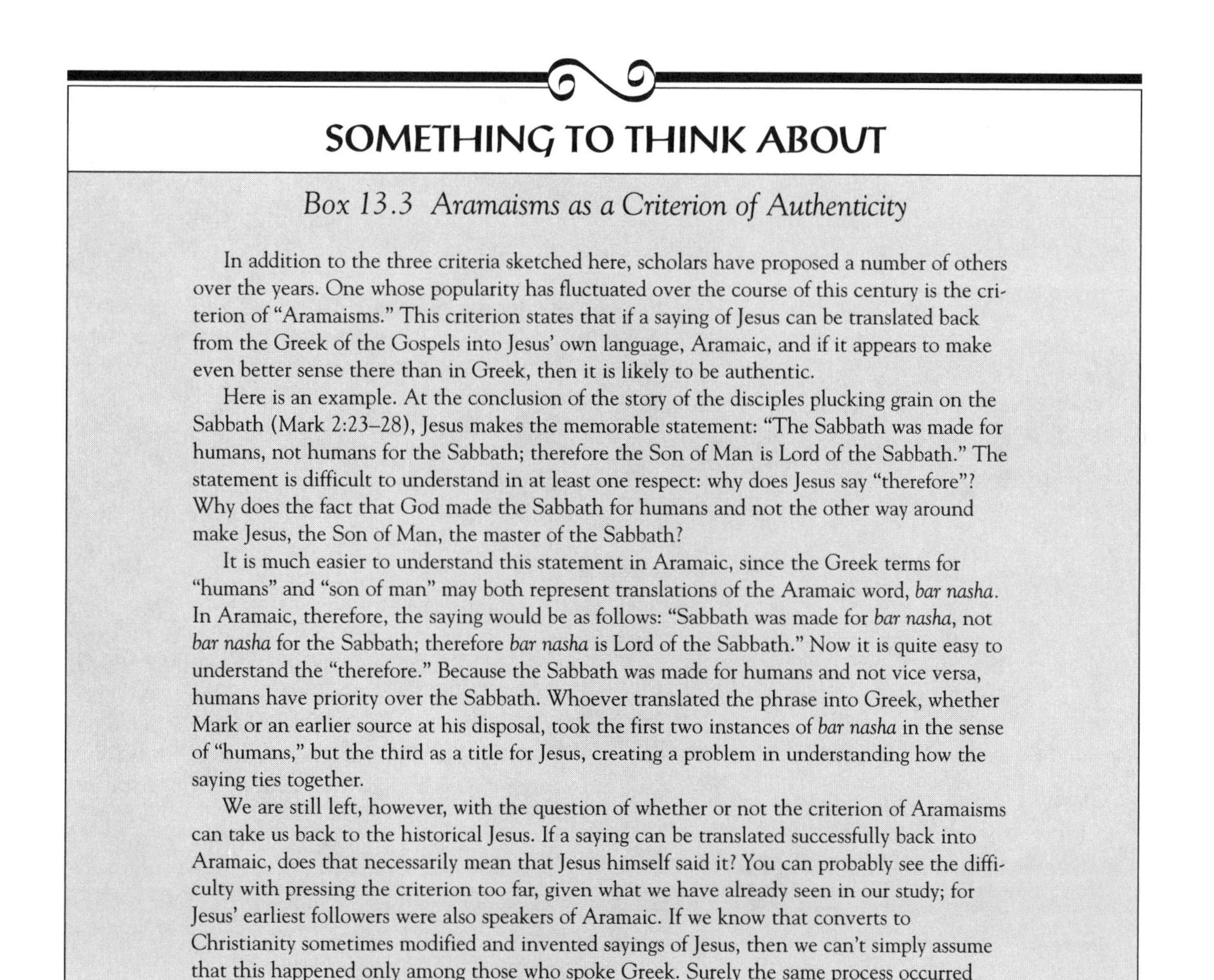

SOMETHING TO THINK ABOUT

Box 13.3 Aramaisms as a Criterion of Authenticity

In addition to the three criteria sketched here, scholars have proposed a number of others over the years. One whose popularity has fluctuated over the course of this century is the criterion of "Aramaisms." This criterion states that if a saying of Jesus can be translated back from the Greek of the Gospels into Jesus' own language, Aramaic, and if it appears to make even better sense there than in Greek, then it is likely to be authentic.

Here is an example. At the conclusion of the story of the disciples plucking grain on the Sabbath (Mark 2:23–28), Jesus makes the memorable statement: "The Sabbath was made for humans, not humans for the Sabbath; therefore the Son of Man is Lord of the Sabbath." The statement is difficult to understand in at least one respect: why does Jesus say "therefore"? Why does the fact that God made the Sabbath for humans and not the other way around make Jesus, the Son of Man, the master of the Sabbath?

It is much easier to understand this statement in Aramaic, since the Greek terms for "humans" and "son of man" may both represent translations of the Aramaic word, *bar nasha*. In Aramaic, therefore, the saying would be as follows: "Sabbath was made for *bar nasha*, not *bar nasha* for the Sabbath; therefore *bar nasha* is Lord of the Sabbath." Now it is quite easy to understand the "therefore." Because the Sabbath was made for humans and not vice versa, humans have priority over the Sabbath. Whoever translated the phrase into Greek, whether Mark or an earlier source at his disposal, took the first two instances of *bar nasha* in the sense of "humans," but the third as a title for Jesus, creating a problem in understanding how the saying ties together.

We are still left, however, with the question of whether or not the criterion of Aramaisms can take us back to the historical Jesus. If a saying can be translated successfully back into Aramaic, does that necessarily mean that Jesus himself said it? You can probably see the difficulty with pressing the criterion too far, given what we have already seen in our study; for Jesus' earliest followers were also speakers of Aramaic. If we know that converts to Christianity sometimes modified and invented sayings of Jesus, then we can't simply assume that this happened only among those who spoke Greek. Surely the same process occurred among Aramaic-speaking Christians as well.

but the facts. When examining ancient sources, however, the historian must always be alert to the perspective of the witness.

We know that early Christians modified and invented stories about Jesus. If we did not think so, we would have to say that Jesus really did make clay sparrows come to life when he was a five-year old and zap his young playmates when they irritated him, that he really did come forth from his tomb at his resurrection with his head reaching above the clouds, supported by angels as tall as skyscrapers, and that he really did reveal the secret Gnostic doctrines to his disciples months and years after his resurrection. No one believes that all of these events actually happened. How, then, did they come to be written down? Somebody made them up, and told them to other people, and eventually they came into the hands of an author—unless he made them up himself.

How can we know which stories were made up and which ones are historically accurate? The surest way is to determine the sorts of things Christians were saying about Jesus in other sources and then ascertain whether the stories told about his sayings and deeds clearly support these Christian views. If they do, then there is at least a theoretical possibility that these sayings and deeds were made up to advance the views that some Christians held dear.

Figure 13.2 Portrayal of Jesus as the Good Shepherd, from one of the oldest surviving Christian sarcophagi.

On the other hand, sometimes a saying or deed attributed to Jesus does not obviously support a Christian cause. A tradition of this kind would likely not have been made up by a Christian. Why then would it be preserved in the tradition? Perhaps because it really happened that way. Dissimilar traditions, that is, those that do not support a clear Christian agenda, are difficult to explain unless they are authentic; they are therefore more likely to be historical.

This criterion too has limitations. Just because a saying or deed of Jesus happens to conform to what Christians were saying about him does not mean that it cannot be accurate. Obviously, the earliest followers of Jesus, who must have appreciated the things that he said did, would have told stories that included such things. Thus, the criterion may do no more than cast a shadow of doubt on certain traditions. For example, the story of the young Jesus withering his playmates and then raising them from the dead, looks like something drawn from later Christian imagination, and the story of his revealing the secret doctrines of gnosis to a handful of followers is too closely aligned with the Gnostic theology to be beyond doubt. The criterion of dissimilarity is best used, however, not in a negative way to establish what Jesus did *not* say or do, but in the positive way to show what he *did* say or do.

This criterion can be clarified by a couple of brief examples. As we have seen, Jesus' association with John the Baptist at the beginning of his ministry is multiply attested. In some traditions, Jesus is actually said to have been baptized by John. Is this a tradition that a Christian would have made up? Most Christians appear to have understood that when a person was baptized, he or she was spiritually inferior to the one who was doing the baptizing. This view is suggested in the Gospel of Matthew, where we find John protesting that he is the one who should be baptized by Jesus, not the other way around. It is hard to imagine a Christian inventing the story of Jesus' baptism since this could be taken to mean that he was John's subordinate. It is more likely that the baptism is something that actually happened. The story that John initially refused to baptize Jesus, on the other hand, is not multiply attested (it is found only in Matthew) and appears to serve a clear Christian agenda. On these grounds, even though the story of John's reluctance cannot be proven to be a Christianized form of the account, it may be suspect.

Consider another example. According to all four canonical Gospels, and perhaps Paul, at the end of Jesus' life he was betrayed by one of his own followers. Is this a story that a Christian believer would invent? Would Christians want to admit that Jesus was turned in by one of his closest friends and allies? It seems unlikely; surely Jesus would have had a commanding presence over those closest to him. Why, then, do we have this tradition of betrayal, which is independently attested? Perhaps the betrayal is something that really happened.

A final, fairly obvious example. The earliest Christians put a good deal of effort into convincing non-Christian Jews that the messiah had to suffer and die, that Jesus' crucifixion was according to the divine plan. It was difficult for them to persuade others in part because, prior to the Christian proclamation of Jesus, there were no Jews, so far as we know, who believed that the messiah was going to be crucified; on the contrary, the messiah was to be the great and powerful leader who delivered Israel from its oppressive overlords. Christians who wanted to proclaim Jesus as messiah would not have invented the notion that he was crucified because his crucifixion created such a scandal. Indeed, the apostle Paul calls it the chief "stumbling block" for Jews (1 Cor 1:23). Where then did the tradition come from? It must have actually happened.

Other sayings and deeds of Jesus do not pass the criterion of dissimilarity. In Mark's Gospel, for example, when Jesus predicts that he is to go to Jerusalem and that he will be rejected by the scribes and elders, crucified, and then in three days raised from the dead, he is proclaiming precisely what the early Christian preachers were saying about him. The Passion predictions cannot pass the criterion of dissimilarity. Does that mean that Jesus did not predict his own death? Not necessarily. It means that we can't *show* that he did so through the use of this criterion. Also, in John's Gospel Jesus claims to be equal with God, a claim that coincides perfectly with what John's community was saying about him. Does that mean that Jesus did not really make this claim? Not necessarily. It means that the claim cannot pass this criterion.

Historians have to evaluate all of the traditions about Jesus to determine whether they coincide with the beliefs and practices of the early Christians who were proclaiming them before they can render a judgment concerning their historical reliability. One of the problems inherent in the criterion of dissimilarity, as you might have guessed, is that we do not know as much about what the early Christians believed and practiced as we would like; moreover, what we do know indicates that they believed and practiced a whole range of things. For these reasons, it is easier to make a judgment concerning a particular tradition when it passes both of the criteria we have discussed. The judgment can be made even more easily when a tradition passes a third criterion as well.

Contextual Credibility

For the testimony of a witness in a court of law to be judged trustworthy, it has to conform with what is otherwise known about the facts of the case. The same applies to historical documents. If a document was unearthed that claimed to be a firsthand account of "Joshua Harrison, explorer of the western territories of the *United States*" and was dated A.D. 1728, you would know that you have a problem.

In ancient documents, reliable traditions must conform with the historical and social contexts to which they relate. In the case of the Gospels, the sayings, deeds, and experiences of Jesus must be able to be plausibly situated in the historical context of first-century Palestine in order to be trusted as reliable. Any saying or deed of Jesus that does not make sense in this context is automatically to be suspected. The sayings of the *Gospel of Philip*, for example, give Gnostic interpretations of the Christian sacraments of baptism and the Eucharist. It is much easier to situate these particular interpretations in the later second or early third century, when we know that Gnosticism was thriving and working out its theology, rather than in the days of Jesus. Something similar may be said of many of the Gnostic sayings of the *Gospel of Thomas*.

Some of the traditions of the New Testament Gospels do not fare well by the contextual criterion either. For example, in Jesus' conversation with Nicodemus in John 3, there is a play on words that creates a certain confusion in Nicodemus's mind. Jesus says "You must be born from above," but Nicodemus misunderstands him to mean "You must be born again." The misunderstanding arises from the fact that the Greek word for "from above" also means "again." Nicodemus has to ask for clarification, which leads Jesus to enter into an extended discourse. From a historical point of view the problem is that the confusion makes sense in Greek, the language of the Fourth Gospel, but it cannot be replicated in Aramaic, the language spoken by Jesus himself (in which the word for "from above" does not also mean "again"). Thus, if this conversation did take place (it passes neither of our other criteria either), it could not have occurred exactly in the way described by John's account.

A somewhat different problem of contextual credibility occurs in John 9:22, where "the Jews" are said to have agreed that anyone who professed belief in Jesus as the messiah was to be "put out of the synagogue." We have good reason for thinking that something of this sort did happen later in the first century, but not during the days of Jesus; at that time Jewish leaders had not yet passed any legislation concerning Jesus or his followers. It is likely, then, that the story as narrated in the Fourth Gospel cannot be historically accurate.

Unlike the other two criteria, the criterion of contextual credibility serves a strictly negative function. The others are used to argue *for* a tradition, on the grounds that it is attested by two or more independent sources and that it is a story that Christians would not have invented. This third criterion is used to argue *against* a tradition, on the grounds that it does not conform to what we know about the historical and social context of Jesus' life.

CONCLUSION: RECONSTRUCTING THE LIFE OF JESUS

To sum up. We know that Christians were modifying and inventing stories about Jesus and that our written sources preserve both historically reliable information and theologically motivated accounts. In light of this situation, the traditions that we can most rely on as historically accurate are those that are independently attested in a number of sources, that do not appear to have been created to fulfill a need in the early Christian community, and that make sense in light of a first-century Palestinian context.

In addition, as a general rule, earlier sources are preferred to later ones in reconstructing the life of Jesus. For this reason, the Synoptic Gospels, and the sources lying behind them, are generally more useful for the task than the later Gospels of John, Thomas, and Peter. Our preference for the Synoptics, however, cannot be absolute, since these Gospels also preserve traditions modified by Christians over the decades in which they were being retold. Moreover, even later sources can occasionally preserve older forms of the tradition.

Finally, with respect to Jesus, or indeed any historical person, the historian can do no more than establish probabilities. In no case can we reconstruct the past with absolute certitude. All that we can do is take the evidence that happens to survive and determine to the best of our abilities what probably happened. Thus, scholars will always disagree about the end results of these labors. But nothing can be done about this: the past cannot ever be empirically proved, it can only be reconstructed.

It is this situation that creates the final methodological problem that I want to address, the problem of how the historian who can establish only what probably has happened in the past can (or cannot!) deal with miracles that are alleged to have occurred. As this is a special problem for the historian interested in knowing what Jesus actually said and did, I have devoted the following chapter to the issue.

SUGGESTIONS FOR FURTHER READING

Meier, John. *A Marginal Jew: Rethinking the Historical Jesus*, Vol 1. New York: Doubleday, 1991. Includes the clearest and most up-to-date discussion of the extracanonical sources for Jesus' life and the methods used by scholars to determine which traditions about Jesus in the New Testament are historically accurate.

Perrin, Norman. *Rediscovering the Teachings of Jesus*. London: SCM Press; New York: Harper & Row, 1967. A classic statement of the criteria used by critical scholars to ascertain the actual teachings of Jesus.

Tatum, W. Barnes. *In Quest of Jesus: A Guidebook*. Atlanta: John Knox, 1982. An excellent introduction that includes discussions of the problems involved in establishing historically reliable traditions in the Gospels and the criteria that can be used to do so.

CHAPTER 14

Excursus: The Historian and the Problem of Miracle

Since the Enlightenment some philosophers have denied that miracles can happen; others have vehemently objected to such denials. This is not the issue that I want to address in this excursus. For the sake of the argument, I will concede the theoretical possibility that miracles can happen. I do so in order to address the historical, as opposed to the philosophical, problem of miracle. The historical problem is not whether miracles can happen; it is whether they can be shown to have happened, even if they have happened.

This is a particularly pressing problem when dealing with the Gospel narratives about Jesus, since miracles crop up on virtually every page. Jesus is born miraculously: his mother has never had sexual intercourse. As an adult he does one miracle after the other: casting out demons, walking on water, calming the storm, feeding the multitudes, healing the sick, raising the dead. At the end comes the biggest miracle of all: Jesus dies and is buried, but three days later he is raised from the dead, never to die again. This return to life is not like those narrated elsewhere in the Gospels; presumably, Jairus's daughter and Lazarus died again when their time came. Jesus' time never was to come; he actually conquered death.

How can we know whether or not any of these Gospel miracles actually happened? I am going to argue that we cannot know, so long as we stick strictly to the canons of historical inquiry for we are limited by the nature of historical evidence and the ways that we have access to events of the past. In making this argument, I am not saying that Jesus' miracles—or those of Apollonius of Tyana or Hanina ben Dosa or anyone else—did not happen. I am saying that even if they did happen, there is no way for the historian to demonstrate it.

MIRACLES IN THE MODERN WORLD AND IN ANTIQUITY

People today typically think of miracles as supernatural violations of natural law, divine interventions into the natural course of events. This popular idea does not fit particularly well into modern scientific understandings of nature, to the extent that scientists today are less confident of the entire category of "natural law" than they were, say, in the nineteenth century. For this reason, it is probably better to think of miracles, not as supernatural violations of natural laws, but as events that contradict the normal workings of nature in such a way as to be virtually beyond belief and to require an acknowledgment that supernatural forces have been at work.

This understanding is itself the major stumbling block for a historical demonstration of miracles, because historians have no access to supernatural forces. They must rely on the public record, that is, events that can be observed and interpreted by any reasonable person, of any religious persuasion. If accepting the occurrence of a miracle requires belief in the supernatural realm, and historians by the very nature of their craft can speak only about events of the natural world (which are accessible to observers of every kind) how can they ever certify the occurrence of a miracle?

Before pursuing this question, we should realize that in the ancient world miracles were not understood in the quasi-scientific terms that we use today. These terms have been available to us only since the advent of the natural sciences during the Enlightenment. To be sure, even in antiquity people understood that nature worked in certain ways. Everyone knew, for example, that iron ax-heads would sink in water, as would the human body if it tried to walk upright on it. But in the ancient world, almost no one thought that there were inviolable "laws" of nature, or even highly consistent workings of nature whose chances of being violated were infinitesimally remote. The question was not whether things happened in relatively fixed ways; it was who had the power to do the things that happened.

For people in Greco-Roman times, the universe was made up of the material world, divine beings, humans, and animals, with everyone and everything having a place and a sphere of authority. A tree could not build a house, but a person could. A person could not make it rain, but a god could. A normal human being could not heal the sick with a word or a touch, or cast out an evil demon, or bring the dead back to life, but a divine human could. Such a person stood in a special relation to the gods, like Jesus or Apollonius. For someone like this to heal the sick or raise the dead was not a miracle in the sense that it violated the natural order; rather, it was "spectacular" in the sense that such things did not happen very often, since few people had the requisite power. When they did happen, they were a marvel to behold.

For most ancients the question was thus not whether miracles were possible. Spectacular events happened all the time. It was spectacular when the sun came up or the lightning struck or the crops put forth their fruit. It was also spectacular when a divine man healed the blind or cured the lame or raised the dead. These occurrences did not involve an intrusion from outside of the natural world into an established nexus of cause and effect that governed the way things work. For ancient people there was no closed system of cause and effect. The natural world was not apart from a supernatural realm. Thus, when spectacular events (which people today might call miracles) occurred, the only questions for most ancient persons were (a) who was able to perform these deeds and (b) what was the source of their power? Was a person like Jesus, for example, empowered by a god or by black magic?

Figure 14.1 Marble statue of Asclepius, son of the god Apollo and known throughout the Greco-Roman world as a great god of healing.

To agree with an ancient person that Jesus healed the sick, walked on water, cast out a demon, or raised the dead is to agree, first, that there were divine persons (or magicians) walking the earth who could do such things and, second, that Jesus was one of them. In other words, from a historian's perspective, anyone who thinks that Jesus did these miracles has to be willing in principle to concede that other people did them as well, including the pagan holy man Apollonius of Tyana, the emperor Vespasian, and the Jewish miracle worker Hanina ben Dosa. The evidence that is admitted in any one of these cases must be admitted in the others as well. But what evidence could there be?

THE HISTORIAN AND HISTORICAL METHOD

For historians who are interested in establishing what probably happened in the past but who are not required either to embrace or to deny particular religious beliefs, what would count as evidence that a miracle had ever taken place? One way to approach the question is by reflecting for a moment on the way in which historians engage in their craft, in contrast, say, to the ways natural scientists engage in theirs. The natural sciences use repeated experimentation to establish predictive probabilities based on past occurrences. To illustrate on the simplest level, suppose I wanted to demonstrate that a bar of iron will sink in a tub of lukewarm water whereas a bar of Ivory soap will float. I could perform a relatively simple experiment by getting several hundred tubs of lukewarm water, several hundred bars of iron, and several hundred bars of Ivory soap. By tossing the bars of iron and soap into the tubs of water, I could demonstrate beyond reasonable doubt that one will sink and the other will float, since the same result will occur in every instance. This experiment does not prove that in the future every bar of iron thrown into a tub of lukewarm water will sink, but it does provide an extremely high level of presumptive probability. In common parlance, a miracle would involve a violation of this known working of nature; it would be a miracle, for example, if a preacher prayed over a bar of iron and thereby made it float.

Unlike the natural sciences, the historical disciplines are concerned with establishing what has happened in the past, as opposed to predicting what will happen in the future. Thus they cannot operate through repeated experimentation. A historical occurrence is a one-time proposition; once it has happened, it is over and done with. Since historians cannot repeat the past in order to establish what has probably happened, there will always be less certainty in their conclusions. It is much harder to convince people that John F. Kennedy was the victim of a lone assassin than to convince them that a bar of Ivory soap will float.

The farther back you go in history, the harder it is to mount a convincing case. For events in the ancient world, even events of earth-shattering importance, there is often scant evidence to go on. All the historian can do is establish what probably happened on the basis of whatever supporting evidence happens to survive.

Of course, every event that occurs is to some extent improbable. Suppose you were in a minor car accident last night. The chances of that happening were probably not very great. Nevertheless, if some people fifteen years from now wanted to show that you had that accident last night, they could appeal to certain kinds of evidence—newspaper articles, police reports, eyewitness accounts—and demonstrate their historical claim to most peoples' satisfaction. They can do this because there is nothing about the event itself that is improbable. People have accidents all the time, and the only issue would be whether or not you had one on the night in question.

As events that defy all probability, however, miracles do not happen all the time. Thus they create an inescapable problem for historians. Since historians can only establish what probably happened in the past, and the chances of a miracle happening, by definition, are infinitesimally remote, they can never demonstrate that a miracle *probably* happened.

This is not a problem for only one kind of historian, for atheists or agnostics or Buddhists or Roman Catholics or Baptists or Jews or Muslims; it is a problem for all historians of every stripe. Even if there are otherwise good sources for a miraculous event, the very nature of the historical discipline prevents the historian from arguing for its probability. Let me illustrate with a hypotheti-

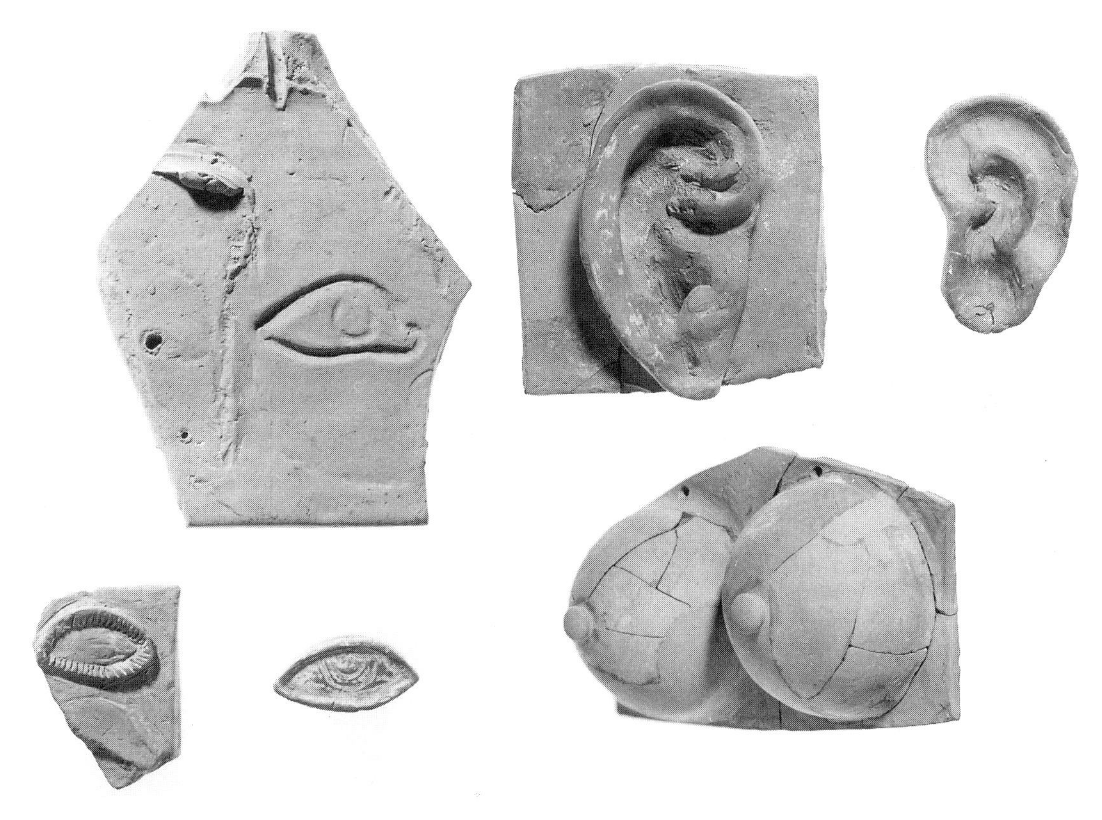

Figure 14.2 Patients who believed they were healed by the god Asclepius would commonly dedicate terra-cotta replicas of their restored body parts to him, hanging them on the walls of his temple. This picture shows some of the offerings found in the temple precincts of Asclepius in the city of Corinth, evidently from people who had previously been deaf (partially?), blind, and, possibly, suffering from breast cancer.

cal example. Suppose that three otherwise credible eyewitnesses claimed to see Reverend Jones of the Plymouth Baptist Church walk across his parishioner's pond in 1926. The historian can certainly discuss what can be known about the case: who the eyewitnesses were, what they claimed they saw, what can be known about the body of water in question, and so forth. What the historian cannot claim, however, at least when discussing the matter as a historian, is that Reverend Jones actually did it. This is more than we can know using the canons of historical knowledge. The problem of historical probabilities restrains our conclusion. We all know several thousand people, none of whom can walk across pools of water, but all of whom at one time or another have been mistaken about what they thought they saw, or have been misquoted, or have exaggerated, or have flat out lied. To be sure, such activities may not be probable, especially for the upstanding members of the Plymouth Baptist Church. But they would be more probable than a miracle that defies the normal workings of nature. Thus, if we as historians can only say what probably happened, we cannot say—as historians—that the good Reverend probably performed a miracle.

I should emphasize that historians do not have to deny the possibility of miracles or deny that miracles have actually happened in the past. Many historians, for example, committed Christians, observant Jews, and practicing Muslims, believe that they have in fact happened. When they think or say this, however, they do so not in the capacity of the historian but in the capacity of the believer. In the sketch of the historical Jesus that follows in Chapter 15, I am not taking the position of the believer, nor am I saying that one should or should not take such a position. I am taking the position of the historian, who on the basis of a limited number of problematic sources has to determine to the best of his or her ability what the historical Jesus actually said, did, and experienced. As a result, in reconstructing Jesus' activities, I will not be able to affirm or deny the miracles that he is reported to have done.

SUGGESTIONS FOR FURTHER READING

Fuller, Reginald. *Interpreting the Miracles*. London: SCM, 1963. A somewhat older study that examines the meaning of the stories of Jesus' miracles for early Christians and the function of these stories in each of the New Testament Gospels.

Kee, Howard Clark. *Miracle in the Early Christian World: A Study in Socio-Historical Method*. New Haven, Conn.: Yale University Press, 1983. A sociological study that tries to explain the function of the miracle stories for the different New Testament authors and to situate the early accounts of Jesus' miracles in the broader context of the understandings of miracles and miracle workers in the Greco-Roman world; for more advanced students.

Meier, John. A Marginal Jew: *Rethinking the Historical Jesus*, Vol 2. New York: Doubleday, 1994. Includes a systematic and careful discussion of the problem that miracles pose for the historian and a detailed evaluation of the traditions of Jesus' miracles as found in the New Testament.

Sanders, E. P. *The Historical Figure of Jesus*. London: Penguin, 1993. Chapter 10 of this very fine introductory study deals with the problems posed for the modern historian by ancient accounts of the miraculous and evaluates the New Testament stories of Jesus' miracles.

CHAPTER 15

Jesus, the Apocalyptic Prophet

In Chapter 13 we saw why it is so difficult, given the nature of our ancient sources, to reconstruct the life of the historical Jesus. If we uncritically accepted whatever our ancient accounts of Jesus happen to say about him, the resulting picture would be hopelessly and endlessly contradictory. We should not throw our hands up in despair, however, as if we can know nothing at all about the things Jesus said and did. On the contrary, when we approach our sources critically, using the kinds of criteria we have discussed, they can indeed supply us with reliable historical information.

In one chapter we will not be able to discuss every facet of Jesus' life. We can however, apply the criteria that I have mapped out to discover the kind of person Jesus was, as revealed by the sorts of things that he taught and did. Since the life of Jesus is a hotly debated area of research among New Testament scholars, I cannot simply describe "the consensus" among present-day historians. Despite what some scholars claim (especially when they want everyone else to agree with them), there is no consensus. Instead, I will make a case for the position that strikes me as the most compelling.

The best place for us to begin is with the one negative criterion discussed in Chapter 13: contextual credibility. If something that Jesus allegedly said or did cannot be credibly situated in his own social and historical context, then it cannot be regarded as authentic. This criterion is similar to a principle that I have emphasized throughout our study, the importance of context for understanding events of the past. Up to this point I have merely touched upon the social and political context of first-century Palestine, principally because I think it is more relevant for understanding the historical Jesus than for understanding the traditions that circulated about him in other parts of the Mediterranean some decades later. To be sure, even to study the Gospels one must understand certain aspects of Judaism, but the precise nature of life in first-century Palestine is chiefly relevant to the study of someone who happened to live there. Jesus was a Jewish man living in the first century of the Common Era in the Roman territory of Galilee. If we want to know about his life, we have to learn about his world.

POLITICAL CRISES IN PALESTINE AND THEIR RAMIFICATIONS

The ancient history of Palestine is long and complex. Here we will consider only the minute aspect of it that had a direct bearing on the context of Jesus' adult life in the 20s of the Common Era. In a nutshell, the political history of the land had not been happy for some eight hundred years; during this time it experienced periodic wars and virtually permanent foreign domination. The northern part of the land, the kingdom of Israel, was overthrown by the Assyrians in 721 B.C.E.; then, about a century and a half later, in 587–86 B.C.E., the southern kingdom of Judah was conquered by the Babylonians. Jerusalem was leveled, the Temple destroyed, and the leaders of the people taken into exile. Some fifty years later, the Babylonian empire was overrun by the Persians, who brought an end to the forced exile and allowed the Judean leaders to return home. The Temple was rebuilt, and the priest in charge of the Temple, the high priest, was given jurisdiction as a local ruler of the people. This man was from an ancient family that traced its line back hundreds of years to a priest named Zadok. Ultimately, of course, the Persian king was the final authority over the land and its people.

This state of affairs continued for nearly two centuries, until the conquests of Alexander the Great, ruler of Macedonia (see box 2.1). Alexander overthrew the Persian empire, conquering most of the lands around the Eastern Mediterranean as far as modern-day India. He brought Greek culture with him into the various regions he conquered, building Greek cities and schools and gymnasia (centers of Greek culture), encouraging the acceptance of Greek culture and religion, and promoting the use of the Greek language. Alexander died a young man in 323 B.C.E. The generals of his army divided up his realm, and Palestine fell under the rule of Ptolemy, the general in charge of Egypt. During all of this time, the Jewish high priest remained the local ruler of the land of Judea. This did not change when the ruler of Syria wrested control of Palestine from the Ptolemaeans in 198 B.C.E.

It is hard to know how widespread or intense the antagonism toward foreign rule was throughout most of this period, given our sparse sources. No doubt many Jews resented the idea that their rulers were answerable to a foreign power. They were, after all, the chosen people of the one true God of Israel, who had agreed to protect and defend them in exchange for their devotion. Judea was the land that he had promised them, and for many Jews it must have been distressing, both politically and religiously, to know that ultimately someone else was in charge.

In any event, there is no doubt that the situation became greatly exacerbated under the Syrian monarchs. Over the century and a half or so since Alexander's death, Greek culture had become more and more prominent throughout the entire Mediterranean region. One Syrian ruler in particular, Antiochus Epiphanes, decided to bring greater cultural unity to his empire by requiring his subjects to adopt aspects of Greek civilization. Some of the Jews living in Palestine welcomed these innovations. Indeed, some men were enthused enough to undergo surgery to remove the marks of their circumcision, allowing them to exercise in the Jerusalem gymnasium without being recognized as Jewish. Others, however, found this process of Hellenization, or imposition of Greek culture, absolutely offensive to their religion. In response to their protests, Antiochus tightened the screws even further, making it illegal for Jews to circumcise their baby boys and to maintain their Jewish identity, converting the Jewish Temple into a pagan sanctuary, and requiring Jews to sacrifice to the pagan gods.

A revolt broke out, started by a family of Jewish priests known to history both as the Maccabeans, based on the name given to one of its powerful leaders, Judas Maccabeus ("the Hammerer"), and also as the Hasmoneans, based on the name of a distant ancestor. The Maccabean revolt began as a small guerrilla skirmish in 167 B.C.E.; soon much of the country was in armed rebellion against its Syrian overlords. In less than 25 years, the Maccabeans had successfully driven the Syrian army out of the land and assumed full and total control of its governance, creating the first sovereign Jewish state in over four centuries. They rededicated the Temple (one of their first acts, in 164 B.C.E., commemorated still in the Hanukkah celebration) and appointed a high priest as supreme ruler of the land. To the dismay of many Jews in Palestine, however, the high priest was not from the ancient line of Zadok but from the Hasmonean family.

The Hasmoneans ruled the land as an autonomous state for some eighty years, until 63 B.C.E., when the Roman general Pompey conquered it. The Romans allowed the high priest to remain in office, using him as an administrative liaison with the local Jewish leadership, but there was no doubt who controlled the land. Eventually, in 40 B.C.E., Rome appointed a king to rule the Jews of Palestine, Herod the Great, renowned both for his ruthless exercise of power and for his magnificent building projects, which served not only to beautify the cities but also to elevate the status of Judea and employ massive

Figure 15.1 Silver coin from Antioch with a portrait of Antiochus Epiphanes and the inscription, in Greek, "King Antiochus, a god made manifest."

numbers of workers. Many Jews, however, castigated Herod as an opportunistic collaborator with the Romans, a traitorous half-Jew at best. The latter charge was based in part on his lineage: his parents were from the neighboring country of Idumea and had been forced to convert to Judaism before his birth.

During the days of Jesus, after Herod's death, Galilee, the northern region of the land, was ruled by Herod's son Antipas; and starting when Jesus was a boy, Judea, the southern region, was governed by Roman administrators known as prefects. Pontius Pilate was prefect during the whole of Jesus' ministry and for some years after his death. His headquarters were in Caesarea, but he came to the capital city Jerusalem, with troops, whenever the need arose.

The point of this brief sketch is not to indicate what Jewish children learned in their fifth-grade history classes; indeed, there is no way for us to know whether a boy like Jesus would ever have even heard of such important figures from the remote past as Alexander the Great or Ptolemy. Rather, the historical events leading up to his time are significant for understanding Jesus' life because they had social and intellectual ramifications for all Palestinian Jews. It was in response to the social, political, and religious crises of the Maccabean period that the Jewish sects of Jesus' day (e.g., the Pharisees, Sadducees, and Essenes) were formed, and it was the Roman occupation that led to numerous nonviolent and violent Jewish uprisings during Jesus' time. For many of Jews, any foreign domination of the Promised Land was both politically and religiously unacceptable. Moreover, it was the overall sense of inequity and the experience of suffering during these times that inspired the ideology of resistance known as apocalypticism, a worldview that was shared by a number of Jews in first-century Palestine.

THE FORMATION OF JEWISH SECTS

It was during the rule of the Hasmoneans, and evidently in large measure in reaction to it, that various Jewish sects emerged. The Jewish historian Josephus mentions four of these groups; the New Testament makes explicit reference to three. In one way or another, all of them play a significant role in our understanding of the life of the historical Jesus.

I should emphasize at the outset that most Jews in Palestine did not belong to any of these groups. We know this much from Josephus, who indicates

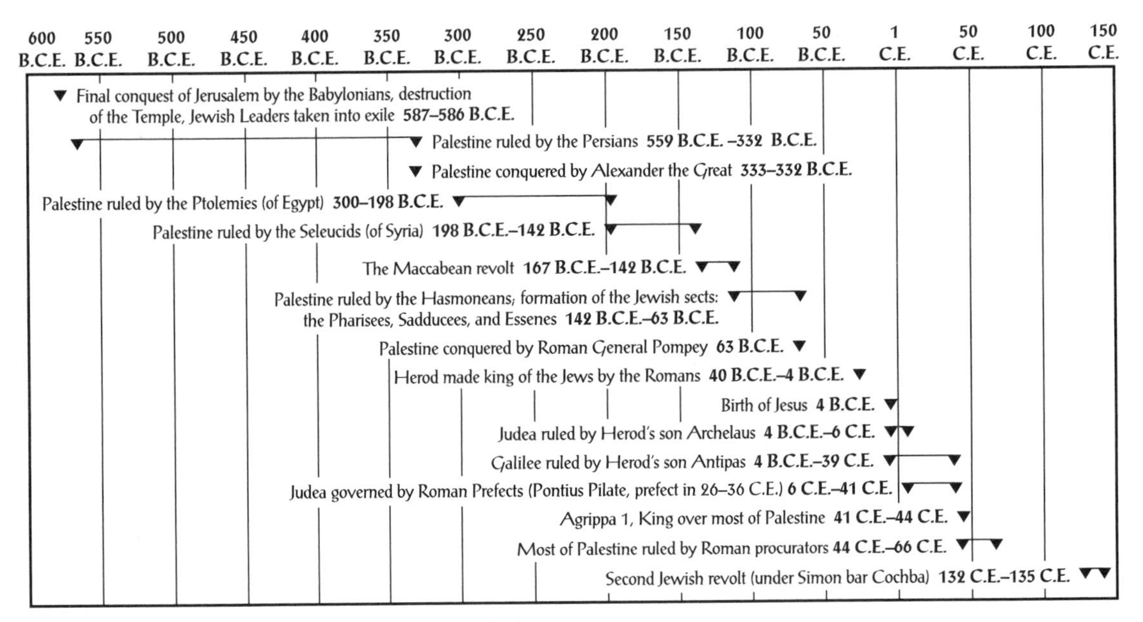

Figure 15.2 Time Line of Key Events in the History of Palestine.

that the largest sect, the Pharisees, claimed 6,000 members and that the Essenes claimed 4,000. The Sadducees probably had far fewer. These numbers should be considered in light of the overall Jewish population in the world at the time; the best estimates put the number at something like three and a half million.

What matters for our purposes here, however, is not the size of these groups, for they were influential despite their small numbers, but the ways in which they understood what it meant to be Jewish, especially in light of the political crises that they had to face. Members of all of the sects, of course, would have subscribed to the basic principles of the religion, as sketched earlier in Chapter 2: each believed in the one true God, the creator of all things, who was revealed in the Hebrew Scriptures, who had chosen the people of Israel, and who had promised to protect and defend them in exchange for their committed devotion to him through following his laws. The groups differed in significant ways, however, in their understanding of what obedience to God's laws required and in how they responded to the rule of a foreign power and to the presence of a high priest from a line other than Zadok's.

Pharisees

The Pharisees represent probably the best-known and least understood Jewish sect. Because of the way they are attacked in parts of the New Testament, especially in Matthew, Christians through the ages have wrongly considered the Pharisees' chief attribute to be hypocrisy.

It appears that this sect began during the Maccabean period as a group of devout Jews intent above all else on keeping the Torah of God. Rather than accepting the culture and religion of the Greeks, these Jews insisted on knowing and obeying the Law of their own God to the fullest extent possible. One of the difficulties with the Torah of Moses, though, is that in many places it is ambiguous. For example, Jews are told in the Ten Commandments to keep the Sabbath day holy, but nowhere does the Torah indicate precisely how this is to be done. Pharisees devised rules and regulations to assist them in keeping this and all the other laws of Moses. These rules eventually became very detailed. Among many other things, they stated what a person could and could not do on the Sabbath day in order to keep it holy, or set apart from all other days. Thus, when it was eventually determined that a faithful Jew should not go on a long journey on the Sabbath, it had to be decided what a "long" journey was, and consequently what distance a Jew could travel on this day without violating its holiness. Likewise, a worker who believed that he or she should not labor on the Sabbath had to know what constituted "work" and what therefore could and could not be done.

The rules and regulations that developed among the Pharisees came to have a status of their own and were known in some circles as the "oral" Law, which was set alongside the "written" Law of Moses. It appears that Pharisees generally believed that anyone who kept the oral law would be almost certain to keep the written law as a consequence. The intent was not to be legalistic but to be obedient to what God had commanded.

The Pharisees may have been a relatively closed society in Jesus' day, to the extent that they stayed together as a group, eating meals and having fellowship only with one another, that is, with those who were like-minded in seeing the need to maintain a high level of obedience before God. Those who did not show this obedience were thought to be unclean.

It is important to recognize that the Pharisees were not the "power players" in Palestine in Jesus' day. That is to say, they appear to have had some popular appeal but no real political clout. In some ways they are best seen as a kind of separatist group; they wanted to maintain their own purity and did so in isolation from other Jews. Many scholars think that the term "Pharisee" itself originally came from a Persian word that means "separated ones." Eventually, however, some decades after Jesus' execution, the Pharisees did become powerful in the political sense. This was after the Jewish War (which I will describe more fully below), which culminated in the destruction of Jerusalem and the Temple in the year 70 C.E. With this calamity the other groups passed from the scene for a variety of reasons, and the Pharisees were given greater authority by the Roman overlords. The oral tradition continued to grow and eventually took on the status of divinely revealed law. It was eventually written down around the year 200 C.E. and is today known as the Mishnah, the heart of the Jewish sacred collection of texts, the Talmud.

The Pharisees are important for understanding the historical Jesus, in part because he set his message over against theirs. As we will see, Jesus did not think that scrupulous and detailed adherence to the laws of Torah were the most important aspect of a Jew's relationship with God, especially as these laws were interpreted by the Pharisees.

Sadducees

It is difficult to reconstruct exactly what the Sadducees stood for because not a single literary work survives from the pen of a Sadducee, in contrast to the Pharisees, who are represented to some extent by the later traditions of the Talmud, by Josephus, who was a Pharisee, and by the one Pharisee who left us writings before the destruction of the Temple (after he had converted to Christianity), the apostle Paul. To understand the Sadducees, however, we must turn to what is said about them in other sources, such as Josephus and the New Testament.

During Jesus' own day, the Sadducees were evidently the real power players in Palestine. They appear to have been, by and large, members of the Jewish aristocracy in Jerusalem who were closely connected with the Jewish priesthood in charge of the Temple cult. Most of the Sadducees were themselves priests (though not all priests were Sadducees). As members of the aristocracy, granted some limited power by their Roman overlords, Sadducees appear to have been conciliatory toward the civil authorities, that is, cooperative with the Roman governor. The local Jewish council, commonly called the Sanhedrin, which was occasionally called together to decide local affairs, was evidently made up principally of Sadducees. With their close connection with the Temple, Sadducees emphasized the need for Jews to be properly involved in the cultic worship of God as prescribed in the Torah. Indeed, it appears that the Torah itself, that is, the five books of Moses, was the only authoritative text that the Sadducees accepted. In any event, we know that they did not accept the oral traditions formulated by the Pharisees. Less concerned with personal purity and the regulation of daily affairs such as eating, travel, and work, the Sadducees focused their religious attention on the sacrifices in the Temple and expended their political energy on working out their relations with the Romans so that these sacrifices could continue.

It may have been their rejection of all written authority outside of the five books of Moses that led the Sadducees to reject several doctrines that later became characteristic of other groups of Jews. They denied, for example, the existence of angels and disavowed the notion of the future resurrection of the dead. Their views of the afterlife may well have conformed, essentially, with those of most non-Jews throughout the empire: either the "soul" perishes with the body, or it continues on in a kind of shadowy netherworld, regardless of the quality of its life here on earth.

The Sadducees are of importance for understanding the historical Jesus, in part because he roused their anger by predicting that God would soon destroy the locus of their social and religious authority, their beloved Temple. In response, some of their prominent members urged Pontius Pilate to have him executed.

Essenes

The Essenes are the one Jewish sect not explicitly mentioned in the New Testament. Ironically, they are also the group about which we are best informed. This is because the famous Dead Sea Scrolls were evidently produced by a group of Essenes who lived in a community east of Jerusalem in the wilderness area near the Western shore of the Dead Sea, in a place that is today called Qumran. Although the term "Essene" never occurs in the scrolls, we know from other ancient authors such as Josephus that a community of Essenes was located in this area; moreover, the social arrangements and theological views described in the Dead Sea Scrolls correspond to what we know about the Essenes from these other accounts. Most scholars are reasonably certain, therefore, that the scrolls represent a library used by this sect, or at least by the part of it living near Qumran.

As was the case with the Gnostic documents uncovered near Nag Hammadi, Egypt, the discovery of the Dead Sea Scrolls was completely serendipitous. In 1947, a shepherd boy searching for a lost goat in the barren wilderness near the Northwest shore of the Dead Sea happened to toss a stone into a cave and heard it strike something. Entering the cave, he discovered an ancient earthenware jar that contained a number of old scrolls. The books were recovered by bedouin shepherds. When news of the discovery reached antiquities dealers, biblical scholars learned of the find, and a search was conducted both to find more scrolls in the surrounding caves and to retrieve those that had already been found by the bedouin, who cut some of them up to sell one piece at a time.

Some of the caves in the region yielded entire scrolls; others contained thousands of tiny scraps that are virtually impossible to piece back together, given the problem that so many of the pieces are missing. Imagine trying to do an immense jigsaw puzzle, or rather dozens of immense jigsaw puzzles, not knowing what the end product of any of them should look like, when most of the pieces are lost and those that remain are all mixed together! All in all hundreds of documents are represented, many of them only in fragments the size of postage stamps, others, perhaps a couple of dozen, in scrolls of sufficient length to give us a full idea of their contents.

Most of the scrolls are written in Hebrew, but some are in Aramaic. Different kinds of literature are represented here (see box 15.1). There are at least partial copies of every book of the Jewish Bible, with the exception of the book of Esther, and some of them are fairly complete. These are extremely valuable because of their age; they are nearly a thousand years older than the oldest copies of the Hebrew Scriptures that we previously had. We can therefore check to see whether Jewish scribes over the intervening centuries reliably copied their texts. The short answer is that, for the most part, they did. There are also commentaries on some of the biblical books, written principally to show that the predictions of the ancient prophets had come to be fulfilled in the experiences of the Essene believers and in the history of their community. In addition, there are books that contain psalms and hymns composed by members of the community, prophecies that indicate future events that were believed to be ready to transpire in the authors' own day, and rules for the members of the community to follow in their lives together.

Sifting through all of these books, scholars have been able to reconstruct the life and beliefs of the Essenes in considerable detail. It appears that their community at Qumran was started during the early Maccabean period, perhaps around 150 B.C.E., by pious Jews who were convinced that the Hasmoneans had usurped their authority by appointing a non-Zadokite as high priest. Believing that the Jews of Jerusalem had gone astray, this group of Essenes chose to start their own community in which they could keep the Mosaic law rigorously and maintain their own ritual purity in the wilderness. They did so fully expecting the apocalypse of the end of time to be imminent. When it came, there would be a final battle between the forces of good and evil, between the children of light and the children of darkness. The battle would climax with the triumph of God and the entry of his children into the blessed kingdom.

Some of the scrolls indicate that this kingdom would be ruled by two messiahs, one a king and the other a priest. The priestly messiah would lead the faithful in their worship of God in a purified temple, where sacrifices could again be made in accordance with God's will. In the meantime, the true people of God needed to be removed from the impurities of this world, including the impurities prevalent in the Jewish Temple and among the rest of the Jewish people. These Essenes therefore started their own monastic-like community, with strict rules for admission and membership. A two-year initiation was required, after which, if approved, a member was to donate all of his possessions to the community fund and share the common meal with all the other members. Rigorous guidelines dictated the life of the community. Members had fixed hours for work and rest and for their meals, there were required times of fasting, and strict penalties were imposed for unseemly behavior such as interrupting one another, talking at meals, and laughing at inappropriate times.

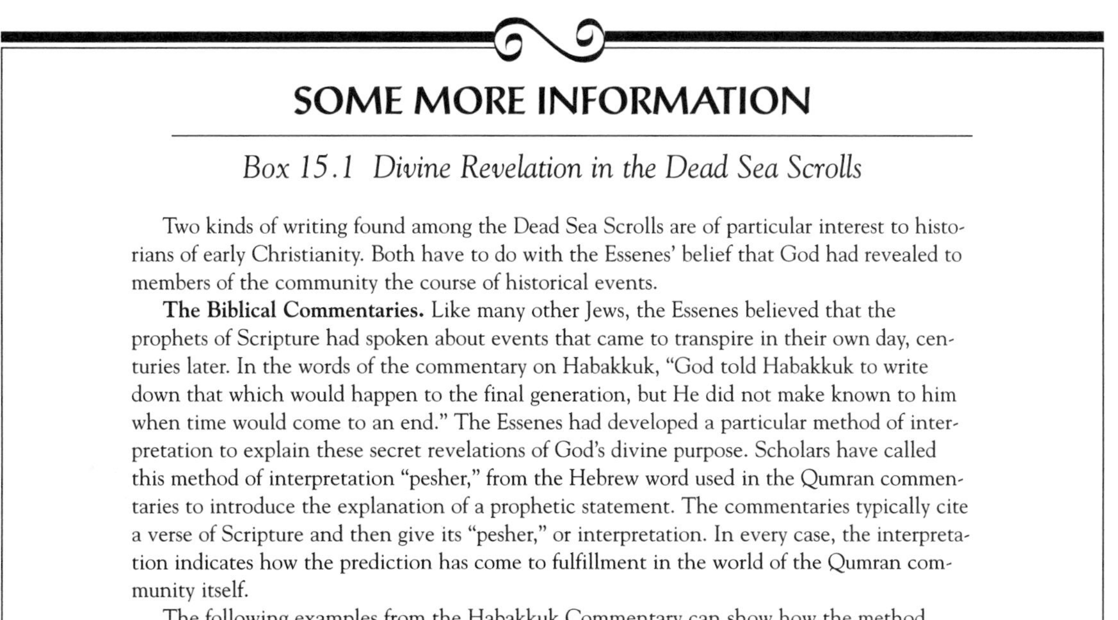

SOME MORE INFORMATION

Box 15.1 Divine Revelation in the Dead Sea Scrolls

Two kinds of writing found among the Dead Sea Scrolls are of particular interest to historians of early Christianity. Both have to do with the Essenes' belief that God had revealed to members of the community the course of historical events.

The Biblical Commentaries. Like many other Jews, the Essenes believed that the prophets of Scripture had spoken about events that came to transpire in their own day, centuries later. In the words of the commentary on Habakkuk, "God told Habakkuk to write down that which would happen to the final generation, but He did not make known to him when time would come to an end." The Essenes had developed a particular method of interpretation to explain these secret revelations of God's divine purpose. Scholars have called this method of interpretation "pesher," from the Hebrew word used in the Qumran commentaries to introduce the explanation of a prophetic statement. The commentaries typically cite a verse of Scripture and then give its "pesher," or interpretation. In every case, the interpretation indicates how the prediction has come to fulfillment in the world of the Qumran community itself.

The following examples from the Habakkuk Commentary can show how the method works. In italics is the passage of Scripture, followed by the pesher. I have placed my own explanatory comments in brackets.

For behold, I rouse the Chaldeans [another name for the Babylonians], *that bitter and hasty nation* (Hab 1:6). Interpreted, this concerns the Kittim [a code name for the Romans] who are quick and valiant in war.

O traitors, why do you stare and stay silent when the wicked swallows up one more righteous than he? (Hab 1:13b). Interpreted, this concerns the House of Absalom [a prominent group of Jews in Jerusalem] and the members of its council who were silent at the time of the chastisement of the Teacher of Righteousness [the leader of the Qumran community at its beginning] and gave him no help against the Liar [the high priest in Jerusalem who was the community's sworn enemy] who flouted the Law in the midst of their whole congregation.

Moreover, the arrogant man seizes wealth without halting... (Hab 2:5). Interpreted, this concerns the Wicked Priest [the same figure as the "Liar" above] who was called by the name of truth when he first arose. But when he ruled over Israel his heart became proud, and he forsook God and betrayed the precepts for the sake of riches.

As you can see simply from these passages, the history of the Qumran community can be read from their own interpretations of the ancient prophecies.

The War Scroll. This scroll details the final war between the forces of good and evil that will take place at the end of time. It sketches the course of the battles, gives regulations for the soldiers who fight, and describes the outcome that is assured by God as the "children of light" (the members of the Essene communities) overcome the "children of darkness" (the Romans, the apostate Jews, and everyone else). The war will take forty years, the first six of which involve overcoming the "Kittim" (the Romans), the rest being devoted to campaigns against the other nations.

This document, then, provides an apocalyptic vision of the final struggle between good and evil, between the forces of God and those of his enemies. While the War Scroll is unique among ancient Jewish literature in its graphic and detailed description of the future battle that will end the age, in general terms, it relates closely to apocalyptic texts written by other Jews in the period, as we will see further in box 15.4.

It appears that when the Jewish war of 66–73 C.E. began the Essenes at Qumran hid some of their sacred writings before joining in the struggle. It may well be that they saw this as the final battle, preliminary to the end of time when God would establish his kingdom and send its messiahs.

The Essenes are important for understanding the historical Jesus, in part because Jesus appears to have shared many of their apocalyptic views, even though he did not belong to their sect. Like them, he believed that the end of time was near and that people had to prepare for the coming onslaught.

The "Fourth Philosophy"

When Josephus writes about Judaism for a Roman audience, he describes each of the sects that we have discussed as a "philosophy," by which he means a group with a distinctive and rational outlook on the world. He never gives a name to the fourth sect that he discusses but simply calls it the "fourth philosophy." The tenets of this philosophy, however, are clear, and they were manifested in several different groups that we know about from various ancient sources. Each of these groups in its own way supported active resistance to Israel's foreign domination.

Figure 15.3 One of the most important of the Dead Sea Scrolls, a Hebrew copy of the book of Isaiah.

The view that characterized these sundry groups was that Israel had a right to its own land, a right that had been granted by God himself. Anyone who usurped that right, and anyone who backed the usurper, was to be opposed, by violent means if necessary. Among those who took this line in the middle of first century were the Sicarii, a group whose name comes from the Latin word for "dagger." These "daggermen" planned and carried out assassinations and kidnappings of high-ranking Jewish officials who were thought to be in league with the Roman authorities. Another group that subscribed to this philosophy, somewhat later in the century, were the Zealots. These were Jews who were "zealous" for the Law and who urged armed rebellion to take back the land God had promised his people. More specifically, based on what we find in Josephus, Zealots were Galilean Jews who fled to Jerusalem during the Jewish revolt around the year 67 C.E., overthrew the priestly aristocracy in the city in a bloody coup, and urged the violent opposition to the Roman legions that ultimately led to the destruction of Jerusalem and the burning of the Temple in 70 C.E.

Such groups are important for understanding the historical Jesus, in part because he too thought that the Romans were to be overthrown. But it was not to be by armed resistance.

POPULAR MODES OF RESISTANCE TO OPPRESSION

As we have seen, Jews in Palestine had been under direct foreign domination for most of the eight centuries prior to the birth of Jesus. The struggles of the Hasmoneans against the Hellenizing policies of their Syrian overlords led to the formation of sects that were active in Jesus' day. Most Jews did not belong to any of these parties, but all Jews, however, were directly affected by the policies of domination enforced by Rome.

As a conquered people, Jews in Palestine were required to pay taxes to the empire. Since the Roman economy was agrarian, taxation involved payment of crops and of monies to fund the armies and infrastructure provided by Rome, including roads, bridges, and public buildings. In monetary terms, the oppression of Jews appears to have been no worse than that of other native populations in the Roman provinces. We have no reliable numbers from ancient sources themselves, but the best estimates by modern scholars suggest that a typical Jewish farmer was taxed on average about 12 or 13 percent of his income to support the Roman presence in the land, on top of taxes to support the Temple and local Jewish administration, which might run an additional 20 percent or so. His total taxes, then, were perhaps a third of his overall income (Sanders, 1992).

These taxes may not appear exorbitant by the standards of today's highly industrialized nations; we should recall, however, that in ancient agrarian societies, without modern means of irrigation, labor-saving machinery, and sophisticated technology, most farmers did well to eke out an existence in the best of circumstances. When one is living close to the edge, having to provide financial support for a foreign oppressor is not a cheery prospect. Paying for Rome's excesses was seen by many Jews, as well as by many others in the empire, as both unmanageable and perverse.

At the same time, the treatment of the Jews in Palestine was better in some respects than that of other inhabitants of the empire. Since the days of Julius Caesar, Jews were not required to supply Rome with soldiers from their ranks (an exemption that was in Rome's best interest as well, since devout Jews refused to soldier every seventh day) or to provide direct support for Roman legions stationed nearby or marching through to the frontiers. In another respect, though, the Jewish situation could be seen as far worse than average, in that many Jews considered it blasphemous to pay taxes to support the Roman administration of the land that God had given them.

Jews reacted to their domination by Rome in different ways. For many Jews, no doubt, the Roman domination was tolerable and had its advantages, for example, protection from hostile nations to the East; but for others, it was a political and religious nightmare. Resistance to Roman power appears to have been widespread, but rarely was it active or violent. Throughout the first century, however, Jews of Palestine did lock horns with their Roman overlords on several occasions. It will suit our purposes to consider the nature of these conflicts

Silent Protests

The population of Jerusalem would swell many times over during the weeklong Passover festival (see Chapter 3), and there is little doubt that those who came to the celebration did not do so for purely commemorative reasons. Jews celebrating the Passover were not simply remembering the past, when God acted on their behalf to save them from their subjugation to the Egyptians; they were also looking to the future, when God would save them yet again, this time from their present overlords, the Romans.

Roman officials appear to have understood full well the potentially subversive nature of the celebration. They typically brought armed troops in just for the occasion, stationing them in the Temple, the locus of all activity. Most Jews did not much appreciate the Roman presence on such sacred occasions.

The tension was particularly evident during a Passover celebration in the 50s of the Common Era, when a Roman governor named Cumanus was procurator of Judea. During the feast, one of the soldiers stationed on the wall of the Temple decided to show his disdain of the Jews and their religion. In the words of Josephus, he "stooped in an indecent attitude, so as to turn his backside to the Jews, and made a noise in keeping with his posture" (*Jewish War* 2.224–27). The worshippers present were not amused. Some picked up stones and began to pelt the soldiers. A report was sped off to Cumanus, who was nearby. When he sent in reinforcements, a riot broke out. According to Josephus (who probably exaggerated the numbers) some 20,000 Jews were killed in the mayhem.

Thus, the Passover feast represented a silent protest against the Roman presence in the Promised Land, but on occasion the event led to violent resistance and death. As a rule, the Romans worked hard at keeping the situation under control, resolving problems before they led to massive uprisings or public riots. You may recall that Jesus was arrested and removed from the public eye during Passover.

Nonviolent Uprisings

Roman administrators would occasionally do or threaten to do something offensive to Jews in Palestine, who would in turn rise up in protest. It appears that for most of the first century these protests were nonviolent. In the year 26 C.E., for example, when Pilate assumed the prefectorship of Judea, he had Roman standards brought into Jerusalem during the night and set up around the city. These standards bore the image of Caesar. Jews in the city erupted in protest and demanded their removal. Pilate refused. According to Josephus, hundreds of the leading citizens staged a kind of sit-in at his residence in Caesarea. After five days, Pilate had the protesters surrounded by soldiers three deep and threatened to have them all put to the sword. The Jews responded by flinging themselves to the ground and stretching out their necks, claiming to prefer death to such a flagrant transgression of their Law. Pilate rescinded his order and had the standards removed.

Something similar happened fourteen years or so later, when the megalomaniacal emperor Caligula required the inhabitants of the empire to worship him as a god, the first Roman emperor to do so. Jews around the world vehemently protested. Some from the diaspora came in delegations to Rome to explain why the act would be offensive and blasphemous for them. Caligula responded with intransigence, ordering that a statue of himself, with the body of Zeus, be set up in the Jerusalem Temple. According to Josephus, tens of thousands of Jews in Palestine appeared in protest before the Roman legate of Syria, Petronius, who had arrived with two full legions to enforce the policy. They vowed not to plant their crops if he carried out his orders and offered themselves as martyrs rather than live to see the desecration of their Temple. Petronius was himself powerless to revoke the emperor's order, although he was impressed both by the strength of the opposition and by the danger to the crops, knowing that Rome could collect no tribute if the land lay fallow. Fortunately for him, he was saved from the consequences of failing to follow the emperor's order; for reasons unrelated to the protest, Caligula was assassinated.

Prophetic Proclamations

One particularly noteworthy form of Jewish protest against foreign domination involved the occasional appearance of self-styled prophets predicting the imminent intervention of God on behalf of his people. This intervention was modeled on earlier acts of salvation as recorded in the Hebrew Scriptures. Some of these prophets gathered a large following among the Jewish masses. For obvious reasons, they were not well-received by the Romans.

Less than fifteen years after Jesus' execution, a prophet named Theudas led a large crowd of Jews to the Jordan River, where he publicly proclaimed that he would make the waters part, allowing his people to cross on dry land. Word of his activities reached the Roman authorities, who evidently knew enough Jewish tradition to recognize the allusion to the Exodus event under Moses, when the children of Israel were delivered from their slavery in Egypt and the Egyptian army was routed during the crossing of the Red Sea. Rather than risk an uprising, the governor sent out the troops; they slaughtered Theudas's followers and brought his head back to Jerusalem for display.

About a decade later another prophet arose, who is called simply "the Egyptian" by Josephus and the book of Acts, the two sources that refer to him. This prophet acquired a large following among the masses (30,000 people according to Josephus), whom he led to the Mount of Olives. There he proclaimed the destruction of the walls of Jerusalem, another transparent reference, this time to the conquest of Jericho, when the children of Israel came into the Promised Land and "the walls came tumbling down." Again, the Roman troops were sent forth to hunt down and slaughter the group.

Other prophets arose and experienced similar fates. Roman administrators of Judea appear to have had no qualms about destroying those whose proclamation of God's intervention on behalf of his people could win them a following and, potentially, lead to riots—especially in Jerusalem.

Violent Insurrections

In Palestine during the first century, there were also violent insurrections in which Jews with forethought and intent engaged in armed revolt against the Romans. These were isolated rather than everyday occurrences, however. One of them occurred around 6 C.E., during Jesus' childhood, when Archelaus, son of Herod the Great, was deposed as ruler of Judea and replaced by a Roman prefect. A census was imposed for tax purposes, and a group of Jews led by a freedom fighter named Judas the son of Hezekiah resisted with the sword. The revolt was crushed, effectively and brutally.

The second, and more disastrous, uprising came 60 years later when Roman atrocities such as the governor's plundering of the Temple treasury led to a widespread revolt. The Romans sent in the legions from the north and within a year subjugated Galilee (this was when Josephus was the commander of the Jewish troops there, prior to surrendering). A group of Galilean Jews who fled from the Roman army arrived in Jerusalem and eventually provoked a bloody civil war against the priestly aristocracy who had been in charge of the Temple and the rest of the city. Once they acquired control, these "Zealots" pressed the fight against the Romans to the end. The result was a horrifying three-year siege of Jerusalem, in which reports of starvation and cannibalism were rampant. The war ended in a bloodbath in which tens of thousands of Jews were slaughtered or enslaved, rebel leaders were crucified, much of the city was leveled, and the Temple was burned to the ground.

In sum, Palestine was under Roman domination in the first century and Jews in the land reacted to the situation in a variety of ways. They sometimes protested in silence, anticipating a deliverance to be wrought by God, they sometimes engaged, when necessary, in acts of nonviolent resistance, and they sometimes became

Figure 15.4 Detail from the arch built in Rome to commemorate the victory of Titus over Jerusalem in the year 70 C.E. This part of the arch shows the menora from the Temple being carried away to Rome.

caught up in spontaneous rioting, provoked by the insensitive treatment from Roman rulers and soldiers. Some publicly proclaimed the imminent end of their suffering through the supernatural intervention of God, while others sought to take matters into their own hands, taking up the sword to engage in violent resistance. The nonviolent protesters had some success in forcing the Romans to back down on particular issues; the violent protesters, whether rioting masses, prophetic figures, or guerrilla warriors, had none whatsoever. In the cases we know of, the Romans effectively and ruthlessly destroyed those who preached or practiced violence against them.

AN IDEOLOGY OF RESISTANCE

Another important aspect of Jesus' historical context involves one of the "worldviews" evident in a number of Jewish writings from around his time. Modern scholars have called this worldview "apocalypticism" from the Greek term *apocalypsis*, which means an "unveiling" or a "revealing." Jews who subscribed to this worldview maintained that God had revealed to them the future, in which he would soon overthrow the forces of evil and establish his kingdom on earth.

We know about Jewish apocalyptic thought from a number of ancient sources. It is first attested in some of the latest writings of the Hebrew Bible, especially the book of Daniel, which scholars date to the time of the Maccabean revolt. It is also prominent in the Dead Sea Scrolls, the writings of the Essene community at Qumran. In addition, it is found in a range of other Jewish writings that did not make it into the Bible. These books are called "apocalypses" because their authors claim that the course of future events had been revealed to them.

The worldview of the apocalypticists originated in the turbulent history of the Jews in Palestine. We have seen that most ancient Jews believed that God had made a covenant with his people to be their divine protector in exchange for their devotion to him through keeping his law. This point of view naturally came to be challenged by political events in Palestine. If God had promised to protect and defend Israel against its enemies, how could one account for its perpetual foreign domination—by the Assyrians, the Babylonians, the Persians, the Greeks, the Syrians, and the Romans?

One of the popular answers was given by ancient Jewish prophets, including those whose writings were later canonized in the Hebrew Bible, prophets such as Isaiah, Jeremiah, Amos, and Hosea. According to these authors, Israel continued to suffer military and political setbacks because it had disobeyed God. He was still their God and he remained the all-powerful ruler of the world, able to dictate the course of human events. But the people of Israel had sinned against him, and their military defeats and economic disasters represented God's punishment for their sins. According to the prophets, if the people would only return to the ways of God, and again become devoted to keeping his Law, he would relent and establish them once more as a sovereign power in their own land.

This basic point of view has always been popular, not only among Jews but also among Christians: people suffer because they have sinned, and this suffering is their punishment. Some Jewish thinkers eventually became dissatisfied with this view, however, because it could not adequately explain historical realities. It was not only the sinners who suffered but people who were righteous as well. Furthermore, matters did not improve when people repented and committed themselves to keeping God's Law. Why would Israel continue to suffer after it returned to God, while other nations that made no effort to please him at all prospered?

Around the time of the Maccabean revolt, when the oppressive policies of Antiochus Epiphanes became too much to bear for many Jews in Palestine, when they were forbidden on pain of death from keeping the Torah, some of them came up with another position. In their view, the suffering of God's people could not be explained as a penalty for their sin. God surely would not punish his people for doing what was right, for keeping his laws, for example. Why, then, did the people suffer? There must be some other supernatural agency, some other superhuman power that was responsible. God was not making his people suffer; his enemy, Satan, was.

According to this new way of thinking, God was still in control of this world in some ultimate sense, but for unknown and mysterious reasons he had temporarily relinquished his control to the forces of evil that opposed him. This state of affairs, however, was not to last forever. Quite soon, God would reassert himself, destroying the forces of evil and establishing his people as rulers over the earth. When this new kingdom came, God would fulfill his promises to his people. This ideology, which tried to make sense of the oppression of the people of God, is commonly called apocalypticism.

Dualism

Jewish apocalypticists were dualists. They maintained that there were two fundamental components to all of reality: the forces of good and the forces of evil. The forces of good were headed by God himself, the forces of evil by his superhuman enemy, sometimes called Satan, Beelzebub, or the Devil. On the side of God were the good angels; on the side of the Devil were the demons. On the side of God were righteousness and life; on the side of the Devil were sin and death. These forces were cosmic powers to which human beings were subject and with which they had to be aligned. No one was in neutral territory. People stood either with God or with Satan, in the light or in darkness, in the truth or in error.

This apocalyptic dualism had clear historical implications. All of history could be divided into two ages, the present age and the age to come. The present age was the age of sin and evil. The powers of darkness were in the ascendancy, and those who sided with God were made to suffer by those in control of this world. Sin, disease, famine, violence, and death were rampant. For some unknown reason, God had relinquished control of this age to the powers of evil—and things were getting worse.

At the end of this age, however, God would reassert himself, intervening in history and destroying the forces of evil. After a cataclysmic break in which all that was opposed to God would be annihilated, God would bring in a new age. In this new age, there would be no more suffering or pain; there would be no more hatred, despair, war, disease, or death. God would be the ruler of all, in a kingdom that would never end.

Pessimism

Even though, in the long run, everything would work out for those who sided with God, in the short term things did not look good. Jewish apocalypticists maintained that those who sided with God were going to suffer in this age, and there was nothing they could do to stop it. The forces of evil were going to grow in power as they attempted to wrest sovereignty over this world away from God. There was no thought of being able to improve the human condition through mass education or advanced technology. The righteous could not make their lives better because the forces of evil were in control, and those who sided with God were opposed by those who were much stronger than they. Things would get worse and worse until the very end, when, quite literally, all hell would break loose.

Vindication

At the end, when the suffering of God's people was at its height, God would finally intervene on their behalf and vindicate his name. In the apocalyptic perspective, God was not only the creator of this world but also its redeemer. His vindication would be universal; it would affect the entire world, not simply the Jewish nation. Jewish apocalypticists maintained that the entire creation had become corrupt because of the presence of sin and the power of Satan. This universal corruption required a universal redemption; God would destroy all that is evil and create a new heaven and a new earth, one in which the forces of evil would have no place.

Different apocalypticists had different views concerning how God would bring about this new creation, even though they all claimed to have received the details in a revelation from God. In some apocalyptic scenarios, God was to send a human messiah to lead the troops of the children of light into battle against the forces of evil. In others, God was to send a kind of cosmic judge of the earth, sometimes also called the messiah or the Son of Man, to bring about a cataclysmic overthrow of the demonic powers that oppressed the children of light.

This final vindication would involve a day of judgment for all people. Those who had aligned themselves with the powers of evil would face the Almighty Judge and render an account of what they had done; those who had remained faithful to the true God would be rewarded and brought into his eternal kingdom. Moreover, this judgment applied not only to people who happened to be living at the time of the end. One could not side with the powers of evil, oppress the people of God, die prosperous and contented, and get away with it. God would allow no one to escape. He was going to raise all people bodily from the dead to receive their reward or punishment: eternal bliss for those who had taken his side, eternal torment for everyone else.

Imminence

According to Jewish apocalypticists, this vindication of God was going to happen very soon. Standing in the tradition of the prophets of the Hebrew Bible, apocalypticists maintained that God had revealed to them the course of history and that the end was almost here. Those who were evil had to repent before it was too late. Those who were good, who were suffering as a result, were to hold on, for it would not be long before God would intervene by sending a savior, possibly on the clouds of heaven, to pass judgment on the people of the earth and bring the good kingdom to those who had remained faithful to his Law. Indeed, the end was right around the corner. In the words of one first-century Jewish apocalypticist: "Truly I tell you, there are some standing here who will not taste death until they see that that kingdom of God has come with power." These, in fact, are the words of Jesus (Mark 9:1). Or as he says elsewhere, "Truly I tell you, this generation will not pass away before all these things have taken place" (Mark 13:30).

Some of the earliest traditions about Jesus portray him as a Jewish apocalypticist who responded to the political and social crises of his day, including the domination of his nation by a foreign power, by proclaiming that his generation was living at the end of the age, and that God would soon intervene on behalf of his people. He would send a cosmic judge, the Son of Man, who would destroy the forces of evil and set up God's kingdom. In preparation for his coming, the people of Israel needed to repent and turn to God, trusting him as a kindly parent and loving one another as his special children. Those who refused to accept this message would be liable to the punishment of God.

JESUS AS A JEWISH APOCALYPTIC PROPHET

Is the ancient portrayal of Jesus as an apocalyptic prophet, which is embodied in a number of our oldest traditions, historically accurate? This is one of the most hotly contested areas of modern scholarship on the New Testament. Some scholars are firmly convinced that the apocalyptic sayings recorded in the Gospels do not go back to Jesus, but were placed on his lips by his later followers, and that as a consequence Jesus is not at all to be understood as an apocalypticist. Let me explain why I think this view is wrong, and why I think an apocalyptic message was not only one component of Jesus' teaching but the very heart and soul of his entire ministry.

First, we should take note of the fact, which you may have already noticed, that the vast majority of apocalyptic sayings attributed to Jesus—for example, those that speak about the imminent appearance of the kingdom of God, the day of judgment, or the coming Son of Man—appear in the earliest accounts, the Synoptic Gospels and their sources, but not in our later ones. There are far fewer such sayings in John, the latest of the canonical Gospels, and virtually none in the Gospel of Thomas, which was later still. This should give us pause. Did Christians who passed along the teachings of Jesus recast them after some time had passed, so that his apocalyptic predictions of the imminent end of the age came to be muted over time and finally silenced?

Given the fact that the end never did come as expected, the reshaping of this tradition would make sense. Recall that even the Gospel of Luke, written perhaps in the 80s of the Common Era, went to some length to alter Jesus' words inherited from Mark, so that in Luke Jesus does not predict that the end will come in his disciples' lifetime, even though the author may have thought that it was going to come in his own. At the same time, we have seen that Christians did not wait until the end of the first century to begin modifying the traditions about Jesus that they retold. To learn whether the apocalyptic message actually comes from Jesus himself and not from the earliest Christians who told stories about him, we need to establish the basic character of Jesus' message and ministry and determine the central aspects of his life.

The Beginning and End as Keys to the Middle

We can begin by isolating traditions from Jesus' life that are virtually beyond dispute to see what they can tell us about the character of his ministry. In particular, there are two key events that can give us insight into his overarching views and concerns. One event occurred at the very outset of Jesus' ministry and the other transpired after its end. My reasoning is that since we can know how Jesus' ministry began and how it ended, we can speak with some confidence about what happened in between.

There is little doubt about how Jesus began his ministry: he was baptized by John. The story is independently attested by multiple sources; Mark, Q, and John all begin with Jesus' associating with the Baptist. Also, it is not a story the early Christians would have been inclined to invent, since it was commonly understood that the one doing the baptizing was spiritually superior to the one being baptized (i.e., the story passes the criterion of dissimilarity). Moreover, in view of our discussion earlier in this chapter, we can see that the event is contextually credible. John appears to have been one of the "prophets" who arose during the first century of the Common Era in Palestine. Somewhat like Theudas and the Egyptian, he predicted that God was about to destroy his enemies and reward his people, as he had done in the days of old.

John the Baptist appears to have preached a message of coming destruction and salvation. Mark portrays him as a prophet in the wilderness who proclaims the fulfillment of the prophecy of Isaiah that God would again bring his people from the wilderness into the Promised Land (Mark 1:2–8). When this happened the first time, according to the Hebrew Scriptures, it meant destruction for the nations already inhabiting the land. In preparation for this imminent event, John baptized those who repented of their sins, that is, those who were ready to enter into this coming kingdom. The Q source gives further information, for here John preaches a clear message of apocalyptic judgment to the crowds that have come out to see him: "Who warned you to flee from the wrath to come? Bear fruits worthy of repentance. . . . Even now the ax is lying at the root of the trees; every tree therefore that does not bear good fruit is cut down and thrown into the fire" (Luke 3:7–9). Judgment is imminent (the ax is at the root of the tree) and it will not be a pretty sight. In preparation, Jews can no longer rely on having a covenantal relationship with God: "Do not begin to say to yourselves, 'We have Abraham as our ancestor'; for I tell you, God is able from these stones to raise up children to Abraham" (Luke 3:8). Instead, they must repent and turn to God anew by doing the things he requires of them.

Jesus went out into the wilderness to be baptized by this prophet. Since nobody compelled him, he must have gone to John, instead of to someone else, because he agreed with John's message. Jesus did not join the Pharisees, who emphasized the scrupulous observance of the Torah, or align himself with the Sadducees, who focused on the worship of God through the Temple cult. He did not associate with the Essenes, who formed monastic communities to maintain their own ritual purity, or subscribe to the teachings of the "fourth philosophy," which advocated a violent rejection of Roman domination. He associated with an apocalyptic prophet in the wilderness who anticipated the imminent end of the age.

So Jesus appears to have begun by embracing the message of a Jewish apocalypticist. Is it possible, though, that he changed his views during the course of his ministry and began to focus on something other than what John preached? This is certainly possible, but it would not explain why so many apocalyptic sayings are found on Jesus' own lips in the earliest sources for his life. Even more seriously, it would not explain what clearly emerged in the aftermath of his ministry. I have argued that we are relatively certain about how Jesus' ministry began; we are even more certain about what happened in its wake. After Jesus' death, those who believed in him established communities of followers throughout the Mediterranean. We have a good idea of what these Christians believed because some of them have left us writings. These earliest writings are imbued with apocalyptic thinking. The earliest Christians were Jews who believed that they were living at the end of the age and that Jesus himself was to return from heaven as a cosmic judge of the earth to punish those who opposed God and to reward the faithful (e.g., see 1 Thess 4:13–18; 1 Cor 15:51–57, writings from the earliest Christian author, Paul). The church that emerged in Jesus' wake was apocalyptic.

Thus, Jesus' ministry began with his association with John the Baptist, an apocalyptic prophet, and ended with the establishment of the Christian church, a community of apocalyptic Jews who believed in him. The fact that Jesus' ministry began apocalyptically and ended apocalyptically gives us the key to interpreting what happened in between. The only connection between the apocalyptic John and the apocalyptic Christian church was Jesus himself. How could both the beginning and the end be apocalyptic if the middle was not as well? Jesus himself must have been a Jewish apocalypticist.

To call Jesus an apocalypticist does not mean that Jesus was saying and doing exactly what every other Jewish apocalypticist was saying and doing. We are still interested in learning specifically what Jesus taught and did during his life. Knowing that his overall message was apocalyptic, however, can help us understand other aspects of the tradition about him that can be established as authentic.

THE APOCALYPTIC DEEDS OF JESUS

The Crucifixion

The most certain element of the tradition about Jesus is that he was crucified on the orders of the Roman prefect of Judea, Pontius Pilate. The crucifixion is independently attested in a wide array of sources and is not the sort of thing that believers would want to make up about the person proclaimed to be the powerful Son of God. Why, historically, was Jesus crucified? This is the question that every reconstruction of the life of Jesus has to answer, and some of the answers proffered over the years have been none too plausible. If, for example, Jesus had simply been a great moral teacher, a gentle rabbi who did nothing more than urge his devoted followers to love God and one another, or an itinerant philosopher who urged them to abandon their possessions and live a simple life, depending on no one but God (see box 15.2), then he would scarcely have been seen as a threat to the social order and nailed to a cross. Great moral teachers were not crucified—unless their teachings were considered subversive. Nor were charismatic leaders with large followings—unless their followers were thought to be dangerous.

The subversive teachers from Jesus' day were labeled as prophets, people who proclaimed the imminent downfall of the social order and the advent of a new kingdom to replace the corrupt ruling powers. According to the traditions recorded in the New Testament and Josephus, John the Baptist was imprisoned and executed because of his preaching; according to the Gospels he directed his words against Herod Antipas, appointed to rule over the Promised Land. Jesus was to fare no better. Those who prophesied the judgment of God were liable to the judgment of Rome.

In the case of Jesus, however, it is not altogether clear that Rome initiated the proceedings. It appears that Jesus' message was directed not only against the Roman powers but also against the Jewish leadership of Jerusalem that supported them, as seen in another tradition that can be established beyond reasonable doubt as authentic.

SOMETHING TO THINK ABOUT

Box 15.2 Was Jesus a Cynic Philosopher?

A number of recent American scholars have proposed that Jesus should be understood not as a Jewish apocalypticist but as a kind of Jewish Cynic philosopher. The term "cynic" in this context does not carry the same connotations that it does for us when we say that someone is "cynical." When referring to the Greco-Roman world, it denotes a particular philosophical position that was advocated by a number of public and well-known characters.

The term "cynic" actually means "dog." It was a designation given to a certain group of philosophers by their opponents, who claimed that they lived like wild mongrels. In some respects, the designation was apt, for Cynics urged people to abandon the trappings of society and live "according to nature." For them, the most important things in life were those over which people could have some control, such as their attitudes toward others, their likes and dislikes, and their opinions. Other things outside of their control were of no importance. Followers of the cynics were therefore admonished not to burden themselves with material possessions, such as nice houses or fine clothes, or to worry about how to earn money or what to eat. To this extent, the Cynics were closely aligned in their views to the Stoic philosophers. They differed, though, in the degree of their social respectability. Cynics rejected most constraints imposed by society, even society's ethical mores, so as to live "naturally." The Cynics who practiced what they preached had virtually no possessions, often lived on the streets, rarely bathed, begged for a living, performed private bodily functions in public places, and spent their days haranguing people to adopt their philosophical views. They were especially renowned for abusing people on street corners and in marketplaces, where they castigated those who thought that the meaning of life could be found in wealth or in any of the other trappings of society.

Scholars who think that Jesus was a Jewish teacher who embraced Cynic values point out that many of his teachings sound remarkably similar to what we hear from the Cynics. Jesus' followers were to abandon all their possessions (Matt 6:19–21; Mark 11:21–22), they were not to be concerned about what to wear or what to eat (Matt 6:25–33), they were to live with the bare essentials and accept whatever was given to them by others (Mark 6:6–13; Luke 10:1-12); they were to condemn those who rejected their message (Luke 10:1–12), and they were to expect to be misunderstood and mistreated (Matt 5:11–12). Was not Jesus, then, a Jewish Cynic?

Other scholars believe that this is taking matters too far. All of our ancient sources portray Jesus as quoting the Hebrew Scriptures to support his perspective, but never does he quote any of the Greek or Roman philosophers or urge his followers to adhere to their teachings. Moreover, the message of his teaching is not, ultimately, about living in accordance with nature. It is about the God of Israel, the true interpretation of his Law and the coming judgment against those who are unrepentant. Thus, while it is true that Jesus' followers were told not to concern themselves with wealth and the trappings of society, these teachings were not rooted in a concern for promoting self-sufficiency in a harsh and capricious world. Rather, his followers were not to be tied to the concerns of this age because it was passing away and a new age was soon to come. Jesus may have appeared to an outsider to be similar in some ways to an itinerant Cynic philosopher, but his message was in fact quite different.

The Temple Incident

We know with relative certainty that Jesus predicted that the Temple was soon to be destroyed by God. Predictions of this sort are contextually credible given what we have learned about other prophets in the days of Jesus. Jesus' own predictions are independently attested in a wide range of sources (cf. Mark 13:1; 14:58; John 2:19; Acts 6:14). Moreover, it is virtually certain that some days before his death Jesus entered the Temple, overturned some of the tables that were set up inside, and generally caused a disturbance.

The account is multiply attested (Mark 11 and John 2) and it is consistent with the predictions scattered throughout the tradition about the coming destruction of the Temple. Therefore, it is unlikely that Christians invented the story, in order to show their own opposition to the Temple, as some scholars have claimed. It is possible, however, that Christians modified the tradition in some ways, as they modified most of the stories that they retold over the years. In the earliest surviving account, Jesus displays a superhuman show of strength, shutting down the entire Temple cult by an act of his will (Mark 11:16). The Temple complex was immense, and there would have been armed guards present to prevent any major disturbances. Mark's account, then, may represent an exaggeration of the effect of Jesus' actions.

It is hard to know whether Jesus' words during this episode should be accepted as authentic. He quotes the prophets Isaiah and Jeremiah to indicate that the Temple cult has become corrupt, calling it "a den of thieves." Indeed, it is possible that Jesus, like the Essenes, believed that the worship of God in the Temple had gotten out of hand and that the Sadducees in control had abused their power and privileges to their own end. But it is also possible that Jesus' actions are to be taken as a kind of enacted parable, comparable to the symbolic actions performed by a number of the prophets in the Hebrew Scriptures (see box 15.3). By overturning the tables and causing a distur-

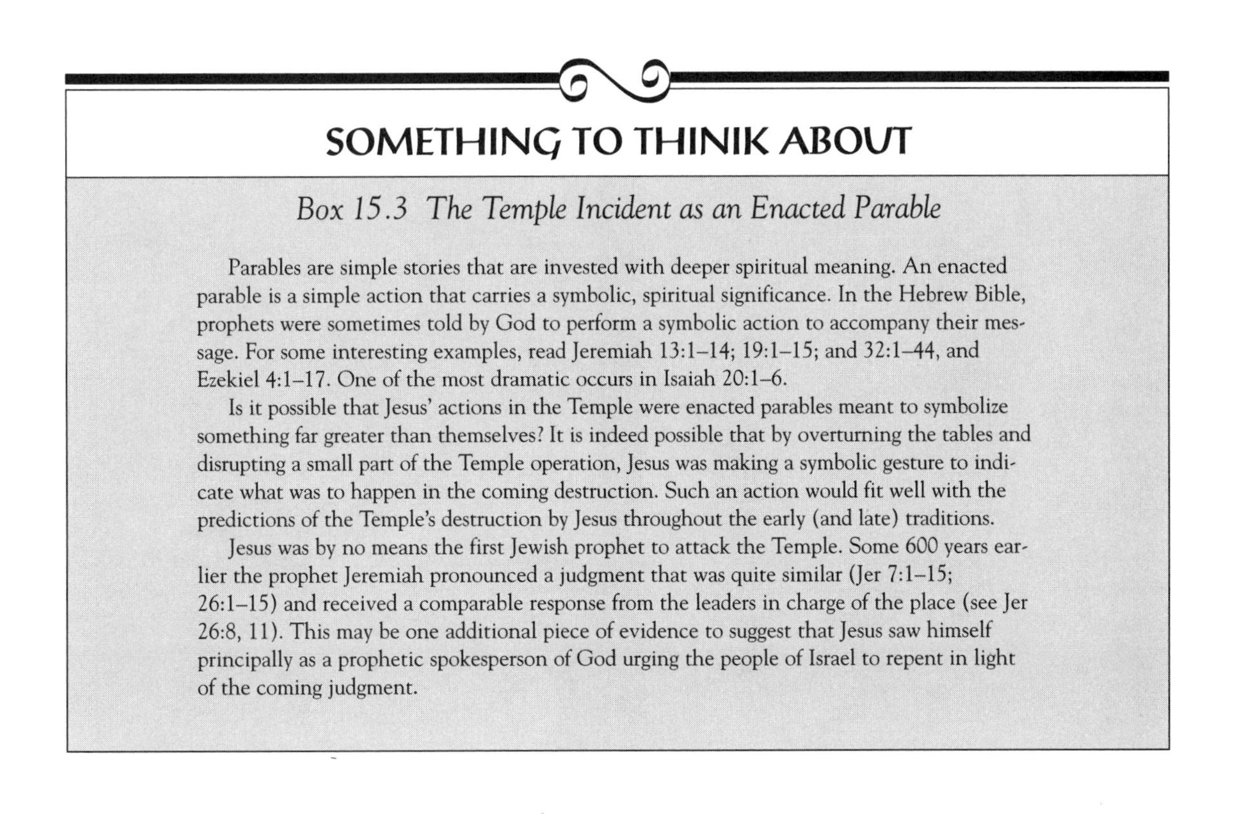

SOMETHING TO THINK ABOUT

Box 15.3 The Temple Incident as an Enacted Parable

Parables are simple stories that are invested with deeper spiritual meaning. An enacted parable is a simple action that carries a symbolic, spiritual significance. In the Hebrew Bible, prophets were sometimes told by God to perform a symbolic action to accompany their message. For some interesting examples, read Jeremiah 13:1–14; 19:1–15; and 32:1–44, and Ezekiel 4:1–17. One of the most dramatic occurs in Isaiah 20:1–6.

Is it possible that Jesus' actions in the Temple were enacted parables meant to symbolize something far greater than themselves? It is indeed possible that by overturning the tables and disrupting a small part of the Temple operation, Jesus was making a symbolic gesture to indicate what was to happen in the coming destruction. Such an action would fit well with the predictions of the Temple's destruction by Jesus throughout the early (and late) traditions.

Jesus was by no means the first Jewish prophet to attack the Temple. Some 600 years earlier the prophet Jeremiah pronounced a judgment that was quite similar (Jer 7:1–15; 26:1–15) and received a comparable response from the leaders in charge of the place (see Jer 26:8, 11). This may be one additional piece of evidence to suggest that Jesus saw himself principally as a prophetic spokesperson of God urging the people of Israel to repent in light of the coming judgment.

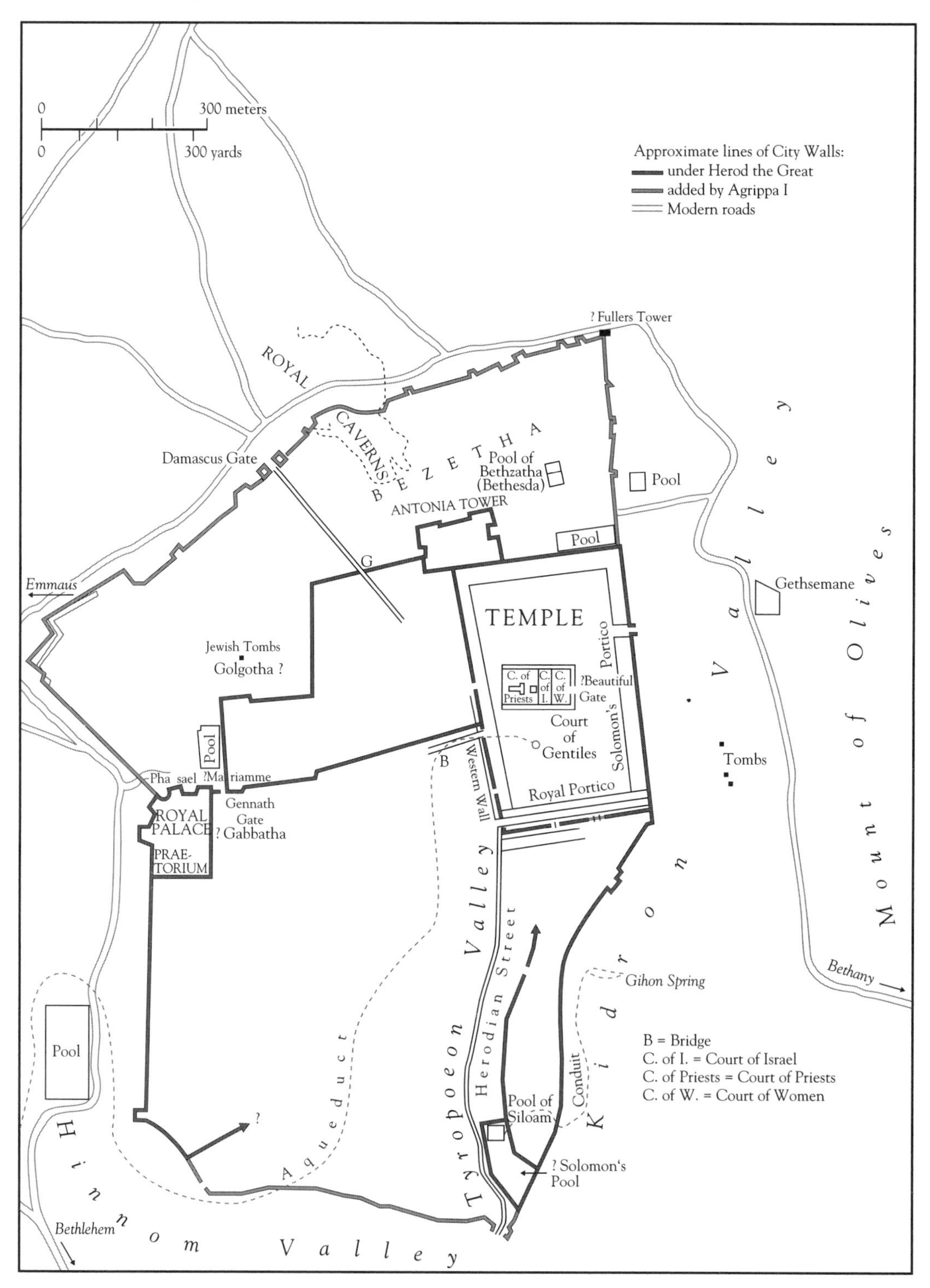

Figure 15.5 Jerusalem in Days of Jesus.

bance, Jesus could have been projecting what was to happen when his words against the Temple came to fruition, foreshadowing the destruction of the Temple that he anticipated was soon to come.

How, though, did Jesus' prediction that the Temple would be destroyed fit into his broader apocalyptic message? Two possible answers suggest themselves. It may be that he believed that in the new age there would be a new Temple, totally sanctified for the worship of God. This was the view of the apocalyptically minded Essenes. Or it may be that Jesus believed there would be no need for a temple at all in the kingdom that was coming, since there would no longer be any evil or sin, and therefore no need for the cultic sacrifice of animals to bring atonement. In either case, the implication of Jesus' actions is clear: for Jesus, the Temple cult and the officials in charge of it were a temporary measure at best and a corruption of God's plan at worst. They would soon be done away with when the kingdom arrived.

This message did not escape the notice of those in charge of the Temple, the chief priests who also had jurisdiction over the local affairs of the people in Jerusalem. These priests, principally Sadducees, were the chief liaison with the Roman officials, in particular, the Roman prefect Pilate. For these reasons, the most plausible scenario for explaining Jesus' death is that Jesus' apocalyptic message, including its enactment in the Temple, angered some of the chief priests on the scene. These priests recognized how explosive the situation could become during the Passover feast, given the tendency of the celebration to become a silent protest that might erupt into something much worse. The Sadducean priests conferred with one another, had Jesus arrested, and put him on trial for his words against the Temple. Knowing that they could not execute Jesus themselves, perhaps because the Romans did not allow the Jewish authorities to execute criminals (the matter is debated among historians), they delivered him over to Pilate, who had no qualms at all about disposing of yet one more troublemaker who might cause a major disturbance.

Jesus' Associations

One other aspect of Jesus' public ministry can be spoken of with confidence by the historian, and here again an apocalyptic context provides some important insights. With whom did Jesus associate? There is no doubt that he had twelve followers whom he chose as his special disciples; the Gospels of Mark (3:16) and John (6:67), and the apostle Paul (1 Cor 15:5) all mention "the Twelve." Curiously, even though the Synoptics give different names for some of these followers (Mark 3:13–19; Matt 10:1–4; and Luke 6:12–16), all three Gospels know that there were twelve of them. But why twelve? Why not eight? Or fourteen?

The number twelve makes sense from an apocalyptic perspective. The present age was coming to an end; God was bringing in his new kingdom for his people. Those who repented and did what God wanted them to do, as revealed in the teachings of Jesus, would enter into that kingdom. This new people of God would arise out of the old. Just as Israel had started out as twelve tribes headed by twelve patriarchs (according to the book of Genesis), so the new people of God would emerge from old Israel with twelve leaders at their head: "Truly I tell you, at the renewal of all things, when the Son of Man is seated on the throne of his glory, you who have followed me will also sit on twelve thrones, judging the twelve tribes of Israel" (Matt 19:28; from Q). Thus the disciples represented the new people of God, those who had repented in anticipation of the kingdom that would come soon, on the day of judgment. This appears to be why Jesus chose twelve of them.

We know that Jesus also associated with two other groups of people, whom early sources designate as "tax collectors" and "sinners." We can accept this tradition as authentic because references to these groups are scattered throughout our sources (e.g., see Mark 2:15; Luke 7:34 [Q]; Luke 15:1–2 [L]); moreover, this is probably not the sort of tradition that a follower of Jesus would be inclined to make up. "Tax collectors" refers to local Jews employed by regional tax corporations to collect the Roman taxes. These persons were unpopular in first-century Palestine because they supported Roman rule and sometimes grew rich through their association with the imperial government. For these reasons, tax collectors had a bad reputation among many of the Jewish subjects of Rome; they were not the sort of people that pious religious leaders were supposed to befriend. "Sinners" does not necessarily refer to prostitutes, as is sometimes thought, although certainly prostitutes and other habitually "sinful" people could be included in their ranks. It refers simply to those who were not scrupulous about observing the law of God. Jesus appears to have spent a good deal of his time with such folk.

From an apocalyptic perspective, these associations make sense. We have numerous teachings of Jesus in which he proclaims that the kingdom is coming not to those who are righteous but to those who are sinful. We have already seen that he does not associate in a friendly way with the religious leaders who scrupulously observe the regulations of the Torah, faithfully attend to the Temple cult, or focus their attention on their own ritual purity. The kingdom that is coming is open to all who are willing to repent of their misdeeds, even the most lowly; they need only turn to God in love and receive his loving acceptance in return. Those who are willing to abandon everything to follow the teachings of Jesus, to turn from their evil ways and love God above all else and their neighbors as themselves—whether they are from the lower social classes, like the impoverished fishermen among the disciples, from the scandalous upper classes, like some of the wealthier tax collectors, or from the ranks of the religious outcasts, like the sinners—all such people will enter into the kingdom of God that is soon to arrive.

In Sum: The Deeds of Jesus

Although historians cannot demonstrate that Jesus performed miracles, they have been able to establish with some degree of certainty a few basic facts about Jesus' life: he was baptized, he associated with tax collectors and sinners, he chose twelve disciples to be his closest companions, he caused a disturbance in the Temple near the end of his life, this disturbance eventuated in his crucifixion at the hands of the Roman prefect Pontius Pilate, and in the wake of his death his followers established vibrant Christian communities. What is striking is that all of these pieces of information add up to a consistent portrayal of Jesus. Jesus was an apocalyptic prophet who anticipated the imminent end of the age, an end that would involve the destruction of Israel, including the Temple and its cult, prior to the establishment of God's kingdom on earth. As we turn now to consider more specifically some of the teachings of Jesus, we can fill out this basic apocalyptic message.

THE APOCALYPTIC TEACHINGS OF JESUS

Scholars have been unable to establish a solid consensus on what the historical Jesus said. Certainly, we cannot uncritically assume that he said many of the things recorded in such Gospels as Thomas or even John. As we have seen, a number of these teachings are not independently attested, and most of them appear to conform to the perspectives on Jesus that developed within the communities that preserved them. Thus, although Jesus makes many self-identifications in John's Gospel—"I am the bread of life," "I am the light of the world," "I am the way, the truth, and the life, no one comes to the Father but through me," "I and the Father are one"—none of these is independently attested in any other early source and all of them coincide with the christology that developed within the Johannine community. Indeed, one interesting piece of evidence that the author of the Fourth Gospel modified his traditions of Jesus' sayings in conformity with his own views is that it is nearly impossible to know who is doing the talking in this narrative, unless we are explicitly told. For John the Baptist, Jesus himself, and the narrator of the story all speak in almost exactly the same way, suggesting that there is only one voice here, that of the Gospel writer.

Is it not possible, though, that the apocalyptic sayings of Jesus were also modified in accordance with the views of the early Christians, who, after all, were apocalypticists? This indeed is a possibility, and one that should be carefully considered, but remember that we have already established *on other grounds* that Jesus was an apocalypticist. It is very hard to explain the basic orientation of his ministry otherwise, given the fact that it began with his decision to associate with the apocalypticist John the Baptist and was followed by the establishment of apocalyptic communities of his followers. Moreover, the deeds and experiences of Jesus that

Figure 15.6 Ancient portrayal of Jesus teaching the apostles, from the catacomb of Domitilla in Rome.

we can establish beyond reasonable doubt are consistent with his identity as an apocalypticist.

Given this orientation, it is not surprising that a large proportion of Jesus' sayings in our earliest sources are teachings about the imminent arrival of the Son of Man, the appearance of the kingdom of God, the coming day of judgment, and the need to repent and live in preparation for that day, the climax of history as we know it. While we cannot assume that every saying in the Gospels that has any tint of apocalypticism in it is authentic, many of the apocalyptic sayings must have come from Jesus himself. Mark's summary of Jesus' teaching appears to be reasonably accurate (Mark 1:15): "The time has been fulfilled, the kingdom of God is near; repent and believe in this good news!"

Here we cannot consider all of the sayings that can be established as authentically from Jesus, but we will explore several of the more characteristic ones. Jesus taught that God's kingdom was soon to arrive on earth. Given Jesus' social context and the apocalyptic character of his ministry, we can assume that he had in mind an actual kingdom, one that would replace the corrupt powers that were presently in control, a kingdom perhaps headed by God's special anointed one, his messiah. This kingdom was going to come in a powerful way (Mark 9:1); people must watch for it and be prepared, for no one could know when exactly it would come and it would strike unexpectedly (Mark 13:32–35; Luke 21:34–36). But Jesus did know that it was to arrive soon—at least within the lifetime of some of his disciples (Mark 9:1; 13:30).

It appears that Jesus expected the kingdom to be brought by one whom he called the Son of Man. Scholars have engaged in long and acrimonious debates about how to understand this designation. Is it a title for a figure that Jews would generally understand, for instance, a reference to the figure mentioned in Daniel 7:13–14? Is it a general description of "a human-like being"? Is it a self-reference, a circumlocution for the pronoun "I"? Moreover, did Jesus actually use the term? Or did the Christians come up with it and attribute it to Jesus? If Jesus did use it, did he actually refer to himself as the Son of Man?

The details of this debate cannot concern us here, but I can indicate what seems to me to be the best way to resolve it. Some of Jesus' sayings mention the Son of Man coming in judgment on the earth (e.g., Mark 8:38; 13:26–27; 14:62; Luke 12:8); these appear to presuppose a knowledge of the passage in Daniel where "one like a son of man" comes and is given the kingdoms of earth. We know of other Jewish apocalypticists who anticipated a cosmic judge of this type, sometimes called the "Son of Man" (see box 15.4). Jesus himself seems to have expected the imminent appearance of such a cosmic judge. In some sayings, such as the ones cited above (especially Mark 8:38 and 14:62), he does not identify himself as this figure but seems, at least on the surface, to be speaking of somebody else. If Christians were to make up a saying of Jesus about the Son of Man, however, they would probably not leave it ambiguous as to whether he was referring to himself. Therefore, on the grounds of dissimilarity (again, hotly debated) such sayings are probably authentic. Jesus anticipated the coming of a cosmic judge from heaven who would bring in God's kingdom.

When he came, there would be cosmic signs and a universal destruction. The messengers of God would gather together those who have been chosen for the kingdom (Mark 13:24–27). On the day of judgment, some people would be accepted into the kingdom, and others cast out. The judge

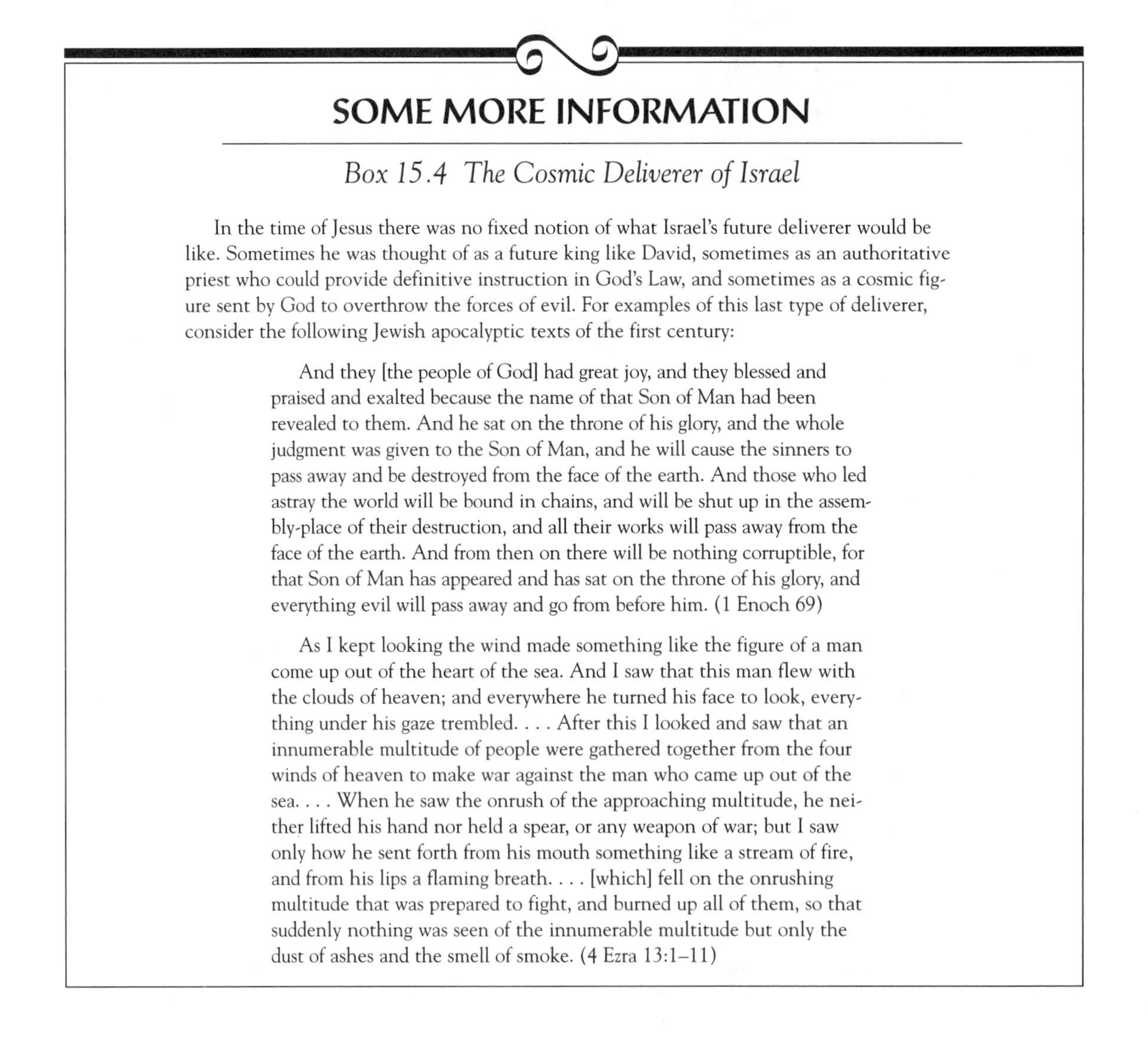

SOME MORE INFORMATION

Box 15.4 The Cosmic Deliverer of Israel

In the time of Jesus there was no fixed notion of what Israel's future deliverer would be like. Sometimes he was thought of as a future king like David, sometimes as an authoritative priest who could provide definitive instruction in God's Law, and sometimes as a cosmic figure sent by God to overthrow the forces of evil. For examples of this last type of deliverer, consider the following Jewish apocalyptic texts of the first century:

> And they [the people of God] had great joy, and they blessed and praised and exalted because the name of that Son of Man had been revealed to them. And he sat on the throne of his glory, and the whole judgment was given to the Son of Man, and he will cause the sinners to pass away and be destroyed from the face of the earth. And those who led astray the world will be bound in chains, and will be shut up in the assembly-place of their destruction, and all their works will pass away from the face of the earth. And from then on there will be nothing corruptible, for that Son of Man has appeared and has sat on the throne of his glory, and everything evil will pass away and go from before him. (1 Enoch 69)

> As I kept looking the wind made something like the figure of a man come up out of the heart of the sea. And I saw that this man flew with the clouds of heaven; and everywhere he turned his face to look, everything under his gaze trembled. . . . After this I looked and saw that an innumerable multitude of people were gathered together from the four winds of heaven to make war against the man who came up out of the sea. . . . When he saw the onrush of the approaching multitude, he neither lifted his hand nor held a spear, or any weapon of war; but I saw only how he sent forth from his mouth something like a stream of fire, and from his lips a flaming breath. . . . [which] fell on the onrushing multitude that was prepared to fight, and burned up all of them, so that suddenly nothing was seen of the innumerable multitude but only the dust of ashes and the smell of smoke. (4 Ezra 13:1–11)

would be like a fisherman who sorts through his fish, taking only the best and disposing of all the others (Matt 13:47–50; *Gosp. Thom.* 8).

This judgment would bring about a total reversal of the social order. Those in positions of power and prestige would be removed, and the oppressed and afflicted would be exalted. It was the forces of evil who were currently in charge of this planet, and those who sided with them were the ones in power. Those who sided with God, however, were the persecuted and downtrodden, who were dominated by the cosmic powers opposed to God. Thus, when God reasserted his control over this planet, all of this would be reversed: "The first shall be last and the last first" (Mark 10:30), and "all those who exalt themselves will be humbled, and those who humble themselves will be exalted" (Luke 14:11; 18:14[Q]) This was not simply a hopeful pipe dream; Jesus expected it actually to happen.

The coming of the Son of Man was not good news for those in power. They would be better served to relinquish their power—to become like children (Mark 10:13–15), to give away their wealth and become poor (Mark 10:23–30), to yield their positions of prestige and become slaves (Mark 10:42–44). Not even the official leaders of the Jewish people would escape, for everyone who lorded it over another would be liable. Indeed, the very locus of power for the influential Sadducees, the Temple of God itself, would be destroyed on judgment day: "There will be not one stone left upon another that will not be destroyed" (Mark 13:2).

On the other hand, those who currently suffered, the oppressed and downtrodden, would be rewarded. This promise is expressed in Jesus' Beatitudes, found in Q: "Blessed are you who are poor, for yours is the kingdom of Heaven [meaning that they will be made rich when it arrives]; Blessed are you who hunger now, for you shall be satisfied [when the kingdom comes]; Blessed are you who weep now, for you shall rejoice; Blessed are you who are hated by others, and reviled . . . for your reward will be great" (Luke 6:20–23; see *Gosp. Thom.* 54, 68–69).

Because there was to be such a dramatic reversal when the Son of Man brought the kingdom, a person should be willing to sacrifice everything in order to enter into it. A person's passion to obtain the kingdom should be like that of a merchant in search of pearls; when he finds one that is perfect, he sells everything that he has to buy it (Matt 13:45–46, *Gosp. Thom.*, 76). People should not, for this reason, be tied to this world or the alluring treasures that it has to offer; instead, they should focus on the kingdom that is coming (Matt 6:19, 33; *Gosp. Thom.*, 63).

At the same time, we should not think that Jesus was maintaining that everyone who happened to be poor or hungry or mistreated would enter into God's kingdom. He expected that people first had to repent and adhere to his teachings. This is what his own disciples had done; they left everything to follow him. As a result, they were promised special places of prominence in the coming kingdom. Similarly, Jesus' association with tax collectors and sinners should not be taken to mean that he approved of any kind of lifestyle. To be sure, he did not insist that his followers keep the detailed traditions of the Pharisees, in part because he appears to have felt that the Torah itself was only a provisional measure: what need would there be of "Law" in a kingdom in which there was no sin or evil? Moreover, he appears to have believed that at the heart of the Torah was the command for people to love God with their entire being and to love their neighbors as themselves (Mark 12:28–31, where he quotes Deut 6:4 and Lev 19:18; see *Gosp. Thom.* 25). Occasionally, in his view, the overly scrupulous attention to the details of the Torah could, perhaps ironically, lead to a violation of these basic principles (Mark 7:1–13). The Sabbath, for example, was created for the sake of humans, not humans for the Sabbath. Human need, therefore, had priority over the punctilious observation of rules for keeping the Sabbath (Mark 2:27–28). For Jesus, then, keeping the Torah was indeed important; this happened, however, not when Jews followed the carefully formulated rulings of the Pharisees but when they repented of their bad behavior and turned to God with their entire being and manifested their love for him in their just and loving treatment of their neighbors.

These examples make it clear that the guidelines for living that Jesus gave, that is, his ethics, were grounded in his apocalyptic worldview. They are probably misunderstood, therefore, when they are taken as principles for a healthy society. Jesus did teach that people should love one another, but not because he wanted to help them lead happy and productive lives or because he knew that if love were not at the root of their dealings with one another society might fall apart. He was not a teacher of ethics concerned with how people should get along in the future. For Jesus, the end was coming soon, within his own generation. The motivation for ethical behavior, then, sprang from the imminent arrival of the kingdom, to be brought by the Son of Man in judgment.

Those who began to implement the ideals of the kingdom, where there would be no sin, hatred, or evil, had in a sense begun to experience the rule of God here and now. This rule of God that would find its climax in the powerful appearance of the Son of Man. The followers of Jesus who had begun to live the life of the kingdom by loving God and their neighbors as themselves were merely a small prelude; they were like a tiny mustard seed in comparison with the great mustard bush that represented the coming kingdom (Mark 4:30–31; *Gosp. Thom.* 20). Indeed, they were not many in number, since the words of Jesus for the most part fell on deaf ears. But when these words came to those who were chosen for the kingdom, they were like vibrant seed falling on rich soil; they bore fruit of far greater worth and magnitude than one could imagine (Mark 4:1–9, *Gosp. Thom.* 9). For this reason, those who heard the good news of the kingdom were not only to prepare themselves but also to proclaim the message of Jesus to others. As the Gospels express it, no one puts a lamp under a bushel but on a light stand, so that all might see the light and recognize the truth that has now been made clear, the truth of God's coming kingdom (Mark 4:21–22, *Gosp. Thom.* 33).

It is difficult to know what Jesus thought about his own role in this imminent kingdom of God. On occasion he speaks as if he expected to enter into the kingdom himself, and he seems to have anticipated that this was to be soon (e.g., Mark 14:25). As we have seen, the disciples were to be leaders in this new kingdom, but who would lead them? Would it still be Jesus? Would he be the ultimate leader of this new kingdom of God on earth, the one whom God appoints as king? If this is what Jesus thought—and, of course, it is impossible to know what anyone thinks, especially someone who lived 2,000 years ago, whom we know only through such fragmentary sources—then he may have considered himself to be the future messiah—but only in this apocalyptic sense.

THE APOCALYPTIC DEATH OF JESUS

As we have seen, several aspects of the Gospel Passion narratives appear to be historically accurate. Jesus offended members of the Sadducees by his apocalyptic actions in the Temple just prior to the Passover feast. They decided to have him taken out of the way. Perhaps they were afraid that his followers would swell as the feast progressed and that the gathering might lead to a riot; or perhaps they simply found his views offensive and considered his attack on the Temple of God blasphemous. In either case, they appear to have arranged with one of his own disciples to betray him. Jesus was arrested and questioned by a Jewish Sanhedrin called for the occasion, possibly headed up by the high priest Caiaphas;. He was then delivered over to the Roman prefect Pontius Pilate, who condemned him to be crucified. The time between his arrest and his crucifixion may have been no more than twelve hours; he was sent off to his execution before anyone knew what was happening.

What else can we know about Jesus' last days? Here we will look at some of the more intriguing questions that have occurred to scholars over the years. One of these is, why was Jesus in Jerusalem in the first place? The theologian might say that Jesus went to Jerusalem to die for the sins of the world; this view, however, is based on Gospel sayings of Jesus that cannot pass the criterion of dissimilarity (e.g., his three passion predictions in Mark). In making judgments about why this itinerant teacher from Galilee went to Jerusalem, we should stick to our historical criteria.

It is possible that Jesus simply wanted to celebrate the Passover in Jerusalem, as did so many thousands of other Jews every year. But Jesus' actions there appear to have been well thought out. When he arrived, he entered the Temple and caused a disturbance. Afterwards he evidently spent sever-

Figure 15.7 A portrayal of Jesus' triumphal entry, found on the famous sarcophagus of a Christian named Junius Bassus.

al days in and out of the Temple, teaching his message of the coming kingdom. Given Jesus' understanding that this kingdom was imminent and the urgency with which he taught others that they needed to repent in preparation for it, we should perhaps conclude that he had come to Jerusalem to bring his message to the center of Israel itself, to the Temple in the holy city, where faithful Jews from around the world would be gathered to worship the God who saved them from their oppressors in the past and who was expected to do so once more. Jesus came to the Temple to tell his people how this salvation would occur and to urge them to prepare for it by repenting of their sins and accepting his teachings. He proclaimed that judgment was coming and that it would involve a massive destruction, including the destruction of the Temple.

Did Jesus realize that he was about to be arrested and executed? Again, there is simply no way to know for certain what Jesus thought. It is not hard to imagine, though, that anyone with any knowledge at all of how prophets of doom were generally received, both in ancient times and more recently, might anticipate receiving similar treatment. Moreover, Jesus would probably have known that the leaders in Jerusalem did not take kindly to his message, and he certainly would have known about their civil power. According to the traditions, of course, Jesus knew that his time had come on the night of his arrest. There are a number of difficulties with accepting the accounts of the Last Supper as historically accurate, especially when Jesus indicates that his death will be for the forgiveness of sins, a clearly Christian notion that cannot pass the criterion of dissimilarity. Still, we have two independent accounts of the event (Mark 14:22–26 and 1 Cor 11:23–26), the earliest of which was written in the mid-50s by Paul, who claims to have received the tradition from others. Did he learn it from someone who was present at the event, or from a Christian who knew someone who was there? In any case, the basic notion that at his last meal Jesus explained that he would not last long in the face of his powerful opposition is not at all implausible.

Why did Judas betray Jesus, and what did he betray? These again are extraordinarily difficult questions to answer. That Judas did betray Jesus is almost certain; it is multiply attested and is not a tradition that a Christian would have likely invented. Why he did so, however, will always remain a mystery. Some of our accounts intimate that he did it simply for the money (Matt 26:14–15; cf. John 12:4–6). This is possibly the case, but the "thirty pieces of silver" is a reference to a fulfillment of a prophecy in the Hebrew Bible (Zech 11:12) and may not be historically accurate. Other theories have been proposed over the years, some of which are more plausible than others (see box 15.5).

What appears certain is that Jesus was eventually handed over to the Roman authorities, who tried him on the charge that he called himself king of the Jews. That this was the legal case against him is multiply attested in independent sources. Moreover, as has often been noted, in the early Gospels the designation of Jesus as king of the Jews is found only in the accounts of his trial (Mark 15, Matt 27, Luke 23, John 18–19); nowhere do his followers actually call him this. Since the early Christians did not generally favor, or even use, the designation "king of the Jews" for Jesus, they probably would not have made it up as the official charge against him. This must, therefore, be a historically accurate tradition.

Claiming to be king of the Jews was a political charge that amounted to insurrection or treason against the state. That is why Jesus was executed by the Romans under Pontius Pilate, not by the Jewish authorities, who may not have been granted the power of capital punishment in any case. That the Romans actually did the deed is attested by a wide range of sources, including even Josephus and Tacitus.

Why, though, did the Roman authorities execute Jesus if it was the Jewish authorities who had him arrested in the first place? We know that Jesus must have offended powerful members of the Sadducees by his action in the Temple. Through the high priest Caiaphas, the chief authority over local affairs, these leaders arranged to have Jesus arrested. Once he was taken, he was brought in for questioning. We cannot know for certain how the interrogation proceeded; none of Jesus' disciples was present, and our earliest account, Mark's, is historically problematic (see box 5.4). Perhaps we can best regard it as a fact-finding interrogation. The Sanhedrin evidently decided to have Jesus taken out of the way. Using the information (given by Judas? see box 15.5) that he had been called the messiah, they sent Jesus before the prefect Pilate. We do not know exactly what happened at this trial. Possibly Pilate was as eager to be rid of a potential troublemaker during these turbulent times as the chief priests were.

When Pilate chose to have someone executed, he could do so on the spur of the moment. There was no imperial legal code that had to be followed, no requirements for a trial by jury, no need to call witnesses or to establish guilt beyond reasonable doubt, no need for anything that we ourselves might consider due process. Roman governors were given virtually free reign to do whatever was required to keep the peace and collect the tribute (see Chapter 25). Pilate is known to history as a ruthless administrator, insensitive to the needs

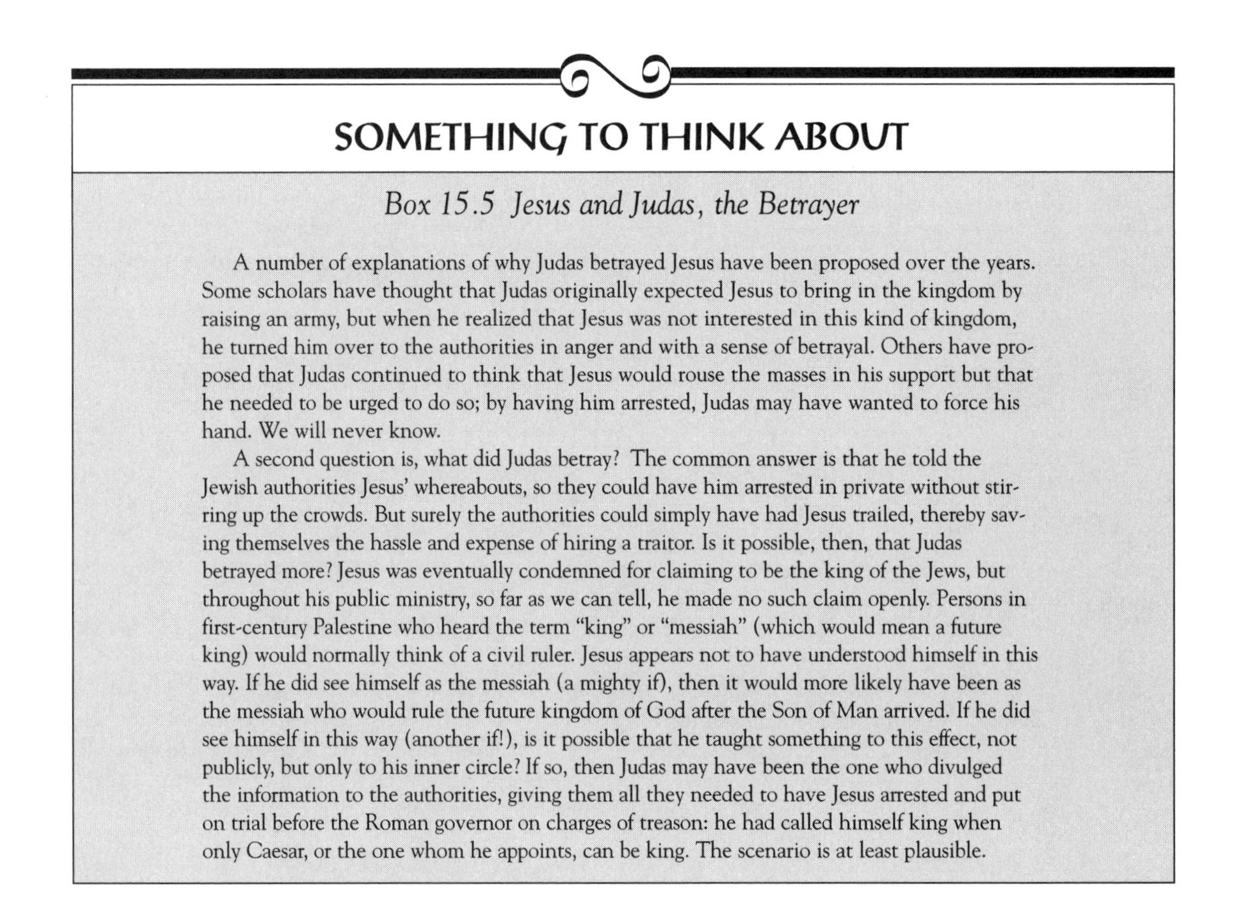

SOMETHING TO THINK ABOUT

Box 15.5 Jesus and Judas, the Betrayer

A number of explanations of why Judas betrayed Jesus have been proposed over the years. Some scholars have thought that Judas originally expected Jesus to bring in the kingdom by raising an army, but when he realized that Jesus was not interested in this kind of kingdom, he turned him over to the authorities in anger and with a sense of betrayal. Others have proposed that Judas continued to think that Jesus would rouse the masses in his support but that he needed to be urged to do so; by having him arrested, Judas may have wanted to force his hand. We will never know.

A second question is, what did Judas betray? The common answer is that he told the Jewish authorities Jesus' whereabouts, so they could have him arrested in private without stirring up the crowds. But surely the authorities could simply have had Jesus trailed, thereby saving themselves the hassle and expense of hiring a traitor. Is it possible, then, that Judas betrayed more? Jesus was eventually condemned for claiming to be the king of the Jews, but throughout his public ministry, so far as we can tell, he made no such claim openly. Persons in first-century Palestine who heard the term "king" or "messiah" (which would mean a future king) would normally think of a civil ruler. Jesus appears not to have understood himself in this way. If he did see himself as the messiah (a mighty if), then it would more likely have been as the messiah who would rule the future kingdom of God after the Son of Man arrived. If he did see himself in this way (another if!), is it possible that he taught something to this effect, not publicly, but only to his inner circle? If so, then Judas may have been the one who divulged the information to the authorities, giving them all they needed to have Jesus arrested and put on trial before the Roman governor on charges of treason: he had called himself king when only Caesar, or the one whom he appoints, can be king. The scenario is at least plausible.

and concerns of the people he governed, willing to exercise brutal force whenever it served Rome's best interests. So, perhaps on the basis of a brief hearing in which he asked a question or two, Pilate decided to have Jesus executed. It was probably one of several items on a crowded morning agenda; it may have taken only a couple of minutes. Two other persons were charged with sedition the same morning. All three were taken outside of the city gates to be crucified.

According to the Gospel traditions, Jesus was first flogged. It is hard to say whether this is a Christian addition to show how much Jesus suffered or a historical account. In any event, he and the others would have been taken by soldiers outside the city gates and forced to carry their crossbeams to the upright stakes kept at the site of execution. The uprights were reused, possibly every day. There the condemned would have been nailed to the crossbeams, or possibly to the uprights themselves, through the wrists and possibly the ankles. There may have been a small ledge attached to the upright on which they could sit to rest.

The death itself would have been slow and painful. Crucifixion was reserved for the worst offenders of the lowest classes: slaves, common thieves, and insurrectionists. It was a death by suffocation. As the body hung on the cross, the lung cavity would distend beyond the point at which one could breathe. To relieve the pain on the chest, one had to raise the body up, either by pulling on the stakes through the wrists or by pushing on those through the feet, or both. Death came only when the victim lacked the strength to continue. Sometimes it took days.

In Jesus' case, death came within several hours, in the late afternoon, on a Friday during Passover week. He was taken from his cross and given a quick burial sometime before sunset on the day before Sabbath.

SUGGESTIONS FOR FURTHER READING

The World of Early Judaism

Barrett, C. K., ed. *The New Testament Background: Selected Documents*. 2d. ed. New York: Harper & Row, 1989. A standard collection of Jewish and pagan texts relevant for the study of the New Testament.

Collins, John. *The Apocalyptic Imagination: An Introduction to the Matrix of Christianity*. New York: Crossroad, 1984. A fine treatment of Jewish apocalypticism as the context for the proclamation of Jesus and his followers.

Fitzmyer, Joseph A. *Responses to 101 Questions on the Dead Sea Scrolls*. New York: Paulist, 1992. Probably the best and easiest way for a beginning student to learn the most important information about the Dead Sea Scrolls, from a renowned expert.

Horsley, Richard A., and John S. Hanson. *Bandits, Prophets, and Messiahs: Popular Movements at the Time of Jesus*. Minneapolis, Minn.: Winston Press, 1985. An enlightening discussion of the sociopolitical context of Jesus' world.

Sanders, E. P. *Judaism Practice and Belief, 63 B.C.E.–66 C.E.* London: SCM Press; Philadelphia: Trinity Press International, 1992. A full, detailed, and authoritative account of what it meant to be a Jew immediately before and during the time of the New Testament.

Schürer, Emil. *The History of Jesus People in the Age of Jesus Christ*. Rev. and ed. Geza Vermes and Fergus Millar. Edinburgh: T & T Clark, 1973. A standard work on the major political, economic, and religious aspects of first-century Judaism; for advanced students.

The Historical Jesus

Evans, Craig, and Bruce Chilton, eds. *Studying the Historical Jesus: Evaluations of the Current Stage of Research*. Leiden: Brill, 1994. Essays on important aspects of the historical Jesus; some of these essays take exception to the view of Jesus as an apocalypticist, preferring instead to see him as a kind of first-century Jewish Cynic; for more advanced students.

Green, Joel, et al., eds. *Dictionary of Jesus and the Gospels*. Downers Grove, Ill.: Intervarsity Press, 1994. A Bible Dictionary with in-depth articles on a wide range of topics pertaining to the historical Jesus and the Gospels, written by prominent evangelical Christian scholars who by and large represent different perspectives from those presented in this chapter.

Meier, John. *A Marginal Jew: Rethinking the Historical Jesus*. Vols. 1–2. New York: Doubleday, 1991, 1994. Written at an introductory level. but filled with erudite documentation in the end notes, this is one of the finest treatments of the historical Jesus of the twentieth century.

Sanders, E. P. *The Historical Figure of Jesus*. London: Penguin, 1993. All in all, perhaps the best single-volume sketch of the historical Jesus for beginning students.

Schweitzer, Albert. *Quest of the Historical Jesus*. Trans. W. Montgomery. New York: Macmillan, 1968. The classic study of the major attempts to write a biography of Jesus up to the first part of the twentieth century, and one of the first and perhaps the most important attempt to portray Jesus as a Jewish apocalypticist.

Vermes, Geza. *Jesus the Jew: A Historian's Reading of the Gospels*. New York: Macmillan, 1973. A readable but learned study of Jesus in light of traditions of other Jewish holy men from his time, by a prominent Jewish scholar.

CHAPTER 16

From Jesus to the Gospels

We began our study of the New Testament with the oral traditions about Jesus that were in circulation in the early Christian churches and saw how the stories that eventually made it into the Gospels were modified and sometimes, perhaps, created by Christians who narrated them in order to convert others to faith and to educate, encourage, and admonish those who had already been converted. We moved from there to a study of our earliest written accounts of Jesus—books that were not the first to be produced by Christians (the letters of Paul were earlier) but were the first to portray the most important figure of early Christianity, Jesus himself. We initially examined these works as literary documents, trying to uncover their distinctive portrayals of Jesus. We then moved behind these portrayals to reconstruct the life of the man himself by applying a variety of historical criteria to uncover what Jesus actually said and did.

We have now come full circle back to where we began. This is an ideal stage for us to pause and reexamine the original point of entry into our study in light of what we have learned en route. Here we will discuss with somewhat greater sophistication (and brevity) the development of the traditions about Jesus that circulated in the early decades of the Christian movement.

THE BEGINNING OF CHRISTIANITY

Hypothetically speaking, every religious and philosophical movement has a point of origin. When did Christianity begin? There are several possibilities.

We might say that it began with Jesus' ministry. Obviously, without the words and deeds of Jesus there would have been no religion based on him. At the same time, Christianity has traditionally been much more than a religion that espouses Jesus' teachings. Indeed, if Jesus was the apocalyptic prophet that he appears to have been, then the Christianity that emerged after his death represents a somewhat different religion from the one he himself proclaimed. In the simplest terms, Christianity is a religion rooted in a belief in the death of Jesus for sin and in his resurrection from the dead. This, however, does not appear to have been the religion that Jesus preached to the Jews of Galilee and Judea. To use a formulation that scholars have tossed about for years, Christianity is not so much the religion of Jesus (the religion that he himself proclaimed) as the religion about Jesus (the religion that is based on his death and resurrection).

Should we say, then, that Christianity began with Jesus' death? This too may contain some element of truth, but it also is somewhat problematic. If Jesus had died and no one had come to believe that he had been raised from the dead, then his death would perhaps have been seen as yet another tragic incident in a long history of tragedies experienced by the Jewish people, as the death of yet another prophet of God, another holy man dedicated to proclaiming God's will to his people. But it would not have been recognized as an act of God for the salvation of the world, and a new religion would probably not have emerged as a result.

Did Christianity begin with Jesus' resurrection? Historians would have difficulty making this judgment, since it would require them to subscribe to faith in the miraculous working of God. Yet even if historians were able to speak of the resurrection as a historically probable event, it could not, in and of itself, be considered the beginning of Christianity, for Christianity is not the resurrection of Jesus but the *belief* in the resurrection of Jesus. Historians, of course, have no difficulty speaking about the belief in Jesus' resurrection, since this is a matter of public record. It is a historical fact that some of Jesus' followers came to believe that he had been raised from the dead soon after his execution. We know some of these believers by name; one of them, the apostle Paul, claims quite plainly to have seen Jesus alive after his death. Thus, for the historian, Christianity begins after the death of Jesus, not with the resurrection itself, but with the belief in the resurrection.

JESUS' RESURRECTION FROM AN APOCALYPTIC PERSPECTIVE

How did belief in Jesus' resurrection eventually lead to the Gospels we have studied? Or to put the question somewhat differently, how does one understand the movement from Jesus, the Jewish prophet who proclaimed the imminent judgment of the world through the coming Son of Man, to the Christians who believed in him, who maintained that Jesus himself was the divine man whose death and resurrection represented God's ultimate act of salvation? To answer this question, we must look at who the first believers in Jesus' resurrection actually were.

The Gospels provide somewhat different accounts about who discovered Jesus' empty tomb and about whom they encountered, what they learned, and how they reacted. But all four canonical Gospels and the *Gospel of Peter* agree that the empty tomb was discovered by a woman or a group of women, who were the first of Jesus' followers to realize that he had been raised. Interestingly, the earliest author to discuss Jesus' resurrection, the apostle Paul, does not mention the circumstance that Jesus' tomb was empty, nor does he name any women among those who first believed in Jesus' resurrection (1 Cor 15:3–8). On one important point, however, Paul does stand in agreement with the early Gospel accounts: those who initially came to understand that God had raised Jesus from the dead were some of Jesus' closest followers, who had associated with him during his lifetime.

It is probably safe to say that all of these followers had accepted Jesus' basic apocalyptic message while he was still alive. Otherwise they would not have followed him. Thus, the first persons to believe in Jesus' resurrection would have been apocalyptically minded Jews. For them, Jesus' resurrection was not a miracle that some other holy person had performed on his behalf. Jesus' followers believed that God had raised Jesus from the dead. Moreover, he had not been raised for a brief period of time, only to die a second time. Jesus had been raised from the dead never to die again. What conclusions would be drawn by these Jewish apocalypticists, the earliest Christians?

We have already seen that apocalypticists believed that at the end of this age the powers of evil would be destroyed. These powers included the devil, his demons, and the cosmic forces aligned with them, the forces of sin and death. When these powers were destroyed, there would be a resurrection of the dead, in which the good would receive an eternal reward and the evil would face eternal punishment. Many Jewish apocalypticists, like Jesus himself, believed that this end would be brought by one specially chosen by God and sent from heaven as a cosmic judge of the earth. Given this basic apocalyptic scenario, there is little doubt as to how the first persons who believed in Jesus' resurrection would have interpreted the event. Since the resurrection of the dead was to come at the end of the age and since somebody had now been raised (as they believed), then the end must have already begun. It had begun with the resurrection of a particular person, the great teacher and holy man Jesus, who had overcome death, the greatest of the cosmic powers aligned against God. Thus, Jesus was the personal agent through whom God had decided to defeat the forces of evil. He had been exalted to heaven, where he now lived until he would return to finish God's work. For this reason, people were to repent and await his second coming.

Sometime after Jesus' resurrection—it is impossible to say how soon (since our sources were written decades later)—these earliest apocalyptic believers began to say things about Jesus that reflected their belief in who he was now that he had been raised. These early reflections on Jesus' significance strongly influenced the beliefs that came to be discussed, developed, and modified for centuries to follow, principally among people who were not apocalyptic Jews to begin with. For example, the earliest Christians believed that Jesus had been exalted to heaven; that is, God had bestowed a unique position upon him. Even during his lifetime, they knew, Jesus had addressed God as Father and taught his disciples that they should trust God as a kindly parent. Those who came to believe in his resurrection realized that he must have had a relationship with God that was truly unique. In a distinctive way, for them, he was *the* Son of God.

Moreover, these Christians knew that Jesus had spent a good deal of time talking about one who was soon to come from heaven in judgment over the earth. For them, Jesus himself was now exalted to heaven; clearly, he must be the judge about whom he had spoken. Therefore, in their view, Jesus was soon to return in judgment as the Son of Man.

Jesus also spoke of the kingdom of God that was to arrive with the coming of the Son of Man. As we have seen, he may have thought that he would be given a position of prominence in that kingdom. For these early Christians, that was precisely what would happen: Jesus would reign over the kingdom that was soon to appear. For them, he was the king to come, the king of the Jews, the messiah (see box 16.1).

Jesus also taught that in some sense this kingdom had already been inaugurated. He therefore taught his followers to implement the values of the kingdom and adopt its ways in the here and now by loving one another as themselves. Those who believed in his resurrection maintained that

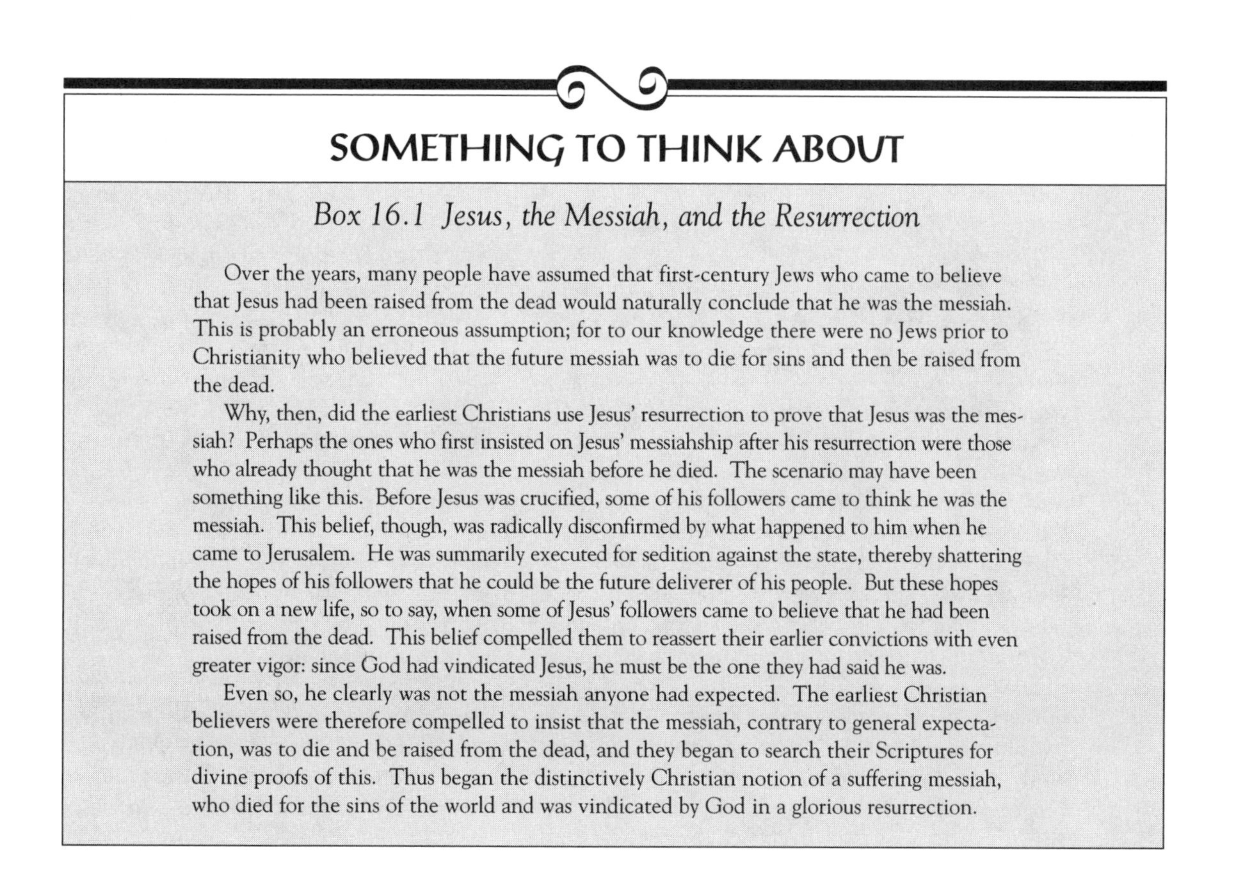

SOMETHING TO THINK ABOUT

Box 16.1 Jesus, the Messiah, and the Resurrection

Over the years, many people have assumed that first-century Jews who came to believe that Jesus had been raised from the dead would naturally conclude that he was the messiah. This is probably an erroneous assumption; for to our knowledge there were no Jews prior to Christianity who believed that the future messiah was to die for sins and then be raised from the dead.

Why, then, did the earliest Christians use Jesus' resurrection to prove that Jesus was the messiah? Perhaps the ones who first insisted on Jesus' messiahship after his resurrection were those who already thought that he was the messiah before he died. The scenario may have been something like this. Before Jesus was crucified, some of his followers came to think he was the messiah. This belief, though, was radically disconfirmed by what happened to him when he came to Jerusalem. He was summarily executed for sedition against the state, thereby shattering the hopes of his followers that he could be the future deliverer of his people. But these hopes took on a new life, so to say, when some of Jesus' followers came to believe that he had been raised from the dead. This belief compelled them to reassert their earlier convictions with even greater vigor: since God had vindicated Jesus, he must be the one they had said he was.

Even so, he clearly was not the messiah anyone had expected. The earliest Christian believers were therefore compelled to insist that the messiah, contrary to general expectation, was to die and be raised from the dead, and they began to search their Scriptures for divine proofs of this. Thus began the distinctively Christian notion of a suffering messiah, who died for the sins of the world and was vindicated by God in a glorious resurrection.

the kingdom proclaimed by Jesus had indeed already begun. As the exalted one, he was already its ruler. He was, in fact exalted above all of creation; for believers, Jesus was the Lord of all that is, in heaven and on earth.

These new and important ways of understanding Jesus came to prominence quickly and naturally. Within several years after his death he was proclaimed in small communities scattered throughout the Eastern Mediterranean as the unique Son of God, the coming Son of Man, the Jewish messiah, and the Lord of all. Christians who understood Jesus in these ways naturally told stories about him that reflected their understanding. For example, when they mentioned Jesus' teaching about the Son of Man, they sometimes changed what he said so that instead of speaking about this other one to come, he was said to be speaking of himself, using the first person singular: "Whoever acknowledges me before others, I will acknowledge before my Father who is in heaven" (Matt 10:32; contrast Mark 8:38). Likewise, when Jesus spoke about himself, they sometimes changed words given in the first person singular ("I") to the title "Son of Man." Thus Matthew's form of Jesus' question to his disciples is, "Who do people say that the Son of Man is?" (Matt 16:13; contrast Mark 8:27).

JESUS' DEATH, ACCORDING TO THE SCRIPTURES

As we have seen, the earliest Christians had an obvious problem when they tried to convince their fellow Jews that Jesus was the one upon whom God had shown his special favor. For non-Christian Jews who were anticipating a messiah figure were not looking for anyone remotely like Jesus. To be sure, the Jewish messianic expectations reflected in the surviving sources are quite disparate. But they all had one thing in common: they all expected the messiah to be a powerful figure who would command the respect of friend and foe alike and lead the Jewish people into a new world that overcame the injustices of the old (see box 5.1). Jesus, on the other hand, was a relatively obscure teacher who was crucified for sedition against the empire. How could a convicted criminal be God's messiah? Jesus never overthrew the state; he was mocked, beaten, and executed by the state. For most Jews, to call Jesus the messiah, let alone Lord of the universe, was preposterous, even blasphemous. To our knowledge, prior to the advent of Christianity, there were no Jews who believed that the messiah to come would suffer and die for the sins of the world and then return again in glory.

Christians today, of course, believe that this is precisely what the messiah was supposed to do. The reason they think so, however, is that the earliest Christians came to believe that the Jewish Bible anticipated the coming of a suffering messiah (see box 16.1). Recall: these earliest Christians were Jews who believed that God spoke to them through their sacred writings. For them, the Scriptures were not simply the records of past events; they were the words of God, directed to them, in their own situation. Not only the earliest Christians, but most Jews that we know about from this period understood the Scriptures in a personal way, as a revelation of meaning for their own times (see box 15.2). Thus, even though the Hebrew Bible never specifically speaks of the messiah as one who is to suffer, there are passages, in the Psalms, for example, that speak of a righteous man who suffers at the hands of God's enemies and who comes to be vindicated by God. Originally, these "Psalms of Lament" may have been written by Jews who were undergoing particularly difficult times of oppression and who found relief in airing their complaints against the evil persons who attacked them and expressing their hopes that God would intervene on their behalf (e.g., see, Pss 22, 35, and 69). Christians who read such Psalms, however, saw in them not the expressions of oppressed, righteous Jews from the distant past but the embodiments of the pain, suffering, and ultimate vindication of the one truly righteous Jew who had recently been unjustly condemned and executed.

As they reflected on what had happened to Jesus, these Jewish Christians saw in his suffering and death a fulfillment of the words of the righteous sufferer described in the Psalms. In turn, these words shaped the ways Christians understood and described the events of Jesus' own Passion. They took the words of Psalm 22, for example, as expressive of the events surrounding Jesus' execution:

> "My God, my God, why have you forsaken me" (v. 1); "All who see me mock at me, they make mouths at me, they shake their heads (v. 7); "I am poured out like water, and all my bones are out of joint . . . my mouth is dried up like a potsherd, and my tongue sticks to my jaws" (vv. 14–15); "A company of evildoers encircles me. My hands and feet have shriveled; I can count all my bones—they stare and gloat over me; they divide my garments among them, and for my clothing they cast lots" (vv. 16–18).

For the early Christians, the sufferings of the righteous Jesus were foreshadowed by the sufferings of the righteous Jew of the Psalms. His sufferings were therefore no mere miscarriage of justice; they were the plan of God.

Other portions of Scripture explained why this suffering was God's plan. Again, these were passages that did not mention the messiah, but Christians nonetheless took them to refer to Jesus, whom they believed to be the messiah. Most important were passages found in the writings of the prophet Isaiah, who also speaks of the suffering of God's righteous one, whom he calls the "Servant of the Lord." According to the "Songs of the Suffering Servant," as scholars have labeled four different passages in Isaiah, the most important of which is Isaiah 52:13–53:12, this servant of God was one who suffered a heinous and shameful fate: he was despised and rejected (53:3), he was wounded and bruised (53:4–5), he was oppressed and afflicted, he suffered in silence and was eventually killed (53:7–8). This is one who suffered and died to atone for the sins of the people (53:4–5)

The interpretation of the original meaning of this passage is difficult, but the widely held view among scholars is that it was originally speaking of the suffering of the nation of Israel during the Babylonian captivity (see Isa 49:3). We have no indication that any Jew, prior to Christianity, ever took the passage as a reference to the future Jewish messiah. You may notice in reading it that the author refers to the Servant's suffering as already having taken place in the past (although his vindication is in the future). Christians, however, understood Jesus' own suffering in light of this and similar passages. For them, these ancient words described well what Jesus went through. Moreover, for them, Jesus clearly was the chosen one, given his resurrection and exaltation (see box 16.1). Their conclusion: God's messiah had to suffer, as a sacrifice for the sins of the world. (see box 16.2)

The crucifixion, then, was turned from a stumbling block for Jews into a foundation stone for Christians (see 1 Cor 1:23). In reflecting upon their Scriptures, the earliest Jewish Christians concluded that Jesus was meant to suffer and die. His death was no mere miscarriage of justice; it was the eternal plan of God. Jesus faithfully carried out his mission, bringing salvation to the world. God therefore exalted him to heaven, making him the Lord of all and setting in motion the sequence of events that would lead to his return in fiery judgment on the earth.

THE EMERGENCE OF DIFFERENT UNDERSTANDINGS OF JESUS

Not all the Christian communities that sprang up around the Mediterranean were completely unified in the ways they understood their belief in Jesus as the one who had died for the sins of the world. We have seen numerous differences that emerged among these groups, particularly as the religion spread from the small group of apocalyptically minded Jews who followed Jesus in Galilee and Jerusalem to other regions and different types of people. This variety can be seen, on its most basic level, in the ways that different believers in the first decades of Christianity would have understood the descriptions of Jesus that we have already examined.

The term "Son of Man," for example, might have made sense to Jews familiar with the prediction of Daniel 7:13–14 that "one like a son of man" was to come on the clouds of heaven. For such an audience, the identification of Jesus as the Son of Man would have meant that he was destined to be the cosmic judge of the earth. A pagan audience,

SOME MORE INFORMATION

Box 16.2 Vicarious Suffering in Jewish Martyrologies and Other Greco-Roman Literature

The idea that someone would suffer and die in order to save others, a notion called vicarious suffering, was not invented by the Christians. Prior to Christianity the notion is found, for example, in a number of stories of Jewish martyrs. Is it possible that these tales affected the ways Christians narrated their stories about Jesus?

In the account of the Maccabean revolt known as 1 Maccabees, we find a Jewish warrior named Eleazar who single-handedly attacks an elephant thought to be bearing the king of Syria, the enemy of God. Eleazar ends up beneath the beast, crushed for his efforts. In the words of the author, "So he gave his life to save his people" (1 Macc 6:44).

A later account of martyrs from the Maccabean period, known as 4 Maccabees, describes in graphic detail the tortures that faithful Jews underwent because they refused to forsake the Law of Moses. The author claims that God accepted their deaths as a sacrifice on behalf of the people of Israel: "Because of them our enemies did not rule over our nation, the tyrant was punished, and the homeland purified—they having become, as it were, a ransom for the sin of our nation. And through the blood of those devout ones and their death as an atoning sacrifice, divine Providence preserved Israel that previously had been mistreated" (4 Macc 17:20–22). In these writings, the death of the faithful martyr brings salvation to others.

Literary portrayals of vicarious suffering can be found in ancient pagan literature as well. One of the most interesting instances occurs in the moving play of Euripides titled *Alcestis*. Alcestis is the beautiful wife of Admetus. He is fated to die at a young age, but the god Apollo, who had earlier befriended him, has worked out a special arrangement with the Fates: someone else can voluntarily die in his stead. Admetus tries in vain to persuade his parents to undertake the task as a familial duty. As a last resort, Alcestis agrees to perform the deed. After her death, Admetus is understandably stricken by grief, although, perversely enough, he is more upset that people will think badly of him than that he has actually made his wife sacrifice her own life for his. But he is comforted by the god Heracles, who goes down into Hades in order to rescue Alcestis from the throes of death and brings her back alive to her stricken husband. Euripides's story is thus about a person who voluntarily dies in someone else's stead and is then honored by a god who conquers death by raising the victim back to life. Sound familiar?

on the other hand, would have to have been told about the book of Daniel, or, as sometimes happened, they would have tried to understand the phrase as best they could, perhaps by taking it to mean that since Jesus was the son of a man, he was a real human being. This is the way many Christians today understand the term, even though it probably would not have meant this either to Jesus or to his apocalyptically minded followers.

The term "Son of God" would have meant something quite different to Jews, who could have taken it as a reference to the king of Israel (as in 2 Sam 7:14 and Psalm 2), than to Gentiles, for whom it would probably mean a divine man. The term "messiah" may have made no sense at all to Gentiles who were not familiar with its special significance in Jewish circles. Literally it would have designated someone who had been anointed or oiled (e.g., an athlete after a hard workout)—scarcely a term of reverence for a religious leader, let alone for the Savior of the world!

Even communities that agreed on the basic meaning of these various titles may have disagreed on their significance as applied to Jesus. Take, for

Figure 16.1 Portrayal of Hercules (= Heracles; with the club on the left) leading Alcestes (middle) back to life away from the god of the underworld, Pluto (seated), after she had voluntarily died in the place of her husband. The scene is found on a pagan sarcophagus of the second century C.E. (SEE BOX 16.2).

instance, the title Son of God. If, in the general sense, the title refers to Jesus' unique standing before God, the question naturally arises, when did Jesus receive this special status? Some early communities appear to have thought that he attained it at his resurrection when he was "begotten" by God as his son. This belief is reflected, for example, in the old traditions preserved in Acts 13:33–34 and Romans 1:3–4. Other communities, perhaps somewhat later, came to think that Jesus must have been God's special son not only after his death but also during his entire ministry. For these believers, Jesus became the Son of God at his baptism, when a voice from heaven proclaimed, "You are my son, today I have begotten you," as the story is preserved in some manuscripts of Luke and among Ebionite Christians. Others came to think that Jesus must have been the Son of God not only for his ministry but for his entire life. Thus, in some of the later Gospels, we have accounts that show that Jesus had no human father, so that he literally was the Son of God (e.g., see Luke 1:35). Still other Christians came to believe that Jesus must have been the Son of God not simply from his birth but from eternity past. By the end of the first century, Christians in some circles had already proclaimed that Jesus was himself divine, that he existed prior to his birth, that he created the world and all that is in it, and that he came into the world on a divine mission as God himself. This is a far cry from the humble beginnings of Jesus as an apocalyptic prophet. Perhaps these beginnings can be likened to a mustard seed, the smallest of all seeds. . . .

The various notions of who Jesus was, and the diverse interpretations of the significance of what he had said and done, came to be embodied in the various written accounts of his life. This, in my judgment, is a certainty. Otherwise, there is no way to explain the radically different portrayals of

Jesus that we find, for instance, in the Gospels of Mark, John, Thomas, and Peter. It was only later, when Christians decided to collect several of these Gospels into a canon of Scripture that the differences came to be smoothed over. From that time on, Matthew, Mark, Luke, and John were all acclaimed as authoritative and interpreted in light of one another. Their placement in the Christian canon thus led to a homogenization, rather than illumination, of their distinctive emphases.

SUGGESTIONS FOR FURTHER READING

Bousset, Wilhelm. *Kurios Christos: A History of the Belief in Christ from the Beginnings of Christianity to Irenaeus*. Trans. John E. Steely. New York: Abingdon, 1970. A classic study that tries to show how the view of Jesus as the Lord developed very early in Christianity among Gentiles living in a polytheistic environment where there were numerous other competing "Lords."

Brown, Raymond. *The Death of the Messiah: From Gethsemane to the Grave*. 2 vols. London: Doubleday, 1994. A massive and exhaustive discussion of the Passion narratives of the four Gospels; among other things, Brown discusses how the Jewish Scriptures influenced early Christian reflections on Jesus' death and resurrection.

Dunn, James D. G. *Christology in the Making: A New Testament Inquiry into the Origins of the Doctrine of the Incarnation*. Philadelphia: Westminster, 1980. A systematic attempt to understand how Christians developed their exalted views of Jesus through the New Testament period.

Frederiksen, Paula. *From Jesus to Christ: The Origins of the New Testament Images of Jesus*. New Haven, Conn.: Yale University Press, 1988. An important study of the earliest Christian views of Jesus and the ways they developed as Christianity moved away from its Jewish roots.

Fuller, Reginald. *Foundations of New Testament Christology*. New York: Scribners, 1965. A classic study that explores the use of christological titles in the New Testament (e.g., Son of Man, Son of God, Messiah, Lord) to learn how early Christians developed their thinking about Jesus.

Hengel, Martin. *Between Jesus and Paul: Studies in the Earliest History of Christianity*. Trans. J. Bowden London: SCM; Philadelphia: Fortress, 1983. A collection of interesting and provocative essays that show the continuities between the apostle Paul's views and those that he inherited from the tradition before him, going back to Jesus; for more advanced students.

Hurtado, Larry. *One God, One Lord: Early Christian Devotion and Ancient Jewish Monotheism*. Philadelphia: Fortress, 1988. A valuable study that argues that the source of conflict between early Christians and non-Christian Jews was not, strictly speaking, over whether Jesus could be thought of as divine but whether he was to be worshipped.

CHAPTER 17

Paul the Apostle: The Man and His Mission

The importance of the apostle Paul in the Christian movement was not universally recognized in his own day. Indeed, Paul appears to have been a highly controversial figure among his contemporaries. From his own letters it is clear that he had at least as many enemies as friends. Nonetheless, for the entire history of Christianity from the first century to our own, no figure except Jesus has proved to be more important.

Consider the New Testament itself. Thirteen of its twenty-seven books claim to be written by Paul. One other book, the Epistle to the Hebrews, was accepted into the canon only after Christians came to believe that Paul had written it, even though it makes no such claim for itself. Yet another book, the Acts of the Apostles, sketches a history of early Christianity with Paul as the principal character. Thus, well over half of the books of the New Testament, fifteen out of twenty-seven, are directly or indirectly related to Paul.

Consider next the spread of Christianity after its inauspicious beginnings among a handful of Jesus' followers in Jerusalem. By the beginning of the second century the religion had grown into an interconnected network of believing communities scattered throughout major urban areas of the empire. Paul was instrumental in this Christian mission. He did not, of course, accomplish it single-handedly. As he himself admits, at the outset he was violently and actively hostile to the spreading Christian church. But in one of the most dramatic turnabouts in history, Paul converted to the faith that he had previously persecuted and became one of its leading spokespersons, preaching the gospel in cities and towns of Syria, Cilicia, Asia Minor, Macedonia, and Achaia (modern-day Syria, Turkey, and Greece), which were significant areas of growth for Christianity in its first few decades.

As important as his role in the geographical spread of the faith—in some respects, far more important—was Paul's contribution to its spread across ethnic lines. More than anyone else that we know about from earliest Christianity, Paul emphasized that faith in Jesus as the messiah who died for sins and was raised from the dead was not to be restricted to those who were born Jews. Moreover, it was not to be restricted to Gentiles who converted to Judaism. The salvation brought by Christ was available to everyone, Jew or Gentile, on an equal basis.

This may not sound like a radical claim in our day, when very few people who believe in Jesus are Jewish and it would be nonsensical to argue that a person must convert to Judaism before becoming a Christian, but people like Paul had to argue the point vehemently in antiquity. For Paul, even though faith in Jesus was in complete conformity with the plan of the Jewish God as found in the Jewish Scriptures, it was a faith for all persons, Jews and Gentiles alike.

At first, Paul probably stood in the minority on this issue. To most of the earliest followers of Jesus, who were born and raised Jewish, it was Paul's claim that a person did not have to be a Jew to be counted among the people of God that would have made no sense. These early Christians maintained that Jesus had been sent by the Jewish God to the Jewish people in fulfillment of the Jewish Law to be the Jewish messiah. Jesus himself had followed Jewish customs, gathered Jewish disciples, and interpreted the Jewish Law. The religion he founded was Jewish. People who wanted to follow Jesus had to become Jews first. This seemed fairly obvious to most early Christians. But not to Paul. The kind of Christianity that was defined and advocated by the apostle Paul was open to both Jews and Gentiles and was rooted in the belief that Jesus had died and been raised for the salvation of the world, not just of Israel.

Before we can begin to examine Paul's views in greater depth, we need to engage in two preliminary tasks. First, we must explore the methodological difficulties that this kind of study involves. Second, we must set our investigation into a somewhat broader context by considering some of the major aspects of Paul's own life, insofar as these can be deduced from his surviving writings.

THE STUDY OF PAUL: METHODOLOGICAL DIFFICULTIES

The problems of reconstructing the life and teachings of the historical Paul are in some ways analogous to the problems of reconstructing the life and teachings of the historical Jesus, in that they relate to the character of our sources. But there is one significant difference: Jesus left us no writings, whereas Paul did. Indeed, thirteen letters in the New Testament fall under Paul's name. A major problem involved in studying these letters, however, is that scholars have good reasons for thinking that some of them were not written by Paul, but by later members of his churches writing in his name.

The Problem of Pauline Pseudepigrapha

The fact that some ancient authors would falsely attribute their writings to a famous person (like Paul) comes as no shock to historians. Writings under a false name are known as "pseudepigrapha." We know of numerous pseudepigrapha produced by pagan, Jewish, and Christian writers of the ancient world (see Chapter 12). Indeed, letters allegedly written by Paul proliferated in the second and later centuries. Among those that still survive are a third letter to the Corinthians, a letter addressed to the church in the town of Laodicea (cf. Col 4:16), and an exchange of correspondence between Paul and the famous Greek philosopher Seneca (see box 17.1). Interestingly, we learn from the church father Tertullian that one second-century Christian was caught in the act of forging writings in Paul's name and confessed to the deed. The question of why authors in antiquity would forge documents in someone else's name is intriguing, and we will take it up later in Chapter 22.

Is it conceivable, though, that some of the letters that made it into the New Testament are this kind of literature, pseudonymous writings in the name of Paul? For most scholars, this is not only conceivable but almost certain; they have, as a consequence, grouped the letters attributed to Paul into three categories (see box 17.1). (In later chapters I will discuss the arguments that have proven persuasive to most historians and allow you to weigh their merits for yourself.)

First there are the three Pastoral epistles. These are the letters allegedly written to the pastors Timothy (1 and 2 Timothy) and Titus, that provide instruction on how these companions of Paul should engage in their pastoral duties in their churches. For a variety of reasons, most critical scholars are persuaded that these letters were written not by Paul but by a later member of one of Paul's churches who wanted to appeal to his authority in dealing with a situation that had arisen after his death. As we will see, the arguments revolve around whether the writing style, vocabulary, and theology of these letters coincides with what we find in the letters that we are reasonably certain Paul wrote, and whether Paul's own historical context can make sense of the issues that the letters address (see Chapter 22).

Next, there are the three epistles of Ephesians, Colossians, and 2 Thessalonians, called the "Deutero-Pauline" epistles because each of them is thought by many scholars to have been written by a "second Paul," a later author (or rather three later authors) who was heavily influenced by

SOME MORE INFORMATION

Box 17.1 The Pauline Corpus

Undisputed Pauline Epistles (almost certainly authentic)
- Romans
- 1 Corinthians
- 2 Corinthians
- Galatians
- Philippians
- 1Thessalonians
- Philemon

Deutero-Pauline Epistles (possibly pseudonymous)
- Ephesians
- Colossians
- 2 Thessalonians

Pastoral Epistles (probably pseudonymous)
- 1 Timothy
- 2 Timothy
- Titus

Paul's teachings (the term "Deutero-" means "second"). Scholars continue to debate the authorship of these books. Most continue to think that Paul did not write Ephesians and probably not Colossians; the case for 2 Thessalonians has proved somewhat more difficult to resolve (see Chapter 22).

Finally, there are seven letters that virtually all scholars agree were written by Paul himself: Romans, 1 and 2 Corinthians, Galatians, Philippians, 1 Thessalonians, and Philemon. These "undisputed" epistles are similar in terms of writing style, vocabulary, and theology. In addition, the issues that they address can plausibly be situated in the early Christian movement of the 40s and 50s of the Common Era, when Paul was active as an apostle and missionary.

The significance of this threefold classification of the Pauline epistles should be obvious. If scholars are right that the Pastorals and the Deutero-Paulines stem from authors living after Paul rather than from Paul himself, then despite the importance of these letters for understanding how Pauline Christianity developed in later years, they cannot be used as certain guides to what Paul himself taught. For methodological reasons a study of Paul has to restrict itself to letters that we can be confident he wrote, namely, the seven undisputed epistles.

The Problem of Acts

What, though, about the book of Acts, Luke's account of the history of the early church, which features Paul as one of its chief protagonists? For a historically accurate account of what Paul said and did, can we rely on Luke's narrative?

Different scholars will answer this question differently. Some trust the book of Acts with no qualms, others take its accounts with a grain of salt, and still others discount the narrative altogether (that is, they discount its historical credibility for establishing what Paul said and did, not necessarily its importance as a piece of literature).

My own position is that Acts is about as reliable for Paul as Luke is for Jesus. Just as Luke modified aspects of Jesus' words to reflect his own theological point of view, for instance, with respect to when the end was to arrive, and similarly changed some of the traditions concerning his actions, for instance, with respect to what occurred during his Passion, so too in the book of Acts Paul's words and deeds have been modified in accordance with Luke's own perspective. Thus, Acts can tell us a great deal about how Luke understood Paul, but less about what Paul himself actually said and did.

In our discussion of Acts I have already indicated why I do not think that the book was written by one of Paul's traveling companions. Even if it were, we would still have to ask whether its portrayal of Paul is historically accurate, for even eyewitnesses have their own perspectives. In any event, in evaluating the reliability of Acts we are fortunate that Paul and Luke sometimes both describe the same event and indicate Paul's teachings on the same issues, making it possible to see whether they stand in basic agreement.

Events of Paul's Life. In virtually every instance in which the book of Acts can be compared with Paul's letters in terms of biographical detail, differences emerge. Sometimes these differences involve minor disagreements concerning where Paul was at a certain time and with whom. As one example, the book of Acts states that when Paul went to Athens he left Timothy and Silas behind in Berea (Acts 17:10–15) and did not meet up with them again until after he left Athens and arrived in Corinth (18:5). In 1 Thessalonians Paul himself narrates the same sequence of events and indicates just as clearly that he was not in Athens alone but that Timothy was with him (and possibly Silas as well). It was from Athens that he sent Timothy back to Thessalonica in order to see how the church was doing there (1 Thess 3:1–3).

Although this discrepancy concerns a minor detail, it shows something about the historical reliability of Acts. The narrative coincides with what Paul himself indicates about some matters (he did establish the church in Thesssalonica and then leave from there for Athens), but it stands at odds with him on some of the specifics.

Other differences are of greater importance. For example, Paul is quite emphatic in the epistle to the Galatians that after he had his vision of Jesus and came to believe in him, he did *not* go to Jerusalem to consult with the apostles (1:15–18). This is an important issue for him because he wants to prove to the Galatians that his gospel message did not come from Jesus' followers in Jerusalem (the original disciples and the church around them) but from Jesus himself. His point is that he has not corrupted a message that he received from someone else; his gospel came straight from God, with no human intervention. The book of Acts, of course, provides its own narrative of Paul's conversion. In this account, however, Paul does exactly what he claims not to have done in Galatians: after leaving Damascus some days after his conversion, he goes directly to Jerusalem and meets with the apostles (Acts 9:10–30).

It is possible, of course, that Paul himself has altered the real course of events to show that he couldn't have received his gospel message from other apostles because he never consulted with them. If he did stretch the truth on this matter, though, his statement of Galatians—"In what I am writing to you, before God, I do not lie"—takes on new poignancy, for his lie in this case would have been bald-faced. More likely the discrepancy derives from Luke, whose own agenda affected the way he told the tale. For him, as we have seen, it was important to show that Paul stood in close continuity with the views of the original followers of Jesus, because all the apostles were unified in their perspectives. Thus, he portrays Paul as consulting with the Jerusalem apostles and representing the same faith that they proclaimed.

As we saw in our discussion of Acts, Luke portrays Paul as standing in harmony not only with the original apostles of Jesus but also with all of the essentials of Judaism. Throughout this narrative, Paul maintains his absolute devotion to the Jewish Law. To be sure, he proclaims that Gentiles do not need to keep this Law, since for them it would be an unnecessary burden. He himself, however, remains a good Jew to the end, keeping the Law in every respect. When Paul is arrested for violating the Law, Luke goes out of his way to show that the charges are trumped up (chaps. 21–22). As Paul himself repeatedly asserts throughout his apologetic speeches in Acts, he has done nothing contrary to the Law (e.g., 28:17).

In his own writings, Paul's view of the Law is extremely complicated. Several points, however, are reasonably clear. First, in contrast to the account in Acts, Paul appears to have had no qualms about violating the Jewish Law when the situation required him to do so. In Paul's words, he could live not only "like a Jew" when it served his purposes but also "like a Gentile," for example, when it was necessary for him to convert Gentiles (1 Cor 9:21). On one occasion, he attacked the apostle Cephas for failing to do so himself (Gal 2:11–14). In addition, Paul did not see the Law merely as an unnecessary burden for Gentiles, something that they didn't have to follow but could if they chose. For Paul, it was an absolute and total affront to God for Gentiles to follow the Law, a complete violation of his gospel message. In his view, Gentiles who did so were in jeopardy of falling from God's grace, for if doing what the Law required could contribute to a person's salvation, then Christ died completely in vain (Gal 2:21, 5:4). This is scarcely the conciliatory view attributed to Paul in Acts.

Paul's Teaching. Paul's teachings in Acts differ in significant ways from what he says in his own letters. Here we look at just one important example.

Almost all of Paul's evangelistic sermons mentioned in Acts are addressed to Jewish audiences. This itself should strike us as odd given Paul's repeated claim that his mission was to the Gentiles. In any event, the most famous exception is his speech to a group of philosophers on the Areopagus in Athens (chap. 17). In this speech, Paul explains that the Jewish God is in fact the God of all, pagan and Jew alike, even though the pagans have been ignorant of him. Paul's understanding of pagan polytheism is reasonably clear here: pagans have simply not known that there is only One God, the creator of all, and thus cannot be held accountable for failing to worship him. Since they have been ignorant of the true God, rather than willfully disobedient to him, he has overlooked their false religions until now. With the coming of Jesus, though, he is calling all people to repent in preparation for the coming judgment (Acts 17:23–31).

This perspective contrasts sharply with the views about pagan idolatry that Paul sets forth in his own letters. In the letter to the Romans, for example, Paul claims that pagan idolaters are *not* ignorant of the one true God, that all along they have known of his existence and power by seeing the things that he has made. Here the worship of idols is said to be a willful act of disobedience. Pagans have rejected their knowledge of the one true God, the maker of all, and chosen of their own free will to worship the creation rather than the creator. As a result of their rejection of God, he has punished them in his wrath (Rom 1:18–32).

These passages appear to be at odds with one another on a number of points. Do pagans know that there is only one God? (Acts, no; Romans, yes.) Have they acted in ignorance or disobedience? (Acts: ignorance; Romans: disobedience.) Does God overlook their error or punish it? (Acts, overlooks; Romans: punishes.)

Some scholars think that the two passages can be reconciled by considering the different audiences that are being addressed. In Acts Paul is trying to win converts, and so he doesn't want to be offensive, whereas in Romans he is addressing the converted, so he doesn't mind saying what he really thinks. To be sure, it is possible that Paul would say the opposite of what he believed in order to convert people or tell a white lie intended to bring about a greater good; but another explanation is that Luke, rather than Paul, is the author of the speech on the Areopagus, just as he is the author of all the other speeches in his account, as we saw in Chapter 9. This explanation goes a long way toward showing why so many of the speeches in Acts sound similar to one another, regardless of who the speaker is—Paul sounding like Peter, for example, and Peter like Paul (compare the speeches of Acts 2 and 13). Rather than embodying Paul's view of the pagan religions, then, the Areopagus speech may embody Luke's view, and thus represent the kind of evangelistic address that he imagines would have been appropriate to the occasion.

SOME MORE INFORMATION

Box 17.2 Other Sources for the Life of Paul

Just as a number of legendary accounts of Jesus sprang up from the first century through the Middle Ages, so too a number of pseudepigraphal accounts of Paul and the other apostles appeared. We will look at one of the earliest and most interesting of these narratives, *The Acts of Paul and Thecla*, in Chapter 21. There we will see how Paul came to be portrayed as a proponent of the gospel of the ascetic life, who deprecated sexual relations of every kind, both within and outside of marriage.

As was the case with the apocryphal tales about Jesus, these stories about Paul are less important for what they tell us about the man Paul himself than for what they reveal about Christianity in the years during which they were written. The same is true of the interesting set of correspondence forged by a third-century Christian in the names of Paul and Seneca, the famous philosopher and mentor of the emperor Nero. Written some two hundred years after both parties were dead (both of them killed, according to tradition, by order of Nero), these fourteen letters were meant to show that Paul's significance as an author was recognized by one of the greatest philosophical minds of his day. In the second letter that "Seneca" addresses to Paul, he claims to be particularly impressed with Paul's writings and expresses his desire to make them known to the emperor himself:

> I have arranged some scrolls [of your letters] and have brought them into a definite order corresponding to their several divisions. Also I have decided to read them to the emperor. If only fate ordains it favourably that he show some interest, then perhaps you too will be present; otherwise I shall fix a day for you at another time when together we may examine this work. And if only it could be done safely, I would not read this writing to him before meeting you. You may then be certain that you are not being overlooked. Farewell, most beloved Paul.

What then are we left with? The book of Acts appears to contain a number of discrepancies with the writings of Paul himself, with respect both to the events of his life and to the nature of his teachings. If this is so, then it cannot be accepted uncritically as a historically accurate portrayal of Paul, any more than the Gospel of Luke can be accepted uncritically as a historically accurate portrayal of Jesus. To gain a historical understanding of Paul, however, we are at least able to proceed on the basis of his own writings, for we have seven other New Testament books that stem from his pen. Our study of Paul and his teachings will therefore rely principally on the undisputed Pauline epistles. Even the use of these letters, however, is not without its problems.

The Occasional Nature of Paul's Letters

Probably the most important insight into the Pauline epistles in modern scholarship is that all of them are "occasional." Paul's letters are not essays written on set themes or systematic treatises that discuss important issues of theology. They are actual communications to particular individuals and communities, sent through the ancient equivalent of the mail. With all but one exception, Paul wrote these letters to address problems that arose in the Christian communities he established. In every case, they are occasioned by situations that he felt compelled to address as an apostle of Christ.

Because of the occasional nature of these letters, they do not contain everything that we may want to know about Paul and his views. Since he

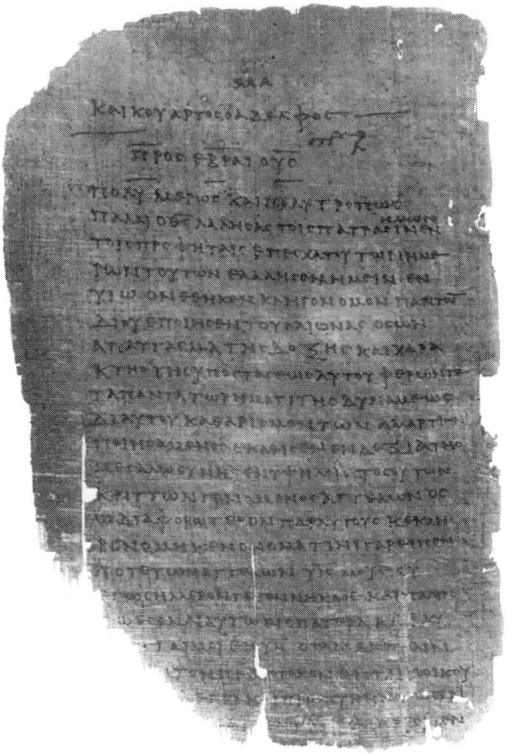

Figure 17.1 A page of P^{46}, the earliest surviving manuscript of Paul's letters, from around the year 200 C.E.

is addressing issues that have come up in the communities that he founded, then beliefs, practices, and perspectives that are not at issue will not be addressed, even when these were of central importance to Paul. As numerous scholars have noted, if Paul had not taken exception to the way the Corinthians were celebrating the Lord's Supper, we would never have known that he even supported (or knew of) the practice.

Another implication of the occasional nature of Paul's letters is that if we want to approach them from a historical perspective, then we need to learn about the occasions that lie behind them. Each of these books has a specific historical setting, a real-life context. If we misconstrue the context, or pretend that it never existed, we change what the books mean. For this reason, we will be applying the contextual method to the Pauline epistles, as we did with the Johannine letters (Chapter 11). For each writing, we will begin by looking for clues as to the historical circumstances that prompted Paul to produce it, or at least the circumstances as he appears to have perceived them. Of course, in every case we have only Paul's side of the argument, but the contextual method will help us understand what he says in light of the way he appears to have construed the context. We should not assume, however, that his perception of the situation was necessarily shared by the people he addressed.

THE LIFE OF PAUL

Paul's letters are chiefly concerned with problems that have arisen in his churches, not with events that transpired in his life. On occasion, however, Paul has reason to mention his past, for instance, when he is trying to establish his credentials as a true apostle of Christ. It appears from such self-references as Galatians 1:11–2:14 and Philippians 3:4–10 that Paul visualized his past in three stages: his life as a Pharisee prior to faith in Christ, his conversion experience itself, and his activities as an apostle afterwards.

Paul the Pharisee

We can say very little for certain about Paul prior to his conversion. He does tell us that he was a Jew born to Jewish parents and that he was zealous for the Law, adhering strictly to the traditions endorsed by the Pharisees (Gal 1:13–14; Phil 3:4–6). He does not tell us when he was born, where he was raised, or how he was educated. The book of Acts, however, does provide some information along these lines. There Paul is said to have been from the Greek city of Tarsus (21:39) in Cilicia, in the southeastern part of Asia Minor, and to have been educated in Jerusalem under the renowned rabbi Gamaliel (22:3). Since Paul himself makes neither claim, a historian might suspect Luke of attempting to provide superior credentials for his protagonist. Tarsus was the location of a famous school of Greek rhetoric, that is, a school of higher learning reserved for the social and intellectual elite, something like an Ivy League University. Jerusalem, of course, was the center of all Jewish life, and Gamaliel was one of its most revered teachers.

Paul's own letters give little indication of the extent of his formal education. Simply his ability to read and write shows that he was better educated than most people of his day; recent studies indicate that some 85–90 percent of the population in the empire could do neither. Moreover, Paul writes on a fairly sophisticated level, showing that he must have had at least some formal training in rhetoric, the main focus of higher education at the time. He is certainly not one of the highest of the literary elite, but he just as certainly had some advanced schooling. It is not altogether implausible, then, that he grew up in a place like Tarsus, if not Tarsus itself. In any event, Paul's native tongue was almost without question Greek, and he gives no indication at all of knowing Aramaic, the language more widely used in Palestine. This is probably an indication that Luke is right in situating him in the Jewish diaspora.

Although Paul gives no indication that he studied in Jerusalem, he clearly did study the Jewish Scriptures extensively, perhaps in some kind of formal setting (comparable, perhaps, to a later rabbinic school?). He appears to be able to quote the Scriptures extensively from memory and to have meditated and reflected on their meaning at a fairly deep level. He knows these Scriptures in their Greek translation, the Septuagint. Since his letters are all addressed to Greek-speaking Christians, it is difficult to know whether he quoted the text in this way in order to accommodate his readers or whether this was the only form of the text that he knew. That is to say, it is hard to know whether or not he could also read the Scriptures in their original Hebrew.

What is certain is that prior to becoming a believer in Jesus Paul was an avid Pharisee (Phil 3:5). In fact, Paul's letters are the only writings to survive from the pen of a Pharisee, or former Pharisee, prior to the destruction of the Temple in 70 C.E. Paul claims that he rigorously followed the "traditions of the fathers" (Gal 1:14). These are usually understood to be the Pharisaic "oral laws" that were in circulation in Paul's youth, nearly two centuries before these laws, or ones like them, were written down in the Mishnah. We get a picture, then, of a devout and intelligent Jewish young man totally committed to understanding and practicing his religion according to the strictest standards available.

As a Pharisee, Paul's religion would have centered around the Law of God, the Torah of Moses, the greatest gift of God to Israel, the exact and thorough adherence to which was the ultimate goal of devotion. Looking back on his early life, Paul could later claim that he had been "blameless" with respect to the righteousness that the Law demands (Phil 3:6). It is hard to know exactly what he meant by that. Did he mean that he never violated a solitary commandment of God? This seems unlikely given his insistence elsewhere that no one has kept the Law in all its particulars (e.g., Rom 3:10–18), a view that he claimed is taught by the Law itself (Rom 3:19–20). Did he mean that he did his best to keep the Law, so he could not be faulted for effort? This interpretation seems more likely. But he may also have meant that he was blameless because the Law itself makes provision for those who sin, in the sacrifices that it requires. These sacrifices were explicitly given for those who inadvertently broke the Law, as a way to restore them to a right standing before God. If Paul did his utmost to keep the Law and performed the required sacrifices for his sins when he failed (perhaps on pilgrimages to Jerusalem), he may well have considered himself "blameless" with respect to the righteousness that the Law demands. In that case, not even the Law could blame him, since he had done what it requires.

Paul's view of himself before the Law is but one of the many issues that have perplexed his interpreters through the years. Somewhat less perplexing is the general view of the world that he must have had as a devoted Pharisee. As we have seen, one of the salient features of the Pharisees, which distinguished them from the Sadducees, for example, was their fervent expectation of a future resurrection of the dead. It appears that Pharisees of the first century, along with other groups such as the Essenes, were by and large Jewish apocalypticists, who anticipated the intervention of God in the world and the destruction of the forces of evil that oppose him. At the end of the age, which would be imminent, God would send a deliverer for his people, who would set up God's kingdom on earth; the dead would be raised, and all would face judgment. Paul almost certainly held these views prior to his conversion to Christianity.

What else can we say about the life of this upright Jewish Pharisee? The one aspect of his former life that Paul himself chose to emphasize in his autobiographical statements in Galatians 1 and Philippians 3 is that it was precisely as a law-abiding, zealous Jew that he persecuted the followers of Jesus. Far from adhering to the gospel, he violently opposed it, setting himself on destroying the church, and he interpreted this opposition as part of his devotion to the one true God.

Why was Paul so opposed to Jesus' followers, and how exactly did he go about persecuting them? Unfortunately, Paul never tells us, but we can make some intelligent guesses, especially with regard to the reasons for his opposition. We have already seen how the Christian proclamation of Jesus as the messiah would have struck most Jews as ludicrous. Various Jews had different expectations of what the messiah would be like. He might be a warrior-king who would establish Israel as a sovereign state, an inspired priest who would rule God's people through his authoritative interpretation of God's Law, or a cosmic judge who would come to destroy the forces of evil. They all, however, anticipated a messiah who would be glorious and powerful. Jesus, on the other hand, was commonly viewed as nothing more than an itinerant preacher with a small following who was opposed by the Jewish leaders and executed by the Romans for sedition against the state. For most faithful Jews, to call him God's messiah was an affront to God.

For Paul, there appears to have been an additional problem, relating to the precise manner of Jesus' execution. Jesus was crucified; that is, he was killed by being attached to a stake of wood. Paul, well versed in the Scriptures, recognized what this meant for Jesus' standing before God, for the Torah states, "Cursed is anyone who hangs on a tree" (Deut 27:26, quoted in Gal 3:13). Far from being the Christ of God, the one who enjoyed divine favor, Jesus was the cursed of God, the one who incurred divine wrath. For Paul the Pharisee, to call him the messiah was probably blasphemous.

This problem would have given Paul sufficient grounds for persecuting the Christian church. How exactly he went about doing so cannot be known. According to the book of Acts, he received authorization from the high priest in Jerusalem to capture and imprison Christians. Paul himself says nothing of the sort, and the fact that churches in Judea had never seen him before he visited them as a Christian argues against it (see Gal 1:22). At the same time, whatever he did to the Christians as a Jewish persecutor, and on whatever authority, he apparently gained something of a reputation for it. He later acknowledges his reputation among the Christian churches as a sworn enemy (Gal 1:13, 23).

All of this changed, of course, when the greatest persecutor of the church became its greatest proponent. The turning point in Paul's life came with his encounter with the risen Jesus. Both Acts and Paul intimate that this happened when Paul was a relatively young man.

Paul's Conversion and Its Implications

It is difficult for historians to evaluate what actually happened to make Paul "turn around," the literal meaning of "convert." Both Acts and Paul attribute his conversion to the direct intervention of God, and this kind of supernatural act, by its very nature, is outside the purview of the historian (see Chapter 14). The historian can, of course, talk about a person's descriptions of divine acts, since narratives of this kind are a matter of the public record. So we will restrict ourselves to what Paul claims to have happened at his conversion and how he says he understood its significance. Even here there are problems, however. Some of these are easily disposed of, because they relate less to Paul than to widespread misperceptions about him, as found, for example, in historical novels about his life that can be picked up in used bookstores. In these accounts, the pre-Christian Paul is a guilt-ridden legalist who felt bound to follow a set of picayune laws that were impossible to keep and whose remorse over his own failings drove him both to insist with increasing vehemence that the Law had to be followed at all costs and to hate those who experienced a personal freedom like the one that Christ reputedly brought. In this version of his life, Paul saw the light when he realized that the solution to his guilt was not to intensify his efforts but to find forgiveness of his sins in Christ, who died to set him free from the Law. Paul, in this view, converted from a religion of guilt to a religion of love, and so became Jesus' faithful follower, bringing the good news of release from sins to those burdened with guilt complexes like his own.

It is with good reason that accounts like this are found in the fiction section of a bookstore. Paul himself does not indicate that he experienced a profound sense of guilt over his inability to keep God's commandments before becoming a Christian, even though after becoming a Christian he came to recognize that God's Law was nearly impossible to keep (see Rom 7:14–24). Prior to his faith in Christ, however, he considered himself to be blameless before the Law (Phil 3:4–6). Thus, he did not convert because he was burdened by a Law that he knew he could not keep. In some sense, this popular view of Paul derives more from a kind of implicit anti-Semitism—the Jews are burdened with an impossible Law and don't do a good job in keeping it—than from Paul himself.

Why, then, did Paul convert, and what did his conversion mean? The book of Acts provides a detailed account of the event, or, rather, it provides three accounts (chaps. 9, 22, and 26) that mention details not found in Paul (e.g., that he was on the "road to Damascus" and that he was "blinded by the light"). These accounts, however, are difficult to reconcile with one another (see box 17.3). Even

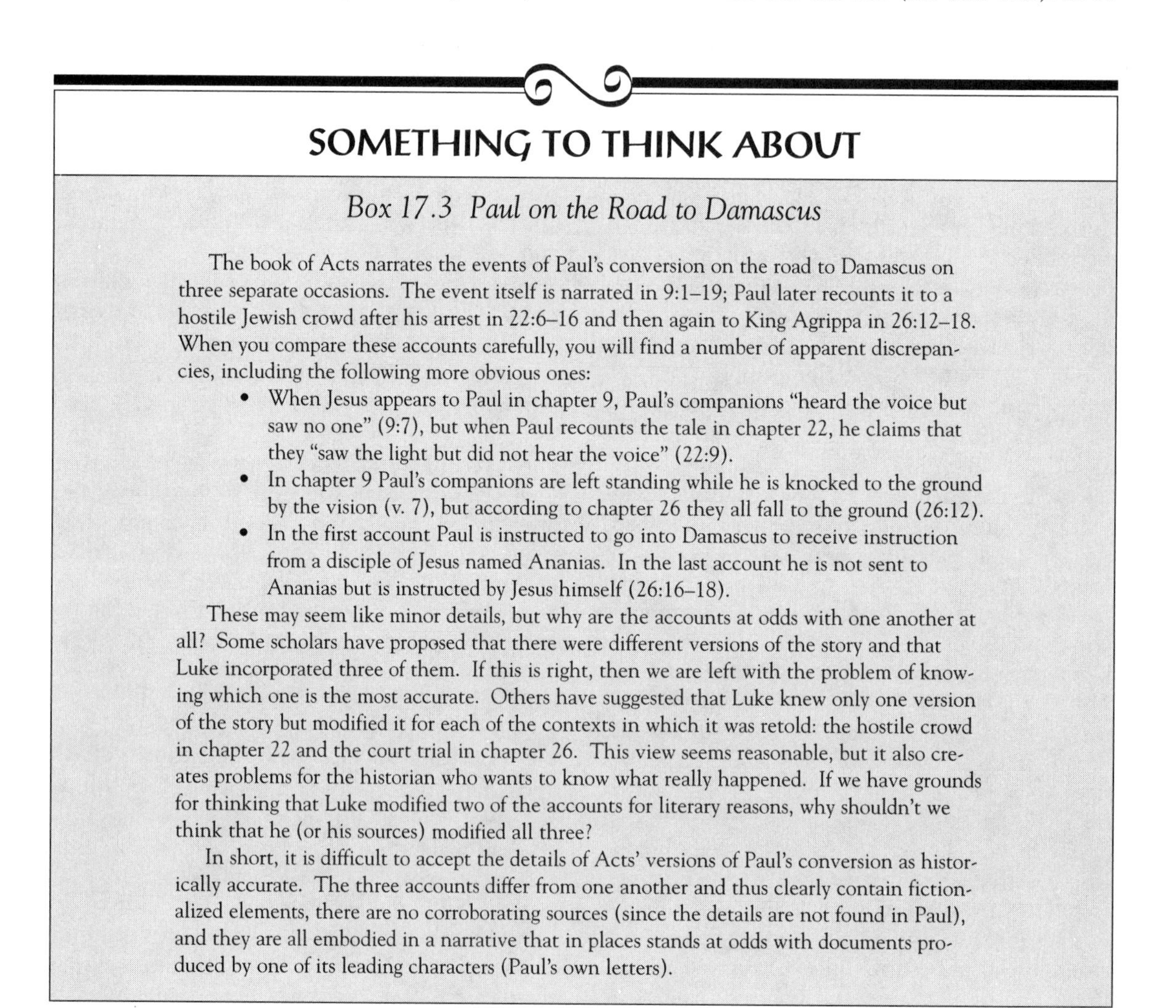

SOMETHING TO THINK ABOUT

Box 17.3 Paul on the Road to Damascus

The book of Acts narrates the events of Paul's conversion on the road to Damascus on three separate occasions. The event itself is narrated in 9:1–19; Paul later recounts it to a hostile Jewish crowd after his arrest in 22:6–16 and then again to King Agrippa in 26:12–18. When you compare these accounts carefully, you will find a number of apparent discrepancies, including the following more obvious ones:

- When Jesus appears to Paul in chapter 9, Paul's companions "heard the voice but saw no one" (9:7), but when Paul recounts the tale in chapter 22, he claims that they "saw the light but did not hear the voice" (22:9).
- In chapter 9 Paul's companions are left standing while he is knocked to the ground by the vision (v. 7), but according to chapter 26 they all fall to the ground (26:12).
- In the first account Paul is instructed to go into Damascus to receive instruction from a disciple of Jesus named Ananias. In the last account he is not sent to Ananias but is instructed by Jesus himself (26:16–18).

These may seem like minor details, but why are the accounts at odds with one another at all? Some scholars have proposed that there were different versions of the story and that Luke incorporated three of them. If this is right, then we are left with the problem of knowing which one is the most accurate. Others have suggested that Luke knew only one version of the story but modified it for each of the contexts in which it was retold: the hostile crowd in chapter 22 and the court trial in chapter 26. This view seems reasonable, but it also creates problems for the historian who wants to know what really happened. If we have grounds for thinking that Luke modified two of the accounts for literary reasons, why shouldn't we think that he (or his sources) modified all three?

In short, it is difficult to accept the details of Acts' versions of Paul's conversion as historically accurate. The three accounts differ from one another and thus clearly contain fictionalized elements, there are no corroborating sources (since the details are not found in Paul), and they are all embodied in a narrative that in places stands at odds with documents produced by one of its leading characters (Paul's own letters).

Paul's own references to the event are somewhat problematic because he is remembering the event long afterwards and is reflecting on it in light of his later experiences.

The first thing to observe about Paul's conversion is that he traces it back to an encounter with the resurrected Jesus. In 1 Cor 15:8–11 he names himself as the last person to have seen Jesus raised from the dead and marks this as the beginning of his change from persecutor to apostle. He appears to be referring to the same event in Gal 1:16, where he indicates that at a predetermined point in time, God "was pleased to reveal his son to me." When Paul experienced this revelation from God, he became convinced, then and there, according to his later perspective, that he was to preach the good news of Christ to the Gentiles.

Whatever Paul experienced at this moment, he interpreted it as an actual appearance of Jesus himself. We don't know how long this was after Jesus' death (several months? several years?) or how Paul, when he saw whatever he saw, knew it to be Jesus, but there is no doubt that he believed that he saw Jesus' real but glorified body raised from the dead. Indeed, as we will see later, one of the reasons that he believed Christians would eventually experience a bodily resurrection from the dead is because he "knew" that Jesus did. For him, Jesus was the "first-fruit" of those who would be raised (1 Cor 15:20).

Did this experience, then, lead Paul to reject his Judaism in favor of a religion for the Gentiles? Was this a conversion to a completely different and contrary set of beliefs? What exactly did his vision of the resurrected Jesus mean for Paul? As we have seen, Paul was probably an apocalyptic Jew prior to coming to believe in Jesus. If it is true that we can understand something new only in light of what we already know, we can ask how Paul would have understood this "new" event of Jesus' resurrection in light of his "old" worldview of Jewish apocalypticism. We can approach the question by considering two related matters: aspects of Paul's worldview that would have been confirmed by an encounter with a man raised from the dead and aspects that would have been reformulated in light of such an experience.

The Confirmation of Paul's Views in Light of Jesus' Resurrection. Apocalypticists maintained that at the end of the age God was going to intervene in history to overthrow the forces of evil and establish his good reign on earth, and that at that time the dead would be raised to face judgment. What would an apocalyptic Jew conclude if he or she came to believe that God had now raised someone? Clearly, for such a person, the end had already begun.

Paul drew exactly this conclusion. As we will see in greater detail later, he believed that he was living in the end of time and that he would be alive when Jesus returned from heaven (see 1 Thess 4:13–18 and 1 Cor 15:51–57). Thus, he speaks of Jesus as the "first-fruit of the resurrection," evoking an agricultural image that refers to the celebration that comes at the conclusion of the first day of the harvest. On the following day, the workers go to the fields and continue their labor. Jesus was the first-fruit of the resurrection in the sense that all the others would also soon be gathered in.

Other agricultural metaphors were common in Jewish apocalyptic circles. The end of the age would be like a great harvest, in which the fruit was gathered and the chaff was destroyed. As an apocalyptic Jew, Paul probably already believed that at the end of the age God would intervene to reward the faithful and punish the sinner and overthrow the forces of evil that plague this world, the demonic rulers and the wicked powers of sin and death. Jesus' resurrection must have confirmed these views, for one of the reasons that there will be a resurrection at the end of time is that death is God's enemy, and when it is destroyed there will be no more dying and no more death. Those who have died will therefore return to life.

For Paul, Jesus has already returned to life, which means that God has begun to defeat the power of death in him. This much Paul "knows," for if Jesus died but is dead no longer, as Paul believes because he has seen him alive after his death, then he has conquered this most dread of God's enemies. The cosmic destruction of the forces of evil has therefore begun.

The Reformulation of Paul's Views in Light of Jesus' Resurrection. Whereas some of Paul's views were confirmed by his belief in Jesus' resurrection, others had to be reconsidered.

1. *Paul's View of Jesus.* First and foremost, of course, Paul's understanding of Jesus himself changed. Rather than being the cursed of God (Paul's original view), Jesus must be the one specially blessed by God, for he was the one God raised from the dead to conquer the cosmic forces of sin and death. Jesus, the conqueror, was thus indeed the messiah, the one appointed by God as Lord (see Chapter 16). He was presently in heaven, awaiting the moment of his return in glory when he would finish the deed that he had begun.

   Once Paul came to believe that Jesus was raised from the dead, the crucifixion itself must have begun to make more sense. Paul appears to have turned to the Jewish Scriptures to understand how Jesus' death was according to the plan of God, evidently knowing that it had to be, since the resurrection showed that he was under God's particular blessing. From the Scriptures, of course, Paul knew of the suffering of the Righteous One of God, whom God ultimately vindicated. Since Jesus was the one whom God vindicated, for Paul he must have been that righteous one who suffered, not as a punishment for his own actions, but for the sake of others. That is to say, even though Jesus was cursed, given his death on the cross, the curse could not have been deserved since he was God's righteous one. He must, then, have borne the curse that was meant for others. As the righteous servant of God, Jesus took the punishment that others deserved and bore it on the cross. God vindicated this faithful act by raising him from the dead.

   By raising Jesus, God showed that his death was meaningful rather than meaningless. It was meaningful because it served as a sacrifice for the sins of others (see box 16.2). More than that, it was a death that actually conquered the cosmic power of sin. Paul "knew" that Jesus conquered sin because he had obviously conquered death. Otherwise he would have remained dead. In Jesus himself, then, God had worked to conquer the evil forces that until now had been in control of this world.

   This new belief in Jesus raised an obvious problem for Paul, the upright Jewish Pharisee whose upbringing and commitments were centered on the Jewish Law. If salvation from sins and the defeat of the powers of sin and death came through Jesus, what was the role of the Law of God, God's greatest gift to his people?

2. *Paul's View of the Law.* Paul's understanding of the Law in light of his faith in Christ is extremely complicated. Some scholars have wondered, given the variety of things Paul says about the Law, whether he ever managed to construct an entirely consistent view. At the very least, it seems clear that Paul came to believe that a person could not be put into a right standing before God by keeping the Law; only faith in Christ could do this. Moreover, he maintained that this view was not contrary to the Law but, perhaps ironically, was precisely what the Law itself taught (Rom 3:31). As we will see, he devotes most of the letter to the Romans to making these points.

   It appears that after his conversion Paul began to think that the Jewish Law, even though in itself an obviously good thing (see Rom 7:12), had led to some bad consequences. The problem for Paul, however, was not the Law per se, but the people to whom it was given.

   Those who had received the good Law of God, according to Paul, had come to misuse it. Rather than seeing the Law as a guide for their actions as the covenant people of God, they began keeping the Law as a way to establish a right standing before God, as if by keeping its various injunctions they could earn God's favor (e.g., Rom 4:4–5; 10:2–4). It is not clear whether Paul thought that Jews *intentionally* used the Law in this way. Moreover, Paul does not appear to have held this view of the Law prior to his conversion but only afterwards. Indeed, this view is found in virtually no other Jewish writing from the ancient world.

In any event, after his conversion Paul came to think that his fellow Jews had attempted to use the Law to bring about a right standing before God. For him this was a misuse of the Law. Instead of making people right before God, the Law shows that everyone is alienated from God: "For no human being will be justified in God's sight by deeds prescribed by the law, for through the law comes the knowledge of sin" (Rom 3:20).

What Paul means by this statement is debated among scholars. On the one hand, he is almost certainly thinking about the repeated insistence in the Jewish Scriptures themselves that God's people have fallen short of his righteous demands (Rom 3:10–20). In addition, he may have been reflecting on the sacrificial system that is provided by the Torah as a way of dealing with human sins (although Paul never mentions it directly), for why would God require sacrifices for sin if people didn't need them? Whatever Paul's precise logic was, it appears certain that as a Christian, he came to believe that the Law points to the problem of human sinfulness against God on the one hand but does not provide the power necessary to overcome that sinfulness on the other. (Why the divinely ordained sacrifices are not sufficient to overcome sin is an issue that he never addresses.) The problem for Paul the Christian apocalypticist is that humans are enslaved by powers opposed to God, specifically the cosmic powers of sin and death, and the Law can do nothing to bring about their release. Since the problem is enslavement to an alien force, people cannot be liberated simply by renewing their efforts to keep the Law of God. It is Christ alone who brings liberation, for Paul, in that he alone has broken the power of death, as proved by his resurrection. Christ has also, therefore, conquered the power of sin.

The Law, then, cannot bring about a right standing before God for those who observe it. Since everyone is enslaved to sin, they are all alienated from God. Only the one who has defeated sin can bring deliverance from sin.

3. *Paul's View of Jews and Gentiles*. As an apocalyptic Jew prior to his conversion, Paul probably believed that at the end of time God would intervene not only on behalf of his people Israel but on behalf of the entire world, since everyone was enslaved to the cosmic forces that opposed God. In other words, Paul would have been particularly attuned to the Jewish Scriptures that spoke of all the nations coming to worship the true God, after turning from their vain devotion to pagan idols and acknowledging that the God of Israel was the one true God (e.g., in Isa 40–66). Once he had decided that the death of Jesus, rather than the Law, was the way to a right relationship with God, he came to believe that the other nations would become God's people not through converting to the Law but through converting to Christ.

In reading the Scriptures, Paul recognized that God had made more than one covenant with the Jewish patriarchs. The first covenant was not with Moses (see Exod 19–20) but with the father of the Jews, Abraham (see Gen 17). God promised Abraham that he would be a blessing for all nations, not just Israel (Gen 12:3). Abraham believed God's promise and was rewarded with a right standing before God, or, as Paul calls it, "righteousness." In Paul's view, this promise was fulfilled in Jesus, not only for the Jew who later inherited the covenant given to Moses but also for the Gentile who trusted that God had fulfilled his promise in the person of Jesus. In other words, the original covenant was for all people, not just the Jews, and it was bestowed before and apart from the Law of Moses, which was given specifically to the Jews. Gentiles, therefore, did not have to follow this Law in order to be heirs of the original covenant.

In short, Paul came to believe, on the basis of his experience of the resurrected Jesus, that all people, both Jews and Gentiles, could have a right standing with God through Christ. Faith in Jesus' death and resurrection was the only way to achieve this standing. The Law was not an alternative way, because the Law brings the

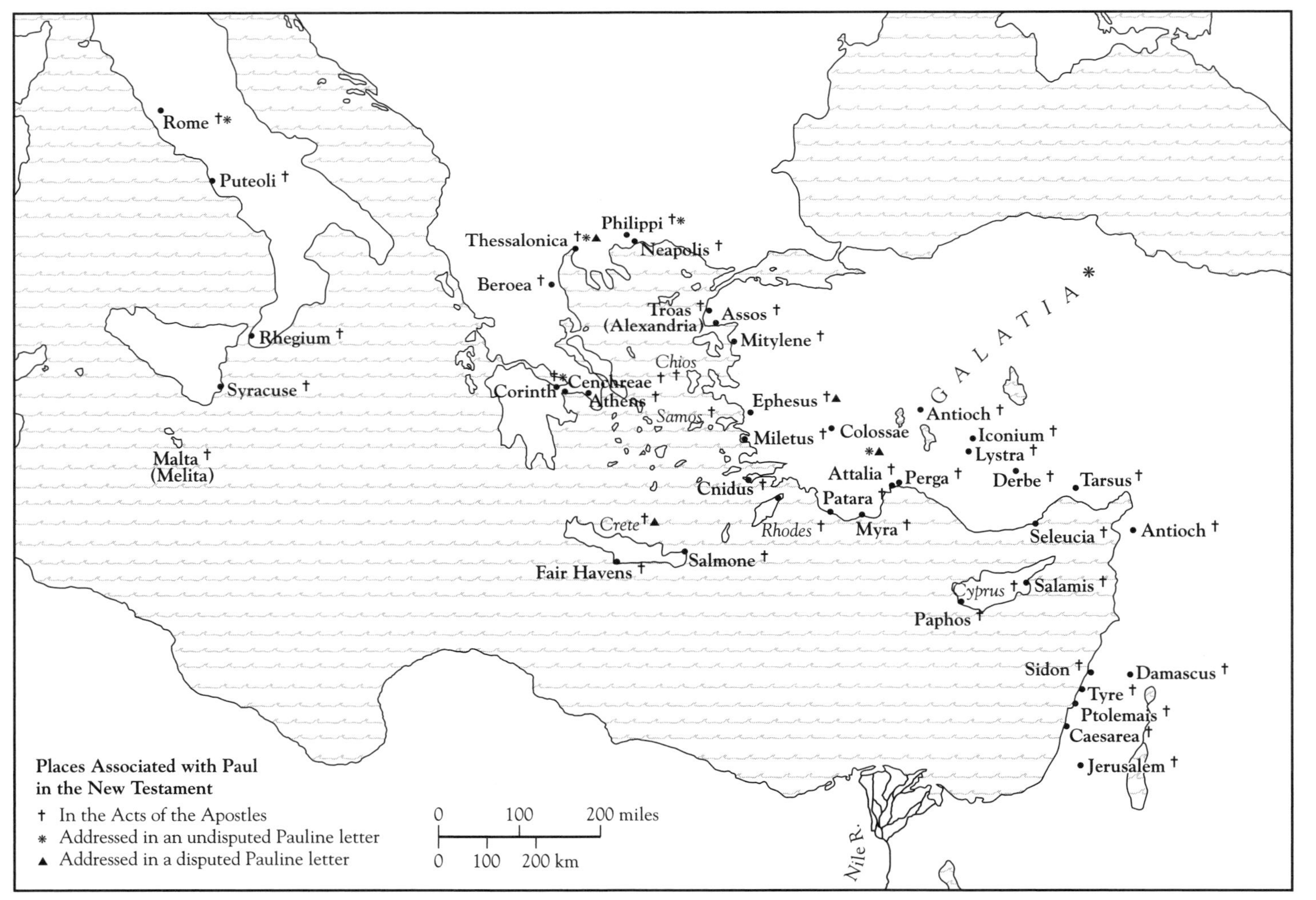

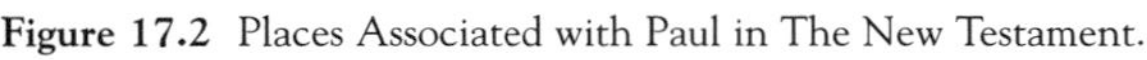

Figure 17.2 Places Associated with Paul in The New Testament.

knowledge of sin but not the power to conquer it. Christ conquered sin, however, and whoever believes in him and accepts his work on the cross will participate in his victory.

Our brief exploration of Paul's theology here has given some indication of how his conversion affected his understanding of Christ, the Law, salvation, faith, and the relationship between Jews and Gentiles. This background will help you in your own reading of Paul's letters. As you will see, the letters, themselves, for the most part presuppose these points of view rather than describe them. Except for a few places that can be tough going, these books are not heavy-duty theological treatises.

Paul the Apostle

After his conversion, Paul spent several years in Arabia and Damascus (Gal 1:17). He doesn't tell us what he did there. After a brief trip to Jerusalem, he then went into Syria and Cilicia and eventually became involved with the church of Antioch. It is not altogether clear when he began his missionary activities further west, in Asia Minor, Macedonia, and Achaia, but in one of his final surviving letters he claims that he was actively involved in spreading the gospel all the way from Jerusalem to Illyricum, north of modern-day Greece (Rom 15:19).

Throughout his career as a preacher of the gospel, Paul saw himself as the "apostle to the Gentiles." By this he meant that he had been appointed by God to bring the good news of salvation through faith in Christ to those who were not Jews. Paul's normal practice appears to have been to establish a Christian community in cities that had previously been untouched by a Christian presence (we will explore his methods in the next chapter). After staying with the new church for some time and providing it with some rudimentary instruction, he would move on to another city and start from scratch. In his wake, evidently, other Christian missionaries would commonly arrive. These sometimes presented a different version of the gospel from the one Paul preached. Some of Paul's letters warn against such people. Moreover, problems frequently arose within the congregations themselves, problems of immorality, infighting, confusion over Paul's teachings, or opposition from outsiders who took exception to this new faith. When Paul learned of such problems, he fired off a letter to warn, admonish, encourage, instruct, or congratulate the church. As we will see, in some instances he was himself the problem.

The letters that we have from Paul's hand represent only some of this correspondence. We can probably assume that there were dozens of other letters that for one reason or another have been lost. Paul mentions one of them in 1 Corinthians 5:9. The authentic letters that have survived are all included within the New Testament. In the chapters that follow we will examine these letters, beginning with a relatively detailed assessment of the earliest one, 1 Thessalonians. In this first instance, we will be looking for information concerning Paul's modus operandi as an apostle, to learn (a) how Paul went about establishing a church and communicating with it after he had left, (b) the nature of his message when he worked to convert people to faith in Christ and when he wrote to resolve problems that had arisen in his absence, and (c) the actual constituency of his churches and the character of their interactions with one another and with the world around them. Having thus set the stage, we will move on in the following chapter to examine five of the other letters, 1 and 2 Corinthians, Galatians, Philippians, and Philemon. There we will apply the contextual method to reconstruct each situation that Paul addresses and assess his response to the problems that he perceives. Finally, an entire chapter will be devoted to the letter to the Romans, the most influential of Paul's writings. There we will explore further some of the important ideas of this apostle, a figure of paramount importance in the history of Christianity down to our own day.

SUGGESTIONS FOR FURTHER READING

In addition to the books mentioned here, see the suggestions for Chapters 18–21.

Aune, David. *The New Testament in its Literary Environment*. Philadelphia: Westminster, 1987. Includes a superb discussion of the practices of letter-writing in Greco-Roman antiquity as the social context for Paul's epistles.

Beker, J. Christiaan. *Paul the Apostle: The Triumph of God in Life and Thought*. Philadelphia: Fortress, 1980. A sophisticated and astute discussion of the apocalyptic character of Paul's theology and its various forms of expression in different situations that the apostle confronted; for advanced students.

Bruce , F. F. *Apostle of the Heart Set Free*. Grand Rapids, Mich.: Eerdmans, 1977. A full study of Paul's life and teachings by a major evangelical Christian scholar.

Fitzmyer, Joseph. *Pauline Theology: A Brief Sketch*. 2d ed. Englewood Cliffs, N.J.: Prentice Hall, 1989. An excellent overview of Paul's teachings for beginning students.

Hawthorne, Gerald, and Ralph Martin. *Dictionary of Paul and His Letters*. Downers Grove, Ill.: Intervarsity, 1993. A Bible dictionary that contains over 200 articles on various topics relating to the life and writings of Paul, written by prominent evangelical scholars who on several major issues take a different perspective from the one presented here (such as the authorship of the Deutero-Pauline and Pastoral epistles).

Keck, Leander. *Paul and His Letters*. Philadelphia: Fortress, 1979. An insightful overview of Paul's theology as expressed in his letters.

Meeks, Wayne. *The First Urban Christians. The Social World of the Apostle Paul*. New Haven, Conn.: Yale University Press, 1983. An impressive and highly influential study that explores the Pauline epistles from a socio-historical rather than theological perspective; for more advanced students.

Meeks, Wayne, ed. *The Writings of St. Paul*. New York: Norton, 1972. A very useful annotated edition of Paul's letters that includes a number of classic essays on various aspects of Paul's thought and significance.

Roetzel, Calvin. *The Letters of Paul: Conversations in Context*. 3d ed. Atlanta: John Knox, 1991. Perhaps the best introductory discussion of each of the Pauline epistles.

Sanders, E. P. *Paul and Palestinian Judaism*. Philadelphia: Fortress, 1977. An enormously influential and erudite study that situates Paul in the context of early Judaism; for advanced students.

Segal, Alan. *Paul the Convert: The Apostolate and Apostasy of Saul the Pharisee*. New Haven, Conn.: Yale University Press, 1990. A very interesting study by a Jewish scholar who examines the importance of Paul's conversion for his theology and practice.

CHAPTER 18

Paul and His Apostolic Mission: 1 Thessalonians as a Test Case

First Thessalonians is a particularly good place to begin a study of Paul's letters. Scholars are almost unanimous in thinking that it was the first of his surviving works to be written, which also means that it is the oldest book of the New Testament and consequently the earliest surviving Christian writing of any kind. It is usually dated to about the year 49 C.E., that is, some twenty years after Jesus' death. It is written to a congregation for which Paul has real affection and in which no major problems have arisen, at least in comparison with what we will find in the letters to the Corinthians and the Galatians. As a consequence, Paul spends most of the letter renewing his bonds of friendship with the congregation, largely by recounting aspects of their past relationship. Since he has just recently left the community, memories of this relationship are still fresh.

Given the nature of the letter, we can learn a good deal about how Paul established this church and about what the people who composed it were like. We can also learn about the difficulties they experienced in light of their conversion, the problems that emerged in their community soon thereafter, and the approach that Paul took to dealing with these problems. To be sure, we are not provided with as much information as we would like about such things; Paul after all was not writing to us, but to people who were already intimately familiar with him. Nonetheless, for historians interested in knowing how the Christian mission was conducted and how the Christian converts fared in their world, 1 Thessalonians provides ample food for thought.

We will examine this particular letter, therefore, not only to learn about its immediate occasion (i.e., the reasons that Paul wrote it) and to uncover its principal themes but also to find clues about various social and historical aspects of Paul's apostolic mission to the Gentiles. This kind of socio-historical investigation will then set the stage for our study of the other Pauline letters.

THE FOUNDING OF THE CHURCH IN THESSALONICA

Thessalonica was a major port city, the capital of the Roman province of Macedonia, where the Roman governor kept his residence, and one of the principal targets chosen by Paul for his mission in the region. This choice appears to be consistent with Paul's missionary strategy. So far as we can tell, he generally chose to stay in relatively large urban areas where he would have the greatest opportunity to meet and address potential converts.

How, then did Paul go about converting people to faith in Christ? That is, how did a Christian missionary like Paul, after arriving in a new city where he had no contacts, actually go about meeting people and talking to them about religion in an effort to convert them? First Thessalonians provides some interesting insights concerning Paul's missionary tactics, or his apostolic modus operandi.

Paul's Modus Operandi

One might imagine that when Paul arrived in town as a complete stranger, he would simply stand on a crowded street corner and preach to those passing by, hoping to win converts by his sincerity and charisma and by the appeal of his message. As we will see, there was a precedent for this kind of proselytizing activity among some of the philosophers in the Greco-Roman world, but Paul gives no indication that this is how he proceeded.

Nor does the book of Acts. In Acts, Paul invariably makes new contacts by going to the local synagogue, where as a traveling Jew he would be quite welcome, and using the worship service there as an occasion to speak of his belief in Jesus as the messiah come in fulfillment of the Scriptures. This tactic seems reasonable, and Acts is quite explicit in saying that this is how Paul did evangelize the people of Thessalonica, winning converts among the Jews and the "devout Gentiles" who joined them in their worship of the God of Israel (Acts 17:2–4). Luke sometimes calls this latter group "[God]-fearers," by which he seems to mean non-Jews who have abandoned their idolatry to worship the Jewish God, without, however, keeping every aspect of the Torah, including circumcision if they were men. According to Acts, Paul converted a number of such people in Thessalonica over a period of three weeks, after which a group of antagonistic Jews rose up to run him out of town (17:2–10).

This portrayal in Acts, however, stands in sharp contrast with Paul's own reminiscences of his Thessalonian mission. Curiously, Paul says nothing about the Jewish synagogue in his letter; indeed, he never mentions the presence of any Jews, either among his Christian converts or among their opponents in town. On the contrary, he indicates that the Christians that he brought to the faith were former pagans, whom he himself converted from worshipping "dead idols to serve the living and true God" (i.e., the Jewish God, whom Paul himself continues to worship through Jesus; 1:9). These converts, in other words, were neither Jews nor God-fearers. How then do we explain the account in Acts 17? It may be that Luke knew in general that Paul had preached in Thessalonica but did not know how he had proceeded or whom he had converted.

If Paul did not preach from the street corner or work through the synagogue, how did he go about making contacts and, eventually, converts? In the course of his letter, Paul reflects on the time he had spent among the Thessalonians, recalling with great pride how he and his Christian companions had worked "night and day so that we might not burden any of you while we proclaimed to you the gospel of God" (2:9). Recent scholars have realized that Paul literally means that he had been working full time and had used his place of business as a point of contact with people to proclaim the gospel. Paul preached while on the job.

Paul's emphasis on the burdens of his toil (2:9) makes it reasonably clear that his job involved some kind of manual labor. The book of Acts indicates that he worked with leather goods (18:3). Sometimes this is interpreted to mean that he was a tentmaker, although the term used can refer to a number of occupations involving animal skins. Paul himself doesn't indicate the precise nature of his employment (presumably the Thessalonians would already know). What he does indicate is that he was not alone in his labors but was accompanied in Thessalonica by two others, Timothy and Silvanus. The three arrived in town in active pursuit of converts; they all, evidently, engaged in the same form of manual labor and all preach-ed their faith to those with whom they came in contact.

Before we try to imagine how this mission took place, we should review the historical context. In our earlier discussion of Greco-Roman religions, we saw that none of the religions of the empire was exclusive; that is, none of them claimed that if you worshipped any one of the gods, it was inappropriate to worship others as well. Perhaps because of their inclusive character, none of these religions was missionary, none of them urged their devotees to pursue converts to participate in their cult and their cult alone. Thus, when Paul and his fellows were working to make converts, they were not modeling themselves on what representatives of other sacred cults in their day were doing.

On the other hand, some of the Greco-Roman philosophical schools were missionary, in that they had leading spokespersons actively engaged in

Figure 18.1 A reconstructed model of a Roman insula, with shops on the lower level and living quarters above, similar to one that Paul may have worked and lived in while engaged in his missionary endeavors in such places as Thessalonica and Corinth.

winning converts to their way of looking at the world. In particular, Stoic and Cynic philosophers were involved in these kinds of activities. They tried to convince people to change their notions about life and their ways of living to conform to the philosophical views that alone could bring personal well-being. More specifically, Stoic and Cynic philosophers urged people to give up their attachments to the things of this world and to make their overarching concerns those aspects of their lives that they themselves were able to control. The Stoic theory was that people who were ultimately committed to matters outside of their control, such as wealth, health, careers, or lovers, were constantly in danger of forfeiting their well-being through the vicissitudes of bad fortune. What happens if you base your happiness on material goods or personal relationships, but then they are lost or destroyed? The solution to this problem is not to take measures to protect what you have, since this may not be within your power; it is, instead, to redirect your affections so your happiness is based on things that cannot be taken away, such as your freedom to think whatever you like, your honor, and your sense of duty. Since these are things that can never be lost, they should lie at the root of your personal well-being and so be the objects of your greatest concern.

Proselytizers for such philosophies could be found in a variety of urban settings throughout the empire. Cynics, those who took the Stoic doctrine to an extreme by abandoning all social conventions, including decent clothing, lodging, bathing, and privacy for bodily functions (see box 15.3), sometimes frequented crowded public places, where they urged their views on passers-by, maligned those who turned away, and badgered people for money (since they rejected social convention, they could scarcely be expected to work for a living). More socially respectable philosophers were often connected with wealthy households, somewhat like scholars-in-residence, and had wealthy patrons who provided for their physical needs in exchange for services rendered towards the family's intellectual and spiritual needs. A few Greco-Roman philosophers believed in working for a living to keep from depending on the support of others for their needs and becoming subservient to the so-called "nicer things in life."

So far as we can tell, this last kind of philosopher was somewhat rare in the empire, but Paul and his companions may have been identified as such persons by outsiders in Thessalonica. They were missionaries with a particular worldview who were trying to convert others to their ideas; they worked hard to support themselves and refused to take funds from others (e.g., 1 Thess 2:9).

Perhaps their mission proceeded something like this. Paul and his two companions arrived in the city and as a first step rented out a room in a downtown insula. Insula were the ancient equivalents of modern apartment buildings, packed close together in urban areas. They had a ground floor containing rooms that faced the street for small businesses (grocers, potters, tailors, cobblers, metal workers, engravers, scribes, and so forth), while the upper two or three stories served as living quarters for the people who worked below and for anyone else who could afford the rent. Shops were places not only of commerce but of social interaction, as customers, friends, and neighbors would stop by to talk. Given the long workdays and the absence of weekends (Jews, of course, took the Sabbath off; and everyone else closed up for special religious celebrations), the workplace was much more an arena of social intercourse than most modern business establishments are today. Contacts could be made, plans could be laid, ideas could be discussed—all over the potter's wheel or the tailor's table or the cobbler's bench.

Did Paul and his companions set up a small business, a kind of Christian leather goods shop, in the cities they visited? If so, this would explain a good deal of what Paul recounts concerning his interaction with the Thessalonian Christians in the early days. He and his companions toiled night and day while preaching the gospel to them (2:9). Like philosophers in that world, they exhorted, encouraged, and pleaded with those who dropped by, urging them to change their lives and adhere to the Christian message (2:12). Like some of the Stoics, they refused to be a burden on any of their converts, choosing to work with their own hands rather than rely on the resources of others (2:9–10).

Paul's Message

Paul obviously could not launch into a heavy exposition of his theology with people who were just stopping by. This was not simply because of the setting but even more because of the nature of his typical encounter. Even though Paul was engaged in manual labor, he was not an ordinary "blue-collar" worker. He was highly educated, far more so than most of the people that he would meet during a workday, and his theological reflections would be enough to befuddle the average

Figure 18.2 The Remains of an Insula in the City of Ostia, near Rome.

Figure 18.3 A shoemaker and cordmaker at work, from an ancient sarcophagus. These were manual laborers like Paul, who according to Acts 18:3 was a leather-worker.

person on the street. Moreover, most people stopping by the shop were almost certainly pagans, worshippers of Greco-Roman deities, who believed that there were lots of gods, all of whom deserved devotion and cult.

How would Paul begin to talk about his gospel with people like this? We are again fortunate to have some indications in Paul's letter. The critical passage is 1:9–10, where Paul reminds his recent converts about what he originally taught them:

> [To turn] to God from idols, to serve a living and true God, and to wait for his Son from heaven, whom he raised from the dead—Jesus, who rescues us from that wrath that is coming.

This appears to have been the core of Paul's proclamation to his potential converts. His first step was to have them realize that the many gods they worshipped were "dead" and "false" and that there was only one "living" and "true" God. In other words, before Paul could begin to talk about Jesus, he first had to win converts to the God of Israel, the one creator of heaven and earth, who chose his people and promised to bless all the nations of earth through them. Thus, Paul's proclamation began with an argument against the existence and reality of the deities worshipped in the local cults.

We have no way of knowing how Paul actually persuaded people that there was only one true God. Quite possibly he recounted tales of how this one God had proven himself in the past, for example, in the stories found in the Jewish Scriptures or in tales of Jesus' apostles, who were said to have done miracles. It is likely that these converts had at least heard of the Jewish God before, so Paul's initial task appears to have been to convince them that this was the only God worthy of their devotion, and that their own gods had no power but were dead and lifeless. It may be that some of these people were already inclined to accept the belief in one God in view of the increasingly widespread notion even in non-Jewish circles that ultimately there was one deity in control of human affairs (see Chapter 2). If so, then Paul's success lay in his ability to convince them, somehow, that this one God was the God that he proclaimed to them.

Once Paul's listeners accepted the notion of the one true God, Paul pressed upon them his belief that Jesus was this one God's Son. Again, it is hard to know how he elaborated this view. There are reasons to doubt that he proceeded by describing Jesus' earthly life, narrating tales of what he said and did prior to his crucifixion, for even though he constantly reminds his Thessalonian audience of what he taught them, he says nary a word about Jesus' sayings or deeds (recall that none of our Gospels was yet in existence; see further Chapter 21). What, then, did he teach them?

Later in the letter we learn that a central component of the converts' faith was the belief that Jesus died "for them" (5:10) and that he was raised from the dead (4:14). From this we can surmise that Paul taught his potential converts that Jesus was a person who was specially connected with the one true God (the "Son of God," as he calls him in 1:9), whose death and resurrection were necessary to put them into a right relation with God. What appears to have been the most important belief about Jesus to the Thessalonians, however, was that he was soon to return from heaven in judgment on the earth. The first reference to this belief is here in 1:10, where Paul reminds his readers that he taught them to "wait for his Son from heaven—Jesus, who rescues us from the wrath that is coming." Further references to the notion of Jesus' return are found in every chapter of the letter (e.g., see 2:19; 3:13; 4:13–18; 5:1–11).

The Thessalonian congregation was also acquainted with the reason that Jesus was soon to return. On this point Paul is unequivocal: Jesus was going to come for his followers to save them from God's wrath. Paul, in other words, had taught his Thessalonian converts a strongly apocalyptic message. This world was soon to end, when the God who created it returned to judge it; those who sided with this God would be delivered , and those who did not would experience his wrath. Moreover, the way to side with this God, the creator and judge of all, was by believing in his Son, Jesus, who had died and been raised for the sins of the world and who would return soon for those who believe in him, to rescue them from the impending wrath.

This appears to have been the burden of Paul's preaching. From beginning to end it was rooted in a worldview that Paul appears to have embraced as a Jewish apocalypticist even prior to his conversion. Thus, to some extent his preaching to the Thessalonians involved convincing them to accept such basic apocalyptic notions as the end of the age, the coming of God's judgment, the need for redemption, and the salvation of the godly. It is striking, in this connection, how much apocalyptic imagery Paul uses throughout the letter. Consider, for example, 5:1–11, where Paul indicates that the end will come suddenly, like a woman's labor pains, that it will come like a thief in the night, that the children of light will escape but not the children of

Figure 18.4 Statue of Artemis (the goddess Diana) from Ephesus. The almost grotesque portrayal of her many breasts emphasizes her role as a fertility goddess, one who gives life in abundance. For Paul, though, she (along with all other pagan deities) was nothing but a "dead idol" (see 1 Thess 1:8–10).

darkness, and that the faithful need to be awake and sober. All of these images can be found in other Jewish apocalyptic texts as well. Moreover, many of Paul's allusive comments throughout the letter make sense only within a Jewish apocalyptic framework; among these are his reference to Satan, the great enemy of God and his people (2:18) and his assurance that suffering is necessary for God's people here at the end of time (3:3–4). Thus, in its simplest terms, Paul's proclamation was designed to transform the Thessalonian pagans into Jewish apocalypticists, who believed that Jesus was the key to the end of the world.

THE BEGINNINGS OF THE THESSALONIAN CHURCH: A SOCIO-HISTORICAL PERSPECTIVE

To some extent, Paul succeeded in his mission. We have no idea how many people he and his companions converted, but there were clearly some. Here we will explore the nature of this group of converts from the perspective of a social historian, asking not so much what they came to believe but rather who they were and how they functioned as a social group.

It is nearly impossible to gauge what kind of people Paul's Gentile converts in Thessalonica were. If they were in regular contact with manual laborers like Paul and his companions in their insula, and if it would have been an excessive burden for them to provide financial support for the missionaries, then we might suppose that for the most part the converts were not among the wealthy and the social elite in town, although certainly some may have been drawn from among the upper classes. If this sketch is correct, then the Thessalonian Christians, as a social group, may have been roughly comparable to the people Paul was later to convert in the city of Corinth farther to the south, the majority of whom were not well educated, influential, or from among the upper social classes, according to 1 Corinthians 1:26 (presumably some were, or Paul would have not have said that "not many of you are").

It seems plausible that the people Paul converted began meeting together periodically, perhaps weekly, for fellowship and worship. This appears to have been the pattern of Paul's churches, as you will see from his other letters (e.g., 1 Cor. 11:17–26; 16:1), and it would make sense of his decision to send a letter to "the church" rather than individual converts. Most scholars think that churches like this would have met in private homes, and so call them "house churches" (e.g., see Philem 2). We have no hard evidence of actual church buildings being constructed by Christians for another two centuries (see box 11.3).

It appears that people in this kind of group experienced unusual cohesion as a social unit. There were, of course, other kinds of social groups in the Greco-Roman world that met periodically for worship and socializing. We are especially well informed about ancient trade organizations and funeral societies. The church in Thessalonica may have been roughly organized like one of these groups (see box 18.1). On the other hand, given its central commitment to a religious purpose, it may have had some close organizational affinities with the Jewish synagogue as well, although the synagogue may have been much larger than the Christian group. It appears that some of the local converts became leaders in the Christian congregation and that they organized their meetings, distributed the funds they collected, and guided the thinking of the group about religious matters (5:12–13).

From a socio-historical point of view, certain features of these converts' new religion provided strong bonds with the group. For one thing, they appear to have understood themselves as a closed group. Not just anyone could come off the street to join; membership was restricted to those who accepted Paul's message of the apocalyptic judgment that was soon to come and the salvation that could be obtained only through faith in Jesus, who died and was raised from the dead. The Thessalonian church had a unified commitment to this teaching, and it made them distinct from everyone else that they came in contact with.

This distinctiveness was evidently known to outsiders as well. Throughout 1 Thessalonians Paul refers to the persecution that the community

SOME MORE INFORMATION

Box 18.1 Rules for a Private Association

Christian house churches may have appeared to outsiders to be like other kinds of voluntary associations found in the Greco-Roman world. Associations were privately organized small groups that met periodically to socialize and share a good meal together; they would often perform cultic acts of worship together; many of them were concerned with providing appropriate burial for their members (a kind of life-insurance arrangement that covered expenses hard to manage on an individual basis). The social activities of such groups were sometimes underwritten by one or more of their wealthier members who served as patrons for the body.

Voluntary associations had rules for membership, some of which we know from surviving inscriptions. To see the close connections of such societies with the early Christian communities, consider the following set of by-laws of a burial society in Lanuvium Italy, a group that met at the temple of the divine man Antinoüs. These bylaws come to us from an inscription dated to 136 C.E. [A sesterce was a coin worth about one-quarter of an average worker's daily wage.]

> It was voted unanimously that whoever desires to enter this society shall pay an initiation fee of 100 sesterces and an amphora of good wine, and shall pay monthly dues of [2 sesterces]. . . . It was voted further that upon the decease of a paid-up member of our body there will be due him from the treasury 300 sesterces, from which sum will be deducted a funeral fee of 50 sesterces to be distributed at the pyre [among those attending]; the obsequies, furthermore, will be performed on foot. . . .
>
> Masters of the dinners in the order of the membership list, appointed four at a time in turn, shall be required to provide an amphora of good wine each, and for as many members as the society has, a [loaf of] bread costing [1 sesterce], sardines to the number of four, a setting, and warm water with service.
>
> It was voted further that any member who has [served as chief officer] honestly shall [thereafter] receive a share and a half of everything as a mark of honor, so that other [chief officers] will also hope for the same by properly discharging their duties.
>
> It was voted further that if any member desires to make any complaint or bring up any business, he is to bring it up at a business meeting, so that we may banquet in peace and good cheer on festive days.
>
> It was voted further that any member . . . who speaks abusively of another or causes an uproar shall be fined 12 sesterces. Any member who uses any abusive or insolent language to a [chief officer] at a banquet shall be fined 20 sesterces.
>
> It was voted further that on the festive days of his term of office, each [chief officer] is to conduct worship with incense and wine and is to perform his other functions clothed in white, and that on the birthdays of [the goddess] Diana and [the divine] Antinoüs he is to provide oil for the society in the public bath before they banquet. (Taken from Naphtali Lewis and Meyer Rheinhold, *Roman Civilization*, 3rd ed. [New York: Columbia University Press, 1990] 2.186-88.)

experienced from those who did not belong. As an apostle who proclaimed the gospel in the face of malicious opposition, Paul himself had suffered in some undisclosed way in the city of Philippi before arriving in Thessalonica (2:1–2). His statement is consistent with Luke's account of the founding of the Philippian church in Acts 16:19–40, although Paul does not corroborate any of Luke's details. In any event, he instructs his Thessalonian converts that they too should expect to suffer (3:3–4). He does not say why they should expect this, but perhaps it is because he believed that the forces of evil were out in full strength here at the end of time (cf. 2:18; 5:1–11). Moreover, he indicates that the Thessalonians had already experienced persecution from their compatriots, just as the earlier Christian communities had been persecuted by the non-Christian Judeans, who had always served as a thorn in the side of the church, in Paul's opinion, from the days of Jesus onward (2:14–16).

A shared experience of suffering can help to consolidate a social group that is already unified by a common set of beliefs and commitments. That is to say, suffering for the cause can function to emphasize and sharpen the boundaries that separate those who "live according to the truth" from those who "live in error." Moreover, the Christian believers in Thessalonica shared their insider status with similar groups of believers throughout their world. Thus Paul emphasizes that their faithfulness to the gospel had become well known to Christian communities throughout the provinces of Macedonia and Achaia (1:7–9) and that they were linked to the communities of Judea as well.

Paul never indicates directly why he mentions the churches of Judea, but he may have done so because of his cherished notion that his message did not represent a new religion but the religion of the Jews come now to fulfillment in Jesus (see Chapter 17). Paul did not teach these converts that they had to become Jews, but he did teach them that the one true God whom they now worshipped was the God of Israel, who in fulfillment of his promises had sent his messiah to die for the sins of the world. This was Jesus, the Son of the Jewish God, who was now prepared to return to deliver his people from the wrath that was to come.

The group of believers in Thessalonica thus understood itself to be part of a much broader social and historical network of the faithful, a network stretching across broad tracks of land and reaching back into the misty ages of history. They were brothers and sisters (1:4) bonded together for a common purpose, standing against a common enemy, partaking of a common destiny—and connected with other communities of like purpose and destiny who all shared the history of the people of God, as recorded in the traditions of the Jewish Scriptures.

The exhortations and instructions that Paul gives serve further to unify the group as rules, guidelines, beliefs, and practices that they share in common. He gives them these instructions, of course, in response to situations that have arisen in the community.

THE CHURCH AT THESSALONICA AFTER PAUL'S DEPARTURE

First Thessalonians 3:1 indicates that after Paul and his companions left Thessalonica they journeyed to Athens, perhaps again to set up shop. After a while, feeling anxious about the young church, they sent Timothy back to check on the situation, and possibly to provide additional instruction and support. When Timothy rejoined his colleagues (either in Athens or in Corinth, which was evidently their next stop; Acts indicates the latter but Paul says nothing of it), he filled them in on the situation (3:6). First Thessalonians represents a kind of follow-up letter. Even though, technically speaking, it was co-authored by Paul, Silvanus, and Timothy (1:1), Paul himself was evidently the real author (e.g., see 2:18).

The most obvious piece of information that Timothy brought back to his colleagues was that the congregation was still strong and deeply grateful for the work they had done among them. The letter is remarkably personable, with professions of heartfelt gratitude and affection flowing from nearly every page, especially in the first three chapters.

Although Paul's epistles generally follow the form of most Greco-Roman letters (see Chapter 11), they are, as a rule, much longer and tend to have a shape of their own. They typically begin with a prescript that names the sender(s) and the addressees, followed by a prayer or blessing ("Grace to you and peace . . ."), and then an expression of thanksgiving to God for the congregation. In most of Paul's letters, the body of the letter, where the main business at hand is addressed, comes next, followed by closing admonitions and greetings to people in the congregation, some references to Paul's future travel plans, and a final blessing and farewell. In 1 Thessalonians, however, the majority of the letter is taken up by the thanksgiving (1:2–3:13). This is clearly a letter that Paul was happy to write, in contrast, say, to Galatians, where the thanksgiving is replaced by a reprimand!

The closest analogy to 1 Thessalonians from elsewhere in Greco-Roman antiquity is a kind of correspondence that modern scholars have labeled the "friendship letter." This is a letter sent to renew an acquaintance and to extend friendly good wishes, sometimes with a few requests or admonitions. Paul's letter also contains some requests and admonitions, based on the news that he has received from Timothy. The congregation has not experienced any major problems, but one important issue has arisen in the interim since Paul's departure. Paul writes to resolve the issue and to address other matters that are important for the ongoing life of the community.

Before considering the major issue that has arisen, we should examine another aspect of life in the Thessalonian church—the community's persecution. We do not know exactly what this persecution entailed. We do know that in a somewhat later period, some sixty years after 1 Thessalonians was written, Roman provincial authorities occasionally prosecuted Christian believers simply for being Christian (see Chapter 25). At least during the New Testament period, however, there was no official opposition to Christianity, in the sense of an established governmental policy or legislation outlawing the religion. People could be Christian or anything else so long as they didn't disturb the peace.

Christians sometimes did disturb the peace, however, and when they did there could be reprisals. Paul himself indicates that over the course of his career he had been beaten with "rods," a standard form of Roman corporal punishment, on three occasions (2 Cor 11:25). Were the Christians of Thessalonica, the capital of the Roman province of Macedonia, being condemned to punishment by the governor who resided there?

In later times, the case against the Christians was taken up by governors at the instigation of the populace, who feared that this new religion was offensive to the Roman gods. Other non-Roman religions were generally not seen as offensive because they did not prohibit their adherents from participating in the state cult. Judaism did not participate, of course, but they were granted an exemption because of the great antiquity of their traditions (in this world, if something was old, it was venerable). Christianity, on the other hand, was not at all ancient; moreover, the Christians not only refused to worship the state gods, they also insisted that their God was the only true God and that all other gods were demonic. For the most part, this notion did not sit well with those who believed not only that the gods existed but also that they could terrorize those who refused to acknowledge them in their cults. Some decades after Paul, cities that experienced disaster would sometimes blame the false religion of the Christians; when that happened, Christian believers were well advised to keep a low profile.

Had something like this happened in Thessalonica? While it is possible that the governor of the province had sent out the troops at the instigation of the masses, Paul says nothing to indicate that the situation was so grave or dramatic. It could be, then, that the Christians were opposed not by the government but by other people (organized groups?) who found their religion offensive to their sense of right and duty—duty to the gods who bring peace and prosperity and duty to the state, which was the prime beneficiary of the gods' kindnesses. It commonly happens that closed, secret societies bring out the worst in their neighbors, and it may be that the Thessalonian Christians, with their bizarre teachings about the end of the world and the return of a divine man from heaven, along with their inflammatory rhetoric (for example, against other local cults), proved to be too much for others. These others could have included families and former friends of the converts, who knew enough to be suspicious and were not themselves inclined to join up. Perhaps they maligned the group or abused it in other ways (physical attacks? graffiti on the walls of its house church? organized protests?).

If something like this scenario is at all plausible, it would help explain some of the other things Paul says in this letter by way of exhortation. He begins the body of the letter (4:1–5:11) by urging his converts not to engage in sexual immorality. The meaning of his words is hotly debated by scholars, to the extent that translators of the New Testament cannot even agree about how to render them into English. This is especially true of verses 4–6: is Paul urging the Thessalonian men to be careful in treating their wives or in handling their genitals? Whichever meaning is preferred, Paul clearly wants the community to behave in socially acceptable ways. Whether or not he is responding to a specific problem of sexual immorality that he wants to nip in the bud is difficult to judge. Given his lack of specificity in the matter, it may be that Paul simply wants the Thessalonian Christians to keep their image pure before the outside world, just in case they are suspected of vile activities commonly attributed to secret societies in the ancient world (see box 18.2). After all, there is no reason to give outsiders additional grounds to malign your group when they may already have all the grounds they need.

The same logic may underlie the exhortations in 4:9–12. The believers are urged to love one another, in what we might call the platonic sense, not to make waves in society ("mind your own affairs"), and to be good citizens ("work with your own hands"). These admonitions serve both to promote group cohesion and to project an acceptable image of the group to those who are outside.

The Major Issue in the Congregation

In 4:13 Paul finally comes to the one serious issue that the Thessalonians themselves have raised. Perhaps not surprisingly, given what we have seen about the character of Paul's message when he converted and instructed these people, it is a question pertaining to the events at the end of time. Paul had earlier instructed the Thessalonians about the imminent end of the world, which would bring sudden suffering to those who were not prepared, like the birthpangs of a woman in labor (see 5:1–3). He had warned them that they must be ready, for the day was coming soon and was almost upon them; they must be awake and sober lest it catch them unawares (5:4–9). His converts had presumably taken his teaching to heart; they were eagerly awaiting the return of Jesus to deliver them from the wrath that was coming. But Jesus hadn't returned and something troubling had happened: some of the members of the congregation had died.

These deaths caused a major disturbance among some of the survivors. The Thessalonians had thought that the end was going to come before they passed off the face of the earth. Had they been wrong? Even more troubling, had those who died missed their chance to enter into the heavenly kingdom when Jesus returned?

Paul writes to respond to their concern. You will notice that the response of 4:14–17 is bracketed by two exhortations to have hope and be comforted in light of what will happen when Jesus appears. At his return in glory those who have died will be the first to meet him; only then will those who are alive join up with them in the air "to be with the Lord forever" (4:17). In other words, there will not simply be a resurrection of the dead for judgment at the end of time; there will also be a removal of the followers of Jesus, both dead and alive, from this world prior to the coming of the divine wrath. The Thessalonians are to be comforted by this scenario. Those who have already died have not at all lost out; rather, they will precede the living as they enter into the presence of the Lord at the end of time.

There are two further points of interest about this passage. First, it is clear that Paul expects that he and some of the Thessalonians will be alive when this apocalyptic drama comes to be played out. He contrasts "those who have died" with "we who are alive, who are left until the coming of the Lord" (v. 15; also see v. 17). He appears to have no idea that his words would be discussed after his death, let alone read and studied some nineteen centuries later. For him, the end of time was imminent.

SOMETHING TO THINK ABOUT

Box 18.2 Charges Against the Christians

There is no solid evidence to suggest that specific allegations of wrongdoing were being made against the church in Thessalonica at the time of Paul's writing, but we do know that other secret societies were widely viewed with suspicion and that certain standard kinds of slander were leveled against them. The logic of these slanders is plain: if people meet together in secret or under the cloak of darkness, they must have something to hide.

It is possible that Paul was aware of such charges and wanted the Thessalonian Christians to go out of their way to avoid them. Such a concern would make sense of his injunctions to maintain pure sexual conduct and to keep a good reputation among outsiders.

As an example of the kinds of accusations that were later leveled against the Christians, consider the comments of Fronto, the tutor of the emperor Marcus Aurelius and one of the most highly respected scholars of the mid second century:

> They [the Christians] recognize each other by secret marks and signs; hardly have they met when they love each other, throughout the world uniting in the practice of a veritable religion of lusts. Indiscriminately they call each other brother and sister, thus turning even ordinary fornication into incest. . . . It is also reported that they worship the genitals of their pontiff and priest, adoring, it appears, the sex of their "father.". . . The notoriety of the stories told of the initiation of new recruits is matched by their ghastly horror. A young baby is covered over with flour, the object being to deceive the unwary. It is then served before the person to be admitted into their rites. The recruit is urged to inflict blows onto it—they appear to be harmless because of the covering of flour. Thus the baby is killed with wounds that remain unseen and concealed. It is the blood of this infant—I shudder to mention it—it is this blood that they lick with thirsty lips; these are the limbs they distribute eagerly; this is the victim by which they seal their covenant; it is by complicity in this crime that they are pledged to mutual silence; these are their rites, more foul than all sacrileges combined. . . . On a special day they gather for a feast with all their children, sisters, mothers—all sexes and all ages. There, flushed with the banquet after such feasting and drinking, they begin to burn with incestuous passions. They provoke a dog tied to the lampstand to leap and bound towards a scrap of food which they have tossed outside the reach of his chain. By this means the light is overturned and extinguished, and with it common knowledge of their actions; in the shameless dark with unspeakable lust they copulate in random unions, all equally being guilty of incest. (Minucius Felix, *Octavius* 9:2–6)

Second, Paul's scenario presupposes a three-storied universe, in which the world consists of an "up" (where God is, and now Jesus), a "here" (where we are), and a "down" (where those who have died are). According to this scenario, Jesus was here with us; he died and so went down to the place of the dead; then God raised him up to where he is. Soon he is going to come back down to earth on the clouds (i.e., from heaven above the sky) to raise up both those who are here and those who are down below, elevating them to the clouds to live with him forever.

SOMETHING TO THINK ABOUT

Box 18.3 The Thessalonians' Perplexity

The occasion of 1 Thessalonians raises some intriguing historical questions. Why were the Thessalonian Christians surprised that some of their members had died, and why didn't they know that at Jesus' return he would raise the dead to be with him forever? Had Paul simply neglected to tell them that part? Morever, why was Timothy unable to answer their question? Why did he have to return to ask Paul about it and leave them in uncertainty for some weeks at the least? Didn't Timothy know what was supposed to happen at the end?

One possibility is that when Paul was with the Thessalonians his own views were in a state of flux. If he himself didn't realize how long it would be before Jesus returned, he might not have discussed the matter with either the Thessalonians or his own close companions, Silvanus and Timothy.

This scenario is based on an ancient way of looking at the world where there actually was an up and a down in the universe. It stands in stark contrast, obviously, to our modern understanding of the earth as the third planet of a solar system formed around a minor star, just one of the billions of stars that make up our galaxy, which itself is just one of billions of galaxies in a universe, in other words, in which there is no such thing as up and down, no "heaven" above our heads or "place of the dead" below. This is simply reminder that Paul's world, and consequently his worldview, is not ours.

CONCLUSION: PAUL THE APOSTLE

It is clear that Paul's self-acclaimed title "apostle of the Gentiles" was no empty phrase. His converts, at least in Thessalonica, were former pagans, whom he contacted from his place of employment and convinced to abandon their traditional cults to worship the one true God, the creator of the world. Moreover, he and his colleagues couched their proclamation in apocalyptic terms: the creator of the world was also its judge, and his day of reckoning was imminent. Soon he was to send his Son, Jesus, who had died and had been raised from the dead and exalted to heaven and who would deliver his followers from the wrath that was soon to come.

Those who accepted this message formed a social group, a church, that met periodically in one of the member's home (or in several homes, depending on its size). The members of the group had unusually strong bonds of cohesion, reinforced by several factors: (1) the insider information they had as those who understood the course of history here at the end of time, (2) the mutual love and support that they showed one another, (3) the common front they projected in the face of external opposition from those who did not know the "truth," and (4) the rules that governed their lives together. Moreover, they understood themselves to stand in unity with other groups similarly organized throughout the provinces of Macedonia and Achaia and reaching all the way to Judea. These groups were unified by their common faith and common commitment to the God of Israel, who now in the end of time had fulfilled his promises to his people through Jesus, and through him to all peoples of the earth, both Jews and Gentiles.

Difficulties had arisen in this community, and Paul wrote a letter to help resolve them. In this the Thessalonians were probably like most of Paul's churches, communities that he established in major urban areas throughout the Mediterranean, each of which experienced problems that required the apostle's intervention and advice.

SUGGESTIONS FOR FURTHER READING

See also the general bibliography for Chapter 17

Hock, Ronald. *The Social Context of Paul's Ministry: Tentmaking and Apostleship*. Philadelphia: Fortress, 1980. An interesting investigation of Paul's apostolic modus operandi in light of representatives of other philosophies in the Greco-Roman world, who worked to support themselves and used their workplace as a forum to propagate their views.

Malherbe, Abraham. *Paul and the Thessalonians: The Philosophic Tradition of Pastoral Care*. Philadelphia: Fortress, 1987. An insightful study of Paul's interaction with the Thessalonians from a socio-historical perspective.

CHAPTER 19

Paul and the Crises of His Churches: 1 and 2 Corinthians, Galatians, Philippians, and Philemon

The thirteen New Testament epistles attributed to Paul are arranged roughly according to length, with the longest (Romans) coming first and the shortest (Philemon) last. As we have seen, this arrangement does not coincide with the actual sequence in which the letters were written; 1 Thessalonians is Paul's earliest surviving letter and Romans the latest. Of the five undisputed letters that remain, however, a case can be made that their canonical sequence also happens to be their chronological. For this reason, we can deal with each of these remaining letters in their canonical order: 1 and 2 Corinthians, Galatians, Philippians, and Philemon.

1 CORINTHIANS

Corinth was a large and prosperous city south of Thessalonica, in the Roman province of Achaia, of which it was the capital. Located on the isthmus dividing the northern and southern parts of modern-day Greece, it was a major center of trade and communication, served by two major ports within walking distance. The city was destroyed in 146 B.C.E. by the Romans but was refounded a century later as a Roman colony. In Paul's time, it was a cosmopolitan place, the home of a wide range of religious and philosophical movements.

Corinth is perhaps best remembered today for the image problem it suffered throughout much of its checkered history, at least among those who advocated the ancient equivalent of "family values." Its economy was based not only on trade and industry but also on commercialized pleasures for the well-to-do. It is not certain that Corinth's loose reputation was altogether deserved, however; some modern historians have suggested that its image was intentionally tarnished by the citizens of Athens, one of its nearby rivals and the intellectual center of ancient Greece. It was an Athenian, the comic poet Aristophanes, who invented the verb "Corinthianize," which meant to engage in sexually promiscuous activities. In any event, many people today know about the city only through the letter of 1 Corinthians, a document that has done little to enhance its reputation.

The congregation that Paul addresses appears to have been riddled with problems involving interpersonal conflicts and ethical improprieties. His letter indicates that some of its members were at each others' throats, claiming spiritual superiority over one another and trying to establish it through ecstatic acts during the course of their worship services. Different members of the community would speak prophecies and make proclamations in languages that no one else (including themselves) knew, trying to surpass one another in demonstrating their abilities to speak in tongues. This one-upmanship had evidently manifested itself outside the worship service as well. Some people had grown embittered enough to take others to court (over what, we are not told). In addition, the personal conduct of community members was not at all what Paul had in mind when he led them away from what he viewed as their degenerate pasts into the church of Christ. At their periodic community meals, some had been gorging themselves and getting drunk while others had been arriving late to find nothing to eat. Some of the men in the congregation had been frequenting prostitutes and didn't see why this should be a problem; one of them was sleeping with his stepmother. This is the community that Paul addresses as the "saints who are in Corinth" (1:2). One wonders what the Corinthian sinners looked like.

The Beginnings of the Church

After leaving Thessalonica, Paul and his companions, Timothy and Silvanus, arrived in Corinth and began, again, to preach the gospel in an effort to win converts (2 Cor 1:19). Possibly they proceeded as they had in the capital of Macedonia, coming into town, renting out a shop in an insula, setting up a business, and using the workplace as a forum to speak to those who stopped by. In this instance, the book of Acts provides some corroborating evidence. Luke indicates that Paul did, in fact, work in a kind of leather goods shop in Corinth, having made contact with a Jewish couple named Aquila and Priscilla who shared his profession in both senses of the term; they had the same career and the same faith in Jesus.

In other respects, however, the narrative of Acts contrasts with what Paul himself says about his sojourn in Corinth. For one thing, Luke indicates that Paul devoted himself chiefly to evangelizing the Jews in the local synagogue until he was dismissed with the left foot of fellowship. Even after leaving the synagogue, according to Luke, Paul principally converted Jews (18:4–11). Paul's own letter gives an entirely different impression. Most of his converts, as one would expect, given his claim to be the apostle to the Gentiles, appear to have been non-Jews. "You know that when you were pagans, you were enticed and led astray to idols that could not speak" (12:2). Here, as in Thessalonica, Paul and his companions worked primarily with Gentiles to convince them both that there was only one God worthy of devotion and worship (the God of Israel) and that Jesus was his Son.

The majority of Paul's converts were evidently from the lower classes, as he himself reminds them: "Not many of you were wise by human standards [highly educated], not many were powerful, [influential in the community], not many were of noble birth" [in the upper classes] (1:26). Recent scholars have observed, however, that at least some of the Corinthian converts must have been well-educated, powerful, and well born, or else Paul would not have said that "not many" of them were. Indeed, if we assume that some members of the community came from the upper classes, we can make better sense of some of the problems that they experienced as a group. It would explain, for example, why some of those coming together for the communal meal (a bring-your-own-supper kind of affair) could come early and enjoy lots of food and good drink; these were comparatively wealthy Christians who and didn't have to work long hours. Others, however, had to come late and had scarcely anything to eat; these were the poorer members, possibly slaves, who had to put in a full day's work. The presence of some upper-class Christians would also explain why some members of this community were perturbed that Paul would not allow them to support him, that is, become his patrons and care for all of his financial needs so as to free him up to preach the gospel (9:7–18, cf. especially 2 Cor 12:13). One of the common ways for a philosopher to make a living in the Greco-Roman world was to be taken into a wealthy household to serve as a kind of scholar-in-residence in exchange for room, board, and other niceties (depending on the wealth of the patron). Paul had reasons for wanting none of this arrangement—he saw it as putting his gospel up for sale—but some of the influential members of the congregation found his attitude puzzling and even offensive, as will become yet clearer in 2 Corinthians.

Other problems in the congregation may also have related to the differing socioeconomic levels of its members. If we can assume that the upper classes in antiquity would have been relatively well educated, it may be that the "knowledge" of some of these people in the Corinthian church allowed them to see things differently from the lower classes and that this led to some differences of opinion in the community. For example, some

Figure 19.1 Picture of an ancient philosopher leaning on his walking stick, from a wall painting of the first century B.C.E. Paul himself would have appeared to many people in his world as an itinerant philosopher.

members may have thought that eating meat offered to idols was a real and present danger, in view of the demonic character of the pagan gods (possibly a lower-class view), while others took such scruples as baseless superstition (possibly the view of some of the more highly educated). This is one of the major issues that Paul addresses in the letter (chaps. 8–10).

During their stay in Corinth, Paul and his companions appear to have converted a sizable number (dozens?) of pagans to the faith. The book of Acts indicates that they spent a year and a half there, in contrast to just three weeks in Thessalonica. Paul himself makes no clear statement concerning the length of his stay, but there are indications throughout his letter that the Christians in Corinth, or at least some of them, had a much more sophisticated understanding of the faith than those in Thessalonica—even if they had, from Paul's perspective, gotten it wrong at points. Indeed, unlike the Thessalonians, who understood their new religion at a fairly rudimentary level, some of the Corinthians had so much knowledge of their faith that they took Paul's gospel simply as a starting point and developed their views in vastly different directions.

What can we say about the message that Paul originally preached to these people? Again, he evidently instructed them in the need to worship the one true God and to await his Son from heaven. As we will see, however, the second part of this message ("to await his Son") made significantly less impact on the converts in Corinth than on those in Thessalonica. It is difficult to know exactly what else he taught these people. It does appear, though, that Paul devoted little if any effort to narrating tales about what Jesus said and did during his public ministry (at a later stage, we will consider whether Paul himself knew very much about this ministry; remember, he was writing long before the Gospels were written). He does summarize a couple of sayings of Jesus, to the effect that Christians shouldn't get divorced (7:10–11) and they should pay their preacher (9:14), and he does narrate the incident of Jesus' institution of the Lord's Supper (11:24–28). But he says nary a word about Jesus' baptism, temptation, transfiguration, preaching of the coming kingdom of God, encounters with demons, appearance before Pontius Pilate, and so on—all of which would have been directly germane to the problems that the Corinthians appear to have experienced. What he does say, and says emphatically, is that the only thing he "knew" among the Corinthians was "Jesus Christ, and him crucified" (2:2).

In other words, Paul's principal message was about Jesus as the crucified Christ. It appears to be a message that the Corinthians, or at least a good portion of them, didn't absorb, at least in Paul's opinion. We will see why momentarily. First, we should consider in some detail Paul's own brief recollection of what he taught the Corinthians about Jesus. In 15:1–2, he reminds his converts of "the good news that I proclaimed to you, which you in turn received, in which also you stand, through which also you are being saved, if you hold firmly to the message that I proclaimed to you." He then summarizes this message:

> For I handed on to you as of first importance what I in turn had received: that Christ died for our sins in accordance with the scriptures, and that he was buried, and that he was raised on the third day in accordance with the scriptures, and that he appeared to Cephas and then to the twelve. (15:3–5).

Thus, of primary importance in Paul's preaching to the Corinthians was the message of Christ's death and resurrection. Jesus died, fulfilling the Jewish Scriptures, and there's proof: he was buried. Moreover, God raised him from the dead, fulfilling the Scriptures. Again there's proof: he was later seen alive. Paul preached a similar message in Thessalonica as well, but with two differences, one in the message and the other in the way that it was received.

With regard to the message itself, we find subtle indications in 1 Thessalonians that Paul directly linked his gospel message with the Jewish religion, but never does he quote the Jewish Scriptures or assume that his followers are personally conversant with them. The situation is quite different with the Corinthians. From the outset, Paul had taught them that Jesus' death and resurrection were both anticipated in the Scriptures (see Chapter 17); moreover, throughout this letter he appeals to the Scriptures in order to make his points. Strikingly, when he does so he emphasizes that the Scriptures were not written only, or even especially, for Jews in times past, but even more

Figure 19.2 One of the earliest visual representations of Jesus' crucifixion, from a cyprus panel door in the church of Saint Sabina in Rome, nearly 350 years after Paul's day. Earlier Christians were reluctant to portray the crucifixion (contrast Paul in 1 Cor 2:2).

particularly for Christians in the present (e.g., 1 Cor 9:9–10; 10:1–13). If the Thessalonians had insider knowledge, the Corinthians have even more; all of God's interactions with his people have been leading up to the present time. The Christian community is God's ultimate concern, and always has been.

This is heady stuff, and there is some indication that it had in fact gone to the heads of some of Paul's converts. This can be seen in a second difference between the Thessalonians and the Corinthians. The former group saw Jesus' resurrection as the beginning of the major climax of history, when he would return and remove the Christians from this world before God's wrath destroys all his enemies. Some of the Corinthians, on the other hand, appear to have interpreted Jesus' resurrection in a more personal sense as his exaltation to glory that they themselves, as those who have participated in his victory, have come to share. Despite Paul's protests, some (or perhaps most?) of the Corinthians came to believe that they had already begun to enjoy the full benefits of salvation in the here and now, as members of Christ's resurrected and exalted body. In Paul's words (which must be taken as a sarcastic echoing of their views, given everything else he says in this letter), "Already you have all you want! Already you have become rich! Quite apart from us you have become kings!" (4:8).

For Paul himself, the Corinthians' notion that they were already enjoying an exalted status couldn't be further from the truth. In his view, the forces of evil were to remain in power in this world until the end came and Christ returned. Until then, life would be a struggle full of pain and suffering, comparable to the pain and suffering experienced by the crucified Christ himself. Those who believed that they had already experienced a full and complete share of the blessings of eternity had simply deceived themselves, creating immense problems for the church and misconstruing the real meaning of the gospel.

The Subsequent History of the Community

There is nothing to indicate that the problems addressed in this epistle had come to a head during Paul's original stay in Corinth. Eventually, he and his companions left to proclaim their gospel elsewhere, leaving the Christians behind to continue the mission for themselves. Soon thereafter, an acquaintance of Paul named Apollos came to Corinth and proved instrumental in providing additional instruction to the Christians there. According to the book of Acts, Apollos was a skilled speaker (18:24–28), and it is clear from Paul's letter that he acquired a considerable following in the congregation (1:12; 3:4–6).

We are not certain of the precise course of Paul's journeys, but he evidently ended up in the city of Ephesus not long after leaving Corinth. Ephesus, another large urban area, was in the western portion of Asia Minor (modern-day Turkey). From there Paul wrote the letter of 1 Corinthians (see 16:8). Timothy and Silvanus had apparently departed from him already, for he wrote the letter not with them but with someone named Sosthenes, who is otherwise mentioned in the New Testament only in Acts 18:17 as the ruler of the Jewish synagogue in Corinth and a convert to Paul's gospel. Paul obviously wrote the letter of 1 Corinthians to deal with problems that had arisen in the congregation. He indicates that he has heard of these problems from two different sources, one oral and one written.

At the beginning of the letter, after the prescript (1:1–3) and thanksgiving (1:4–9; notice how much shorter it is than the one to the Thessalonians), Paul states that he has learned about the activities of the congregation from "Chloe's people" (1:11). We do not know who this Chloe was; the name occurs nowhere else in the letter or in the rest of the New Testament. We do know that it was the name of a woman, and the reference to her "people" is usually taken to mean her slaves or former slaves who had come to Ephesus, perhaps on her business, and had met with Paul to pass along some news. Since Chloe owned slaves who managed her business affairs, she must have been a wealthy woman in Corinth; whether she herself was a member of the Christian community is difficult to judge. In any event, her unnamed "people" must have been active in the congregation, given the inside information that they passed along to Paul.

The news was not good. The church was divided against itself, with different factions claiming different leaders, each of whom, from Paul's perspective, was seeking to usurp the claims of others by demonstrating their own spiritual superiority and claiming to represent the true faith as expounded by one or another famous authority (Paul, Cephas, Apollos, and Christ himself; 1:12). The conflicts had gotten nasty at times, with some of the members taking others to court over their differences (not their differences over inner church politics, of course, but over matters that the civil law courts could decide). Moreover, immorality was evidently rampant. Generally, this was not the happy community of the faithful that Paul had envisioned, especially compared to the model church of the Thessalonians.

The information from Paul's other source was equally troubling. It appears that he had received a letter from some of the Corinthians (probably not all of them; as we will see, not everyone felt beholden to him) in which they expressed their different opinions on some critical matters and sought Paul's judgment (e.g., see 7:1). The letter had been brought by three members of the church—Stephanas, Fortunatus, and Achaicus—who evidently had waited for Paul to pen a reply (16:15–18). The issues were of some moment; there were members of the congregation, just to take one example, who had been teaching that it was not right even for married couples to have sex. One can sense the urgency of their query.

Paul wrote 1 Corinthians to deal with the various problems and issues that had arisen. Giving fairly straightforward answers, he deals with each problem in turn, in the sequence shown in box 19.1. From Paul's perspective, however, one big problem evidently underlay all of these specific problems.

Paul's Response to the Situation: The End as the Key to the Middle

Paul's perspective is best seen toward the conclusion of the letter. In good rhetorical style (i.e., following the instructions of those who taught rhetoric in his day), Paul provides at the end the key to what has come before. We saw earlier that Paul begins chapter 15 by summarizing the content of the gospel message that he preached to the Corinthians, the message of Christ's death and resurrection; he then draws out the implications of this message. Sometimes this chapter is misread as an attempt to prove that Jesus was raised from the dead, for example, by citing a group of "witnesses" in verses 5–8. In fact, Paul is not trying to demonstrate to the Corinthians something they don't believe, he is reminding them of something they already know (see vv. 1 and 3), that Jesus was raised bodily from the dead.

SOME MORE INFORMATION

Box 19.1 Overview of 1 Corinthians

Chapter	Subject Matter
1–4	Divisions in the church
5	Man living with his stepmother
6	Lawsuits among the believers
	Men going to prostitutes
7	Sex and marriage
8–10	Meat offered to idols
11	Women's use of headcoverings in the church
	Abuse of the Lord's Supper
12–14	Spiritual gifts in the congregation
15	Belief in the resurrection

SOMETHING TO THINK ABOUT

Box 19.2 Possibilities of Existence in the Afterlife

Some scholars have thought that Paul and his Corinthian opponents disagreed about the resurrection because they had fundamentally different understandings about the nature of human existence, both now and in the afterlife. Perhaps it would be useful to reflect on different ways that one might conceive of life after death.

Annihilation. One possibility is that a person who dies ceases to exist. This appears to have been a popular notion in the Greco-Roman world, as evidenced by a number of inscriptions on tombstones that bemoan the brevity of life which ends in nonexistence. One of the most widely used Latin inscriptions was so popular that it was normally abbreviated (like our own R.I.P. for "Rest in Peace") as N.F.N.S.N.C.: "I was not, I am not, I care not."

Disembodied Existence. Another possibility is that life after death is life apart from the body. In some strands of Greek thought influenced above all by Plato, the body itself was thought to be the bane of human existence, because it brought pain, finitude, and death to the soul that lived within it. These people did not think of the soul as immaterial; it was thought to be a "substance," but a much more refined substance than the clunker of a shell that we call the body (cf. the Gnostics; see Chapter 11). The catchy Greek phrase sometimes used to express the notion that the coarse material of the body is the prison or tomb for the more refined substance of the soul was "sōma—sēma," literally, "the body—a prison." For people who thought such things, the afterlife involved a liberation of the soul from its bodily entombment.

Bodily Resurrection. A third possibility is that the body is not inherently evil or problematic but has simply become subject to the ravages of evil and death. For many Jews, for example, the human body was created by God, as were all things, and so is inherently good. And what God has created he will also redeem. Thus, the body will not ultimately perish but will live on in the afterlife. How can this be, given the indisputable fact that bodies eventually decay and disappear? In this view, God will transform the physical body into a spiritual body that will never experience the ravages of evil and death, a glorified body that will never get sick and never die. As a Jewish apocalypticist Paul maintained this third view of the nature of human existence, whereas his opponents in Corinth, like many Christians after them down to our own day, appear to have subscribed to the second.

For Paul, Jesus' resurrected body was a glorified spiritual body, not like the paltry mortal flesh that we ourselves are stuck with; but just as importantly, it was an actual body that could be seen and recognized (15:5–8, 35–41). Paul's point is that the exalted existence that Jesus entered involved the total transformation of his body (15:42–49, 53–54). It was not some kind of ethereal existence in which his disembodied soul was elevated to the realm of divinity; his was a bodily resurrection (see box 19.2). The reason this matters becomes clear in the context of Paul's response. There were some in Corinth who were saying that there was no such thing as the resurrection of bodies from the dead (15:12).

Paul spends most of chapter 15 demonstrating that since Christ was raised bodily from the dead—and since he is the "first fruit" of the resurrection, as all of the Corinthians came to believe when they accepted his gospel message—then there is going to be a future resurrection of the dead when Christians come to participate in Christ's exalted status, that is, when they themselves are raised in glorious immortal bodies (15:12–23, 50–55). It is then that Christian believers will enjoy the full benefits of their salvation. For Paul, the end has not come yet. Despite the claims of some, presumably some of the most "spiritual" among the Corinthian leaders, Christians do not yet have the full benefits of salvation; they are not yet exalted to a heavenly status. Even the elect are living in a world of sin and evil, and they will continue to do so until the end comes.

This basic message underlies not just chapter 15 but all of 1 Corinthians. To some extent, each of the problems experienced by this congregation is related to the basic failure to recognize the limitations and dangers of Christian existence in the age before the end. The first problem that Paul attacks (in chaps. 1–4) is the divisions within the church that were caused, evidently, by leaders claiming to be spiritually superior to one another and to adhere to the teachings of various predecessors (Paul, Cephas, Apollos, or Christ; 1:12). One might expect Paul to take a side in this argument, that is, to insist that the faction that had the good sense to line up with him was right. Instead, he insists that all of the sides (even his) are in error. They are in error because they have elevated the status of individual leaders on the basis of their superior wisdom and superhuman power (1:18–25), perhaps thinking that these characteristics could be transferred from one person to the next in the act of baptism (as suggested, possibly, in 1:14–17). The leaders themselves, who are left unnamed, have apparently agreed on one major point, that wisdom and power indicate the superior standing of those who have already been exalted to enjoy the privileges and benefits of the exalted life in Christ.

For Paul, though, a high evaluation of wisdom and power represents a fundamental misunderstanding of the gospel. The gospel is not about human wisdom and human power, things that may be impressive and attractive by normal standards. Instead, and somewhat ironically, God works not through what appears to be wise and powerful but through what appears to be foolish and weak. What could be more (apparently) foolish and weak than the plan to save the world through a crucified man (1:18–25)? According to Paul's gospel, that is precisely what God has done; and by so doing, he has shown that human power and wisdom have no part to play in the salvation of the world. Paul goes on to note that the congregation as a whole, and he himself, are scarcely powerful and wise by normal standards (1:26–2:5). God does not work in human ways.

Paul points out that the very existence of several of the Corinthians' problems shows that the Corinthian believers have not been exalted to the heavenly heights. The "wise and powerful" leaders of the community, for example, have been unable to deal with the most rudimentary issues. They have not recognized how shameful it is for a man to sleep with his stepmother (5:1–3) or for others to visit prostitutes (6:15–20) or for others to rely on civil law courts instead of the "wise" judgment of those in the community (6:1–9). Moreover, by foolishly thinking that they are already exalted and ruling with Christ, these believers overlook the real and present dangers in their daily existence. They do not see that there are still evil forces in the world, which will infect the congregation if allowed to enter. They do not see, to take one of the most complicated of Paul's discussions, that if women fail to wear head coverings during church services they are susceptible to the invasion of evil angels who might pollute the entire body of believers (11:10; see Chapter 23); nor do they realize that those who have been united with Christ can infect the entire body when they become united with a prostitute (6:15–20).

In addition, the Corinthians' sense of self-exaltation, in Paul's judgment, has made them ultimately unconcerned about how to treat one another in this sinful and fallen world. Many have engaged in uncontrolled acts of ecstasy in their services of worship, prophesying and speaking in tongues not to benefit others who are in attendance but, in Paul's view, simply to elevate themselves in the eyes of others (chaps. 12–14). From their own vantage point they may have understood their worship activities as signs of their participation in the heavenly resurrected existence that was theirs in Christ. But Paul believes these

Figure 19.3 Painting of the Christian celebration of the Lord's Supper, from the catacomb of Priscilla (see 1 Cor 11:23–26).

activities reveal something else. Those who engage in them have forgotten that the Spirit gives gifts to members of the congregation so they can benefit and serve others, not exalt themselves (especially chap. 12). Anyone who has all of the gifts that can be given by the Spirit but who fails to love the brothers and sisters in Christ is still in total poverty. This is the message of 1 Corinthians 13, the famous "love chapter," which is a favorite passage even today, especially at Christian weddings. The passage, however, does not speak of love in the abstract, and certainly not to modern notions of sentiment and sexual passion. Specifically it is about the use of spiritual gifts in the church. If the gifts are not used to benefit others, then they are of no use.

Paul's notion that Christian love is to guide ethical behavior in this evil age explains a number of positions that he takes in this letter. One prominent example is his position on meat offered to idols. In rough outline, the historical situation is reasonably clear. Meat that was sold at the pagan temples could be purchased at a discount. We are not altogether certain why. Possibly the meat was considered as already used, since it had been offered to a god, or possibly it was left over from a pagan festival. In any event, some of the Corinthian Christians (those who were less educated, in the lower classes?) thought that to eat such meat was tantamount to sharing in idolatry; they would not touch it on any condition. Others (more highly educated, in the upper classes?) claimed superior knowledge in this case, pointing out that idols had no real existence since there were no gods other than the one true God. Eating such meat could therefore do no harm and could actually save on much needed resources.

Paul, oddly enough, agrees that the other gods don't exist, but he disagrees that it is proper to eat the meat (chaps. 8–9). His reasoning is that those who see a Christian eating such meat may be encouraged to do so themselves, even while thinking that the gods do exist. They would be encouraged, that is, to do something that they themselves think is wrong, and this could harm their conscience (8:7–10). Rather than behaving in ways that might eventually hurt somebody, then, believers should do everything to help others, even if it involves avoiding something that in itself is not wrong (8:11–13).

Ultimately, this is an apocalyptic view. The need to love one another and to behave in ways that are most useful to them is directly related to the fact that evil still prevails in this world. Since

Christians continue to live in an age dominated by the forces of evil, they are not yet exalted and are not altogether free to do whatever their superior knowledge permits them to do.

Paul's apocalyptic notions appear to affect his entire view of life in this world. In another example drawn from this letter, Paul maintains that married couples should not pretend that they already live as angels, "who neither marry nor are given in marriage" (to quote another famous person; see Mark 12:25). Sexual temptations are great in this age, and marriage is a legitimate way to overcome them in God's eyes. Spouses should therefore grant one another their conjugal rights (7:1–6). Those who are able to withstand such temptations, however—like Paul himself, who says that he has the "gift" (7:7)—should not go to the trouble of becoming married in the first place. In Paul's view, his generation is living at the very end of time, and much work needs to be done before Christ returns. Those who are married are obligated to take time for their spouses and tend to their needs; those who are not can be fully committed to Christ (7:25–38). Thus, it is better to remain single, but if one cannot stand the heat, it is better to marry than to burn (7:8–9).

In Sum: Paul's Gospel Message to the Corinthians

While we have not been able to explore the Corinthians' questions and problems or Paul's responses in depth, we have seen what the big problem was from Paul's perspective and how it manifested itself in so many ways in his Corinthian congregation. Overall, the message that Paul had for the Corinthians was not so different from the message that he had for the Thessalonians. Jesus was soon to return when God entered into judgment of this world. When he did so, his followers would experience a glorious salvation. Until then, however, believers were compelled to live in this world. Their exaltation was a future event, not a present reality, however much it was prefigured in their community, the church.

The church in Corinth appears not to have been a happy place. Paul saw a community that was divided against itself and that tolerated immoral and scandalous behavior while claiming (ironically, in Paul's eyes) to enjoy an exalted standing with Christ. One can sense Paul's exasperation and disbelief: You are living a heavenly existence? You??? Even more, one can sense his concern. This was a major church in his mission field that had by and large gone astray from the basic intent of his gospel message. He treated the Corinthians as friends (e.g., see the prescript and closing) but realized that he was at odds with a number of them on significant issues. As we are going to see, the situation did not much improve once they received his letter.

2 CORINTHIANS

One of the reasons that Paul's letters to the Corinthians are so fascinating is that they allow us to trace his relationship with the congregation over a period of time. In no other instance do we have undisputed letters addressed to the same community at different times (with the possible exception of the church in Philippi). Paul's relationship with the Corinthians continued to ebb and flow in light of events that transpired after the writing of 1 Corinthians. By the time he came to write 2 Corinthians his tone had changed, though his tune had not.

The Unity of the Letter

Paul's tone changes even within his second letter, and rather severely. Indeed, many scholars are convinced that 2 Corinthians does not represent a solitary letter that Paul sat down one day and wrote but a combination of at least two letters that he penned at different times for different occasions. According to this theory, someone else, possibly a member of the Corinthian congregation itself, later edited these letters with "scissors and paste." The result was one longer letter, possibly designed for broader circulation among Paul's churches.

When you read through the letter carefully yourself, you may be struck by the change of tone that begins with chapter 10 and continues to the end. In chapters 1–9 Paul appears to be on very good terms with this congregation. He is overflowing with joy for them, almost as much as he was for the Thessalonians, even though he acknowledges that their relationship has been more than a little stormy in the past (see especially 2:1–11 and 7:5–12). He gives us some of the details. Some time before (but after the writing of 1 Corinthians) he had paid a second visit to Corinth (the first being when he converted them; 1:19). For some undisclosed reason, over some undisclosed issue, someone in the congregation publicly insulted him and he departed in humiliation. He indicates that he had been one angry fellow when he left. Soon thereafter he wrote a harsh letter that caused him great pain, in which he upbraided the congregation severely for their conduct and views and threatened to come to them again in judgment. But now, just prior to the writing of 2 Corinthians itself (or at least prior to the writing of chaps. 1–9), the bearer of the painful letter, Titus, had returned and given him the good news that the Corinthians have repented of their poor judgment and behavior, disciplined the person who had caused Paul's pain, and committed themselves once more to Paul as their spiritual father in Christ (7:5–12).

Paul's reaction could not be more appreciative: "He [Titus] was consoled about you, as he told us of your longing, your mourning, your zeal for me, so that I rejoiced still more" (7:7). Thanks to this good news, Paul now bubbles with joy for their renewed relationship, despite the hardships that he himself continues to experience: "I often boast about you; I have great pride in you; I am filled with consolation; I am overjoyed in all our affliction" (7:4). Paul is writing this conciliatory letter to express his gratitude for their about-face (1:15–2:4) and to explain why he was not fickle when he changed his travel plans: he had chosen not to visit them a third time simply to avoid causing anyone any more pain (2:1–2).

But then, in chapters 10–13, everything seems to change, or rather, to revert. No longer is Paul joyful in this congregation that has returned to him. Now he is bitter and incensed that they have come to question his authority and to badmouth his person (10:2, 10–11);. He threatens to come to them a "third" time in judgment, in which he will not be lenient (13:1–2), and he warns the congregation against those who oppose him, newcomers in their midst whom he sarcastically calls "superapostles" (11:5). He admits that these superapostles can perform miraculous deeds and spectacular signs, but he nonetheless sees them as false apostles, ministers of Satan who prey on the minds of the Corinthians (11:12–14) and lead them into all sorts of disorder and disobedience (12:19–21).

Is it possible that Paul could gush with joy over this congregation and at the same time threaten fierce retribution against it? Of course it is possible, but it doesn't seem likely. How, then, might we explain this change of tone?

One detail of the summary above may have struck you: in chapters 10–13 Paul threatens to make a third visit in judgment against the congregation, whereas in chapters 1–9 he indicates that he had canceled his visit because he did not want to cause further pain. Indeed, he intimates that there was no longer any need to make it. The congregation received his angry and painful letter, and it had its desired effect (or Titus, the bearer of the letter, had this effect). They have come to grieve over how they mistreated him and have now returned to his good graces.

Based on the differences between the two parts of the letter, many scholars believe that chapters 10–13 represent a portion of the earlier "painful" letter mentioned in 2:4, that is, the letter that was written soon after Paul's public humiliation and before his reconciliation with the Corinthians, a reconciliation gratefully discussed in chapters 1–9. If so, then a later editor has combined the two letters by eliminating the closing of one of them (the "thankful" or "conciliatory" letter of chapters 1–9, which was written second) and the prescript of the other (the "painful" letter of chapters 10–13, written first). By doing so, the editor created one longer letter that embodies the ebb and flow of Paul's relationship with the Corinthians over a relatively long period of time. Some scholars go even further, and maintain that more than two letters are embodied here, based on the uneven flow of Paul's argument throughout chapters 1–9 (see box 19.3).

The History of Paul's Relationship with the Community

We can map out the history of Paul's interaction with the Corinthians in terms of a sequence of visits and letters. There is, of course, a good deal of information that we do not have; but what we do have, including the bits and pieces that come from 1 Corinthians, falls out along the following lines.

Paul's First Visit. This was when Paul and Silvanus and Timothy first arrived in Corinth, set up shop, preached the gospel, won a number of converts, and provided them with some rudimentary instruction before leaving for other areas ripe for mission (2 Cor 1:19).

Paul's First Letter. Paul evidently wrote a letter to the Corinthians that has been lost. He refers to it in 1 Corinthians 5.9. It appears to have dealt, at least in part, with ethical issues that had arisen in the community.

The Corinthians' First Letter to Paul. Some of the Corinthians, either in response to Paul's first letter or independently of it, wrote Paul to inquire further about ethical matters, for example, about whether Christians should have sex with their spouses (1 Cor 7:1).

Paul's Second Letter: 1 Corinthians. In response to the Corinthians' queries and in reaction to information that he received from "Chloe's people," Paul wrote 1 Corinthians from Ephesus. In it he announced his plans to travel through Macedonia south to Corinth, where he hoped to spend the winter (1 Cor 16:5–7). He apparently sent the letter back with Stephanas and his two companions, who were members of the Corinthian church (1 Cor 16:15–17).

Paul's Second Visit. In 2 Cor 2:1–4 Paul indicates that he does not want to make "another" painful visit; this suggests that his most recent visit had been painful. It appears, then, that after the writing of 1 Corinthians, Paul fulfilled his promise to come to Corinth for a second time. But he was not well received. Someone in the congregation did something to cause him pain and possibly public humiliation (2 Cor 2:5–11). He left, uttering dire threats that he would return in judgment against them (2 Cor 13:2).

The Arrival of the Superapostles. Either prior to Paul's departure or soon thereafter, other apostles of Christ arrived in town, claiming to be true spokespersons of the gospel. These "superapostles" (as Paul calls them; 2 Cor 11:5) were of Jewish ancestry (11:22) and appear to have appealed precisely to that aspect of the Corinthians' views that Paul found most repugnant, their notion that life in Christ was already an exalted, glorified existence. For these superapostles it was; that was why they could do the spectacular deeds that established their credentials as apostles. Clearly they and Paul did not see eye to eye. At some point the attacks became personal: the superapostles evidently maligned Paul for his clear lack of power and charismatic presence ("his bodily presence is weak and his speech contemptible," 10:10); he in turn claimed that they were ministers of Satan rather than apostles of Christ (11:13–15). Paul argued that his gospel message would be totally compromised if the Corinthians accepted the claims of his opponents (11:4).

Paul's Third Letter (the Painful letter, partly embodied in 2 Cor 10–13). After his second visit, Paul wrote a letter in which he went on the attack against the superapostles. He continued to insist that the life of the believer is not the glorified, exalted existence that Christ presently enjoys. Believers live in an age of evil and suffering, in which God's enemy Satan is still active and in control; those who boast of their power and wisdom do not understand that the end has not yet come, that this is an age of weakness in which God's wisdom appears foolish. Apostles, in particular, suffer in this age, since they are the chief opponents of the cosmic powers of evil who are in charge (11:20–31). Even though apostles may have had a glimpse of the glory to come (12:1–4), they are still subject to pain and suffering, which keeps them from boasting of their own merits and forces them to rely totally on the grace of God for what they can accomplish (12:5–10). In light of these criteria, the superapostles are not apostles at all. Paul also used this letter to attack the person who had publicly humiliated him and to warn the congregation to deal with him prior to his arrival in judgment, for Paul himself would not be lenient when he came (13:1–2).

Part of this letter, principally the part that dealt with the superapostles, is found in what is now 2 Corinthians 10–13. The letter was sent with Paul's companion Titus, and it evidently had its desired

SOMETHING TO THINK ABOUT

Box 19.3 The Partitioning of 2 Corinthians

A number of New Testament scholars believe that 2 Corinthians comprises not just two of Paul's letters but four or five of them, all edited together into one larger composition for distribution among the Pauline churches. Most of the "partition theories," as they are called (since they partition the one letter into a number of others), maintain that chapters 1–9 are not a unity but are made up of several letters spliced together. Read the chapters for yourself and answer the following questions:

- Does the beginning of chapter 8 appear to shift abruptly to a new subject, away from the good news Titus has just brought Paul (about the reconciliatory attitude of the Corinthians) to Paul's decision to send Titus to collect money for the needy among the Christians? There is no transition to this new subject, and 8:1 sounds like the beginning of the body of a letter. Could it have been taken from a different writing?
- Do the words of 9:1 seem strange after what Paul has said in all of chapter 8? He has been talking for twenty-four verses about the collection for the saints, and then in 9:1 he begins to talk about it again as if it were a new subject that had not yet been broached. Could chapter 9 also, then, have come from a separate letter?
- Does the paragraph found in 6:14–7:1 seem odd in its context? The verse immediately preceding it (6:13) urges the Corinthians to be open to Paul, as does the verse immediately following it (7:2). But the paragraph itself is on an entirely different and unannounced topic: Christians should not associate with nonbelievers. Moreover, there are aspects of this passage that appear unlike anything Paul himself says anywhere else in his writings. Nowhere else, for example, does he call the Devil "Beliar" (v. 15). Has this passage come from some other piece of correpondence (possibly one that Paul didn't write) and been inserted in the midst of Paul's warm admonition to the Corinthians to think kindly of him?

If you answered yes to all three of these questions, then you agree with those scholars who see fragments of at least five letters in 2 Corinthians: (a) 1:1–6:13; 7:2–16 (part of the conciliatory letter); (b) 6:14–7:1 (part of a non-Pauline letter?); (c) 8:1–24 (a letter for the collection, to the Corinthians) (d) 9:1–15 (a letter for the collection, to some other church?); and (e) 10:1–13:13 (part of the painful letter).

affect. The Corinthians punished the one who had insulted Paul (2 Cor 2:5–11), repented of the pain they had caused him, and returned to his fold (2 Cor 7:5–12). Paul in the meantime canceled his plan to make another visit to the congregation (2 Cor 1:15–2:2).

Paul's Fourth Letter (the Conciliatory letter, partly embodied in 2 Cor 1–9). After hearing the good news from Titus, Paul wrote a friendly letter to express his pleasure at the Corinthians' change of heart (2 Cor 2:5–11; 7:5–16). He also wanted to explain why he had not come for another visit, to assure them that he was not simply being fickle in making and revising his plans (1:15–2:4). Part of this letter (without, at least, its closing) is found in 2 Corinthians 1–9, or possibly only chapters 1–7, since some scholars think that chapters 8–9 are part of another letter, or possibly even two letters (see box 19.3).

The Overarching Points of the Letter

After someone edited the two (or three or four or five) letters into the one book that we call 2 Corinthians, we lose sight of Paul's relationship with this congregation. Thus, we can never know whether all the problems were solved, or whether any more stormy incidents occurred. Nor can we determine whether the Corinthians decided to adopt Paul's point of view and reject the perspectives brought in by others from the outside.

Clearly, though, the basic message that Paul tried to convey in 1 Corinthians is very much in evidence in the collection of letters we are investigating here. Consider first the fragment of the painful letter (chaps. 10–13), written in part to address the claims of superiority made by the superapostles. Rather than simply attacking them on their own terms, for example, by arguing that he could do better miracles than they, Paul dismisses their very grounds for considering themselves apostles. This is reminiscent of the way he treated the leaders of the divisive factions in 1 Corinthians 1–4, where he denies that earthly wisdom and power are signs of the divine. For him, the credentials of an apostle are not the glorious acts that he or she can perform, as if this were an age of exaltation and splendor. The true apostle will suffer, much as Christ suffered. For the end has not yet come, and those who rely on spectacular acts of power in this age must be suspected of collusion with the cosmic forces that are in charge of this age, namely, Satan and his vile servants (11:12–15).

This is why Paul goes to such lengths to "boast in his weaknesses" in this letter (12:5), principally by detailing all the ways that he has suffered as Christ's apostle (11:17–33). It may not seem like much to boast about—being beaten up regularly, living in constant danger and in fear for one's life—but for Paul these are signs that he is the true apostle of Christ, who himself suffered the ignominious fate of crucifixion. In particular, Paul claims that God has kept him weak so that he would be unable to boast about any work that he himself has performed. Anything good that comes of his ministry has necessarily been performed by God (12:6–10). The same cannot be said of the superapostles.

Paul's apocalyptic message stresses in the strongest terms that believers are not yet glorified with Christ. They live in a world of sin and evil and must contend with forces greater than themselves, until the end comes and Christ's followers are raised into immortal bodies to be exalted with him. For reasons that are ultimately unknown, the Corinthians came to agree with Paul on precisely this point. It is hard to imagine what changed their minds. Was Paul (or his representative Titus) simply too persuasive to refute? Were the superapostles discredited in some other way? We will never know.

We do know that after their reconciliation Paul wrote another letter in which, along with his gratitude for the church's change of heart, he expressed in somewhat more subdued fashion his basically apocalyptic view of life in this world. He begins the letter, now embodied in 2 Cor 1–9 (or 1–7), by stressing his own suffering and the grace of God that was manifest through it (1:3–11). This is to some extent the message of the entire epistle. The gospel is an invaluable treasure, even though it has not been fully manifested in this age of pain and suffering. The body has not yet been glorified and believers are not yet exalted. As a result, "we have this treasure in clay jars" (4:7). Believers themselves are lowly and their bodies of little worth, but the gospel message that they proclaim is a treasure for the ages. As Paul puts it later, in the body the believer groans, longing to be clothed with a heavenly, glorified body (5:1–10). The present age is therefore one of suffering and of longing for a better age to come.

With this longing, however, comes the assurance that in the future the hoped-for glory will become a reality for those who have been reconciled to God through Christ (5:16–21). Until this future reality makes itself known, life in this world is characterized by affliction and hardship. The suffering of the present age, however, is not enough to tarnish the hope of the true believer, for "this momentary affliction is preparing us for an eternal weight of glory beyond all measure" (4:17). This, above all else, is the apocalyptic message that Paul seeks to convey to his Corinthian converts.

GALATIANS

With the letter to the Galatians we enter into an entirely different set of issues from those evident so far of Paul's correspondence. On the one hand, there is no question concerning the unity of this epistle; it is just one letter, written completely at one time, to address one problem. But the problem itself was quite unlike anything that had arisen among the Thessalonians and the Corinthians. In brief, the occasion of the letter was as follows. After Paul converted a number of Gentiles to faith in Christ in the region of Galatia, other missionaries arrived on the scene, insisting that believers must follow parts of the Jewish Law in order to be fully right before God. Specifically, the men in these congregations had to accept the Jewish rite of circumcision.

Paul was absolutely outraged at this proposal. Whereas other apostles to the Gentiles may have looked upon circumcision as merely unnecessary, as a painful operation that Gentiles would have no reason to undergo unless they really wanted to, for Paul the matter was far more serious. For him, Gentiles who underwent circumcision showed a complete and absolute misunderstanding of the meaning of the gospel. In his view, for a Gentile to be circumcised was not simply a superfluous act; it was an affront to God and a rejection of the justification he has provided through Christ. Those who propose such a thing have perverted the gospel (1:7) and are cursed by God (1:8). Paul's anger in this letter is transparent at the outset. It is the only letter that he does not begin by thanking God for the congregation.

The Occasion and Purpose of the Epistle

Paul addresses the letter to "the churches of Galatia" (1:2). Unfortunately, we do not know, specifically, where the letter was sent. Before the Roman conquests, Galatia was a region in the north-central portion of Asia Minor (modern-day Turkey), a sparsely populated territory that was eventually linked by the Romans with the more populous region of the south, which included the cities of Lystra, Derbe, Iconium, and Pisidian Antioch. The Romans called this entire province Galatia, even though the name had earlier been used only to refer to its northern portion.

To what then, is Paul referring when he speaks of the churches of Galatia? Does he mean churches throughout the entire Roman province, comparable to the churches of Achaia and Macedonia that he refers to elsewhere (e.g., 1 Thess 1:7)? Or is he referring only to churches in the northernmost region, the region inhabited by people who would, unlike the southerners, refer to themselves as Galatians (see Gal 3:1)? The problem is complicated by the fact that the book of Acts indicates that Paul established churches in the southern region, in the cities that I have just named. Paul himself, however, never mentions these cities, in Galatians or anywhere else. Moreover, he claims that he founded the Galatian churches in somewhat unusual circumstances: he had taken seriously ill and was nursed back to health by the Galatians (at least by some of them). In this context, he preached the gospel and converted them (4:13–17). He does not appear, then, to have established these churches as he passed through the region preaching in the local synagogues, as is recorded in Acts.

Although we do not know to which churches Paul sent the letter, we do know that newcomers had arrived in Galatia preaching a gospel that Paul sees as standing at odds with his own, and the Galatian Christians appear to have been persuaded by them (1:6–9). We cannot be certain what these opponents actually preached. All we have is Paul's description of their message, and we have no guarantee that he knows, understands, or presents it accurately. It is clear, however, that he sees as the major point of contention the newcomers' insistence that (male) Gentile converts to Christianity have to be circumcised in order to be fully right before God (see e.g., 5:2–6). Paul interprets his opponents to mean that a person has to perform the works prescribed by the Jewish Law to have salvation. This message is totally unaccept-

SOMETHING TO THINK ABOUT

Box 19.4 The Logic of the Opponents' Position in Galatia

Paul's Galatian opponents may well have appealed to the Jewish Scriptures to argue their position. For both Paul and his opponents, Gentiles had been allowed to enter into the covenant that God had made with the Jewish people. They too could stand in a unique relationship with this one who created the world and chose his people. But the Scriptures were quite clear concerning what this covenantal relationship had involved from the beginning, when God first established it with the father of the Jews, Abraham:

> God said to Abraham, "As for you, you shall keep my covenant, you and your offspring after you throughout their generations. This is my covenant which you shall keep, between me and you and your offspring after you: Every male among you shall be circumcised . . . including the slave born in your house and the one bought with your money from any foreigner who is not of your offspring. So shall my covenant be in your flesh an everlasting covenant. Any uncircumcised male who is not circumcised in the flesh of his foreskin shall be cut off from his people; he has broken my covenant." (Gen 17:9–14)

Paul's opponents may simply have argued that while the covenant was now open to all who believed in Christ, God had not rescinded the rules of the covenant itself: it was an "everlasting" covenant, that is, one that would not be changed. Those who wished to belong to it must be circumcised, as God had said from the very beginning.

able from his point of view. According to the gospel that he preaches—and this, as he points out, is the message that led the Galatians to faith in Christ in the first place—a person is "justified" (made right with God) not by doing the works of the Jewish Law but by having faith in Christ (2:16). In Paul's view, the newcomers' message completely contradicts his own.

What else might these newcomers have taught? It is possible that they actually took the offensive against Paul himself (or at least that he thought they did) by questioning not only his views but also his authorization to proclaim them. This would explain the opening part of Paul's response, in which he vehemently denies that he has perverted the message of the gospel that he received from the apostles who came before him (e.g., Jesus' disciples in Jerusalem), because in fact his message didn't come originally from these apostles, or from any human at all. It came from God, in a direct revelation. It is also possible that Paul's Galatian opponents insisted that their message was truer to the Scriptures than his; they may have argued that since the Jewish Bible portrays circumcision as the sign of the covenant, any man who wants to become a full member of this covenant must first be circumcised.

In basic outline, the message of Paul's Galatian opponents appears similar to that proclaimed by other early Christians. The implicit logic behind it may have been that God is totally consistent and does not "change the rules." This is the Jewish God who gave the Jewish Law, who sent the Jewish Jesus as the Jewish messiah to the Jewish people in fulfillment of the Jewish Scriptures. Those who want to enjoy the full benefits of salvation, according to this view, must obviously join the Jewish people by being circumcised if they are men and by practicing the Law whether they are men or women (see box 19.4).

Scholars dispute whether these newcomers were Jews from birth or Gentiles who had converted to Judaism. Galatians 5:12 may suggest the latter: Paul hopes that when they perform the operation of circumcision on themselves, the knife slips. In either case, they were almost certainly believers in Jesus who taught others to adhere to some, or all, of the dictates of the Jewish Law. Paul finds this view offensive both to his person (since his authority is being questioned) and to his message (since his gospel is being compromised).

Paul's Response

Paul begins to make his case against his opponents already in the prescript of his letter; he is an apostle who has been "sent neither by human commission nor from human authorities, but through Jesus Christ and God the Father" (1:1). That is to say, he neither dreamt up his apostolic mission nor received it from any other human. He has been commissioned by God himself. That this self-defense is occasioned by the Galatians' acceptance of a contrary message becomes clear as Paul moves into the body of the letter. Instead of thanking God for these churches, Paul begins with a rebuke: the Galatians have deserted God by adopting a gospel that differs from the one that Paul preached to them (1:6–9). Anyone who affirms a different gospel, however, stands under God's curse.

In this early stage of the letter, Paul does not indicate what this other gospel entails. He evidently can assume that the Galatians know perfectly well what he is referring to, even though we as outsiders do not find out until somewhat later. Rather than launching directly into a theological refutation, he begins his counterattack by raising the question of authorization. Quite apart from what his message is, what authority stands behind it? Did he invent his gospel message? Or did he receive it from someone else and then change some of its details? Paul insists that his message comes directly from a revelation of Christ. Consider the ominous implications: what if someone disagrees with it?

To establish his point, Paul devotes nearly two chapters to an autobiographical sketch of his earlier life. The sketch might seem odd to a reader who is familiar with Paul's general reluctance to reminisce about his past, but the autobiography bears directly on the question at hand, the reliability of his gospel message. It shows that "the gospel that was proclaimed by me is not of human origin; for I did not receive it from a human source, nor was I taught it, but I received it through a revelation of Jesus Christ" (1:11–12).

To demonstrate his point, Paul recounts his conversion, in which he switched from being a persecutor of the church to being a preacher of its gospel. This conversion occurred through a direct act of God, who "was pleased to reveal his Son to me, so that I might proclaim him among the Gentiles" (1:15). Thus, the revelation of who Jesus really was, as opposed to who Paul had earlier thought he was, came directly from God and for a clear purpose: so Paul could take the message to the Gentiles, that is, to non-Jews like the Galatians.

This message was not given by the Jerusalem apostles or by anyone else: "I did not confer with any human being, nor did I go up to Jerusalem to those who were already apostles before me" (1:16–17; contrast Acts 9:19–30). Why is Paul so emphatic on this point? It may be that he suspects that his Galatian opponents have claimed that he modified the gospel that he originally learned from Jesus' earliest followers, the Jerusalem apostles. If so, then his autobiographical sketch shows that the claim is simply not true ("before God, I do not lie!" 1:20). On the other hand, he may know that his opponents have claimed superior authorization for themselves, by pointing to the Jerusalem apostles as the source of their own message. If so, then his sketch shows that whatever the source of his opponents' message, his own came straight from God.

To be sure, Paul does not deny that he has had some contact with the Jerusalem apostles. He admits that three years after his conversion (i.e., long after his views were set) he went to visit Cephas for fifteen days. He does not, however, indicate precisely why he went. Indeed, the term that he uses, which is sometimes simply translated as "to visit" (Gal 1:18), can mean either that he went "to learn something" or "to convey some information." It may be that he went to keep Cephas, the chief apostle in Jerusalem at the time, apprised of his actions (see box 19.5).

SOMETHING TO THINK ABOUT

Box 19.5 Cephas and Peter

Most people naturally assume that when Paul says that he went to Jerusalem to visit Cephas, he is referring to Peter, Jesus' closest disciple. This is because neither "Peter" nor "Cephas" was a proper name in the Greco-Roman world, but both are translations of the word "rock" ("Peter" is Greek and "Cephas" is Aramaic). Moreover, according to the Fourth Gospel, this word was the nickname (something like our modern name "Rocky") that Jesus bestowed upon his disciple Simon (John 1:41).

A number of Christian authors from the second to the eighth centuries, however, believed that there were two different persons, one named Peter and the other Cephas, that is, two important followers of Jesus who shared the same unusual nickname. If this ancient tradition is right, then Peter would have been Jesus' original disciple and Cephas would have been the leader of the church in Jerusalem some years later.

Could this tradition be historically accurate? Interestingly, the only surviving author from antiquity who was personally acquainted with Cephas was the apostle Paul. Judge for yourself: when Paul speaks about Cephas, does he mean Peter the disciple? Look especially at Galatians 2:6–9 where he mentions both names, in the same breath, without indicating that he is referring to the same person. Indeed, he appears to assume that these two persons are engaged in two different kinds of activity, Cephas the head of the Jerusalem church and Peter the missionary to the Jews. It may indeed be that Paul (and his Galatian readers) knew two different apostles who went by similar nicknames—Cephas a resident of Jerusalem, who converted to faith in Jesus after seeing him raised from the dead (1 Cor 15:5) and who became prominent among the apostles (like James the brother of Jesus, who is also mentioned in these verses), and Peter, Jesus' disciple who was engaged in missionary work outside of Jerusalem. If so, then Paul did not go to Jerusalem to learn something about the historical Jesus from his closest disciple Peter; he went to consult with the leader of the Jerusalem church, Cephas.

Some fourteen years later Paul met with a larger group of apostles for a similar reason, to inform them of his missionary activities (2:1–10). It was his second trip to Jerusalem (in the book of Acts it happens to be his third), and it represented a critical moment for the Gentile mission. One does not get the sense from Paul that he made this second visit because he wanted to make sure that his gospel message was right, as if he could imagine it being wrong! (Remember, he claimed to have received it from God himself.) Instead, Paul went to convince the Jerusalem apostles that Gentiles were not required to follow the Jewish Law, including circumcision, the "sign of the covenant," in order to be right with God, or "justified" (2:1–5). He met with the leaders privately to persuade them of his views (2:2), and he succeeded without qualification (2:7–10), even though others were present who argued the alternative perspective. Paul calls these other people "false believers" (2:4) and sees them as the predecessors of his opponents in Galatia.

The important point for Paul is that the Jerusalem apostles agreed with him rather than with his adversaries at the conference. Even though these apostles were committed to evangelizing Jews (2:7–9), they conceded that there was no need for Gentile converts to be circumcised. Emblematic of this decision was the fate of the Gentile Titus, who accompanied Paul to the conference and who was not compelled to be circumcised by those who took the opposing perspective (2:3–4). By securing this agreement with the Jerusalem apostles, Paul could rest assured that they would give his mission their full blessing and not try to undermine it. In his words, he knew that he "was not running, or had not run, in vain" (2:2).

Paul provides one other autobiographical detail to secure his point. After his meeting with the Jerusalem apostles, one of them, Cephas, came to spend time with him and his church in Antioch. At first, Cephas joined with Paul and the other Christians of Jewish background in sharing "table fellowship" with the Gentile believers ("he used to eat with the Gentiles"; 2:11–12). But when representatives of the apostle James, the brother of Jesus, arrived on the scene, Cephas withdrew from fellowship with the Gentiles, and the other Jewish-Christians joined with him (2:12–13). Paul saw this withdrawal as an act of hypocrisy and openly rebuked Cephas for it. In Paul's view, Cephas had compromised the earlier decision not to compel Gentiles to obey Jewish laws (2:14).

Scholars have different opinions concerning what this conflict was all about. It may be best to assume that eating with the Gentiles somehow required Cephas and his Jewish-Christian companions to violate kosher food laws. They may have thought that this was acceptable so long as they gave no offense to other believers, but when the representatives of James, that is, Jewish-Christians who perhaps continued to keep kosher, came to town, Cephas and his companions realized that they had to decide with whom they were going to eat. They chose not to give offense to their Jewish brothers and sisters and so ate with them.

For Paul, this was an absolute affront because it suggested that there was a distinction between Jew and Gentile before God, whereas the agreement that had been struck in Jerusalem maintained that there was not. Jew and Gentile were on equal footing before God, and any attempt to suggest Jewish superiority was a compromise of the gospel.

We do not know the outcome of this confrontation, in part because we never hear Cephas's side of the argument. Paul's narration of the incident is important, though, because it introduces the issue that the letter is ultimately about: the relationship of Paul's gospel message to the Jewish Law (2:15). At this stage, Paul begins to mount theoretical and scriptural arguments to show that the Jewish Law has no role in a person's right standing before God and that, as a consequence, his opponents in Galatia are in error not only for doubting his authorization but also for perverting his gospel. These arguments are somewhat intricate, so here I will simply summarize some of the salient points.

What Was the Basic Issue? Paul begins in 2:15–21 with a forceful expression of his views. Even as a good Jew himself, he has come to realize that a person's right standing ("justification") before God does not come through doing the works of the Jewish Law but through faith in Christ (2:16). If a person could be made right with God through the Law, then there would have been no reason for Christ to die (2:21).

Not only is this the right way to understand the Law, according to Paul, it is also the message that the Law itself teaches. Now that he has come to grasp this message of the Law, he can say that "through the Law I died to the Law" (2:19). This is a difficult saying, which might be paraphrased as follows: "Through the correct understanding of the Law that the Law itself has provided, I have given up on the Law as a way of attaining a right standing before God." Once the Law is abandoned as a way to God, then, no one should pretend that it affects one's standing before God; or to use Paul's image, it is wrong to "build up" the importance of the Law for salvation once its importance has already been "torn down" (2:18).

The matter is significant because the Galatians, former pagans who converted to faith in Christ, have begun to adopt the view that Paul opposes, namely that doing works of the Law (in particular, circumcision) is important for one's standing before God. Paul is incensed and incredulous: "You foolish Galatians! Who bewitched you? . . . Did you receive the Spirit by doing the works of the law or by believing what you heard?" (3:1).

Why Does Paul Appeal to the Law to Dispute this View of the Law? One of the most striking things about Paul's response to the Galatians' situation is that he bases a good deal of his argument against his opponents' emphasis on the Law on a careful interpretation of the Jewish Scriptures themselves. This approach may seem ironic to an outside reader—Paul is citing the Jewish Law in order to show that the Law is to play no role in a person's standing before God! For Paul, though, this line of argument is completely sensible. He maintains that the Scriptures themselves teach that the Law was not given in order to bring about a right standing before God. From the very beginning, people have been made right with God by faith, starting with the father of the Jews, Abraham himself, in Genesis, the first book of the Law. For Paul, the true children of Abraham are those who have faith, just as Abraham had faith—whether they are Jews who have the Law or Gentiles who don't (3:6–9).

It is also possible that Paul makes such a lengthy appeal to the teachings of the Torah to show that he himself is quite capable when it comes to interpreting the Jewish Scriptures. Not only was he raised Jewish and zealous for Jewish traditions prior to his conversion to Christ (1:13–14), he continues to explore the Jewish Scriptures and is second to none in his ability to interpret them.

What is the Problem with Gentiles Keeping the Law? Paul claims that those who do not live by faith but by the Law, that is, those who try to attain a right standing before God by keeping the Law, are subject to God's curse rather than his blessing, despite their motivation and desire. On the one hand, the Torah itself curses those who do not "obey all the things written in the book of the law" (2:10). Paul does not explain why everyone is automatically put under this curse, but it may be because in his opinion no one ever does "obey all the things written in the law," as he indicates elsewhere (see Rom 3:9–20). Indeed, even though he does not explicitly mention this issue, Paul may be thinking that the Law itself demonstrates his point, since a good portion of the Torah is devoted to describing the sacrifices that have to be performed by all Jews, even the Jewish high priest, to atone for their sins when they inadvertently violate the Law. If one must obey all of the things in the Law or suffer its curse, and the Law itself indicates that no one does so, where then does that leave us? Clearly everyone who tries to obey the Law stands under the curse that the Law itself pronounces.

Moreover, and this point is more clearly expressed in the passage, the Law cannot place someone in a right standing before God because the Scriptures indicate that a person will find life through having faith (Hab 2:6, quoted in 3:11). Carrying out the Law, though, is not a matter of trusting God (faith); it is a matter of doing something (work). If faith is the way to life, then doing the Law will not satisfy the requirement. Only faith like the faith of Abraham, the father of all believers (not of Jews only), will put one in a right standing before God.

Why Then Did God Give the Law in the First Place? The question naturally arises, then, if practicing the Law does not put a person into a right standing before God, and it was never meant to do so, why was it ever given at all (3:19)? Paul's answer in 3:19–29 has caused interpreters difficulties over the years. Perhaps it is best to understand his comments to mean that the Law was given to provide instruction and guidance to the Jewish people, informing them of God's will and keeping them "in line" until God came to fulfill his promise to Abraham to "bless his offspring" (3:16). This fulfillment would come in Christ, who was himself the offspring of Abraham spoken of in the promise (3:16). Thus the Law served as a "disciplinarian" until the arrival of Christ; it is called a *paidogogos* (to use the Greek term), i.e., one who made sure the children kept on the straight and narrow until they reached maturity. At no point, though, was the Law meant to put a person into a right standing before God. It couldn't do so because justification comes through faith, not action.

Who Then are the True Descendants of Abraham? Paul understands that the Jews and Gentiles who have faith like that of Abraham are his true descendants, as opposed to unbelieving Jews who are simply his physical progeny. This perspective is especially clear in the allegory that Paul gives in 4:21–30. The allegory represents an original and intriguing interpretation of the story of Genesis 21. (You should read the story on your

Figure 19.4 God giving the Law to Moses, from a Hebrew manuscript of the Pentateuch from Bavaria. Unlike in this portrayal (and unlike in the book of Exodus itself), Paul claimed that the Law did not come directly from God but through angelic intermediaries, thereby lessening its divine character and eternal importance (Gal 3:19).

own before examining again Paul's interpretation of it.) In Paul's view, Abraham's son Isaac, born of the promise, represents the Christian church (i.e., all those who believe in God's promise), while his son Ishmael, born of the flesh, represents Jews who do not believe in Christ. In other words, those who have faith in Christ are the legitimate heirs of God's promise. Unbelieving Jews, on the other hand, are children born into slavery (since Hagar, the mother of Ishmael, was a slave). Those who submit to the Jewish Law apart from faith in Christ submit to a yoke of slavery; they correspond to the son of the slave woman. Those who do have faith will never submit to this yoke. An amazing interpretation this: Jews are not the children promised to Abraham, but Christians (whether Jews or Gentiles) are!

Doesn't This Teaching Lead to Lawlessness? Paul concludes this letter by addressing a problem that some might think is implicit in his teaching that all people, Jews and Gentiles, are made right with God through faith apart from performing the works of the Law. If the Law was given in order to provide direction and discipline to God's people, but Gentile believers don't have to keep it, aren't they liable to turn to wild and reckless behavior?

For Paul, nothing could be further from the truth. In perhaps one of the greatest ironies in his thinking, Paul indicates that Gentile believers in

Christ, who are not obligated to keep the Law (and therefore must not be circumcised) are to be totally committed to one another in love because in so doing, they fulfill the Law! Indeed, for Paul, Christians must be enslaved to one another in love (5:13) precisely because "the whole law is summed up in a single commandment, 'You shall love your neighbor as yourself'" (5:14).

His argument raises a number of tantalizing questions. First, how can Paul tell his converts not to follow the Law (You must not be circumcised) and then require them to follow it (You must love one another so as to fulfill the Law)? Evidently—although this is not a point that he makes explicit in any of his writings—Paul thinks that there are different kinds of laws provided in the Jewish Scriptures (compare what we found with respect to the Gospel of Matthew in Chapter 7). There are some laws that are distinctive to being Jewish. These would include circumcision and kosher food laws. Paul insisted that his Gentile converts not keep these laws: indeed he claims here in Galatians that those who do so "have cut yourselves off from Christ; you have fallen from grace" (5:4). At the same time, he urges his converts to keep the principle that summarizes the entire Torah; they should love their neighbors as themselves. It is hard to escape the conclusion that Paul saw some laws as distinctively Jewish (Be circumcised) and others as applicable to all people (Love your neighbor).

Paul seems to imply in Galatians 3, however, that no one is able to keep all of the laws (including, presumably, the law to love one's neighbor). How then can he insist that Christians fulfill the Law? Paul evidently believes that those who receive the Spirit of God through believing in Christ (3:1) are empowered by the Spirit to do what the Law commands. Indeed, their lives will bear fruit in ways that fulfill the law, and they will do those things that no law forbids (5:22–23). Those who do not have the Spirit on the other hand, that is, those who are not believers, are necessarily ruled by their flesh, and by nature engage in activities that are contrary to the Law and will of God (5:16–21). Such persons will never inherit the kingdom of God (5:21). Thus, perhaps ironically, those who have faith in Jesus, not those who are circumcised, are the ones who fulfill the righteous demands of God's Law.

In Sum: Paul and the Law

This question of the relationship of faith in Christ to the Jewish Law is one that continued to perplex Paul throughout his life. Indeed, it is one of the central questions that he had to address as an apostle of Christ, for he taught at one and the same time that Christ was the fulfillment of the Law and that believers did not have to perform the works of the Law—meaning, as we have seen, that they did not have to carry out those aspects of the Law that in outsiders, eyes made Jews Jewish. The question proved to be of ongoing importance because it related to larger ones that Paul's version of the gospel compelled him to address, including the questions of whether God had abandoned his people Israel by making faith in Christ the sole means of salvation and whether God had as a consequence proved himself to be unfaithful and untrustworthy by not staying true to his promise always to be the God of Israel. These are some of the issues that Paul would explore in the fuller, and somewhat less heated, exposition of his views of the gospel in his letter to the Romans (see Chapter 20).

PHILIPPIANS

We do not know very much about the Christian community in Philippi because Paul does not provide as many explicit reminders of their past relationship as he does, for example, for the Thessalonians and Corinthians. There is some information provided in Acts 16; unfortunately, little of it can be corroborated from Paul's letter itself. Paul never mentions, for example, the principal characters of Luke's account, Lydia and the Philippian jailer.

The city of Philippi was in eastern Macedonia, northeast of Thessalonica, along one of the major trade routes through the region. Paul speaks in 1 Thessalonians of being shamefully treated in Philippi prior to taking his mission to Thessalonica (1 Thess 2:1–2). We should probably assume that he is referring to his initial visit to the city when he founded the church there. In view of their rough treatment, Paul and his companions may not have spent much time there, perhaps only enough to make some converts, instruct them in the rudiments of the faith, and get out of town while the getting was good.

We have little information about the converts themselves. We can probably assume that the Philippian church, like the other congregations Paul established, consisted chiefly of converted pagans who had been taught to worship the one true God of Israel and to expect the return of his Son, Jesus. References to these teachings can be found throughout the epistle (e.g., 1:6, 10–11; 2:5–11; 3:20–21). Why, though, did Paul write it? The answer to this question is somewhat complicated, more complicated, for example, than in the case of Galatians, for it appears to many scholars that different parts of this letter presuppose different occasions. As was the case with 2 Corinthians, Philippians may represent a combination of two or more pieces of correspondence.

The Unity of the Letter

The first two chapters of Philippians sound very much like a friendship letter written by Paul to his converts. The occasion of the letter is reasonably evident (see especially 2:25–30). The Philippians had sent to Paul one of their stalwart members, a man named Epaphroditus, for some reason that is not disclosed (until chap. 4). While there ministering to Paul, Epaphroditus was taken ill; the Philippians had heard of his illness and grew concerned. Epaphroditus in turn learned of their concern and became distraught over the anxiety that he had caused. Fortunately, his health returned, and he was now set to make his journey back home to Philippi. Paul wrote this letter to keep the Philippians informed of his situation and to express his pleasure that all had turned out well.

Paul sent the letter from prison (1:7). We do not know where he is imprisoned or why, except that it was in connection with his preaching of the gospel. He uses the letter to comment on his adversity and to reassure his congregation that it has turned out for the good: as a result of his bonds, others have become emboldened to preach (1:12–18). Paul uses his own situation to explain that suffering is the destiny of Christians in the present age (1:29–30)—a message comparable to that which he proclaimed in the Corinthian correspondence. He continues by providing some general words of admonition (as was common in friendship letters): the Philippians are to be unified, serving one another rather than themselves, and thereby following the example of Christ (2:1–11).

One of the most striking features of this letter comes after these general exhortations. For the friendly and joyful tone that characterizes the letter's first two chapters shifts almost without warning at the beginning of chapter three. Indeed, if one didn't know that there were two more chapters left in the book, it would appear that the letter was drawing to a close at the end of chapter two. Paul has explained his own situation, given some admonitions, stated the purpose of his writing, and provided his concluding exhortation: "Finally, brothers and sisters, rejoice in the Lord" (3:1). Why does he say "finally" but then change the subject completely and continue writing for another two chapters? Indeed, the words that follow are hard to understand in the immediate context: "To write the same things to you is not troublesome to me, and for you it is a safeguard" (3:1). Why would anyone find his exhortation to rejoice troubling? Paul immediately launches into a vitriolic attack on people who are his enemies, presumably in Philippi,, people whom he calls "dogs," "evil workers," and "those who mutilate the flesh" (3:2). He then defends his own understanding of the gospel against these false teachers (3:3–11). A peaceful letter of friendship has now become a harsh letter of warning.

Moreover, the issue of unity within the Christian community takes on an additional twist in these chapters. We learn that there are two women in particular, Euodia and Syntyche, who are at odds with one another and causing something of a disturbance in the community (4:2–3). No longer does Paul deal in the abstract with the need for unity; now he actually puts some names on the problem. What is particularly interesting is that Epaphroditus is again mentioned in these closing chapters. If you didn't know better, though, you would think that he had just arrived, not that he had been with Paul already for an extended period of time (e.g., see 4:18, "I am fully satisfied, now that I have received from Epaphroditus the gifts you sent"). In any case, it is now clear why Epaphroditus has come and why

Paul is penning this letter. The Philippians have sent him to bring a financial contribution, and Paul is writing a thank-you note.

The timing of his response is puzzling. If Epaphroditus has been with Paul for such a long period of time—long enough for him to become deathly ill, for the Philippians to get word of it, for him to learn that they were distressed, and for him then to recover—why is Paul only now writing to tell them that he has received the gift? Surely he was in communication with them before this (since they have heard that Epaphroditus arrived and that he later became deathly ill).

Scholars differ on how to evaluate the various pieces of this contextual puzzle. One solution is that there are two or possibly even three letters that have been edited together here, letters that come from different times and were written for different occasions. For simplicity's sake, I'll assume that there are two letters and explain how the theory works.

After Paul established the Philippian church, he left to pursue his apostolic work elsewhere. We don't know exactly where he was when he was writing this letter, or series of letters (Rome? Ephesus?), only that he was in jail. The Philippians learned of his needs and sent him a gift of money through the agency of one of their leading members, Epaphroditus. Paul thankfully received the gift and learned (from Epaphroditus himself?) about two major problems in the community: some false teachers had begun to stress the need to keep the Jewish Law (see 3:3–6), and two women in the congregation had argued over something in public (4:2–3). He wrote the Philippians a letter, partially embodied now in chapters 3–4, thanking them for the gift, warning against the false teachers, and urging Euodia and Syntyche to get along.

After Paul sent this letter, Epaphroditus became ill, the Philippians learned of it and became concerned, Epaphroditus heard of their concern and became distraught, and finally he recovered. In the course of the communication that was obviously going back and forth, Paul learned of the improved situation in Philippi. When Epaphroditus became well enough to travel, Paul sent another letter back with him, a friendship letter explaining how things now fared with him and providing some renewed (but general) exhortations to the community to maintain their unity in Christ. Most of this letter is now found in Philippians 1–2. Some such scenario would explain why there are such differences between the first and second parts of the letter.

The Overarching Points of the Letter

Some of the issues that we have seen Paul address in other letters are found here as well. Throughout the Thessalonian and Corinthian correspondence, for example, we saw Paul emphasize that prior to the return of Christ in judgment suffering was the lot of the Christian. This is part and parcel of his apocalyptic message, that even though the powers of evil have begun to be defeated through the cross of Christ, the end has not yet come. This continues to be an age under the dominion of the cosmic powers opposed to God, and those who stand against them will bear the brunt of their wrath. Christians will necessarily suffer, but all will be redeemed when Christ returns. This message continues to find expression here in Philippians, where Paul again portrays himself as one who suffers for the sake of Christ (e.g., 1:7, 17), where he again emphasizes that it is the call of the Christian to suffer (1:29), and where he again stresses that at Christ's return all will be made right (3:20–21).

One other motif that holds the two parts of the letter together is the need for these Christians to maintain their unity by practicing self-giving love for one another. The message finds its most pointed expression in the request in chapter 4 for the two women Euodia and Syntiche to stop fighting, but it is expounded at greatest length in chapter 2. Here Paul recounts the actions of Christ on behalf of believers, in a passage that scholars have come to call the "Christ hymn" of Philippians (2:6–11; see box 19.6). This is one of the most poetic and beloved portions of all of Paul's letters; readers have long observed the striking cadences of the passage, its balanced rhythms and exalted views. It has all the marks of an early hymn sung in worship to Christ, and Paul quotes it in full because it makes an important point for his Philippian readers (cf. the prologue of the Fourth Gospel; see

SOME MORE INFORMATION

Box 19.6 The Christ Hymn of Philippians

One of the first things any pagan author said about the early Christians was that they "sang hymns to Christ as if to a god" (Pliny the Younger's Letter X to Trajan). Many scholars believe that several of the earliest hymns to Christ have been inserted by the authors of the New Testament in appropriate places of their writings (e.g., John 1:1–18). There are various ways to reconstruct the original form of the hymn that Paul appears to be citing in Philippians 2:6–11. The following reconstruction shows how the hymn can be broken down into two major parts, each comprising three fairly equally balanced stanzas of three lines each; the first indicates the progressive condescension (or self-humbling) of Christ, the second his subsequent exaltation by God.

The Condescension of Christ

Though he was in the form of God,
he did not regard equality with God
as something to be exploited.

But he emptied himself
taking the form of a slave
being born in human likeness.

And being found in human form,
he humbled himself,
and became obedient unto death.

The Exaltation of Christ

Therefore God also highly exalted him,
and gave him the name
that is above every name.

So that at the name of Jesus,
every knee should bend,
in heaven and on earth and under the earth.

And every tongue should confess
that Jesus Christ is Lord,
to the glory of God the Father.

Chapter 10). Even though many of the details of the hymn are hotly disputed, its basic message is reasonably clear. Rather than striving to be equal with God, Christ humbled himself, becoming human and submitting to a death on the cross. God responded to this humble act of obedience by exalting Christ above everything else in creation, making him the Lord of all.

Paul does not cite this hymn simply because it is a powerful and moving expression of the work of Christ. Rather, he uses it because Christ's humble obedience provides a model of action for his followers, who should also lower themselves for the sake of others (2:1–4). Rather than seeking their own good and working for their own glory, Christians should seek the good and work for the glory of others. You will notice that Christ is not the only example of self-giving, sacrificial love in this chapter. Paul also claims that he himself is willing to be sacrificed for his Philippian converts (2:17), that his companion Timothy seeks the interests of others rather than his own (2:19–24), and that their own Epaphroditus has risked everything for the sake of others (2:25–31). The Philippians are to follow these worthy examples, living in unity with one another through self-sacrificing love.

Whether this admonition had its desired effect or not is something we will probably never know. After this letter (or this sequence of letters), we hear nothing more from Paul of his relationship with his converts in Philippi.

PHILEMON

The letter to Philemon is a little gem hidden away in the inner recesses of the New Testament. Merely a single page in length, the size of an average Greco-Roman letter, it is the only undisputed epistle of Paul addressed to an individual. Rather than dealing with major crises that have arisen in the church, the letter concerns a single man, the runaway slave Onesimus, and his fate at the hands of his master, Philemon.

The Occasion and Purpose of the Letter

On first reading, there may be some confusion concerning the recipient of the letter since it is addressed to three individuals and a church: "To Philemon our dear friend and co-worker, to Apphia our sister, to Archippus our fellow soldier, and to the church in your house" (v. 2). It is clear, however, that the letter is really addressed to a solitary individual because Paul speaks to a single person in the body of the letter ("you" singular in Greek, starting with v. 4 and continuing through v. 24). Evidently, the principal recipient is Philemon, since he is the first one to be named, just as Paul names himself first as the sender of the letter, prior to mentioning his "co-author," Timothy.

Our only clues about who Philemon was come from the letter itself. To begin with, he must have been a relatively wealthy Christian. He had a private home large enough to accommodate a church (i.e., a private gathering of Christians) and he owned slaves. Moreover, he evidently had valuable property that could be stolen, as Paul thinks that Onesimus might have run off with some of it, or else embezzled some of the funds entrusted to his charge (v. 18). Tradition holds that Philemon was a leader of the church in the town of Colossae, an identification possibly suggested by the fact that in verse 23 Paul conveys greetings from Epaphras who, according to Colossians 4:12, was a member of that church (although many scholars doubt that Colossians was actually written by Paul).

Wherever Philemon was from, he appears to have stood in Paul's debt, as Paul not so subtly reminds him: "I say nothing about your owing me even your own self" (v. 19). (By claiming to say nothing about it, of course, Paul says all that needs to be said!) For this reason, it appears likely that Philemon was one of Paul's converts. Apart from these things, we cannot say much about the man himself. As for the occasion of Paul's letter to Philemon, we know that Paul writes from prison (v. 1). Again, we don't know where he is or why he is being punished; it does appear, though, that he anticipates being released (v. 22). While in prison, he met and converted Philemon's runaway slave Onesimus. When he speaks of Onesimus in verse 10 as one "whose father I have become," the Greek literally says "whom I begot"—the same phrase that Paul uses in 1 Cor 4:15 to refer to his converts in Corinth. The letter does not explicitly indicate whether Onesimus himself is imprisoned, for example, for having been caught in flight with some of his master's goods (v. 18), or whether he has come to visit Paul in jail as a friend of his master. The former option seems unlikely. The Roman empire was a big place, and to think that Paul and the slave of one of his converts just happened to end up in the same jail cell, whether in a major urban center like Ephesus or in a small rural village, simply defies the imagination. On the other hand, if Onesimus was trying to get away from his master, why would he have gone straight to see one of his friends?

Recent studies of ancient Roman slavery law may provide an answer to this question. It was a legally recognized practice for a slave who had incurred his or her master's wrath to flee to one of the master's trusted associates to plea for his intervention and protection. The associate then served as a kind of official mediator, who would try to smooth out differences that had arisen through misunderstanding or even malfeasance. Malfeasance appears to be the issue here.

A possible scenario, then, would be something like the following. Philemon's slave Onesimus has done something wrong, possible stealing from the household or incurring some other kind of financial loss for his master (v. 18). Rather than stand and face the consequences, he flees to Paul, the apostle who had converted his master to a new religion and who was therefore a known and respected authority for him. While visiting Paul, Onesimus himself becomes converted to faith in Christ, a conversion that proves convenient for the nasty little business at home: Paul can now urge Philemon to receive Onesimus back not only as a slave but as much more, as a brother in Christ (v. 16), one who has been "useful" to Paul and can now be "useful" to Philemon (v. 11). Here Paul is playing with words. Slaves were often given descriptive names, such as the Latin Fortunatus, which means "lucky," or Felix, which means "happy." The Greek name Onesimus means "useful."

In his mediatorial role, Paul urges Philemon not to punish his slave, who has now had a change of heart, and to charge the apostle himself with whatever debt he has incurred. Paul appears to know full well that Philemon will simply write off his loss, given the (spiritual) debt he owes him (vv. 18–19). But is this all that Paul wants Philemon to do? Scholars have long debated the real meaning of his request, some thinking that Paul wants Philemon to manumit Onesimus (i.e., release him from his slavery), and others that he more specifically wants him to free him to engage in missionary work. Unfortunately, there is little in the text that suggests either possibility. Even verse 16, which urges Philemon to receive Onesimus "no longer as a slave but . . . [as] a beloved brother," is concerned with how he reacts to this errant member of his household; it is not telling him to change his status. (Consider an analogy: if I were to say to a female acquaintance, "I love you not as a woman but as a friend," this would not be to deny her gender!) It may be that the modern abhorrence of slavery has led interpreters to find in Paul a man ahead of his time, who also opposed the practice.

Yet Paul may be asking for something else. He emphasizes that Onesimus has been useful to him and states quite plainly that even though he would like to retain his services he doesn't want to do so without the leave of his master (vv. 12–14). Moreover, at the end of his short letter he asks Philemon to provide him with some kind of additional benefit in light of his own debt to Paul (the word "this" in v. 20 is not found in Greek; literally the text says, "Yes, provide me with a benefit"). What exactly is Paul looking for? Although Paul says not a word about Onesimus being set free, it appears that he would like to have him sent back. Perhaps Paul is asking Philemon to present him with a gift in the person of Onesimus, the slave.

Insights into Paul's Apostolic Ministry

The short letter to Philemon can provide us with some important insights into Paul's view of his apostolic ministry. One thing to observe is Paul's reciprocal relationship with his converts in this letter. In his other letters, he occasionally appears to be the all-knowing and all-powerful apostle, who makes his demands and expects people to follow them. On certain points that he feels strongly about, such as what his congregations believes about his apocalyptic message and how they treat the Jewish Law, he is altogether adamant. But on other issues he falls short of making demands. In the present instance, he expresses his desire as a request, although, to be sure, he phrases it in such a way that it would seem impossible for Philemon to turn him down. Even here, that is, while claiming not to assert his apostolic authority, Paul in fact appears to be doing so (cf. vv. 17–19).

A more important point to be gleaned from this letter relates specifically to its subject matter. It may come as a shock to modern readers that Paul did not use this occasion to lambaste the evils of the institution of slavery. Not only does Paul fail to condemn slavery in general, but he does not denounce its practice among Christians in particular. He never commands his convert Philemon to manumit his brother in Christ, Onesimus, let alone set free all of his other slaves. Was Paul not concerned for the plight of the oppressed?

Throughout his letters Paul shows a remarkable lack of concern for the social inequities of his world (a lack, that is, from a modern perspective). Despite his views that all people are equal in Christ—Jew and Gentile, slave and free, men and women (Gal 3:28)—Paul evidently did not see the need to implement this egalitarian ideal in the workings of society at large. He maintained that slaves should stay enslaved, that men should continue to dominate women, and that Christians as a whole should stay in whatever social roles they find themselves (see especially 1 Cor 7:17–24). But isn't this a bit short-sighted?

For us today it may indeed appear short-sighted, but for Paul it was based on the long view. For this evident lack of concern for a person's standing in society was related to his notion that the history of the world as we know it was soon going to come to a crashing halt when God entered into judgment with it. Soon the wrath of God would strike, annihilating the forces of evil and bringing in his kingdom, in which there would be no more pain or suffering or injustice or inequity. The equality that Paul sought was not one to be effected by social change; it was one to be brought by God himself, when he destroyed this evil age and set up his kingdom on earth. Little did Paul know that the faithful would still be around some nineteen centuries later to ponder his words.

SUGGESTIONS FOR FURTHER READING

See the suggestions at the end of Chapter 17.

CHAPTER 20

The Gospel according to Paul: The Letter to the Romans

No book of the New Testament has proven to be more influential in the history of Christian thought than Paul's letter to the Romans. One of the most frequently quoted pieces of Christian literature during the early centuries of the church, it was awarded pride of place in the orthodox canon of Scripture as the first, and longest, of Paul's epistles. At the end of the fourth century it was instrumental in the conversion of Saint Augustine, a man whose own writings, based in large measure on his understanding of Romans, shaped the thinking of theologians throughout the Middle Ages. It stood at the center of the debates between Protestants and Catholics during the sixteenth-century Reformation, when Protestant leaders such as Martin Luther, Philip Melanchton, and John Calvin saw it as the clearest exposition of Christian doctrine in the writings of the apostles. The book continues to influence and inspire Christian readers in many lands and many languages, theologians and laypeople alike, who cherish its words and puzzle over their meaning.

What, then, is this book that has inspired so much reflection and spawned so much controversy? The short answer is that it is a letter by Paul to the Christian congregation in Rome. The historian who comes to the task of interpreting this letter cannot allow him- or herself to be so overawed by its historical significance as to lose sight of this simple fact. This was a letter that Paul wrote to a particular church. As with all of his letters, this one had an occasion and was written for a reason.

THE OCCASION AND PURPOSE OF THE LETTER

In one important respect the letter to the Romans is unlike all of Paul's other letters: it is written to a congregation that Paul did not establish, in a city that he had never visited (see 1:10–15). Given what we have already seen about Paul's sense of his apostolic mission, this circumstance should give us pause. Paul's other letters were written to deal with problems that had arisen among those whom he had converted to faith in Christ. That clearly is not the case here (see box 20.1)

Even more strikingly, Paul does not appear to be writing to resolve problems that he has heard about within the Roman church. The issues that he discusses appear to relate instead to his own understanding of the Christian gospel. This is clearly the case in chapters 1–11, but even his exhortations in chapters 12–15 are general in nature, not explicitly directed to problems specific to the Christians in Rome. Nowhere, for example, does he indicate that he has learned of their struggles and that he is writing to convey his apostolic advice (contrast all of his other letters). Possibly, then, he simply wants to expound some of his views and explain why he holds them, but why would he want to do so for a church that he has never seen?

There may be some clues concerning Paul's motivation at the beginning and end of the letter. At the outset he states that he is eager to visit the church to share his gospel with them (1:10–15).

One might think that Paul is preparing the Romans for his visit, giving them advance notice about what he is up to, but at the end of the letter a fuller agenda becomes more evident. In his closing, Paul indicates that he has completed the work that he has to do where he is—probably Achaia (in Corinth itself?), since according to 16:1 the person carrying the letter, Phoebe, is a deacon of the church in Cenchreae, Corinth's nearby port. Moreover, he says he is eager to extend his mission into the western regions, specifically Spain, and wants to visit Rome on the way:

> But now, with no further place for me in these regions, I desire, as I have for many years, to come to you when I go to Spain. For I do hope to see you on my journey and to be sent on by you, once I have enjoyed your company for a little while. (15:23–24)

In light of these comments, it appears that Paul is interested in more than simply meeting with the Roman Christians. He evidently wants them to provide support, moral and financial, for his westward mission; possibly he would like to use Rome as the base of his operation to the regions

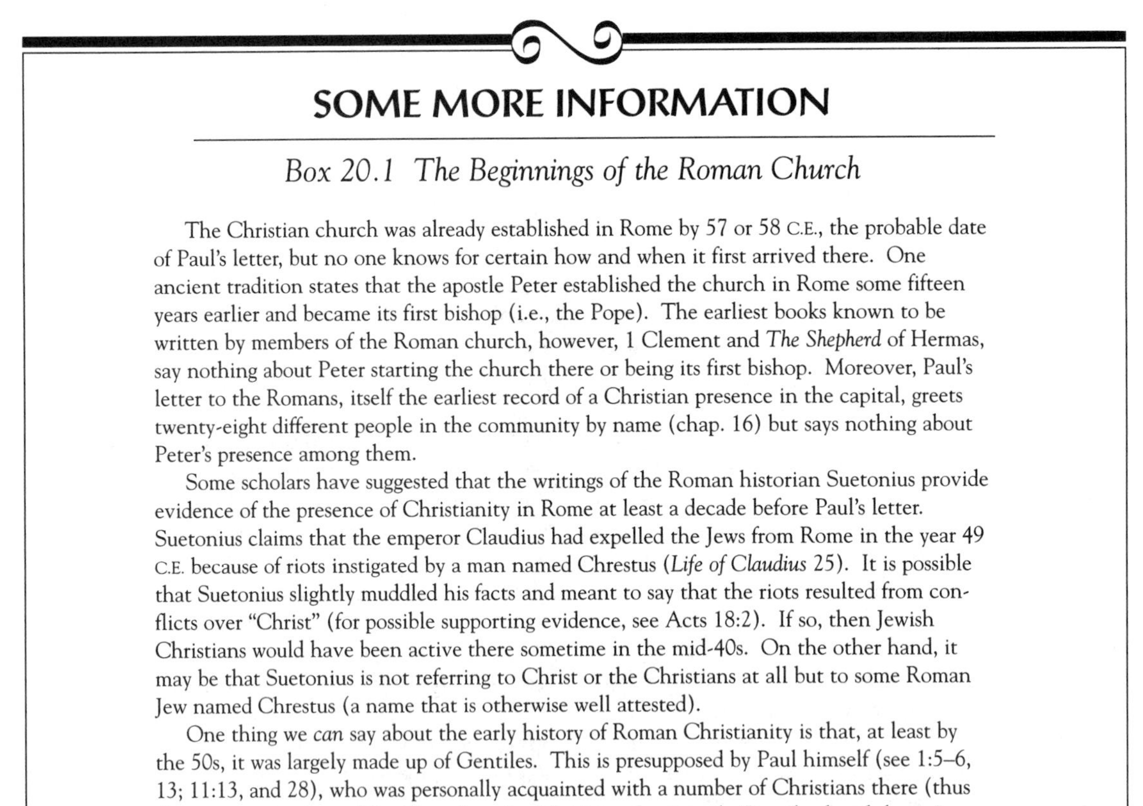

SOME MORE INFORMATION

Box 20.1 The Beginnings of the Roman Church

The Christian church was already established in Rome by 57 or 58 C.E., the probable date of Paul's letter, but no one knows for certain how and when it first arrived there. One ancient tradition states that the apostle Peter established the church in Rome some fifteen years earlier and became its first bishop (i.e., the Pope). The earliest books known to be written by members of the Roman church, however, 1 Clement and *The Shepherd* of Hermas, say nothing about Peter starting the church there or being its first bishop. Moreover, Paul's letter to the Romans, itself the earliest record of a Christian presence in the capital, greets twenty-eight different people in the community by name (chap. 16) but says nothing about Peter's presence among them.

Some scholars have suggested that the writings of the Roman historian Suetonius provide evidence of the presence of Christianity in Rome at least a decade before Paul's letter. Suetonius claims that the emperor Claudius had expelled the Jews from Rome in the year 49 C.E. because of riots instigated by a man named Chrestus (*Life of Claudius* 25). It is possible that Suetonius slightly muddled his facts and meant to say that the riots resulted from conflicts over "Christ" (for possible supporting evidence, see Acts 18:2). If so, then Jewish Christians would have been active there sometime in the mid-40s. On the other hand, it may be that Suetonius is not referring to Christ or the Christians at all but to some Roman Jew named Chrestus (a name that is otherwise well attested).

One thing we *can* say about the early history of Roman Christianity is that, at least by the 50s, it was largely made up of Gentiles. This is presupposed by Paul himself (see 1:5–6, 13; 11:13, and 28), who was personally acquainted with a number of Christians there (thus the greetings in chap. 16). How, though, did this predominantly Gentile church begin? Most scholars, realizing that we can never know for certain, simply assume that Christianity was brought to the imperial capital either by travelers who had converted to the faith while abroad (e.g., see, Acts 2:8–12), or by Christians who decided for one reason or another to relocate there, or by another missionary.

beyond. But why would he need to provide such a lengthy exposition of his views in order to get their support? Don't they already know who he is—the apostle to the Gentiles? And wouldn't they readily undertake to provide him with whatever assistance is needed?

Paul's lengthy discourse suggests that the Romans have only a dim knowledge of who he is or, even more likely, that they have heard a great deal about him and what they have heard has made them suspicious. If this is the case, or at least if Paul believes that it is, then presumably their suspicions would relate to the issues that Paul addresses throughout the letter, issues such as whether Gentiles and Jews can really be thought of as equal before God, and, if they can, (a) whether God has forsaken his promises that the Jews would be his special people and (b) whether Paul's "law-free gospel" to the Gentiles leads to lawless and immoral behavior (cf. Galatians).

The tone and style of this letter support the view that Paul wrote it to explain himself to a congregation whose assistance he was eager to receive. When reading through Romans carefully, one gets the sense that Paul is constantly having to defend himself and to justify his views by making careful and reasoned arguments (e.g., see 3:8; 6:1, 15; 7:1). Moreover, he makes this defense in a neatly crafted way, following a rhetorical style known in antiquity as the "diatribe." This involved advancing an argument by stating a thesis, having an imaginary opponent raise possible objections to it, and then providing answers to these objections. Consider the following rhetorical questions and answers:

> Then what advantage has the Jew? Or what is the value of circumcision? Much in every way. For in the first place the Jews were entrusted with the oracles of God. (3:1–2)

> What then? Are we any better off? No, not at all; for we have already charged that all, both Jews and Greeks, are under the power of sin. (3:9)

> What then are we to say? Should we continue in sin in order that grace may abound? By no means! How can we who died to sin go on living in it? (6:1–2)

Since the author both asks and answers the questions, the diatribe is remarkably effective in showing that he knows what he is talking about and that he is always right. By employing this style, Paul could effectively counter arguments that others had made against his teachings.

It should be noted that Paul's travel plans include not only the trip through Rome to Spain but an earlier jaunt to Jerusalem. Paul has collected funds for the poor Christians of Judea from his Gentile converts in Macedonia and Achaia (15:25–27) and appears uneasy over his upcoming trip to deliver them (15:30–32). He is quite openly fearful of "unbelievers" in Judea (presumably Jews who don't take kindly to his faith in Jesus) and apprehensive of his reception by the "saints" (presumably Jewish-Christians who have not warmed to his law-free gospel to the Gentiles). Some scholars have suspected that his letter to the Romans is a kind of trial run of presenting his views, an attempt to get his thoughts organized on paper before having to present them to a hostile audience in Judea.

There may be some truth in this, but chiefly the letter appears to be directed to the situation that Paul expects to find where he addresses it, in Rome. He wants to use this church as his base of operation and knows (or thinks) that he has some opposition. He writes a letter to persuade this congregation of the truth of his version of the gospel. This gospel insists that Jews and Gentiles are on equal footing before God: both are equally alienated from God and both can be made right with God only through Christ's death and resurrection. Moreover, the salvation that is offered in Christ comes to people apart from adherence to the Jewish Law, even though the Law itself bears witness to this faith as the only means of salvation. Indeed, Christ is the goal of this Law. Above all else, the gospel shows that God has not gone back on his promises to the Jews and has not rejected them as his people. In Christ, all of the promises of God have come to fruition. Furthermore, the Romans can rest assured that this gospel does not lead to moral laxity: Paul is himself no moral reprobate and he does not urge his converts to engage in wild and lawless activities.

Figure 20.1 Reconstruction of central city Rome, roughly as it would have looked in Paul's day.

THE THEME OF THE EPISTLE

Paul begins his letter to the Romans in his usual way, with a prescript naming and describing himself and his addressees, in which he anticipates the central concern of his letter, the meaning of his gospel (1:1–7; see box 20.2). The prescript is followed by a thanksgiving to God for this congregation (1:8–15), in which he announces his plans to visit the congregation order to share his gospel with them. Paul then gives a brief delineation of his gospel in two verses that scholars have long recognized as setting out the theme of the epistle:

> For I am not ashamed of the gospel; it is the power of God for salvation to everyone who has faith, to the Jew first and also to the Greek. For in it the righteousness of God is revealed through faith for faith; as it is written, 'The one who is righteous will live by faith.'" (1:16–17)

As he is occasionally wont to do, Paul has packed a great deal into these two verses. To help us understand the letter as a whole we should spend a few moments unpacking them.

1. **Paul is not ashamed of the gospel.** Paul may be writing the Romans to provide a relatively full and accurate account of the gospel message that he proclaims, perhaps in light of the partial and inaccurate report that he suspects they have already heard. He begins by assuring them that this message brings him no shame.
2. **Paul's gospel is God's powerful means of salvation.** The gospel that Paul preaches represents God's powerful act of salvation to the world, it is the way God has chosen to save those who are headed for destruction. The implication is clear: apart from this gospel, there would be no salvation.

SOMETHING TO THINK ABOUT

Box 20.2 Paul's Gospel to the Romans

Scholars have long maintained that Paul's opening comments in Romans 1:3–4 are not his own words but those of an old Christian creed that he is quoting, perhaps one that was commonly confessed by Christians when they came to be baptized (cf. the Philippians hymn; see box 19.5). One reason for thinking this is that Paul expresses himself here in ways that are quite uncustomary for him, judging from his other undisputed letters. Nowhere else, for example, does he refer to Jesus as "descended from David according to the flesh," nowhere else does he call the Holy Spirit "the spirit of holiness," and nowhere else does he claim that Jesus was "declared to be Son of God" at his resurrection. Why though would Paul begin his letter in such an unusual way?

If it is true that Paul was writing this letter to correct any misunderstanding about his Gospel message, it may be that he wanted to begin by affirming a confessional statement that he knew was familiar to his audience, so that they would recognize that his gospel was not "off base" but was the same gospel they had come to believe when they joined the Christian church. If so, then we have another indication that this is a letter that Paul spent some considerable care in constructing, giving thought to how he might best win over this important church to support his Gentile mission (see 1:5–6).

3. **This salvation comes to those who have faith.** The English noun "faith" (*pistis*) and the verb "believe" (*pisteuein*) are translations of the same Greek root. For Paul faith (or believing) refers to a trusting acceptance of God's act of salvation. It does not refer simply to intellectual assent (as in "I believe you are right") but implies a wholehearted conviction and commitment. Throughout this letter Paul will insist that a person is put into a right relationship with God not by adhering to the dictates of the Jewish Law but by trusting God's act of salvation, that is, by believing in Christ's death and resurrection.
4. **Salvation comes first to the Jew and then to the Greek.** By "Greek" Paul simply means "Gentile" (since it stands in contrast to "Jew"). The salvation given in the gospel comes to both Jews and Gentiles. Jews received it first, since God is the God of the Jews who sent his Son to the Jewish people in fulfillment of the Jewish Scriptures (as Paul indicates both in Romans and throughout his writings); but it also comes to the Gentiles. Indeed, one of Paul's overarching points throughout this letter is that despite the advantages of the Jews (for example, having the Scriptures in which the promises of God are given), Jew and Gentile are on equal footing before God. All have sinned against God and all can be made right with God only by faith in Christ.
5. **The gospel reveals the righteousness of God.** Is it right that God should not give preference to his own people? Paul's gospel insists that God is unequivocally right in the way he brings about salvation; that is, he is "righteous" in the way that he makes all people, Jew and Gentile, "right" with himself. This indeed is a major theme of Romans: God has not gone back on his promises and has not rejected his people the Jews. The death and resurrection of Jesus are the fulfillment of these promises, and faith in him is given first to Jews, and through them to the entire world.

6. **The Scriptures proclaim the Gospel.** Paul claims that God has been perfectly fair and consistent ("righteous") in his treatment of the Jews and of all people because the Scriptures themselves teach that salvation is based completely on faith ("through faith for faith"), rather than on doing the works prescribed in the Jewish Law. Quoting the prophet Habakkuk, Paul emphasizes that a right standing before God, a standing that provides life, comes only through faith: "The one who is righteous will live by faith." To paraphrase: the one who is made right with God through faith will find life."

Paul wants to emphasize that his gospel message is not something that he has made up himself. We saw in Galatians that he claimed to have received it through a revelation from God. We are going to see in Romans (as we saw in Galatians as well) that he also thinks that it is rooted in the Jewish Scriptures. In large measure, Romans is an extended argument that Paul's gospel of salvation, that is, his message of how a person, Jew or Gentile, comes into a right standing before God, derives from these sacred books.

PAULINE MODELS FOR SALVATION

Rather than launching into a passage-by-passage exposition of Romans, it will be more useful for us to reflect in broader terms on what Paul has to say in this letter about his central theme, the gospel. (Remember: Paul is not speaking about a Gospel book that contains a record of Jesus' words and deeds but about his own gospel message.) Paul has a variety of things to say about it, and it is easy at places to become confused and wonder if Paul is being consistent with himself. In most instances (I'm not sure I can vouch for all of them), Paul is not inconsistent and is not himself confused. The difficulty is that he discusses God's act of salvation in a number of different ways and sometimes does not clearly indicate which way he is thinking about. In other words, Paul has various modes of understanding, various conceptual models, of what it means to say that God brought about salvation through Jesus' death and resurrection.

There are at least two major models that Paul uses for understanding the importance of Christ's death in the letter to the Romans (see box 20.3). I will call these the judicial and the participationist models (these are not, of course, Paul's own terms). Paul does not see these as mutually exclusive of one another; on the contrary, he sometimes combines different conceptualities in one statement. For our immediate purposes, however, it will be useful to see how the models work in isolation from one another. Both models understand that human beings are somehow alienated from God and that Christ's death and resurrection somehow work to resolve that problem. The nature of the problem and the way Christ has solved it, however, are expressed differently in the two models.

The Judicial Model

Paul sometimes understands the human problem with respect to God and the divine solution to the problem in legal or judicial terms. In his mind there appears to be a rough analogy between the act of salvation and the human judicial process. The way it works, in simple terms, is as follows.

God is a lawmaker who has made laws for people to follow (all people, not just Jews); everyone, though, has broken these laws. God is also the judge before whom people appear as lawbreakers. The penalty for breaking God's laws is death, and everyone is found to be guilty as charged. This is the human problem. In Paul's words, "everyone has sinned" (i.e., broken God's laws, see Rom 3:23), and "the wages of sin is death" (i.e., death is the penalty for all who have sinned, Rom 6:23).

The divine solution to this problem is again conceived in judicial terms. Jesus is one who does not deserve the death sentence; he dies to pay the penalty for others. God shows that he is satisfied with this payment by raising Jesus from the dead (Rom 3:23–24; 4:24–25). Humans can avail themselves of Christ's payment of their debt simply by trusting that God will find it acceptable. It is not a payment they have either earned or deserved; it is a beneficent act done on their behalf by someone else, an act that can be either accepted or rejected (3:27–28; 4:4–5). Those who accept it are then treated as if they are "not guilty" (even though they are in fact completely guilty), because someone else has accepted their punishment for them.

This, then, is the judicial model for understanding how salvation works. The problem is sin, which is understood to be a transgression of God's law; the solution is Christ's death and resurrection, which are to be received by faith. A person who has faith is restored to a right standing before God. Sometimes this way of looking at things is called Paul's doctrine of justification by faith. In this model the Jewish Law plays no role in salvation. Those who have broken the Law and incurred the sentence of death cannot remove their guilt simply by obeying a number of other statutes, just as a convicted embezzler will not be set free by pleading that he has obeyed all of the traffic laws. The only way to be restored to a right standing before God (to be "justified") is through the death of Jesus, a payment of the penalty owed by others.

The Participationist Model

Most of us today have no trouble understanding how the act of salvation can be seen as analogous to a judicial process. The participationist model, however, is much harder to get our minds around. This is partly because it involves a way of thinking that is no longer prevalent in our culture. Under this second model, the human problem is still called sin, sin is still thought to lead to death, and Christ's death and resurrection still work to resolve the problem; but sin, death, and Jesus' death and resurrection all mean something *different* from what they mean under the judicial model.

Consider the following uses of the word "sin" in the book of Romans:

- Sin is in the world. (5:13)
- Sin rules people. (5:21; 6:12)
- People can serve sin. (6:6)
- People can be enslaved to sin. (6:17)
- People can die to sin. (6:11)
- People can be freed from sin. (6:18)

It should be reasonably clear that sin in these verses is not simply something that a person does, a disobedient action against God, a transgression of his laws. It is instead a kind of cosmic power, an evil force that compels people to live in alienation from God. The human problem under this model is that people are enslaved to this demonic power and are unable to break free from their bondage.

The power of sin is related to another power, the power of death. In the participationist model, death is not simply something that happens when a person stops breathing. It is a cosmic force that is intent on enslaving people; when it succeeds, it totally removes a person from the realm of God. Here again the situation is desperate; all people are subject to the overpowering force of death, and there is nothing that they can do to set themselves free.

As in the judicial model, the solution has to come from God himself, and it takes the form of Jesus' death and resurrection. If the problem is enslavement to alien powers, then the solution must be liberation. Christ's death and resurrection provide freedom from the powers of sin and death that have subjugated the human race. How, though, does this liberation happen?

As an apocalypticist Paul knew that the cosmic force of sin was present in this world, but he came to believe that Christ's death had conquered the power of sin. He evidently came to believe this after he believed that Jesus had been raised from the dead. For Paul, Jesus' resurrection showed beyond any doubt that Jesus was no longer subject to the power of death, the most dreaded of all cosmic forces of evil. Jesus had conquered death through his resurrection; thus, reasoning backwards, at his death he must have defeated the related powers (including the Devil and his agent, sin). Furthermore, Jesus' victory can lead to the salvation of others. That is to say, a person can participate with Christ in his victory (Rom 6:5–8): hence the name I have given this model of conceptualization. A person participates in this victory by being united with Christ in his death and resurrection. According to Paul, this happens when a person is baptized (Rom 6:3–4).

Baptism was a rite that had been practiced among the Christians from the earliest of times. In the early years of the religion, of course, no one was "born" a Christian; new members of the religion converted to it either from Judaism or from loyalty to one of the other cults. Those who converted were initiated into the church through the ritual of baptism. Baptism involved being immersed in water (later sources suggest that running water was to be preferred) while an officiant pronounced sacred words to indicate the significance of the act. For Paul the act was not simply significant as a symbolic statement that a person's sins had been cleansed or that he or she had entered into a new life; the act involved something that really happened. When people were baptized, they actually experienced a union with Christ and participated in the victory brought at his death (in the immersion under the water; see especially Rom 6:1–11).

Although Paul believed that a person who had been baptized had "died" with Christ, that is, had participated fully in Christ's victory over the power of sin, he evidently did not believe that such a person had yet been "raised" with Christ, that is, set completely free from the power of death. Paul knew full well that this had not yet occurred since people, even believers, continued to die! So he is quite emphatic that Christians have died with Christ but that they have not yet been raised with him (6:5, 8). They will be raised only when Christ returns and brings about the resurrection at the end of time. (You may recall that the major problem at Corinth was that some people believed that they had already been raised with Christ, and Paul had to insist that this was simply not so.) Until then, to be sure, Christians live in "newness of life" (Rom 6:4), because they are no longer subject to the power of sin. But their salvation is not yet complete, for the end has not yet come. Only when it does come will they "be united with him in a resurrection like his" (6:5).

Comparison and Contrast of the Two Models

The two models of salvation we have been looking at are ways of understanding something. They are not the thing itself. Paul's gospel is not "justification by faith" or "union with Christ." These are ways of reflecting on or thinking about his gospel. His gospel is God's act of salvation in Christ; the models are ways of conceptualizing how it worked.

The way salvation worked differed according to which model Paul had in mind. In both of them, the problem is "sin," but in one model, sin is an act of disobedience that a person commits, whereas in the other it is a cosmic force that works to enslave people. In both models, the solution is provided by Christ's death and resurrection, but in one Christ's death pays the penalty for human disobedience, and in the other it breaks the cosmic power of sin. In both models a person has to appropriate the benefits of Christ's death, but in one this is done through faith, that is, a trusting acceptance of the payment, whereas in the other it occurs through baptism, a ritual participation in the victory.

As you read through Romans on your own, you can see that Paul does not neatly differentiate between these two models. Even though he uses the judicial model more consistently in chapters 1–4 and the participationist model in chapters 6–8 (to choose the clearest places), he does not ever think of them as conflicting with one another, and he regularly combines the two in the things he says. He would never have thought, for instance (so far as we can tell), that someone could be baptized and so participate in Christ's death without also having faith and so trusting Christ's payment for sin. The two models go hand in hand; they are not so much confused as combined. Their coalescence is clear at a number of points in Paul's discussion. Why, for example, does Paul maintain that everyone is guilty before God? Because everyone has sinned, that is, committed acts of transgression (the judicial model, 3:23). Why has everyone sinned? Because everyone is enslaved to the power of sin (the participationist model, 3:9). Why is everyone enslaved to the power of sin? Because Adam committed an act of disobedience (judicial model), which allowed the power of sin to enter into the world (participationist model; 5:12). And so it goes.

Despite the fact that these two models neatly dovetail in Paul's own thought, it is often useful for readers to keep them conceptually distinct when reading through his letters, especially the letter to the Romans. Therefore, when you find Paul speaking of "sin" in any given verse, you should ask what

SOME MORE INFORMATION

Box 20.3 Judicial and Participationist Models of Salvation in Paul

The Judicial Model	The Participationist Model
Sin—human disobedience that brings a death penalty	Sin—a cosmic power that enslaves people
Jesus' Death—payment of the penalty of sin	Jesus' Death—defeat of the power of sin
Appropriation—acceptance of the payment through faith, apart from works of the Law	Appropriation—participation in Christ's victory through baptism

he means by it. Is he referring to an act of transgression or a cosmic power? When he refers to the effects of Christ's death and resurrection, is he thinking of a payment of a debt or liberation from bondage? In this connection, I should point out that these are not the only two models that Paul uses to conceptualize what Christ has done for salvation (see box 20.4). They are, however, the two models that appear most prominently throughout the book of Romans, as can be seen in the following section-by-section synopsis of the letter.

THE FLOW OF PAUL'S ARGUMENT

- *The Human Dilemma: All Stand Condemned before God (1:18–3:20)*. Paul's gospel follows a "bad news, good news" scheme that is designed to show the reader how desperate the situation is for all people, Gentiles and Jews. Gentiles have abandoned their knowledge of the one true God to worship idols, resulting in wild and rampant immorality (1:18–32). Jews are no better, for even though they have the Law and the sign of circumcision, they do not practice the Law and so also stand condemned (2:1–29). Indeed, all people, Jews and Gentiles, have sinned against God (the judicial notion; 3:1–18), for all are under the power of sin (the participationist notion; 3:9). This view that Jew and Gentile are equally condemned before God does not at all represent a rejection of Judaism, however, for according to Paul it is the teaching of the Jewish Scriptures themselves (3:10–20).
- *The Divine Solution: Salvation through Christ's Death (3:21–31)*. The Jewish Law gives the knowledge of sin but not the solution to sin. The solution comes in the fulfillment of this Law in the death of Jesus, a sacrifice for the sins of others to be received through faith. Performing the works of the Jewish Law does not contribute to this salvation through faith, so Jews have no grounds for boasting of a special standing before God. Jews and Gentiles are on equal footing, all are made right with God through faith in the death of Jesus.
- *The Gospel Message Is Rooted in the Scripture (4:1–25)*. The Father of the Jews, Abraham himself, shows that being made right with God comes through faith rather than by doing the works of the Law. He himself was justified (made right with God) by trusting in God's promise before he was given the

sign of circumcision (a "work" of the Law). His true descendants are those who continue to trust in God and in the fulfillment of his promises, which has now occurred in the death and resurrection of Jesus.

- *Christ's Death and Resurrection Bring Freedom from the Powers Opposed to God (5:1–8:39)*. Those who believe in Christ have been made right with God and will be saved from the wrath of God that is coming upon this world (5:1–11). They will also be delivered from the reign of God's mortal enemy, death, which entered into the world through the disobedience of Adam, Christ's counterpart, but which has now been conquered by Christ's own act of obedience (5:12–21). Moreover, those who have been united with Christ in his death have participated in his victory over the power of sin; they can, therefore, and should, serve the new power that is over them in Christ, the divine power of righteousness (6:1–23). Before a person was united with Christ he or she was compelled by the power of sin to violate the good Law that God had given, so the Law led to condemnation rather than to salvation (7:1–25). But now the part of the self that was subject to sin, the flesh, has been put to death in Christ, so a person no longer needs to submit to its cravings and violate the Law (8:1–17). Those who have been united with Christ will eventually experience the complete salvation that will come when God redeems this fallen world (8:18–39).
- *The Gospel Message Is Consistent with God's Dealings with Israel and Represents a Fulfillment of His Promises (9:1–11:36)*. Paul now deals with the major questions that have been simmering beneath the surface of the letter all along. If what he says is true, that God's act of salvation comes equally to Jew and Gentile alike, with no distinction, hasn't God gone back on his promises to Israel (9:6)? On the contrary, for Paul, God's decision to save Gentiles and Jews by faith is a fulfillment of his promises and is consistent with how he has always worked, as is evident from the Jewish Scriptures themselves. God has always chosen people not on the basis of their actions ("works") but on the basis of his own will (9:6–18). Indeed, the Jewish prophets indicate that God shows mercy on whom he chooses and that he had planned from ages past to make a people who were not his own (the Gentiles) into his own, whereas many of the Jews would be rejected (9:19–29). The failing lies not in God but in the Jews who have not accepted Christ, for they have mistakenly supposed that God gave them the Law as a means for attaining a right standing before him, whereas the Law itself points to Christ (9:30–10:4). A right standing before God therefore comes exclusively through faith in Christ, and many of the Jews have been faithless (10:5–21). God himself, however, is faithful. He has remained true to his promises to the Jews, saving a remnant of them and using the salvation of the Gentiles to bring about his ultimate purpose, the salvation of all of Israel. Gentiles who have been added to the people of God must not therefore vaunt themselves against Jews; Israel is still the people of God's special calling, and he will once again bring them all to faith (11:1–36).
- *The Law-Free Gospel Does Not Lead to Lawless Behavior (12:1–15:13)*. The believer in Christ gives him- or herself to others in self-sacrificing love. Indeed, this is the new cultic act of worship that fulfills the old cultic acts of sacrifice (12:1–21). Believers in Christ are to be obedient to civil authorities (13:1–7), to follow the core of the Torah by loving others as themselves (13:8–10), to lead moral, upright lives in view of their coming salvation (13:11–14), and to refrain from passing judgment or doing things that offend others (14:1–15:6). Paul's law-free gospel, in other words, will not lead to lawless activities.

SOME MORE INFORMATION

Box 20.4 Other Models of Salvation in Paul

In addition to the judicial and participationist models, Paul has other ways of conceptualizing God's act of salvation in Christ, even though he rarely explains how the analogies work in detail. Consider, for instance, the following.

- Sometimes Paul likens salvation to a reconcilation in which two people have had a falling out. A mediator (Christ), at a sacrifice to himself, intervenes and restores their relationship (e.g., see Rom 5:10 and 2 Cor 5:18–20).
- Paul often describes salvation as a redemption, in which a person's life is "purchased" by God through the price of Christ's blood, much as a slave might be purchased by gold (Rom 3:24; 8:23). Never does he explain, however, from whom or what the person is being purchased (the cosmic forces? the devil? sin?).
- Paul sometimes portrays Christ's death as a sacrifice that, like the sacrifices of animals in the Jewish Temple, was designed to bring atonement with God. This view embodies the ancient view that the blood of a sacrifice "covers over" the sins of the people: the technical term for this act of covering is "expiation" (Rom 3:25).
- At other times Paul compares salvation to a rescue from physical danger, in which a person is confronted with peril and certain death only to be saved by someone who heroically intervenes at the cost of his own life (see Rom. 5:7-8).

These models are not at mutually exclusive; sometimes Paul applies several of them even within the same passage. Consider for yourself the theologically packed statement of Romans 3:21–26, where Paul uses the judicial, participationist, redemptive, and sacrificial models at one and the same time!

- *Close of the Letter (15:14–16:27)*. Paul indicates his reasons for writing (15:14–21), discusses his travel plans (15:22–33), and sends greetings to a large number of persons in the congregation (16:1–27). Indeed, he greets so many people by name (twenty-eight altogether) that some scholars have questioned whether this final chapter originally belonged to the letter, since it was written to a congregation Paul had never visited. If the chapter is original to the book, it indicates that a number of people whom Paul had come to know in other contexts had moved to Rome or were known to be visiting there.

CONCLUSION: PAUL AND THE ROMANS

We do not know for certain whether Paul's plans to visit the congregation en route to Spain ever came to fruition. According to the book of Acts, Paul was arrested in Jerusalem before he could make the trip and was then, almost coincidentally, sent to Rome to stand trial before the Roman emperor for his alleged crimes (Acts 21–28). The author of Acts does not seem to know of any contact between Paul and the Christians living in Rome prior to his arrival; indeed, as customarily happens everywhere Paul goes in Acts, he ends up spending his days not with Christian believers but with recalcitrant Jewish leaders and, evidently, with anyone else who would come to hear him preach while under house arrest (Acts 28:16–31). There are later traditions that indicate that Paul was eventually martyred in Rome; a member of the Roman church, writing sometime around 95 C.E., mentions Paul's death during the tyrannical persecution of the Christians during the reign of Nero (ca. 64 C.E.). This writing, traditionally attributed to the bishop of Rome, Clement, may indeed preserve a historical recollection (see Chapter 26).

Even though we cannot gauge whether Paul succeeded in his Western mission, or indeed, whether he ever gained a following among the Christians in Rome, we can say for certain that he succeeded in one respect. Romans is the most closely reasoned letter that survives from his pen, one that continues to intrigue scholars and to inspire believers. It lays out in the clearest terms he could muster important aspects of Paul's gospel, namely God's power that brings salvation for both Jew and Gentile.

SUGGESTIONS FOR FURTHER READING

See also the suggestions at the end of Chapter 17.

Donfried, Karl P. ed. *The Romans Debate*. 2d ed. Peabody, Mass.: Hendrikson, 1991. A collection of significant essays by eminent New Testament scholars, who discuss (and disagree over) the occasion and purpose of Paul's letter to the Romans.

Wedderburn, A. J. M. *The Reasons for Romans*. Edinburgh: T & T Clark, 1988. The most complete book-length discussion of the reasons that Paul wrote his letter to the Romans: it was to explain his law-free gospel to the predominantly Gentile Roman community in light of the tensions between Jews and Gentiles there and in view of his own imminent journey to Jerusalem.

CHAPTER 21

Does the Tradition Miscarry? Paul in Relation to Jesus, James, Thecla, and Theudas

Jesus was a Jewish apocalyptic prophet who proclaimed the coming of the kingdom of God in power and urged his fellow Jews to repent and to keep the Law of God, as he himself interpreted it, in preparation for the imminent appearance of a cosmic judge from heaven, the Son of Man. Paul, on the other hand, was a Christian apostle who proclaimed a salvation apart from the Jewish Law, a missionary who urged Gentiles to put all their trust in Jesus' death and resurrection and all their hope in his imminent return from heaven. Did Jesus and Paul represent the same religion?

The writers of the Gospels maintained that God had brought salvation to this world through the words and deeds of Jesus. The apostle Paul also wrote about salvation, but he said almost nothing about Jesus' words and deeds, apart from the deeds of his death and resurrection. Did the Gospel writers and Paul share the same religion?

Some members of Paul's congregations claimed his support for views that he himself found outrageous (cf. 1 Cor 1:12). After his death, Marcionites, Gnostics, and Proto-orthodox Christians all subscribed to beliefs that they argued came from his writings. Was there one form of Pauline Christianity or several forms? To expand the question yet further: was there one thing that could be called Christianity in the first two centuries of the Common Era or several different things? Should we speak of early Christianity or of early Christianities? Did any of the forms of early Christianity coincide with the religion advocated by Jesus himself? Or at some point, even a number of points, did the tradition miscarry?

These are perplexing and complex questions, but ones that we need to ask if we are to approach the writings of the New Testament from a historical perspective. Having examined all of the early Gospels, the teachings of Jesus himself, and the undisputed writings of Paul, we have arrived at a good stage to take a step back and consider in somewhat broader terms the nature of early Christianity and its diversity. Since we have just completed our study of Paul, we can pursue our questions, using his epistles as a fulcrum, by evaluating how Paul's form of Christianity related to some of what came before and to some of what came after.

PAUL IN RELATION TO WHAT CAME BEFORE

Prior to the writing of the Gospels, Christians throughout the Mediterranean were telling stories about Jesus, about the things that he said, did, and experienced. Did Paul tell these stories?

Paul and the Traditions about Jesus

We can be relatively certain that members of Paul's churches told stories about the earthly Jesus. The author of the book of Acts, after all, belonged to one of these churches (at least we can assume so since Paul was the hero of his narrative), and he also wrote a Gospel. But Luke was writing some thirty years after Paul's active ministry. Did these traditions about Jesus circulate in Paul's churches during his own day? Did Paul teach his converts these stories? Did he know them himself?

These questions themselves may come as a shock—they have never occurred to most people who read the New Testament—but they are a source of endless fascination for the historian of early Christianity. Paul scarcely says anything about the historical Jesus, that is, about the things that Jesus said, did, and experienced between the time of his birth and the time of his death. You can see this for yourself by rereading Paul's letters and listing everything that he says about Jesus' life, up to and including his crucifixion. Part of the surprise is that you won't need an entire sheet of paper.

Paul gives the following information. He says that Jesus was born of a woman (Gal 4:4; this is not a particularly useful datum; one wonders what the alternative may have been!) and that he was born a Jew (Gal 4:4), reputedly from the line of King David (Rom 1:3). He had brothers (1 Cor 9:5), one of whom was named James (Gal 1:19). He had twelve disciples (1 Cor 15:5) and conducted his ministry among Jews (Rom 15:8). He had a last meal with his disciples on the night on which he was betrayed (1 Cor 11:23; it is possible, however, that Paul is not referring here to Judas who "betrayed" Jesus, since the Greek word he uses literally means "handed over" and more commonly refers to God's action of handing Jesus over to his death, as in Rom 4:25 and 8:32). Paul knows what Jesus said at this last meal (1 Cor 11:23–25). Finally, he knows that Jesus died by being crucified (1 Cor 2:2). He also knows of Jesus' resurrection, of course, but here we are interested only in what he tells us about Jesus' life prior to his death.

In addition to the words spoken at the Last Supper, Paul may refer to two of the sayings of Jesus, to the effect that Christians shouldn't get divorced (1 Cor 7:11; cf. Mark 10:11–12) and that they should pay their preacher (1 Cor 9:14; cf. Luke 10:7). Still other teachings of Paul sound similar to sayings of Jesus recorded in the Gospels—for instance, he says that Christians should pay their taxes (Rom 13:7; cf. Mark 12:17) and that they should fulfill the Law by loving their neighbors as themselves (Gal 5:14; cf. Matt 22:39–40)—but Paul gives no indication that he knows that Jesus himself spoke these words.

Paul, of course, has a lot to say about the *importance* of Jesus, especially the importance of his death and resurrection and his imminent return from heaven, but in terms of historical information, what I've listed above is about all that we can glean from his letters. We hear nothing here of the details of Jesus' birth or parents or early life, nothing of his baptism or temptation in the wilderness, nothing of his teaching about the coming kingdom of God. We have no indication that he ever told a parable, that he ever healed anyone, cast out a demon, or raised the dead. We learn nothing of his transfiguration or triumphal entry, of his cleansing of the Temple, of his interrogation by the Sanhedrin or trial before Pilate, of his being rejected in favor of Barabbas, of his being mocked, of his being flogged, and so on. The historian who wants to know about the traditions concerning Jesus, or indeed, about the historical Jesus himself, will not be much helped by the surviving letters of Paul.

Why does Paul not remind his congregations of the things Jesus said and did? Does he think that they are unimportant or irrelevant? Does he assume that his readers already know them? Does *he* know them? How could he *not* know? Let me explore three lines of thinking that scholars have pursued over the years, as a way to stimulate your own thinking on these matters.

Option One. Paul knew a large number of traditions about Jesus but never spoke of them in his surviving letters because he had no occasion to do so. This is perhaps the easiest way to explain why Paul scarcely ever mentions the events of Jesus' life. Someone who takes this line could point out that Paul evidently knew other apostles (cf. Gal 1–2) who must have told him stories about Jesus; moreover, it would make sense that when he founded his churches he must have told them *something* about the man whom he proclaimed as the Son of God who died and was raised from the dead. Who exactly was he? What did he do? What did he teach? How did he die? Surely questions such as these must have occurred to Paul's converts, and surely he must have answered them. If so, then we might conclude that Paul never mentioned these traditions in his letters because he knew that his converts already knew them.

You may, however, detect a flaw in this reasoning. Paul spends a good amount of time in his letters reminding his converts of what he taught them when he was among them. If he had taught them about the historical Jesus, why would he not remind them of these stories also? Moreover, on occasion, though relatively rarely, Paul does use one of the traditions about Jesus to convince his converts of a necessary course of action. For instance, when the Corinthians were celebrating the Lord's Supper in a way that offended Paul, he reminded them of how Jesus instituted it among his disciples. In other words, when the need arose, Paul was inclined to cite stories of Jesus to authorize his views as those promoted by Jesus himself, the ultimate Lord of the community.

If Paul was demonstrably inclined to use the traditions about Jesus in this way, why does he not do so more often? The problem with this first option is that Paul had plenty of occasions to mention traditions about Jesus to buttress his views, but he scarcely ever took the opportunity. When he told the Romans to pay their taxes (Rom 13:6–7), why did he not say: "Remember the words of the Lord Jesus, that we should render unto Caesar the things that are Caesar's"? When he told the Galatians that they should love one another so as to fulfill the Law (Gal 5:13–14), why didn't he point out that this was what Jesus himself had said? When he spoke of the sufferings of the present age to the Corinthians (2 Cor 4:7–18, 11:23–29), why didn't he remind them of the details of Jesus' own passion or of Jesus' call to take up one's cross and follow him? It is hard to explain why if Paul, in fact, knew more than he said.

Option Two. Paul knew more of the traditions of Jesus but considered them irrelevant to his mission. This option is similar to the one preceding with a major difference. In this case, Paul knew many of the traditions about what Jesus said and did, but he did not refer to them extensively either in person or in writing because he considered them irrelevant to his message of Jesus' death and resurrection. Support for this view can be found in a passage like 1 Corinthians 2:2, where Paul insists that the only thing that mattered to him during his entire stay among the Corinthians was "Christ, and him crucified" (cf. 1 Cor 15:3–5). That is to say, what Jesus said and did prior to his death was of little relevance; what mattered was that he died on the cross and that this brought about a right standing before God (as evidenced in his resurrection).

In considering this option, it is not adequate to claim that it can't be right because the words and deeds of Jesus must have been important to Paul. This is like saying that the traditions must have been important to Paul because they must have been important. Rather than simply presupposing our conclusion we have to provide evidence for it. There is, in fact, at least one serious problem with this view. If it were true that Paul did not consider the words and deeds of Jesus to be important, we would be unable to explain why Paul sometimes does appeal to these words and deeds when he is insisting on proper behavior among his congregations (e.g., in 1 Corinthians alone, see 7:11, 9:14, and 11:23–25). Thus, even granting the central importance of Jesus' death and resurrection for Paul, he must have taught his churches something more than the events at the end of Jesus' life—if, that is, he knew more.

Option Three. Paul didn't mention more about Jesus' words and deeds because he didn't know very much more. According to this theory, the life of Jesus was not only unimportant to Paul when he established his churches and addressed their problems, but it was also unimportant to him personally. He never inquired further into the things Jesus said and did, and possibly never even thought about inquiring further, because he simply wasn't interested.

Is this plausible? According to Paul, Jesus himself appeared to him at his conversion; but Paul never indicates that Jesus gave him a crash course in all that he had said and done prior to his death. Also, Paul evidently knew some of Jesus' apostles—his brother James and some of his former disciples in Jerusalem (but see box 19.4)—but he indicates that they spent very little time together and suggests that when they did meet they discussed the future of the Gentile mission rather than the words and deeds of Jesus (Galatians 1–2).

Possibly the other apostles told him *something*, but if so, we are left with the problem that Paul sometimes uses Jesus' words as an authority for his own views but usually does not. If he knew more and taught his congregations more, and if these traditions were of central importance to Paul's

Gospel and his converts' faith, why does he scarcely ever refer to them in his surviving writings or remind his readers that he has told them about them before? I'm afraid that I must leave this dilemma for you to resolve.

Paul and the Historical Jesus

Whereas the preceding problem (did Paul know more about the traditions about Jesus and, if so, why didn't he utilize them in his letters?) was largely a matter of speculation, it is possible to take the question of Paul's relationship to Jesus in a different direction by asking whether the religious points of view that these two men represented were identical, similar, or different. Even this question is not completely straightforward, of course. We do not have any writings from Jesus and therefore have to reconstruct his teachings on the basis of later traditions that are not always historically accurate. Moreover, even though we do have writings from Paul, these are occasional pieces of correspondence, not systematic expressions of his thought. Still, we have devoted some considerable effort to establishing the teachings of Jesus and highlighting the views of Paul, so we have some basis for exploring this question.

The first point to emphasize is perhaps too easily overlooked. Jesus and Paul agreed on a number of very basic issues as two first-century Jewish men. They both subscribed, for example, to the belief in the one God who had created the world, who made a covenant with his people Israel, and who revealed his will through the Jewish Scriptures. Moreover, they were both apocalypticists who thought that they were living at the end of time and that God was soon going to intervene in history by sending a cosmic redeemer from heaven to overthrow the forces of evil that plague this world.

Despite these fundamental similarities, Jesus and Paul also differed on a number of points (see box 21.1). First, while both expected the imminent appearance of a cosmic judge from heaven, for Jesus this divine figure was to be the Son of Man predicted by the prophet Daniel; for Paul it was to be Jesus himself. Both Jesus and Paul maintained that strict adherence to the laws of Torah, particularly as interpreted by the Pharisees, would not contribute to a person's salvation on the day of judgment, but they disagreed on what *would* make a difference. For Jesus, people needed to repent of their sins and keep the central teachings of the Torah by loving God with their entire being and their neighbors as themselves. For Paul, no amount of obedience to the Law would help when God's judgment came; salvation would come only to those who trusted in Christ's death and resurrection as God's act of deliverance from sin.

Both men did understand that Jesus himself was of central significance for those who would be saved on that day, but Jesus appears to have thought that his own importance lay in his teaching about the end time, in his prophetic call for repentance, and in his correct interpretation of the will of God as revealed in the Scriptures. His followers were those who gave up everything to adhere to his teachings. Paul, on the other hand, scarcely mentions any of these things. For him, what ultimately mattered was Jesus' sacrificial death and vindication by God at the resurrection. Those who would be saved were those who had committed themselves in faith to the Christ who died and rose again.

Finally, both Jesus and Paul maintained that in some sense the end had already begun, but they disagreed as to *how* it began. For Jesus it began in the community of his followers, who abandoned everything to live lives of faith in God and of love toward their neighbors. For Paul, it started with Jesus' victory over the powers of sin and death at the cross, the beginning of the defeat of God's cosmic enemies. Christians could participate in this victory by being baptized into Christ's death and sharing in the Spirit of God who now dwelt among his people, prior to the end when Christ returned.

In light of these similarities and differences, do Jesus and Paul represent the same religion? Again, I must leave that for you to decide.

PAUL IN RELATION TO WHAT CAME AFTER

Up to this point we have looked at Paul's relationship to some aspects of the Christian religion that preceded him. It would also be beneficial to consider Paul's relationship to other authors we have considered, for example the Gospel writers who produced their accounts some years later.

SOMETHING TO THINK ABOUT

Box 21.1 Jesus and Paul: Some of the Differences

The Historical Jesus	The Apostle Paul
The coming judge of the earth is the Son of Man.	The coming judge of the earth is Jesus himself.
To escape judgment, a person must keep the central teachings of the Law as Jesus himself interpreted them.	To escape judgment, a person must believe in the death and resurrection of Jesus, and not rely on observance of the Law.
Faith involves trusting God to bring his (future) kingdom to his people.	Faith involves believing in the (past) death and resurrection of Jesus.
Jesus' own importance lies in his proclamation of the coming of the end and in his correct interpretation of the Law.	Jesus' importance lies in his death and resurrection for sins.
The end of the age began in the lives of Jesus' followers, who accepted his teachings and began to implement them in their lives.	The end of the age began with the defeat of the power of sin at the cross of Jesus.

Indeed, you should make such comparisons and contrasts for yourself. Imagine, for instance, comparing Paul with Matthew on the subject of Torah observance: are Jesus' followers required to follow the Law or not?

Here, however, we will consider Paul's relationship to the tradition that he himself, in some sense, started. Just as Jesus began a tradition that eventuated in Gospels that varied both among themselves and from the things that Jesus himself had said (contrast the teachings, for example, in Mark, John, and Thomas), so Paul stood at the head of a tradition of Pauline Christianity, a form of Christianity that developed in ways that some Christian believers found inspiring and others repugnant.

Paul and James

One form of Pauline Christianity appears to lie behind the opinions attacked by the New Testament book of James. This book provides an extended set of admonitions to unnamed Christians living outside of Palestine, who are call "the twelve tribes in the Dispersion" (1:1, which some scholars have taken as a reference to Jewish–Christians but other scholars as a symbolic title of all Christians as the "new Israel"). It is a letter in form, at least partially: it begins with a prescript that names the author and contains a greeting. There is no epistolary conclusion, however, and the "letter" gives no indication of a specific occasion. It is instead a collection of pieces of good advice to those who "believe in our glorious Lord Jesus Christ" (2:1).

There is some question concerning the identity of the book's author. He gives his name as James, which readers over the centuries have taken to refer to the brother of Jesus, but there is little reason to think that the author is claiming to be that particular James. The name was fairly common in the first century; just within the pages of the New Testament, in addition to the brother of Jesus, we encounter James the son of Zebedee (Matt 4:21), James the son of Alphaeus (Matt 10:3), James the son of Mary (Matt 27:56), and James the father of Judas (Luke 6:16). If the author of this epistle was James the brother of Jesus (or was at least claiming to be), it is somewhat strange that he never refers to any personal knowledge of his brother or of his teachings.

The letter that James writes is full of exhortations to his readers, and these strong moral teachings do indeed appear to reflect (though they never quote) traditions of Jesus' own teaching. For instance, believers should not swear oaths, but let their "yes be yes" and their "no be no" (5:12, cf. Matt 5:33–37); loving one's neighbor fulfills the Law (2:8, cf. Matt 22:39–40); and those who are rich should fear the coming judgment (5:1–6, cf. Matt 19:23–24). (I will say a few more words about these admonitions, and the book as a whole, in Chapter 26.) Of particular interest is what James has to say in 2:14–26, a passage that has been much-cited since the Protestant Reformation, when Martin Luther made the unequivocal claim that it contradicts the gospel proclaimed by Paul and so should have only a secondary standing in Scripture.

James, in this passage, and Paul cover much of the same ground. Both discuss justification, both consider the relationship between faith and works, and both use the Old Testament figure of Abraham to establish their points. The points they make, however, are different. For Paul, as we have seen, "a person is justified by faith apart from the works prescribed by the law" (Rom 3:28); for James, however, "a person is justified by works, not by faith alone" (James 2:24). Given their different perspectives, it is odd that both Paul and James appeal to Abraham in support. Paul maintains that "if Abraham was justified by works, he has something to boast about, but not before God. . . . Therefore his faith was reckoned to him as righteousness" (Rom 4:2, 22); James, on the other hand, argues that "our ancestor Abraham was justified by works" (2:21). Yet more peculiarly, each author claims that Genesis 15:6 ("Abraham believed God, and it was reckoned to him as righteousness") supports his own interpretation of the relationship of faith and works to justification (Rom 4:1–5; Gal 3:6; James 2:23).

Thus, at least on the surface, it appears that Paul and James are fundamentally at odds with one another. Paul claims that faith in Christ is all one needs to be justified, and James argues that one needs more than faith. Paul rejects works of the Law as a prerequisite for justification and James insists that works are absolutely necessary.

Nonetheless, most modern scholars have come to think that the differences between James and Paul are only skin deep, because James and Paul do not appear to mean the same things when they speak about "faith" and "works." (If they use the terms in different ways, then they can scarcely be contradicting one another when one of them insists on faith without works and the other on both faith and works.) For Paul, as we have seen, "faith" means a trusting acceptance of Christ's death to put one into a right relationship with God. "Works" for him are the works of the Jewish Law, that is, aspects of the Law that make Jews distinctive as the people of Israel (e.g., circumcision, the Sabbath, kosher food laws). When James, on the other hand, speaks of "faith" in 2:14–26, he appears to mean "intellectual assent to a proposition." He points out, for example, that "even the demons believe" that "God is one . . . and shudder" (2:19). Presumably these demons are not committed to this belief; they simply acknowledge it. This kind of intellectual acknowledgment, according to James, cannot justify anyone. Paul, of course, would not disagree; he simply doesn't mean this when he uses the term "faith."

Moreover, James insists that those who have true faith will do "works," by which he appears to mean "good deeds," such as feeding the hungry and helping the destitute (2:14–16). Those who fail to do such works do not have real faith, or as James himself puts it, their faith is "dead" (2:17). Again, when the matter is put in this way, Paul would scarcely disagree: he too expects believers to behave in certain ways (cf. Gal 5:16–26; 1 Cor 6:9–12).

Paul and James appear, then, to be referring to different things when they speak of faith and works. Yet surely it cannot be a coincidence that they both address the issue of justification by faith and works, that they both use Abraham as an example to prove their points, and that they both quote Genesis 15:6 on this matter. How, then, did this come about?

We don't know exactly when the book of James was written. But if it was produced sometime late in the first century, it is not difficult to imagine a scenario that could explain its strong case against justification by faith alone. It may have happened like this. Paul himself had insisted that a person was justified by a trusting acceptance of Christ's death, not by works of the Law. When Paul passed from the scene, according to this scenario, his words became a kind of catch phrase among his congregations: "faith without works." Some Christians took this to mean that it mattered only what you believed, not what you did. (Indeed, some people may have understood Paul this way even while he was still alive; see Rom 3:8.). Word of this notion got around to an author living in another community who took serious exception to its implications. He wrote a tractate that gave a long series of admonitions to believers, including the admonition to put their faith to work in their lives. Despite what Paul had said, or rather, despite what some people claimed Paul had said, faith needed to be practiced in order to be genuine. For as Abraham himself showed, a "person is justified by works and not by faith alone."

Paul's words thus may have taken on a life of their own as they were used in new contexts, gaining a meaning that was independent of what they originally meant when he proclaimed them to his converts. Interestingly, the distortion of Paul's message is explicitly recognized as a problem even within the pages of the New Testament (2 Pet 3:16).

Paul and Thecla

Something similar seems to have happened in a series of stories that we know were in circulation at the beginning of the second century among other Christians who saw themselves as adherents of the teachings of Paul. Scholars have long known of a letter, written pseudonymously in the name of Paul's companion Titus, that endorses a strict ascetic life involving, among other things, the total renunciation of the joys of sex. In his own letters even Paul urged celibacy for the sake of the gospel. If possible, Christians were to refrain from marriage and the fleeting pleasures of conjugal bliss; it was better for them to devote themselves completely to the Lord, since the time of the end was near (1 Corinthians 7). Never, though, does Paul make salvation contingent upon total abstinence.

The end that Paul anticipated never came, of course, but his teachings concerning celibacy survived, and indeed took on a life of their own. Some of the most interesting pieces of early Christian literature are narratives composed around the person of Paul and modeled, to a limited extent, on the book of Acts, the only narrative about him to be included in the New Testament. Of the noncanonical accounts, perhaps the best known are those that relate the exploits of Paul and his female disciple, Thecla. In these and similar accounts, Paul is portrayed as a hard-core advocate of sexual renunciation, an apostle who preaches the joys of abstinence to audiences eager to escape the drudgeries of arranged marriages and to evade oppressive social arrangements that appear in the guise of established family structures (see further Chapter 23). Not surprisingly, those who take Paul's words to heart are usually women, destined otherwise to live under the oppressive yokes of their future husbands. Thecla's story is typical of these narratives. Engaged to a wealthy man of the upper classes, she hears Paul's disquisition and breaks her engagement. She leaves home to follow the apostle and enjoy the freedom of one liberated from the concerns of the body and the domination of a husband. Her estranged fiancé, as you might imagine, is not amused.

Thecla's exploits are recounted in a second-century novelistic work called *The Acts of Paul and Thecla*. As the plot develops, her fiancé (in cahoots with her mother, who is set to lose a prosperous retirement from the deal) turns on her and prosecutes her, eventually seeking her execution. She is miraculously delivered, however, by the God who protects those who have forsaken all to adhere to his will of sexual renunciation. In several related adventures, this divine protection and Thecla's fidelity to her cause are put to the test. In every instance, the God proclaimed by Paul delivers his faithful servants from those who are determined to make them compromise.

Taking the historian's view, one might ask whether the historical Paul himself would have recognized this version of his own proclamation.

Whatever the apostle would have made of it, the stories about Paul and Thecla enjoyed a wide popularity in certain circles, perhaps chiefly, as some scholars have suggested, among Christian women who, as converts, enjoyed a certain liberation from the constraints of marriage and enforced subservience. This liberation received an apostolic sanction in the ascetic message proclaimed by the missionary to the Gentiles himself (see Chapter 23).

Paul and Theudas

Still other versions of Paul's teachings were in circulation at roughly the same time. In these versions his chief concerns were only indirectly, if at all, related to sexual renunciation. We have already touched on the understanding of Paul promulgated by the second-century Christian Marcion (see Chapter 1), whose views differed on a number of counts from those advanced in the tales of Thecla. They appear to have differed as well from those passed along by a shadowy figure of the early second century by the name of Theudas. We know of this person only because later proto-orthodox Christians maintained that he was the teacher of the infamous Gnostic Valentinus. Valentinus developed a Christian Gnostic theology quite similar to the account that I described in Chapter 1. He evidently claimed to have acquired his knowledge of this theology from Theudas, possibly in the city of Alexandria, where Valentinus was educated. Theudas was said to have been a disciple of Paul.

As we have seen, Gnostics claimed to have secret knowledge about the truths of the universe, knowledge not accessible to just anyone, indeed, not even to ordinary Christians (see Chapter 11). Some Gnostic Christians appealed to Paul as their ultimate authority. Had not Paul himself indicated that he could not speak to some believers "as spiritual people, but rather as people of the flesh" (1 Cor 3:1)? Did he not differentiate between those who were spiritual and those who were not (1 Cor 2:14–15)? Did he not allude to the "mystery" of the gospel that was "hidden" from the rulers of this age and the "wisdom, secret and hidden" that was only for those who were "mature" (1 Cor 2:6–7)? The Gnostics' claim to Paul may strike the historian as odd, since they were polytheists who denied that there was only one God, the creator of heaven and earth. They also typically maintained that Jesus Christ was two persons, one divine and one human and denied that the human body (much less this material world) was to be redeemed at the resurrection. Yet they claimed to stand in the Pauline tradition and to have derived their views from the apostle himself through his faithful disciple Theudas.

CONCLUSION: PAULINE CHRISTIANITIES

We have again moved full circle back to where we began. Whether we consider the traditions that began with the sayings of Jesus or those that began with the teachings of Paul, we discover a wide diversity within early Christianity. This diversity is so pervasive that some scholars prefer to speak of early Christianities rather than early Christianity, and of Pauline Christianity not as one subset of this larger whole (or wholes) but as a number of subsets—Pauline Christianities. We have already seen that a good deal of this diversity, though not nearly all of it, can be found within the pages of the New Testament. We will see more of this diversity now, as we examine several writings that scholars have come to doubt as having come from the pen of their reputed author, the apostle Paul.

SUGGESTIONS FOR FURTHER READING

Davies, Stevan. *The Revolt of the Widows: The Social World of the Apocryphal Acts*. Carbondale, Ill.: Southern Illinois University Press, 1980. An interesting socio-historical investigation that argues that the apocryphal Acts, including the *Acts of Paul and Thecla*, were authored by women in order to counter views that came to be canonized in the New Testament.

Elliott, J. K. *The Apocryphal New Testament: A Collection of Apocryphal Christian Literature in an English Translation*. Oxford: Clarendon, 1993. An excellent one-volume collection of noncanonical works, including the apocryphal Acts, in a readable English translation with nice, brief introductions.

Furnish, Victor Paul. *Jesus according to Paul*. Cambridge: Cambridge University Press, 1993. An introductory discussion of Paul's understanding of Jesus that raises the question of how much Paul actually knew about Jesus' life; ideal for beginning students.

MacDonald, Dennis. *The Legend and the Apostle: The Battle for Paul in Story and Canon*. Philadelphia: Westminster, 1983. A fascinating account that argues that the Pastoral epistles were written pseudonymously in Paul's name to counter views attributed to Paul in the apocryphal Acts.

Pagels, Elaine. *The Gnostic Paul: Gnostic Exegesis of the Pauline Letters*. Philadelphia: Fortress, 1975. A full discussion of the ways Gnostic interpreters understood each of Paul's letters, appropriate for students familiar with the basic issues.

CHAPTER 22

In the Wake of the Apostle: The Deutero-Pauline and Pastoral Epistles

None of the New Testament writings that we have studied to this point can rightly be called pseudonymous. A pseudonymous writing, or "pseudepigraphon," to use the technical term (plural "pseudepigrapha"), is a book whose author writes under a false name, claiming to be someone other than he or she really is. None of the New Testament Gospels or the Johannine epistles or the book of Acts makes any such claim. As we have seen, these books were all written anonymously, only later to be attributed to persons named Matthew, Mark, Luke, and John. The book of James is in a somewhat different category since its author gives his name. If the author had claimed to be James the brother of Jesus, then we could rightly call his book pseudonymous, if we could show that he was not who he said he was. But as we saw in the previous chapter, James was a common name in antiquity, and this particular James does not claim to be Jesus' brother. Rather than being pseudonymous, then, his book is probably better considered "homonymous," that is, written by someone with the same name as a famous person.

We have found examples of pseudonymous writings outside of the New Testament, however, in such works as the Gospels of Thomas and Peter, the Pseudo-Pauline letter of *3 Corinthians*, and *Pseudo-Titus*. Is it conceivable that any books of this sort came to be included in the New Testament canon? The consensus among critical scholars is a resounding yes. Before launching into a discussion of six such books—the three Deutero-Pauline epistles and the three Pastorals—I will set the stage a bit further by discussing the broader phenomenon of pseudonymity in the ancient world.

PSEUDONYMITY IN THE ANCIENT WORLD

In the modern world, there are two kinds of pseudonymous writing. On the one hand, some authors assume a pen name simply to keep their identity secret (sometimes, a transparent secret); this was the case when Samuel Clemens wrote as Mark Twain and when Marian Evans wrote as George Elliot. On the other hand, some authors deceptively claim to be someone famous. This happened, for example, some years ago when the so-called Hitler diaries turned up. These were forged to look like journals kept by Adolf Hitler through the Second World War. At first, the forger's craft fooled the world, but before long experts determined beyond any doubt that the books were not authentic. They were then relegated to the trash heap of historical curiosities.

Thus, in the modern world, a "forgery" is a kind of pseudonymous writing in which an author falsely claims, for one reason or another, to be a famous person. Antecedents for this kind of pseudonymous writing can certainly be found in the ancient world. Indeed, forgery was a relatively common and widely recognized practice in antiquity. This was a world in which there were no copyright laws and, in fact, no legislation of any kind to guarantee literary ownership. Nor were means available for the mass production of literature; authors could not count on the worldwide dissemination of their books or assume that the kind and quality of their work would be widely known. Books were manufactured one at a time, by hand. New copies were ponderously and painstakingly made from old ones and disseminated slowly and sporadically at best. Libraries were rare, and most people could not read in any case. For most people, reading a book meant hearing someone else read it aloud.

We know that forgery was relatively widespread in this world because the ancients themselves say so. Authors throughout Greek and Roman antiquity make numerous references to the practice and issue frequent warnings against it. Some authors even mention books that were falsely written in their own names. One famous author from the second century C.E., the Roman physician Galen, went so far as to write a book explaining how his authentic writings could be distinguished from those forged by others. Sometimes the forger himself was caught in the act, as happened with the author of *3 Corinthians*. More commonly, literary people had to judge whether a book was authentic or not on the basis of its writing style and contents.

A number of factors motivated ancient authors to produce documents in someone else's name. For some forgers, there was the profit motive. If a new library began collecting old books and advertised its willingness to pay good gold for original copies, an amazing number of "originals" could show up (sometimes of works that no one had ever heard of before!). A different motivation was at work in the philosophical schools, where authors sometimes wrote in the name of their teacher, not in order to sell their works for a profit but as an act of humility. In the Pythagorean school, for example, some writers were quite forthright in this view: since everything they thought and believed was ultimately derived from the philosophy of their founder Pythagoras, it would be the height of arrogance for them to lay claim to any originality. Such persons attributed the treatises they wrote to Pythagoras and considered it a virtue.

Perhaps the most common reason to forge a writing in antiquity was to get a hearing for one's own views. Suppose that you as an amateur philosopher wanted to present your ideas to the world, not to make yourself rich or famous but simply because, in your judgment, the world needed to hear them. If you wrote in your own name (Mark Aristides, or whatever), no one would be much intrigued or feel compelled to read what you had to say, but if you signed your treatise "Plato," then it might have a chance.

Someone who wrote in the name of a famous person was therefore not necessarily driven by wicked intent. Sometimes the writer's motive was pure as the driven snow, at least in his or her opinion. For example, the Christian caught red-handed in the act of forging *3 Corinthians* and other "Pauline" works claimed that he had done it out of "love of Paul," according to the church father Tertullian, who recounts the incident (see box 22.1). Presumably he meant that he wanted to show what Paul would have written from beyond the grave, had he been able to address the problems that had arisen in the church. Other Christians and Jews may have been similarly motivated, including, for example, the author of the canonical book of "Daniel," who lived in the second century B.C.E. but wrote in the name of the famous wise man of four centuries earlier.

Ancient forgers used some fairly obvious and standard techniques to convince their readers that they were who they said they were. To begin with, the mere claim to be somebody carries a lot of weight with most readers, ancient and modern. If a book begins with the words "I Moses write to you these words" or "The vision which I, Abraham, had" or "Paul an apostle of Jesus Christ, to the saints who are in Ephesus," then most readers will simply assume that the alleged author is the actual author, barring the presence of something obvious in the text to discourage the assumption. The trick of the forger was to make sure that nothing of the sort could be found. Forgers, therefore, typically tried to imitate the writing style of the author they were claiming to be. Of course, some forgers made a more strenuous effort along these lines than others, and some were more gifted at it. Such imitation was actually an art that was taught in

SOME MORE INFORMATION

Box 22.1 Paul's Third Letter to the Corinthians

We have already seen a sample of a Pauline pseudepigraphon in the forged correspondence between the apostle and the Roman philosopher Seneca. Another example is the third letter that Paul allegedly wrote to the Christians of Corinth to oppose heretics who had arisen in their midst. As the following extract shows, the letter was in fact produced after Paul's death, to attack views that proto-orthodox Christians of the mid second century considered heretical, including the docetic view that Jesus did not have a real fleshly body and the adoptionist view that his mother was not a virgin. Interestingly enough, these are issues that Paul himself never explicitly addresses in his authentic letters.

> Paul, the prisoner of Jesus, Christ, to the brethren in Corinth—greeting! Since I am in many tribulations, I do not wonder that the teachings of the evil one are so quickly gaining ground. For my Lord Jesus Christ will quickly come, since he is rejected by those who falsify his words. For I delivered to you in the beginning what I received from the apostles who were before me, that . . . God, the almighty, who is righteous and would not repudiate his own creation, sent the Holy Spirit through fire into Mary the Galilean, who believed with all her heart, and she received the Holy Spirit in her womb that Jesus might enter into the world, in order that the evil one might be conquered through the same flesh by which he held sway, and convinced that he was not God. For by his own body, Jesus Christ saved all flesh. . . . (3 Cor 1:1–4, 12–14)

the schools of higher learning as part of rhetorical training. Advanced students were regularly required to compose a speech on a set theme imitating the style of a great orator of the past.

Forgers typically added elements of verisimilitude to their works, that is, comments designed to make the writing appear to have come from the pen of its alleged author. In a forged epistle, for example, such comments might include off-the-cuff references to an event that the reader could be expected to recognize as having happened to the alleged author, personal requests of the recipient (why would anyone other than the real author ask his reader to do something for him?), or even an emphatic insistence that he himself really is the author, sometimes making it appear that the author "doth protest too much." One of the most interesting ploys along these lines is when a pseudonymous author insists that his readers not read books that have been written pseudonymously; who would suspect such an author to be a forger himself? An intriguing example occurs in a Christian book of the fourth century called the *Apostolic Constitutions*, a set of church instructions allegedly written by the apostles after Jesus' resurrection. The book admonishes its readers not to read books that falsely claim to be written by the apostles!

This final ploy can tell us something about the attitudes toward forgery among people in antiquity. Some modern scholars have argued that the practice was so widespread that nobody passed judgment on it; others have claimed that forgeries were so easily detected that everyone could see through them and simply accepted them as literary fictions. The ancient sources themselves suggest that both views are wrong. Forgers were commonly successful because people did not always see through them. When they did see through them, they were usually not amused. Indeed, despite its common occurrence, forgery was almost universally condemned by ancient authors (except among members of some of the philosophical schools).

Scholars in the ancient world went about detecting forgeries in much the same way that modern scholars do. They looked to see whether the ideas and writing style of a piece conformed with those used by the author in other writings, and they examined the text for any blatant anachronisms, that is, statements about things that could not have existed at the time the alleged author was writing (like the letter reputedly from an early seventeenth-century American colonist that mentions "the United States"). Arguments of this kind were used by some Christian scholars of the third century to show that Hebrews was not written by Paul or the Book of Revelation by John the son of Zebedee. Modern scholars, as we will see, concur with these judgments. To be sure, neither of these books can be considered a forgery. Hebrews does not claim to be written by Paul (it is anonymous), and the John who wrote Revelation does not claim to be the son of Zebedee (it is therefore homonymous). Are there other books in the New Testament, though, that can be considered forgeries?

The question itself brings us up against a problem of terminology. Many scholars are loath to talk about New Testament "forgeries" because the term seems so loaded and suggestive of ill intent. But the word does not have to be taken that way. It can simply refer to a book written by an author who is not the famous person that he or she claims to be. It is striking that few scholars object to using the term "forgery" for books, even Christian books, that occur outside of the New Testament. This may suggest that the refusal to talk about New Testament forgeries is not based on historical grounds but on faith commitments (either of the scholars or of their audiences), that is, it represents a theological judgment that the canonical books need to be granted a special status. A historical introduction to these books should not, however, be so bashful.

Neither, of course, should it be bashing. When I use the term "forgery," I do not mean it in a derogatory sense. The authors of these forged documents may well have been upright individuals who had good reasons for doing what they did, or at least thought they did. If they wrote in the name of some other famous person, however, they were still producing a forged document. This is no less true for the canonical letter allegedly to Titus than for the noncanonical letter allegedly from Titus.

What now can we say about the Deutero-Pauline epistles of 2 Thessalonians, Colossians, and Ephesians, and the Pastoral epistles of 1 and 2 Timothy and Titus? What are these letters about, and did Paul, their alleged author, really write them?

THE DEUTERO-PAULINE EPISTLES

2 Thessalonians

We can begin with the letter whose authorship remains in greatest doubt, 2 Thessalonians. As was the case with 1 Thessalonians, this letter claims to be written by "Paul, Silvanus, and Timothy to the church of the Thessalonians" (1:1). Whoever the actual author of the letter was, its occasion appears to be reasonably clear. It was written to a group of Christians who were undergoing intense suffering for their faith (1:4–6). We do not know how this suffering manifested itself—whether there was some kind of official governmental opposition to these people, or hostility from the local population, or something else. We do know that the author wrote to assure his readers that if they remained faithful, they would be rewarded when Christ returned in judgment from heaven. At this "parousia" of Jesus, those who opposed them and rejected their message would be punished with "eternal destruction," but the saints would enter into their glorious reward (1:7–12).

A second reason for the letter was that some members of this Christian community had come to believe that the end of time had already come upon them, that is, that the day of judgment was going to happen not in the indefinite future but right away (2:1–2). Some of those who thought this found confirmation in prophecies spoken by members of the congregation and, still more interestingly, in a letter that was reputedly written by Paul (2:2). The author of 2 Thessalonians, claiming to be the real Paul, warns his readers not to be deceived. Whatever an earlier forger may have asserted, the end had not yet come because there were certain events that had to transpire first (2:3).

The author describes these events in an apocalyptic scenario that sounds very much like what we find in the Apocalypse of John. A kind of antichrist figure is to be revealed on earth before Christ returns; this "lawless person" is ultimately "destined for destruction" (2:3). Exalting himself above every other "so-called god or object of worship," he will eventually take his seat in God's Temple in Jerusalem, "declaring himself to be God" (2:4). The author reminds his readers that he fully informed them of this scenario when he was with them (2:5); moreover, it has obviously not yet occurred, since no one has yet come forward to assume the grandiose role of this antichrist. Indeed, the author mysteriously indicates that there is some supernatural force restraining the lawless one for the time being, but once this force is removed, he will make his appearance, setting in motion the final confrontation between Christ and the forces of evil headed by Satan (2:6–12).

In large measure, then, this letter was written to assure this congregation of Christians that the end was not yet upon them. As "Paul" fully instructed them previously (2:5), Christ would not return until this apocalyptic scenario played itself out.

We discover in the final chapter of the book that the problem in the congregation was not simply one of establishing an appropriate timetable for upcoming events. Some members of this church were so persuaded that the end was absolutely imminent that they had quit their jobs and were simply waiting for it to happen (3:6–15). Their decision had grave social implications. Those who kept their jobs were having to feed those who hadn't, and this situation of apocalyptic freeloading was a source of tension in the congregation. In terms quite reminiscent of 1 Thessalonians, the author reminds his readers how he and his companions had lived among them, working for their own meals and refusing to be a burden on others (3:7–10). He insists that they do likewise (3:11–15).

The author of this book certainly sounds like Paul, at least in such places as the prescript, which is very close to the opening of 1 Thessalonians, and the recollection of Paul's toil among the Thessalonians when he was first with them. Also, a number of Pauline themes are sounded throughout the epistle. These include the necessity of suffering, the expectation of ultimate vindication, and the apocalyptic hope that stood at the core of Paul's gospel.

Do these similarities mean that Paul wrote the letter? The problem from a historian's point of view is that someone who had decided to imitate Paul would no doubt try to sound like Paul. If both Paul and an imitator of Paul could sound like Paul, how could we possibly know whether we are dealing with the apostle himself or one of his later followers?

There is a way to resolve this kind of historical whodunit, and it involves looking at the other side of the coin, that is, at the parts of the this letter that do not sound like Paul. These peculiar features provide the best indicators of whether the letter is authentic or was written by a member of one of Paul's churches after the apostle himself had passed from the scene. Such negative evidence is useful because we would expect an imitator to sound like Paul, but we would not expect Paul not to sound like Paul. It is, therefore, the differences from Paul that are most crucial for establishing whether Paul wrote this, or any other, disputed letter.

With respect to 2 Thessalonians, the most intriguing issue is one that I have already alluded to: the author writes to assure his readers that even though the end will be soon it will not come right away. Other things must happen first. They should therefore hold on to their hopes and their jobs, for there is still time left. Does this sound like the same person who urged the readers of his first letter to stay alert so as not to be taken by surprise when Jesus returns (1 Thess 5:3, 6), since the end would come with no advance warning, "like a thief in the night" (1 Thess 5:2), bringing "sudden destruction" (1 Thess 5:3)? According to 2 Thessalonians there will be plenty of advance warning. That which is restraining the man of lawlessness will be removed, then the antichrist figure will reveal himself, exalt himself above all other objects of worship, establish his throne in the Jerusalem Temple, and declare himself to be God. Only then will Christ return. How is this like a thief in the night who comes when people least expect it?

It is particularly interesting that the author claims to have taught the Thessalonians these things while he was with them (2:5). If he had done so, one might wonder why he did not appeal to this knowledge of upcoming events in his first letter, when he answered the Thessalonians' question about "those who have fallen asleep," for example, by pointing out that of course some people would die before the end since it was not imminent. In 1 Thessalonians, however, he does not say, "Remember that the day of the Lord is not already here; first the man of lawlessness must be revealed." Indeed, if the Thessalonians had already been fully apprised of this future course of events at the time of the first letter, one might wonder why they were surprised by the death of some of their members in the first place.

Finally, if the future appearance of the antichrist actually was a central component of Paul's teaching, as intimated in 2 Thess 2:5, it is very strange that he never says a word about it in any of his other letters. These difficulties make it hard to see how Paul could have written both of the letters to the Thessalonians. One of the most interesting things about the second one is how it ends: "I, Paul, write this greeting with my own hand. This is the mark in every letter of mine; it is the way I write" (3:17). This means that "Paul" dictated the letter to a scribe but then added his own signature to it, as he did, for example, in Galatians (see Gal 6:11). What is peculiar is that he claims this to be his invariable practice, even though he does not appear to have ended most of his other letters this way, including, 1 Thessalonians! The words are hard to account for as Paul's, but they make perfect sense as the words of an imitator of Paul who wants his readers to be assured that despite the fact that they have received at least one letter that was forged in Paul's name (2:2), this is not another one.

We obviously don't know who actually wrote this letter if it wasn't Paul and can only speculate about when he was living. We can assume that he wrote sometime after Paul had died, possibly near the end of the first century, when writing letters in Paul's name became both more feasible and, from what we can tell, more popular. Moreover, we know that during the period some Christian groups were beginning to face increased hostilities within their social contexts and that some of them were turning to a renewed hope in the return of Christ in light of these conflicts.

Thus the author must have been a Christian from one of the churches that Paul established, who evidently had read 1 Thessalonians (hence, for example, the similar prescript). He wrote to help resolve the problems that Christians of his day were facing, choosing to do so in the name of Paul, the founder and hero of his church, one whose words would be heard and heeded. Writing as the apostle himself, he urged his readers to keep the faith and to maintain their hope but not to expect the end of the age in the immediate future. God's plan for the end was in the process of being implemented, but believers must not be too eager, living only for tomorrow and not tending to the needs of today. They must suffer boldly and wait faithfully for the day of judgment in which their longings would be fulfilled and their afflictions vindicated.

Colossians

As is the case with 2 Thessalonians, scholars continue to debate the authorship of Colossians, although here there is an entirely different set of problems to consider. There is no real problem, however, in understanding the ostensible occasion of the letter. "Paul" is in prison for preaching the gospel (4:3). While there, he has heard news of the church in Colossae (1:3), a small town in western Asia Minor not far from the larger cities of Hierapolis and Laodicea. "Paul" did not establish this church, but his coworker and companion Epaphras, a citizen of the place, did (1:7–8, 4:3). The news that "Paul" has learned about the Colossians is mixed. On the one hand, he is excited and pleased to learn that they have converted to faith in Christ and have committed themselves to his gospel through the work of Epaphras (1:7–8). On the other hand, he has learned that there are false teachers among them who are trying to lead them into a different kind of religious experience (2:4). He is writing to address the situation.

The author of the letter alludes to his opponents' notions but does not give a detailed description of them, on the assumption, we might suppose, that his readers already knew full well what he was talking about. He labels this new teaching a "philosophy and empty deceit" (2:8) and counters it by indicating that believers have already experienced a "spiritual circumcision" (2:11). Moreover, he insists that since Christ has erased the requirements of the Jewish Law for believers through his death, they need not follow regulations concerning what to eat and concerning what special days to keep as religious festivals (2:13–17). These passages make it appear that the false teachers were advocating some form of Judaism, perhaps like the opponents of Paul in Galatia. But they also insisted on "self-abasement and the worship of angels," basing their appeal on special visions that they have had (2:18–19). This suggests that they advocated an ascetic lifestyle and possibly the ecstatic adoration of higher beings.

Scholars have debated the precise nature of this false teaching for many years. In general terms "Paul's" opponents were evidently promoting some kind of Jewish mysticism, comparable to that known from other ancient texts, in which people were encouraged to experience ecstatic visions of heaven and thereby be transported to the divine realm where they would find themselves filled with the joy and power of divinity. Such people were commonly ascetic, urging that bodily desires must be avoided if one wanted to escape the body and enjoy the pleasures of the spirit. If these persons were Jews, they may well have rooted their asceticism in the Jewish Scriptures and so, perhaps, urged their followers to keep kosher food laws, observe the Sabbath, and if they were males to be circumcised.

In response to these views, the author of Colossians insists that Christ himself is the fullest expression of the divine. In his words, Christ is the very "image of the invisible God, the firstborn of all creation" (1:15). There is little reason for Christian believers to worship angels when they can worship the one "in whom all the fullness of God was pleased to dwell" (1:19). Indeed, the other invisible beings are said to have been both created by and made subservient to Christ himself: "For in him all things in heaven and on earth were created, things visible and invisible, whether thrones or dominions or rulers or powers—all things have been created through him and for him" (1:16). Moreover, Christ alone is responsible for the ultimate benefits bestowed upon the believer. It is Christ who has reconciled all people to God (1:21–22; 2:13–15). When he did so, he destroyed everything that brought alienation, including the Law with all of its "legal demands" (2:14). What sense is there, then, in returning to the adherence to the Law? For this author, Christ destroyed the need to do so, and those who are in Christ can enjoy the full benefits of the divine (2:10, 14–19).

These benefits, which are conferred only through Christ, include an exalted status that is already available to the believer. This author maintains that there is no need for physical circumcision for those who have experienced the real, spiritual circumcision that comes through faith in Christ (2:9–10), or for ecstatic worship of angels for those who have already been raised up to the heavenly places in Christ (2:12; 3:1–3), or for human regulations of what to handle and what to eat, which give only the appearance of piety, for believers in Christ who have a full experience of the divine itself (2:20–23). Indeed, all that the Colossians have sought through their mystical experiences is already theirs in Christ, so long as they do not depart from the gospel message they have heard (2:23).

The Colossians are therefore to enjoy the full experience of the divine as those who have been raised to the heavenly places in Christ (3:1). This does not mean, however, that they can neglect their physical lives in this world or behave as though their bodies no longer matter. Indeed, they must go on living in this world until Christ returns. This means maintaining moral and upright lives. Thus the author gives a number of moral exhortations concerning vices to avoid (fornication, passion, greed, and the like; 3:5–11) and virtues to embrace (compassion, kindness, humility, and the like; 3:12–17). In addition, he gives advice to different social groups within the congregation concerning their interactions with one another, addressing wives and husbands (3:18–19), children and fathers (3:20–21), slaves and masters (3:22–4:1).

The letter closes with some final instructions (4:2–6), greetings to members of the Colossian church, both from "Paul" and those with him (4:7–17), and his own signature and final benediction (4:18). Was this, however, actually Paul's signature?

In a number of ways, this letter looks very much like those that Paul himself wrote. The prescript written in the names of both Paul and Timothy, the basic layout of the letter, and the closing all sound like Paul, and a number of important Pauline themes are sounded throughout: the importance of suffering in this world, Jesus' death as a reconciliation, and the participation of believers in Jesus' death through baptism. Paul may well have written this letter.

Over the past century, however, scholars have put forward a number of arguments against the authenticity of Colossians. Some of these arguments are not very strong. Some scholars, for instance, have claimed that the vocabulary is largely non-Pauline, despite the fact that the number of unusual words here is about the same as in Philippians, an undisputed epistle of comparable size. Others have insisted that there is no trace of Paul's apocalyptic views here, apparently ignoring such passages as 3:1–6. Still others have asserted that Paul would not have written to a congregation that he didn't found himself, overlooking, evidently, his letter to the Romans! The situation is different in Romans, of course, but at least in Colossians "Paul" is writing to a congregation that he could consider his own, in that his companion Epaphras supposedly founded it.

There are, however, more solid grounds for questioning Paul's authorship of this letter. One of the most compelling arguments depends on a detailed knowledge of Greek, for the writing style of Colossians differs markedly from that found in Paul's undisputed letters. Whereas Paul tends to write in short, succinct sentences, the author of Colossians has a more complex, involved style. The difference is not easily conveyed in English translation, in part because the long complicated Greek constructions have to be broken up into smaller sentences to avoid making them appear too convoluted. Colossians 1:3–8, for example, consists of just one sentence in Greek. The problem is not that this is bad or unacceptable Greek but that Paul wrote in a different style (just as Ernest Hemingway and William Faulkner both wrote correct English, but in very different ways). This kind of evidence has convinced a large number of linguistic specialists that Paul did not write the letter.

Other arguments can be more readily evaluated just from the English text. The most striking is one that you may have already surmised: the author believes that Christians have participated with Christ not only in his death but also in his resurrection. The author is quite emphatic on this critical point: believers have already been raised with Christ "in the heavenly places" to enjoy the full benefits of salvation (2:12; 3:1). Paul himself, however, is equally emphatic: even though Christians have "died" with Christ in their baptism, they have not yet been raised with him. They will not be raised until the very end, when Christ returns (see box 22.2). Not only does Paul stress this point in his most explicit discussion of a baptized person's participation with Christ in his death in Romans 6, but he also argues precisely this point against his opponents in Corinth, who claimed already to have experienced the resurrection and so to be ruling with Christ.

How is it that Paul in his undisputed letters can be so emphatic that believers have not yet experienced the resurrection with Christ, whereas the author of Colossians can be equally emphatic that they have? It is certainly possible that Paul changed his mind, either because he genuinely thought better of it later (although this seems unlikely given his vehemence on the point) or because when attacking a different heresy, he had to take a different approach, either consciously misrepresenting his views or forgetting what he had earlier said. It seems more plausible, however, that Paul went to his grave believing, and consistently insisting, that Christians had not yet been raised with Christ. If so, it is hard to accept that he wrote the letter to the Colossians.

This conclusion is supported by the fact that the author of Colossians has a different writing style from Paul's. It also makes sense of other anomalies in the letter, two of which I will mention here. For one thing, the author is particularly

SOMETHING TO THINK ABOUT

Box 22.2 The Resurrection of Believers in Paul and Colossians

If Paul did write Colossians, then his views about the time and significance of the resurrection of Christians changed, for here believers are said already to "have been raised with Christ" (3:1). Recall that 1 Corinthians was written in large measure against those who believed that Christians had already come to enjoy the blessings of the resurrected existence (see 1 Corinthians 15). The contrast in the verb tenses of Romans 6:4 and Colossians 2:12 (see italics) is also telling.

Rom 6:4	Col 2:12
For if we have been united with him in a death like his, we *will* certainly *be united* with him in a resurrection like his. . . . But if we have died with Christ, we believe that we *will* also *be raised* with him.	When you were buried with him in baptism, you *were also raised* with him through faith in the power of God, who raised him from the dead.

concerned with the interactions of believers in their social arrangements, as wives and husbands, children and fathers, slaves and masters. You won't find such things emphasized in Paul's undisputed letters, possibly because Paul did not look upon his churches as being in this world for the long haul (see the discussion of Philemon). For Paul, social arrangements were what they were, and there was no need for Christians to go out of their way to disrupt or sustain them. Since the end was near, believers were to put their energies into preparing for it rather than bothering themselves with the rules and standards of society. The household rules given in Colossians, on the other hand, show that this author expected the church to be around for a long time.

In addition, we should consider the nature of the false teaching in Colossae. If the false teachers there were urging Gentiles to be circumcised and to keep parts of the Law, as suggested by 2:8–19, why isn't "Paul" totally outraged and incensed, as he was in Galatians? Here he is positive and upbeat, trying to show them a better way, portraying the Jewish Law as simply passé and unnecessary. Does this attitude jibe with the rip-roaring, white-hot anger that Paul spewed forth when a similar problem emerged elsewhere?

You will have to evaluate these arguments for yourself. If Paul wrote the letter, then the ostensible occasion set forth at the outset of this discussion was the real occasion, and Paul adopted a different writing style, advocated different views, and assumed a different tone from his other letters. On the other hand, if these changes do not seem plausible, then we must conclude that Paul did not write the letter.

Who wrote the letter if Paul did not? We will never know, but he must have been a member of one of Paul's churches who saw the apostle as an ultimate authority figure. This person wrote a fictitious letter to deal with a real problem that he had come to know about, possibly within his own congregation. If this is what happened, though, then the address to the "Colossians" is itself probably a fiction, for the town, and any church that happened to be there, was destroyed by an earthquake around the year 61 C.E. It may well be that this unknown author had access to one or more of

Paul's other letters, including almost certainly the letter to Philemon, since the same names appear in the greetings of the two letters. Using these other letters as models, he penned an authoritative denunciation of a false philosophy that had begun to spread, putting this pseudonymous writing into circulation as an authentic letter of the apostle Paul.

Ephesians

While the arguments against the Pauline authorship of 2 Thessalonians and especially of Colossians have persuaded a number of scholars, with the letter to the Ephesians the matter is even more clear cut. The majority of critical scholars are convinced that Paul did not write this letter. To say that scholars are convinced of this position, however, is not to say that it has been proven. Scholarly opinion, after all, is still opinion, not fact. For this reason, you will need to evaluate the evidence for yourself (at least as much of the evidence as I can present here) and make your own judgment.

Before jumping to the question of authorship, we should begin once more with the ostensible situation lying behind the epistle. Unlike with the other letters of the Pauline corpus, the occasion for Ephesians is notoriously difficult to determine. We do learn that "Paul" was writing from prison to Gentile Christians (3:1). There is some question, though, concerning where the epistle was sent and for what reason.

Most English translations indicate that the addressees are "the saints who are in Ephesus" (1:1), but the words "in Ephesus" are not found in the earliest and best Greek manuscripts of this letter. Most textual experts think that the words were not in the letter originally but were added by a scribe after it had already been in circulation for a time. If so, then Ephesians was written as a kind of "circular letter," designed to make the circuit of a number of Pauline churches, sent to "the saints who are faithful" but not to the saints of any particular location. Such a letter would have been copied in several of the places that it was received, including the city of Ephesus. It appears that the copyist in Ephesus decided to personalize the letter by adding the words "in Ephesus" to the addressees, so that when the Ephesian Christians read it they would think that it was written particularly to them. Then, both this scribe's copy of the letter and other copies that lacked the words "in Ephesus" were used by later copyists who reproduced the letter. This would explain why some of our surviving manuscripts have the words "in Ephesus" and others don't. (We will discuss the interesting business of how and why Christian scribes changed their texts in Chapter 28.)

Originally, then, the letter may not have been sent to a particular congregation but to a number of congregations, for example, throughout Asia Minor. The overarching purpose of Ephesians is to remind its Gentile readers that even though they were formerly alienated from God and his people, Israel, they have now been made one through the work of Jesus—one with the Jews through Jesus' work of reconciliation and one with God through his work of redemption (2:1–22). More specifically, Jesus' death has torn down the barrier that previously divided Jew and Gentile, that is, the Jewish Law, so that both groups are now absolutely equal; Jews and Gentiles can live in harmony with one another without the divisiveness of the Law (2:11–18). Moreover, Christ has united both Jew and Gentile with God (2:18–22). Believers have not only died with Christ, they have also been raised up with him to enjoy the benefits of a heavenly existence (2:1–10). Thus Jew and Gentile are unified with one another and with God. This is the "mystery" of the gospel that was concealed from earlier generations but has now been revealed to "Paul" and through him to the world (3:1–13).

The second half of the letter (chaps. 4–6) consists of exhortations to live in ways that manifest this unity. It is to be evident in the life of the church (4:1–16), in the distinctiveness of the believers from the rest of society (4:17–5:20), and in the social relations of fellow Christians, that is, in their roles as wives and husbands, children and fathers, slaves and masters (5:21–6:9). The letter closes with an exhortation to continue to fight against the forces of the devil that are trying to disrupt the life of the congregation (6:10–20) and then "Paul's" final closing statement and benediction (6:21–24).

Once again, however, we must ask the critical question: was this letter actually sent by Paul? Broadly speaking, Ephesians may sound like something that Paul could have written. Allowance must be made, of course, for its character as a circular letter, in which the author addresses no specific problem, such as moral improprieties or false teachings, and therefore offers no specific resolutions. Some scholars have argued that Paul would not have written such a letter, but how could we know?

The real difficulty with Ephesians is not with its occasion or broad scope but with the details of what the author actually says and the way in which he says it (as was also the case with 2 Thessalonians and Colossians). While the writing style of Colossians appears to be non-Pauline, the style of Ephesians is even more so. No one who reads this letter in Greek can help being struck by its incredibly long sentences when measured against Paul. In Greek, the opening thanksgiving of 1:3–14 (twelve verses) is one sentence. Again, this is not bad writing style; it simply isn't Paul's.

Some scholars have demonstrated this point in convincing terms (see the article on Ephesians by Victor Furnish in the *Anchor Bible Dictionary II*. 535–42). There are something like a hundred complete sentences in this book, nine of them over fifty words in length. Contrast this with what you find in Paul's undisputed letters. Philippians and Galatians, for example, are roughly the same length as Ephesians; Philippians has 102 sentences, only one of which is over fifty words, and Galatians has 181 sentences, with only one over fifty words. Or consider these portions of the longer undisputed letters: in the first four chapters of Romans there are 581 sentences, only three of which are over fifty words; in the first four chapters of 1 Corinthians, there are 621 sentences, with only one over fifty words. Paul tended to write in a succinct style. The author of Ephesians did not.

In addition, this author uses a total of 116 words that are not found in any of Paul's undisputed letters. To be sure, Paul uses unique words in all of his letters, depending on what he happens to be talking about, but 116 non-Pauline words seems inordinately high compared with what we find

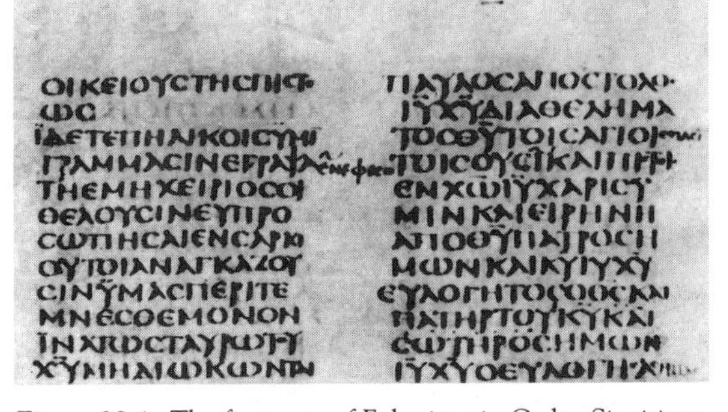

Figure 22.1 The first page of Ephesians in Codex Sinaiticus, the oldest complete manuscript of the New Testament. Notice that the first verse has been corrected in the margin. The letter was originally addressed "to the saints," but a later scribe made the address more specific by inserting the phrase "who are in Ephesus." For a discussion of such scribal changes of our manuscripts, see Chapter 28.

elsewhere. For example, the book of Philippians, a letter of comparable, but slightly shorter, length, has one of the highest number of unique words (in proportion to the total number of words) among Paul's undisputed epistles, but the total there is only 76.

When taken in combination with what the letter of Ephesians actually says, these differences in style and vocabulary suggest that someone other than Paul wrote it, someone imitating the letters of Paul but without complete success. To examine the contents of Ephesians, we can look at one particular passage that is central to the overarching theme of the book and whose ideas appear to resemble those that Paul sets forth in some of his undisputed letters. Once we move beneath the surface, however, these resemblances begin to evaporate.

Ephesians 2:1–10 discusses the conversion of its Gentile readers from their earlier lives to the salvation they have experienced in Christ. There are a number of important Pauline themes here: a person's separation from God before being converted to Christ is spoken of as "death" (vv. 1–2), the cosmic power of the devil is designated as "the ruler of the power of the air" (v. 2), the grace of God brings salvation through faith, not works (vv. 8–9), and the new existence leads to a moral life (v. 10). Surely this is Pauline material.

There are peculiarities here as well, however, as we can see when we dig deeper into the text. The first and most obvious problem concerns the status of the believer, which is described in a way that is strikingly similar to what we found in Colossians. Even though Paul's undisputed letters are quite emphatic that the resurrection of believers (even in a spiritual sense) has not yet happened, the author of Ephesians pronounces that "God . . . made us alive together with Christ . . . and raised us up with him and seated us with him in the heavenly places in Christ Jesus" (vv. 5–6). This view of the Christian believer is even more exalted than the one in Colossians; the words the author uses of the believer's status mirror those he uses of Christ himself:

> God put this power to work in Christ when he raised him from the dead and seated him at his right hand in the heavenly places, far above all rule and authority and power and dominion, and above every name that is named, not only in this age but also in the age to come. And he has put all things under his feet and made him the head over all things for the church. (vv. 20–22)

According to Ephesians 2, believers are seated with Christ in the heavenly places, above everything else. Can this be the same author who castigated the Corinthians for maintaining that they had already come to be exalted with Christ and were therefore already ruling with him?

Another interesting difference from Paul's own letters is the way the author of Ephesians 2:1–10 conceptualizes "works." In Paul's gospel, Gentiles are made right with God not by doing the works of the Law but through faith in Christ's death. Thus, when Paul speaks about works, he is referring to doing those aspects of the Law that make Jews distinctive as the people of Israel (e.g., circumcision and kosher food laws). Ephesians, however, no longer refers to the Jewish Law, but speaks instead of "good deeds" (see 2:8–10). Interestingly, as we found in the previous chapter, the author of James countered a later version of Paul's gospel that insisted that faith without doing good deeds was adequate before God. It appears that the author of Ephesians understands "works" in this later, non-Pauline, sense.

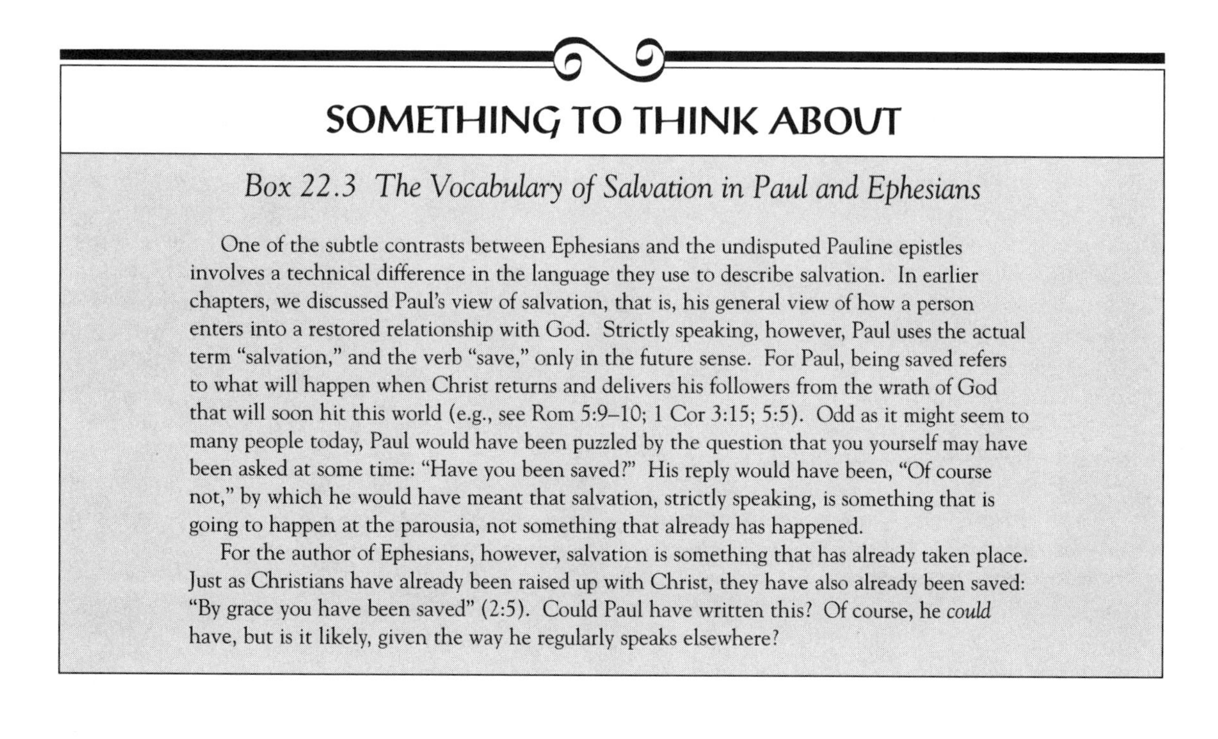

SOMETHING TO THINK ABOUT

Box 22.3 The Vocabulary of Salvation in Paul and Ephesians

One of the subtle contrasts between Ephesians and the undisputed Pauline epistles involves a technical difference in the language they use to describe salvation. In earlier chapters, we discussed Paul's view of salvation, that is, his general view of how a person enters into a restored relationship with God. Strictly speaking, however, Paul uses the actual term "salvation," and the verb "save," only in the future sense. For Paul, being saved refers to what will happen when Christ returns and delivers his followers from the wrath of God that will soon hit this world (e.g., see Rom 5:9–10; 1 Cor 3:15; 5:5). Odd as it might seem to many people today, Paul would have been puzzled by the question that you yourself may have been asked at some time: "Have you been saved?" His reply would have been, "Of course not," by which he would have meant that salvation, strictly speaking, is something that is going to happen at the parousia, not something that already has happened.

For the author of Ephesians, however, salvation is something that has already taken place. Just as Christians have already been raised up with Christ, they have also already been saved: "By grace you have been saved" (2:5). Could Paul have written this? Of course, he *could* have, but is it likely, given the way he regularly speaks elsewhere?

Just as the notion of "works" appears to have lost its specifically Jewish content, so too does the author's own former life in which he engaged in these works. Paul himself spoke proudly of his former life as one in which he had kept the Jewish Law better than the zealous Pharisaic companions of his youth. In his own words, "with respect to the righteousness found in the Law, I was found to be blameless" (Phil 3:6). Paul's conversion was not away from a wild and promiscuous past to an upright and moral present; it was from one form of rigorous religiosity to another. What of the author of Ephesians? Evidently, he did not conceive of Paul's past in this way, for according to him "all of us once lived among them (i.e., the pagans) in the passions of our flesh, following the desires of flesh and senses" (2:3). It is true that Paul himself occasionally speaks of having been subject to the law of sin and of having done the things that he knew he ought not to have done (Romans 7); but in his undisputed letters the extent of his transgression involved such things as "coveting" what he shouldn't have (Rom 7:7–8), not the wild and dissolute lifestyle of the pagans that he sometimes maligned (e.g., see Rom 1:18–32). In terms of his lifestyle, Paul lived "blamelessly." Not so the author of Ephesians.

Who, then, was this author and why did he write the letter? Once again, our historical curiosity is stymied by a lack of evidence. Clearly the author was a member of a church that was committed to Paul's understanding of the gospel, but he evidently lived at a later time, perhaps near the end of the first century, when some of Paul's views had developed in directions that Paul himself had not taken them, for example with respect to what it meant to be saved apart from works. This author may well have had access to other letters written under Paul's name. Scholars have long noted, for example, a number of similarities between Ephesians and Colossians, including their openings and closings, their views of being raised already with Christ, and their instructions to wives and husbands, children and fathers, slaves and masters.

Possibly, then, an unknown author concerned with tensions that had erupted between Gentiles and Jews in the churches that he knew (in Asia Minor?) wrote to reaffirm what he saw to be the core of Paul's message, that Christ brought about a unification of Jew and Gentile and a reconciliation of both with God, and that all members of the Christian church should respond to their new standing in Christ by embracing and promoting the unity provided from above.

THE PASTORAL EPISTLES

Up to this point I have tried to show why scholars continue to debate the authorship of the Deutero-Pauline epistles, but when we come to the Pastoral epistles, 1 and 2 Timothy and Titus, there is greater scholarly unanimity. These three letters are widely regarded by scholars as non-Pauline. In discussing the authorship of the Pauline epistles, we should constantly remember that we are not asking whether or not Christians in the first or second century would have forged documents in Paul's name. We know for a fact that some did: 2 Thessalonians alludes to a forged letter (2:2), and a proto-orthodox Christian confessed to forging *3 Corinthians*. Moreover, everyone agrees that some of the writings that survive in Paul's name are Christian forgeries (e.g., the correspondence between "Paul" and the philosopher Seneca and the apocalypse written by "Paul"). What we are asking, then, is whether any given document that claims to be written by Paul can sustain its claim.

Before addressing the issue of the authorship of the Pastoral epistles, we should note their ostensible occasion and overarching points, both as a group (since most scholars are reasonably certain that they all came from the same pen) and individually. These letters are grouped together as pastoral epistles because each claims to be written by Paul to a person he has appointed to lead one of his churches: Timothy, his young companion left to minister among the Christians in Ephesus, and Titus, his companion left on the island of Crete. Moreover, these epistles contain pastoral advice, that is, advice from the apostle to his appointed representatives concerning how they should tend their Christian flocks.

Each of these epistles presupposes a slightly different situation, but the overarching issues are the same. The problems involve *(a)* false teachers who are creating problems for the congregations and *(b)* the internal organization of the communities and their leaders. "Paul" urges his representatives to take charge, to run a tight ship, to keep everyone in line, and above all to silence those who promote ideas that conflict with the teachings that he himself has endorsed.

1 Timothy

1 Timothy presupposes that Paul and Timothy visited Ephesus on the way to Macedonia (1:3) and that Paul decided to leave Timothy behind to bring the false teachers under control (1:3–11), to bring order to the church (2:1–15), and to appoint moral and upright leaders to keep things running smoothly (3:1–13). Most of the letter consists of instructions concerning Christian living and social interaction, for instance on how Christians ought to pray, on how they ought to behave towards the elderly, the widows, and their leaders, and on what things they ought to avoid, namely, pointlessly ascetic lifestyles, material wealth, and heretics who corrupt the truth.

The nature of the false teaching that the author disparages is somewhat difficult to discern. Some members of the congregation have evidently become enthralled with "myths and endless genealogies" (1:4). This phrase has struck a chord with modern interpreters familiar with various strands of Christian Gnosticism. Recall from our discussion in Chapter 11 that Gnostic Christians developed elaborate mythologies that traced the genealogies of divine beings all the way back to the one true God. Some strands of Gnosticism were deeply rooted in Judaism; the Jewish Scriptures themselves, especially the first chapters of Genesis, proved to be a limitless resource for speculation about how the world and the supernatural beings who rule it came into existence. It is striking in this connection that the author of 1 Timothy goes on to attack those who want to be "teachers of the law" (1:7).

Most of the Gnostic groups that we know about were rigorously ascetic. Wanting to escape the material world, they chose to punish their bodies so as not to be enslaved by them, refraining from sexual relations and insisting on strict and uninteresting diets. The author of 1 Timothy correspondingly lambastes false teachers because they "forbid marriage and demand abstinence from foods" (4:3). Moreover, he concludes his letter with a final exhortation to "avoid the profane chatter and contradictions of what is falsely called knowledge" (6:20). The Greek word for "knowledge," of course, is *gnosis;* those who were gnostics claimed to know what was not available to the general public, not even to their fellow Christians. It seems altogether reasonable, then, to assume that this letter was directed against an early form of Christian Gnosticism.

The author does not attack the views of his opponents head-on but instead urges Timothy not to heed their words and, if possible, to bring them into submission (1:3). As we will see later, many of the instructions that the author gives to the leadership of the church may represent an effort to become organized in order to face these opponents with a unified front. In any event, the qualifications of those who are to be appointed leaders of the church, the bishops and deacons, whose duties are never spelled out, soon take center stage. Only men are allowed to occupy these positions, and they are to be morally upright and strong personalities who can serve as models to the community and command respect in the world outside the church.

The tight organization of the church is important not only for addressing the problems posed by false teachers but also for monitoring the inner workings of the community itself. In particular, the author is concerned about the role women should play in the congregation (not much of one; see especially 1 Tim 2:11–15) and about the position and activities of "widows," who appear to be enrolled by the church and provided with some kind of material support in exchange for their pious deeds (5:4–16). The author evidently thinks that women in general and widows in particular have stirred up problems and are not to be trusted (e.g., 5:11–13; see Chapter 23).

2 Timothy

The second Pastoral epistle presupposes a somewhat different situation. It too is written by "Paul" to Timothy (1:1). Now, however, "Paul" is in prison in Rome (1:16–17; his location in 1 Timothy was not specified), and he is clearly expecting to be put to death soon (4:6–8), after a second judicial proceeding (the first one evidently did not go well; 4:17).

Figure 22.2 Picture of a woman officiating at the Lord's Supper, an activity that the author of the Pastorals would no doubt have disapproved of. (Some scholars have suggested that the participants in the meal look astonished in this painting.)

He writes to Timothy not only to encourage him to continue his pastoral ministry and to root the false teachers out of his church but also to ask him to join him as soon as possible (4:21), bringing along some of his personal belongings (4:13).

In this letter we learn something more about Timothy himself. He is portrayed as a third-generation Christian, having been preceded in the faith by his mother Eunice and grandmother Lois (1:5). He was trained in the Scriptures from his childhood (3:15) and as an adult became a companion of "Paul," collaborating with him in his mission to some of the cities of Asia Minor (3:10–11). He was ordained to Christian ministry through the ritual of laying on of hands (1:6; 4:1–5). As the author's faithful representative in Ephesus (one of the few anywhere, evidently, see 1:16–17; 4:10–18), Timothy is charged to overcome those who lead the saints astray with their idle talk and corrupt lives (2:16–18, 23–26; 3:1–9; 4:3–5).

There is even less evidence concerning the nature of the false teaching here than in 1 Timothy. Two of the opponents are specifically said to have claimed that "the resurrection has already taken place" (2:17), a claim that sounds familiar from other Pauline writings we have examined. But mostly the author attacks his opponents with general slander, providing no specifics concerning what they actually said. Thus, the opponents are called

> lovers of themselves, lovers of money, boasters, arrogant, abusive, disobedient to their parents, ungrateful, unholy, inhuman, implacable, slanderers, profligates, brutes, haters of good, treacherous, reckless, swollen with conceit, lovers of pleasure rather than lovers of God, holding to the outward form of godliness but denying its power. (3:2–5)

They may well have been all these things and more, but the passage provides no clue about what they actually taught or stood for. Timothy, in any event, is to oppose them with all his strength, and to continue the ministry that "Paul" has assigned to him until he comes to see the apostle in his bondage in Rome.

Titus

The book of Titus is far more like the first Pastoral epistle than the second. Indeed, the letter seems something like a Readers' Digest version of 1 Timothy, with its list of qualifications for church leaders and its moral instructions for members of the congregation in their relations with one another.

The presupposed situation is that "Paul" has left his trusted comrade Titus on the island of Crete as an apostolic representative to the church there (1:4–5). In particular, Titus was supposed to appoint elders, or bishops, in the churches of every town (1:5–9). "Paul" is now writing in order to urge Titus to correct the false teachings promoted by Jewish-Christian believers, which appear to involve both complicated "mythologies" that confuse the faithful (1:10–16) and "genealogies and quarrels about the law" (3:9). As in 1 Timothy, the false teaching may therefore involve Gnostic speculation. Titus is not to argue with these people; he is to warn them twice to change their views and afterwards simply ignore them, "since you know that such a person is perverted and sinful, being self-condemned" (3:11). The errant parties themselves, needless to say, probably thought otherwise.

A good portion of the epistle contains the apostle's sage advice to various social groups within the congregation: older men (2:2), older women (2:3), younger women (2:4–5), younger men (2:6–8); and slaves (2:9–10). Near the end, the advice becomes more general in nature, involving basic admonitions to engage in moral behavior in light of the new life for those who have been saved (3:1–7, especially v. 5). The letter concludes with several greetings and a request for Titus to join the apostle in the city of Nicopolis, where he plans to spend the winter (3:12). There were several cities of this name in Asia Minor and elsewhere in the empire; it is not clear to which of these the author refers.

THE HISTORICAL SITUATION AND AUTHORSHIP OF THE PASTORAL EPISTLES

Most scholars are reasonably convinced that all three Pastoral epistles were written by the same author. With 1 Timothy and Titus there can be little doubt. The writing style, subject matter, and specific content are altogether similar. If they were not written by the same person, we would have to suppose that one of them was used by an imitator as the model for the other, but there appears to be no reason to think that this is what happened. The question of 2 Timothy has proven somewhat more complicated since its content is different. Yet even here the vocabulary and writing style are closely aligned with the other two. The salutation of the letter matches that of 1 Timothy: "To Timothy, my . . . child . . . : Grace, mercy, and peace from God the Father and Christ Jesus our Lord" (1 Tim 1:2; 2 Tim 1:2). No other Pauline letter has the same wording. Moreover, many of the same concerns are clearly to the fore in both letters, especially the concern for the administration of the church and the weeding out of false teachers.

Assuming, then, that all three letters come from the same hand (even granting 2 Timothy's different occasion and content), was that hand the apostle Paul's? By pursuing this question, we can learn a good deal about these epistles, particularly about the historical situation that they presuppose. Here I will set forth the arguments that have struck most scholars as decisive in showing that Paul did not write them.

At the outset, we should consider the unusual vocabulary used throughout these letters. Before adducing the data themselves, let me first explain their significance. Suppose (to imagine a relatively bizarre situation) that someone were to uncover a letter allegedly written by Paul that urged its readers to attend mass every Saturday night, to go to confession once a week, and to say three Hail Marys for every unintentional sin they committed. What would you make of such a letter? Some of its words would indicate Christian practices and beliefs that developed long after Paul had died (e.g., mass, Hail Marys). Others were used by Paul, but not in the same way (e.g., confession). With the passage of time, significant words in any language are invested with new meanings and new words are created, which is why Shakespearean English sounds so strange to many people today and why our language would have struck Shakespeare as peculiar. The vocabulary of this hypothetical letter alone would show you that the apostle Paul did not write it.

With the Pastoral epistles, of course, we find nothing so blatant, but we do find an inordinate number of non-Pauline words, most of which do occur in later Christians writings. Sophisticated studies of the Greek text of these books have come up with the following data (see the works cited in the suggestions for further reading): apart from personal names, there are 848 different words found in the Pastorals; of these, 306 occur nowhere else in the Pauline corpus of the New Testament (even including the Deutero-Paulines). This means that over one-third of the vocabulary is not Pauline. Strikingly, over two-thirds of these non-Pauline words are used by Christian authors of the second century. Thus, it appears that the vocabulary represented in these letters is more developed than what we find in the other letters attributed to Paul.

Moreover, some of the words that Paul does use in his own letters take on different meanings in the Pastorals. As brief examples, Paul's word for "having a right standing before God" (literally, "righteous") now means "being a moral individual" (i.e., "upright"; Tit 1:8) and the term "faith," which for Paul refers to a trusting acceptance of the death of Christ for salvation, now refers to the body of teaching that makes up the Christian religion (e.g., Tit 1:13).

Of course, the argument from vocabulary can never be decisive in itself. Everybody uses different words on different occasions, and the Christian vocabulary of Paul himself must have developed over time. The magnitude of these differences must give us pause, however, particularly since they coincide with other features of the letters that suggest they were written after Paul had passed off the scene. To begin with, there is the nature of the problems that the letters address. If the major form of false teaching being attacked was some kind of Christian Gnosticism, then one might ask when this kind of religion can be historically documented. In fact, the first Christian Gnostics that we know by name lived in the early to mid second century. To be sure, the second-century Gnostics may have had some predecessors near the end of the first century (as we discussed in Chapter 11), but there is almost no evidence to suggest that they were spouting "myths and endless genealogies" that sanctioned strictly ascetic lifestyles or that they were otherwise plaguing the Christian congregations during the lifetime of Paul himself. Not even Paul's adversaries in Corinth were this advanced.

Of even greater importance is the way in which these false teachings are attacked in the Pastorals, for the author's basic orientation appears to be very much like what we find developing in second-century proto-orthodox circles. From our earlier discussions, you may have wondered how one form of the widely diversified Christian movement ended up becoming dominant. How did it happen that from all the variety that we have seen within early Christianity, only the Roman Catholic Church emerged, the church from which the Eastern orthodox and Protestant churches of today also derive? The story is far too long to narrate in full here, interesting as it is. For our purposes, it is enough to indicate that the group that I've called the proto-orthodox was successful in countering the claims of other groups, and therefore in attracting more converts to its own perspectives, by forming a unified front that claimed a threefold authorization for its understanding of the religion. This unified front involved (a) developing a rigorous administrative hierarchy that protected and conveyed the truth of the religion (eventuating, for example, in the papacy), (b) insisting that all true Christians profess a set body of doctrines promoted by these leaders (the Christian creeds), and (c) appealing to a set of authoritative books of Scripture as bearers of these inspired doctrinal truths (the "New" Testament; see Chapter 1). Or to put the matter in its simplest and most alliterative terms, the proto-orthodox won these conflicts by insisting on the validity of the clergy, the creed, and the canon.

These forms of authorization were not in place during Paul's day. They are in the process of development, however, in the Pastoral epistles.

The Clergy. The one Pauline community whose inner workings we know in some detail, thanks to the apostle's extended correspondence with it, is the church in Corinth. This was a troubled church, one that was rife with inner turmoil, characterized by what Paul considered to be personal immorality, and subject to what he regarded to be false teaching. How did Paul deal with the problems, or rather, to whom in the church did he appeal when he decided to deal with them? He wrote to the entire church, pleading with them to adhere to his advice. Why didn't he address his concerns to the person in charge, the elder or overseer who could make decisions and run a tighter ship? Quite simply because there was no such person there.

Paul's churches were "charismatic" communities, that is, congregations of people who believed that they had been endowed with God's Spirit and so been given "gifts" (Greek *charismata*) to enable them to minister to one another as teachers, prophets, evangelists, healers, almsgivers, tongues-speakers, tongues-interpreters, and so on. There was nobody ultimately in charge, except the apostle (who wasn't on the scene), because everyone had received an equal endowment of the Spirit, and so no one could lord it over anyone else. At least that is how Paul thought the church ought to be (see 1 Corinthians 12–14).

What happens, though, when everyone feels Spirit-led but not everyone agrees on where the Spirit leads? In such a situation, who is to say that one person's teaching is of the Spirit and another's is not? Who is to decide how the church funds should be used? Who is to reprimand a brother or sister involved in dubious personal activities? At the start, Paul evidently did not find these issues of local leadership pressing, since he believed that the end was soon to arrive and that the Spirit was simply a sort of down-payment of what was to come, a kind of interim guide to how life would be in the kingdom. But what happens when the end does not arrive and there is no one person or group of persons to take charge? Presumably, as happened in the church in Corinth, a fair bit of chaos.

The developments within the Pauline communities appear to have taken place in response to this chaos. With the passing of time, Paul's churches developed a kind of hierarchy of authority in which church leaders emerged and began to take control of what happened within the congregation. To a limited extent, this development began in the later years of Paul's ministry: in the letter to the Philippians, for example, he mentions "overseers and deacons" as among his recipients (1:1). But Paul assigns no special roles to these persons nor does he assume that they can deal directly with the issues that he addresses.

Some fifty years or so after Paul had died, however, these offices had developed considerably in proto-orthodox circles. Each Christian locality had a clear-cut leader called a "bishop" (the Greek word is *episkopos*, literally meaning "overseer," as in Phil 1:1), under whom served "presbyters" (Greek for "elders"), who appear to have tended to the spiritual needs of the communities, and "deacons" (Greek for "ministers"), who may have focused on their material needs. In the early second-century writings of Ignatius, for example, we find churches in Asia Minor with a solitary bishop in charge and a board of presbyters and deacons under him (see box 22.4 and, more fully, Chapter 25). Above all, the bishops were to root out all traces of heretical teaching.

Later on in the second century, when we come to such proto-orthodox authors as Irenaeus and Tertullian, we find explicit arguments for what is sometimes called the "apostolic succession." According to these authors, the apostles established a single bishop over each of the major churches in Christendom; these bishops in turn hand-picked their own successors and ordained them to ministry, and so forth down to the writers' own day. These authors considered the bishops of these churches to be the rightful heirs of the apostles. Needless to say, they were also the bishops who subscribed to the proto-orthodox points of view.

With the passing of time, then, a church hierarchy developed out of the loosely organized, charismatic churches established by Paul and presumably by other missionaries like him. Where do the Pastoral epistles stand in this line of development? In these letters "Paul" writes to his officially designated representatives, ordained by the laying on of hands, instructing them to appoint bishops and deacons who are suitable for the governance of the

SOME MORE INFORMATION

Box 22.4 Church Hierarchy in Ignatius

The undisputed letters of Paul contain nothing like the structured hierarchy that begins to make itself evident in the works of later writers such as Ignatius, who urges that the solitary bishop of the church should hold complete sway over his congregation and that the presbyters and deacons should also be given special places of authority (cf. the Pastorals). As Ignatius says to the Christians of Smyrna:

> Let all of you follow the bishop, as Jesus Christ follows the Father; and follow the presbytery as you would follow the Apostles. And respect the deacons as you respect the commandment of God. Let no one do anything that relates to the church apart from the bishop. The only eucharist that is valid is the one performed by the bishop or by the person that he appoints. Wherever the bishop happens to be, consider this the entire congregation, just as where Jesus Christ is, there you will find the whole church. It is not fitting for anyone to perform a baptism or to celebrate the Lord's Supper if the bishop is not present. But whatever the bishop should approve, this also is pleasing to God. . . . The one who honors the bishop has been honored by God; the one who does anything apart from the knowledge of the bishop serves the devil.
> (Ign. *Smyr.* 8–9)

church and to pass along to them the true teaching that the apostle himself has provided. The clerical structure of these letters appears far removed from what we find in the letters of Paul, but it is closely aligned with what we find in proto-orthodox authors of the second century.

The Creed. Proto-orthodox Christians of the second and third centuries felt a need to develop a set of doctrines that were to be subscribed to by all true believers. As was the case with the proto-orthodox clergy, the proto-orthodox creed was acclaimed as a creation of the apostles themselves; hence the name of the most famous of these statements of faith, devised in the fourth century and known today as the Apostles' Creed.

The proto-orthodox creeds affirmed beliefs that were denied by other groups who claimed to be Christian, and they repudiated beliefs that these other groups affirmed. For example, Gnostic Christians claimed that there were many gods, not just one, and that the true God had never had any contact with the material world, which had been created by a lesser, evil deity. In response, the proto-orthodox creed proclaimed: "We believe in One God, the Father, the Almighty, Maker of Heaven and Earth" (as stated in its somewhat later formulation, the Nicene creed). Many of the Gnostics, moreover, claimed that Jesus was one person and Christ was another. The orthodox creed, however, maintained, "We believe in one Lord Jesus Christ." Other groups of Christians denied that Jesus was a real man who had actually been born, while still others denied that his birth had been at all special or that his mother had been a virgin. In response, the proto-orthodox creed affirmed that he "was born of the Virgin Mary and made man."

The Christians who devised and affirmed these orthodox creeds portrayed Christianity as a religion devoted to a set of doctrinal truth statements, containing ideas or notions that were to be acknowledged by all believers as true. For them, "the faith" referred to the body of Christian teachings that were to be affirmed. As we have seen, this contrasts with Paul's own usage, in which "faith" is not a propositional term but a relational one, signifying a trusting acceptance of the death of Christ to bring about a restored relationship with God. Significantly, in the Pastoral epistles what is of critical importance is "the teaching," that is, the body of knowledge conveyed by the apostle, sometimes simply designated as "the faith" (e.g., see 1 Tim 1:10; Tit 1:9, 13). These epistles, then, appear to represent a form of Christianity that arose in the wake of Paul's own ministry.

The Canon. I have already talked about the development of the Christian canon of Scripture in Chapter 1. We do not find proto-orthodox authors endorsing a specific collection of distinctively Christian books until near the end of the second century. The movement toward a canon was already afoot somewhat earlier, however, in writers who quoted the words of Jesus and the writings of the apostles as authoritative in matters pertaining to doctrine and practice. These words were not understood simply as pieces of good advice; they came to be seen as standing on a par with the Jewish Scriptures themselves, which the Christians continued to revere and study (cf. 2 Tim 3:16).

There is scant evidence that this had already happened by the time the Pastoral epistles were written, but the little that does exist is intriguing. The first book of Timothy quotes a passage from the Torah and sets it next to a saying of Jesus (5:18). Strikingly, the author labels both sayings as Scripture. We appear to be headed down the path that will eventuate in the proto-orthodox canon.

CONCLUSION: THE POST-PAULINE PASTORAL EPISTLES

There are other aspects of the Pastoral epistles that make them appear to date after the death of the apostle Paul: their preoccupation with social arrangements in this world and the Christians' respectability in the eyes of outsiders rather than with the apocalypse that is soon to come, their insistence that the leaders of the church be married rather than single and celibate (which was Paul's own preference for both himself and his converts), their assumption that Timothy is a third-generation Christian preceded in the faith by both his mother and grandmother, and their concern to silence women who have, in the author's view, gotten out of hand (a matter we will explore in the following chapter). But the most compelling reason for thinking that they were written near the end of the first century, or somewhat later, is that their vocabulary and concerns reflect what was transpiring among proto-orthodox Christians a generation or two after Paul's death. These Christians were less concerned with the imminent end of the world than with the problems confronting a church that was to be here for a long time to come. This was a church that needed to strengthen itself through tighter organization and to ward off false teachings that had proliferated with the passing of time.

An unknown author within a church that subscribed to Paul's authority took up his pen, perhaps some thirty or forty years after the apostle himself had died, to do what some Pauline Christians had done before him and what others would do afterwards: compose writings in the name of the apostle to address the crushing problems of his day. Not surprisingly, the stances that this anonymous author took differed not only from those promoted by Paul himself in his undisputed letters but also from those advanced by other Pauline Christians. The differences are particularly evident in the author's attacks on gnosis, on women's involvement in the church, and on strictly ascetic lifestyles. As we have seen, on these subjects the author of the Pastorals stood at odds with what other Christians believed, even though they also appealed to the apostle in supporting their own views (see 2 Pet 3:15–16).

The church that the apostle Paul left behind thus developed in complex and unpredictable ways. As a result, Pauline Christianity, like all other forms of early Christianity, was a remarkably diverse phenomenon, whose manifold forms of expression would not be unified until the triumph of proto-orthodoxy in later centuries.

SUGGESTIONS FOR FURTHER READING

Beker, J. Christiaan. *The Heirs of Paul: Paul's Legacy in the New Testament and in the Church Today*. Philadelphia: Fortress, 1991. A clear assessment of the theology of the Deutero-Pauline and Pastoral epistles, especially in light of the views embodied in the undisputed Paulines.

Donelson, L. R. *Pseudepigraphy and Ethical Argument in the Pastoral Epistles*. Tübingen: Mohr/Siebeck, 1986. A study that sets the Pastoral epistles in the context of ancient practices of pseudepigraphy and that establishes their theological and ethical points of view.

Harrison, P. N. *The Problem of the Pastoral Epistles*. Oxford: Humphrey Milford, 1921. A classic study that provides an authoritative demonstration that Paul did not write the Pastoral epistles in their present form.

Lincoln, Andrew, and A. J. M. Wedderburn. *The Theology of the Later Pauline Letters*. Cambridge: Cambridge University Press, 1993. A clear overview of the major themes of Colossians and Ephesians.

MacDonald, M. Y. *The Pauline Churches: A Socio-historical Study of Institutionalization in the Pauline and Deutero-Pauline Writings*. Cambridge: Cambridge University Press, 1988. A study that uses a social-science approach to help explain various aspects of the institutionalization of churches associated with Paul, especially as evidenced in Ephesians, Colossians, and the Pastoral epistles; for more advanced students.

Roetzel, Calvin. *The Letters of Paul: Conversations in Context*. 3d ed. Atlanta: John Knox, 1991. Perhaps the best introductory discussion of each of the Pauline epistles, including the Deutero-Paulines and Pastorals.

Young, Frances. *The Theology of the Pastoral Epistles*. Cambridge: Cambridge University Press, 1994. A clear overview of the major themes of the Pastoral epistles.

CHAPTER 23

From Paul's Female Colleagues to the Pastor's Intimidated Women: The Oppression of Women in Early Christianity

Women played a prominent role in the earliest Christian churches, including those associated with the apostle Paul. They served as evangelists, pastors, teachers, and prophets. Some were wealthy and provided financial support for the apostle; others served as patrons for entire churches, allowing congregations to meet in their homes and supplying them with the resources necessary for their gatherings. Some women were Paul's coworkers on the mission field. Why, then, do most people today think that all of the early Christian leaders were men?

This question has generated a number of interesting studies in recent years. Here I will present one of the persuasive perspectives that has emerged from these studies. Despite the crucial role that women played in the earliest Christian churches, by the end of the first century they faced serious opposition from those who denied their right to occupy positions of status and authority. This opposition succeeded in pressing Christian women into submission to male authority and obscured the record of their earlier involvement.

WOMEN IN PAUL'S CHURCHES

Despite the impression that one might get from such ancient Christian writings as the Pastoral epistles, women were not always a silent presence in the churches. Consider Paul's letter to the Romans, in which he sends greetings to and from a number of his acquaintances (chap. 16). Although Paul does name more men than women here, the women in the church appear to be in no way inferior to their male counterparts. There is Phoebe, a deacon (or minister) in the church of Cenchreae, entrusted by Paul with the task of carrying the letter to Rome (vv. 1–2). There is Prisca, who along with her husband Aquila, is largely responsible for the Gentile mission and who supports a congregation in her own home (vv. 3–4; notice that she is named ahead of her husband). There is Mary, Paul's colleague who works among the Romans (v. 6). There are Tryphaena, Tryphosa, and Persis, women whom Paul calls his "co-workers" for the gospel (vv. 6, 12). And there are Julia and the mother of Rufus and the sister of Nereus, all of whom appear to have a high profile in this community (vv. 13, 15). Most impressively of all, there is Junia, a woman whom Paul names as "foremost among the apostles" (v. 7). The apostolic band was evidently larger—and more inclusive—than the list of twelve men of common knowledge.

Other Pauline letters provide a similar impression of women's active involvement in the Christian churches. In Corinth women are full members of the body, with spiritual gifts and the right to use them. They actively participate in services of worship, praying and prophesying alongside the men (1 Cor 11:4–6). In Philippians the only two believers worth mentioning by name are two women, Euodia and Syntyche, whose dissension concerns the apostle, evidently because of their prominent standing in the community (Phil 4:2). Indeed, according to the narrative of Acts, the church in Philippi began with the conversion of Lydia, a woman of means whose entire household came to follow her lead in adopting this new faith. She was the head of her household when the apostle first met her and soon became head of the church that met in her home (Acts 16:1–15).

Even after the period of the New Testament, women continued to be prominent in churches connected with Paul. The tales connected with Thecla, recounted in Chapter 21, appear to have struck a resonant chord with such people. Here were stories of women who renounced sexual relations and thereby broke the bonds of patriarchal marriage, that is, the laws and customs that compelled them to serve the desires and dictates of their husbands. Joining the apostle, these women came to experience the freedom provided by an ascetic life dedicated to the gospel. These narratives portray Paul as one who proclaimed that the chaste will inherit the kingdom, with women in particular being drawn to his message.

Even though the stories themselves are fictions, they appear to contain a germ of historical truth. Women who were associated with Paul's churches came to renounce marriage for the sake of the gospel and attained positions of prominence in their communities. Recall that letters later written in Paul's name speak of such women and try to bring them into submission. Some of these women were "widows," that is, women who had no husband overlord (whether they had previously been married or not). Such women are said to go about telling "old wives tales" (1 Tim 4:7 and 5:13), possibly stories like *The Acts of Paul and Thecla* that justified their lifestyles and views. Even in writings that oppose them, such women are acknowledged to be important to the church because of their full-time ministry in its service (1 Tim 5:3–16).

There is still other evidence of women enjoying prestigious positions in churches, well into the late second century. Some of this evidence derives from Gnostic groups that claimed allegiance to Paul and that were known to have women as their leaders and spokespersons. Other evidence comes from groups associated with the prophet Montanus and his two women colleagues, Prisca and Maximillia, who had forsaken their marriages to live ascetic lives, insisting that the end of the age was near and that God had called his people to renounce all fleshly passions in preparation for the final consummation.

How is it that women attained such a high status and assumed such high levels of authority in the early Christian movement? One way to answer the question is by looking at the ministry of Jesus himself, to see whether women enjoyed a high profile from the very outset of the movement.

WOMEN ASSOCIATED WITH JESUS

Most of the studies of women in early Christianity have been less than rigorous when it comes to applying historical criteria to the traditions about Jesus that describe his involvement with women. We ourselves should not fall into the trap of accepting traditions as historical simply because they coincide with an agenda that we happen to share, feminist or otherwise. So we will begin our reflections by applying the historical criteria established at an earlier stage of our study (Chapter 13) to find out what we can know with relative certainty about women in the ministry of Jesus.

To being with, we can say with some confidence that Jesus associated with women and ministered to them in public. To be sure, his twelve closest disciples were almost certainly men (as one would expect of a first-century Jewish rabbi). It is largely for this reason that the principal characters in almost all of the gospel traditions are men. But not all of them are. In fact, the importance of women in Jesus' ministry is multiply attested in the earliest traditions. Both Mark and L (Luke's special source), for example, indicate that Jesus was accompanied by women in his travels (Mark 15:40–41; Luke 8:1–3), a tradition corroborated by the *Gospel of Thomas* (*Gosp. Thom.* 114). Mark and L also indicate that women provided Jesus with financial support during his ministry, evidently serving as his patrons (Mark 15:40–41; Luke 8:1–3). In both Mark and John, Jesus is said to have engaged in public dialogue and debate with women who were not among his immediate followers (John 4:1–42; Mark 7:24–30). Both Gospels also record, independently of one another, the tradition that Jesus had physical contact with a woman who anointed him with oil in public (Mark 14:3–9; John 12:1–8). In Mark's account this is an unnamed woman in the house of Simon, a leper; in John's account it is Mary the sister of Martha and Lazarus, in her own home.

In all four of the canonical Gospels, women are said to have accompanied Jesus from Galilee to Jerusalem during the last week of his life and to have been present at his crucifixion (Matt 27:55; Mark 15:40–41; Luke 23:49; John 19:25). The earliest traditions in Mark suggest that they alone remained faithful to the end: all of his male disciples had fled. Finally, it is clear from the Synoptics, John, and the *Gospel of Peter* that women followers were the first to believe that Jesus' body was no longer in the tomb (Matt 28:1–10; Mark 16:1–8; Luke 23:55–24:10; John 20:1–2; *Gosp. Pet.* 50–57). These women were evidently the first to proclaim that Jesus had been raised.

There are other interesting traditions about Jesus' contact with women that do not pass the criterion of multiple attestation, including the memorable moment found only in Luke's Gospel when Jesus encourages his friend Mary in her decision to attend to his teaching rather than busy herself with "womanly" household duties (Luke 10:38–42). Since Luke, however, appears to be especially concerned with highlighting the prominence of women in Jesus' ministry (see Chapter 8), it is difficult to accept this tradition as historical. Indeed, it is difficult in general to apply the criterion of dissimilarity to the traditions about Jesus' involvement with women. As we have already seen, some early Christians were committed to elevating the status of women in the church; people such as this may have invented some such traditions themselves.

As for the contextual credibility of these traditions, it is true that women were generally viewed as inferior by men in the ancient world, but there were exceptions. Philosophical schools like the Epicureans and the Cynics, for example, advocated equality for women. Of course, there were not many Epicureans or Cynics in Jesus' immediate environment of Palestine, and our limited sources suggest that women, as a rule, were generally even more restricted in that part of the empire with respect to their ability to engage in social activities outside the home and away from the authority of their fathers or husbands. Is it credible, then, that a Jewish teacher would have encouraged and promoted such activities?

We have no solid evidence to suggest that other Jewish rabbis had women followers during Jesus' day, but we do know that the Pharisees were supported and protected by powerful women in the court of King Herod the Great. Unfortunately, the few sources that we have say little about women among the lower classes, who did not have the wealth or standing to make them independent of their fathers or husbands. One consideration that might make the traditions about Jesus' association with women credible, however, is the distinctive burden of his own apocalyptic message. Jesus proclaimed that God was going to intervene in history and bring about a reversal of fortunes: the last would be first and the first last; those who were rich would be impoverished and the poor would be rich; those who were exalted now would be humbled and the humble would be exalted. Jesus associated with the outcasts and downtrodden of society, evidently as an enactment of his proclamation that the kingdom would belong to such as these. If women were generally looked down upon as inferior by the men who made the rules and ran the society, it does not seem implausible that Jesus would have associated freely with them, and that they would have been particularly intrigued by his proclamation of the coming kingdom.

Some recent scholars have proposed that Jesus did much more than this, that he preached a radically egalitarian society. According to this view, he set about to reform society by inventing a new set of rules to govern social relations and aimed to create a society in which men and women would be treated as absolute equals. This, however, may be taking the evidence too far and possibly in the wrong direction. As we have seen, there is little to suggest that Jesus was concerned with transforming society in any fundamental way, let alone in terms of gender relations. In his view, society, with all of its conventions, was soon to come to a screeching halt, when the Son of Man arrived from heaven in judgment on the earth. Far from building a new society, a community of equals, Jesus was preparing people for the destruction and divine recreation of society.

All the same, even though Jesus may not have urged a social revolution in his time, it would be fair to say his message had revolutionary implications. In particular, we should not forget that Jesus urged his followers to begin to implement the ideals of the kingdom in the present in anticipation of the coming Son of Man. For this reason, there may indeed have been some form of equality practiced among the men and women who accompanied Jesus on his itinerant preaching ministry—not as the first step toward reforming society but as a preparation for the new world that was soon to come.

It is possible that the position of women among Jesus' followers while he was alive made an impact on the status of women in the Christian church after his death. This would help explain why women appear to have played significant roles in the churches connected with the apostle Paul, the early Christian churches that we are best informed about. But it would explain these significant roles only in part. For a fuller picture, we should return to Paul to consider not only the roles that women played in his churches but also his own view of these roles.

PAUL'S UNDERSTANDING OF WOMEN IN THE CHURCH

The apostle Paul did not know the man Jesus or, probably, any of his women followers. Moreover, as we have seen at some length, many of the things that Paul proclaimed in light of Jesus' death and resurrection varied from the original message heard by the disciples in Galilee. For one thing, Paul believed that the end had already commenced with the victory over the forces of evil that had been won at Jesus' cross and sealed at his resurrection. The victory was not by any means yet complete, but it had at least begun. This victory brought newness of life, the beginning if not the fulfillment of the new age. For this reason, everyone who was baptized into Christ was "a new creation" (2 Cor 5:16). And a new creation at least implied a new social order: "As many of you as were baptized into Christ have clothed yourself with Christ. There is no longer Jew or Greek, there is no longer slave or free, there is no longer male and female; for all of you are one in Christ Jesus" (Gal 3:27–28).

No male and female in Christ—this was a radical notion in an age in which everyone knew that males and females were inherently different. The notion, though, was deeply rooted in the Pauline churches. Modern scholars have recognized that in Galatians 3:28 Paul is quoting words that were spoken over converts when they were baptized. No wonder there were women leaders in the Pauline churches. Women could well have taken these words to heart and come to realize that, despite widespread opinion, they were not one whit inferior to the men with whom they served.

Like Jesus himself, however, Paul does not seem to have urged a social revolution in light of his theological conviction (recall our discussion of Philemon). To be sure, with respect to one's standing before Christ, it made no difference whether one was a slave or a slave owner; slaves were to be treated no differently from masters in the church. Thus, when believers came together to enjoy the Lord's supper, it was not proper for some to have good food and drink and others to have scarcely enough. In Christ there was to be equality, and failure to observe that equality could lead to disastrous results (1 Cor 11:27–30). Paul's view, however, did not prompt him to urge all Christian masters to free their slaves or Christian slaves to seek their release. On the contrary, since "the time was short," everyone was to be content with the roles they were presently in; they were not to try to change them (1 Cor 7:17–24).

How did this attitude affect Paul's view of women? Whether consistent with his own views of equality in Christ or not, Paul maintained that there was still to be a difference between men and women in this world. To eradicate that difference, in Paul's view, was unnatural and wrong. This attitude is most evident in Paul's insistence that women in Corinth should continue to wear head coverings when they prayed and prophesied in the congregation (1 Cor 11:3–16). A number of the details of Paul's arguments here are difficult to understand and have been the source of endless wrangling among biblical scholars. For example, when he says that women are to have "authority" on their heads (the literal wording of v. 10), does he mean a veil or long hair? Why would having this "authority" on the head affect the angels (v. 10)? Are these good angels or bad? And so on. Despite such ambiguities, it is quite clear from Paul's arguments that women could and did participate openly in the church alongside men—but they were to do so as women, not as men. Nature taught that men should have short hair and women long (at least, that's what nature taught Paul), so women who made themselves look like men were acting in ways contrary to nature and therefore contrary to the will of God.

For Paul, therefore, even though men and women were equal in Christ, this equality had not yet become a full social reality. We might suppose that it was not to become so until Christ returned to bring in the new age. That is to say, men and women had not yet been granted full social equality any more than masters and slaves had been, for Christians had not yet experienced their glorious resurrection unto immortality. While living in this age, men and women were to continue to accept their "natural" social roles, with women subordinate to men just as men were subordinate to Christ and Christ was subordinate to God (1 Cor 11:3).

WOMEN IN THE AFTERMATH OF PAUL

Paul's attitude toward women in the church may strike you as inconsistent, or at least as ambivalent. Women could participate in his churches as ministers, prophets, and even apostles, but they were to maintain their social status as women and not appear to be like men. This apparent ambivalence led to a very interesting historical result. When the dispute over the role of women in the church later came to a head, both sides could appeal to the apostle's authority to support their views. On one side were those who urged complete equality between men and women in the churches. Some such believers told tales of Paul's own female companions, women like Thecla, who renounced marriage and sexual activity, led ascetic lives, and taught male believers in church. On the other side were those who urged women to remain in complete submission to men. Believers like this could combat the tales of Thecla and other women leaders by portraying Paul as an apostle who insisted on marriage, spurned asceticism, and forbade women to teach.

Which side of this dispute produced the books that made it into the canon? Reconsider the Pastoral epistles from this perspective. These letters were allegedly written by Paul to his two male colleagues, Timothy and Titus, urging them to tend to the problems in their churches, including the problem of women. These pastors were to appoint male leaders (bishops, elders, and deacons), all of whom were to be married (e.g., 1 Tim 3:2–5, 12) and who were to keep their households, including of course their wives, in submission (1 Tim 2:4). They were to speak out against those who forbade marriage and urged the ascetic life (1 Tim 4:3). They were to silence the women in their churches; women were not to be allowed to tell old wives' tales and especially not to teach in their congregations (1 Tim 4:7). They were to be silent and submissive and sexually active with their spouses; those who wanted to enjoy the benefits of salvation were to produce babies (1 Tim 2:11–13).

The Pastoral epistles present a stark contrast to the views set forth in *The Acts of Paul and Thecla*. Is it possible that these epistles were written precisely to counteract such views? Whether or not they were, these letters are quite clear on the role to be played by women who are faithful to Paul and his gospel. The clearest statement is found in that most (in)famous of New Testament passages, 1 Timothy 2:11–13. Here we are told that women must not teach men because they were created inferior, as indicated by God himself in the Law. God created Eve second, and for the sake of man; a woman (related to Eve) therefore must not lord it over a man (related to Adam) through her teaching. Furthermore, according to this author, everyone knows what happens when a woman

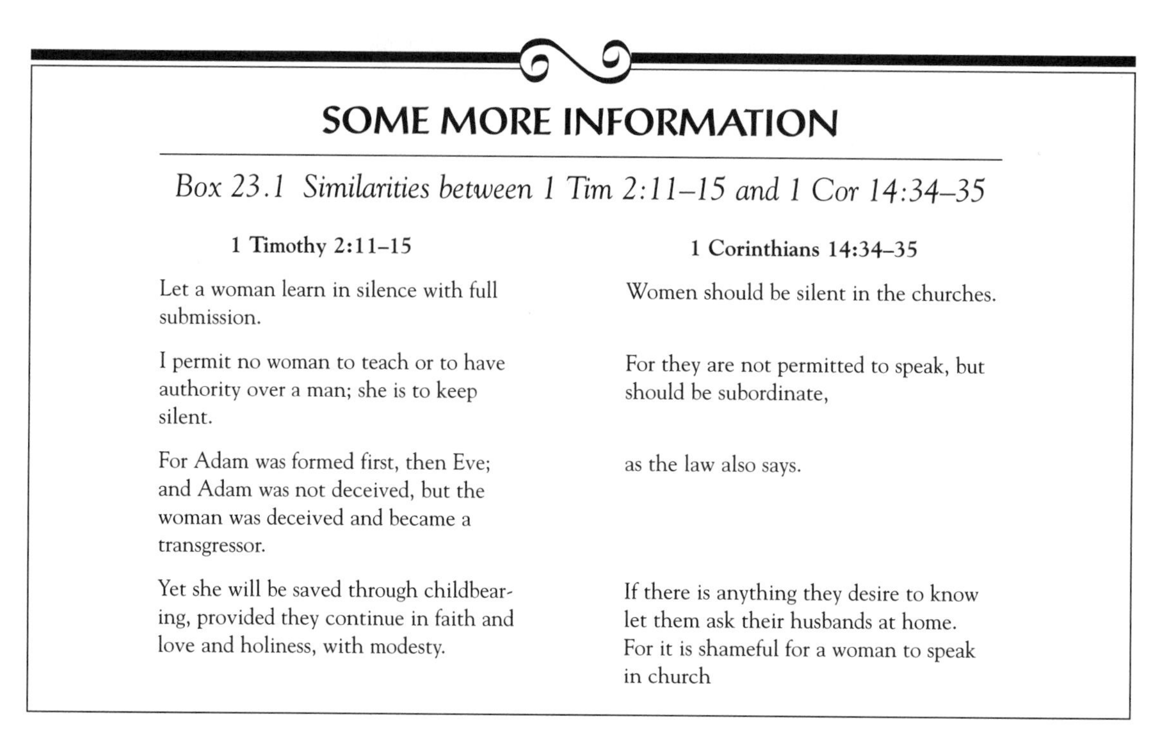

SOME MORE INFORMATION

Box 23.1 Similarities between 1 Tim 2:11–15 and 1 Cor 14:34–35

1 Timothy 2:11–15	1 Corinthians 14:34–35
Let a woman learn in silence with full submission.	Women should be silent in the churches.
I permit no woman to teach or to have authority over a man; she is to keep silent.	For they are not permitted to speak, but should be subordinate,
For Adam was formed first, then Eve; and Adam was not deceived, but the woman was deceived and became a transgressor.	as the law also says.
Yet she will be saved through childbearing, provided they continue in faith and love and holiness, with modesty.	If there is anything they desire to know let them ask their husbands at home. For it is shameful for a woman to speak in church

does assume the role of teacher. She is easily duped (by the Devil) and leads the man astray. So women are to stay at home and maintain the virtues appropriate to women, bearing children for their husbands and preserving their modesty. Largely on the basis of this passage, modern critics sometimes malign the apostle Paul for his misogynist views. The problem, of course, is that he did not write it.

Paul does, however, seem to say something similar in his undisputed letters, in the harsh words of 1 Corinthians 14:34–35. Indeed, this passage is so similar to that of 1 Timothy 2:11–15, and so unlike what Paul says elsewhere, that many scholars are convinced that these too are words that Paul himself never wrote; rather, they were later inserted into the letter of 1 Corinthians by a scribe who wanted to make Paul's views conform to those of the Pastoral epistles. The parallels are obvious when the two passages are placed side by side (see box 23.1).

Both passages stress that women are to keep silent in church and not teach men. This is allegedly something taught by the Law (e.g., in the story of Adam and Eve). Women are therefore to keep their place, that is, in the home, under the authority of their husbands.

It is not absolutely impossible, of course, that Paul himself wrote the passage that is now found in 1 Corinthians, but as scholars have long pointed out, Paul elsewhere talks about women leaders in his churches without giving any indication that they are to be silent. He names a woman minister in Cenchreae, women prophets in Corinth, and a woman apostle in Rome. Even more significantly, he has already indicated in 1 Corinthians itself that women are allowed to speak in church, for example, when praying or prophesying, activities that were almost always performed aloud in antiquity. How could Paul allow women to speak in chapter 11 but disallow it in chapter 14?

Moreover, it is interesting that the harsh words against women in 1 Corinthians 14:34–35 interrupt the flow of what Paul has been saying in the context. Up to verse 34 he has been speaking about prophecy and he does so again in verse 37. It may be, then, that the intervening verses were

Figure 23.1 Statue of a vestal virgin in the Roman forum, circa 70 C.E. The six vestal virgins, among the most prominent women in Roman society, were priestesses who guarded the sacred hearth of Rome and were accorded other special privileges and responsibilities.

not part of the text of 1 Corinthians but originated as a marginal note that later copyists inserted into the text after verse 33 (others inserted it after verse 40). However the verses came to be placed into the text, it does not appear that they were written by Paul. Evidently they were written by someone living later, who was familiar with and sympathetic toward the views of women advanced by the author of the Pastoral epistles.

In Paul's own churches, there may not have been an absolute equality between men and women. Women were to cover their heads when praying and prophesying, showing that as females they were still subject to males. But there was a clear movement toward equality that reflected the movement evident in the ministry of Jesus himself. Moreover, Paul's preference for the celibate life (a view not favored by the author of the Pastorals) may have helped promote that movement toward equality, for women who followed his example would not have had husbands at home who could serve as their religious authorities. Indeed, we know of such women from the second and later centuries—ascetics who preferred the freedom of single life to the restrictive confines of ancient marriage.

ANCIENT IDEOLOGIES OF GENDER

The Pauline churches eventually moved to the position embraced by the Pastoral epistles. They restricted the roles that women could play in the churches, insisted that Christians be married, and made Christian women submit to the dictates of their husbands both at home and in the church. It would be easy to attribute this move simply to male chauvinism, as much alive in antiquity as it is today, but the matter is somewhat more complicated. In particular, we need to consider what male domination might have *meant* in an ancient context; for most people in the ancient Roman world thought about gender relations in terms that are quite foreign to us who live in the modern Western world.

People in our world typically consider males and females to be two different kinds of human beings related to one another like two sides of the same coin. We sometimes refer to "my better half" or to "the other half of the human race." In antiquity, however, most people did not think of men and women as different in *kind* but as different in *degree*. For them there was a single continuum that constituted humanity. Some human beings were more fully developed and perfect specimens along that continuum. Women were on the lower end of the scale for biological reasons: they were "men" who had been only partially formed in the womb, and thus they were undeveloped or imperfect from birth. They differed from real men in that their penises had never grown, their lungs had not fully developed, and the rest of their bodies never would develop to their full potential. Thus, by their very nature, women were the weaker sex.

This biological understanding of the sexes had momentous social implications. Ancient Roman society was somewhat more forthright than ours in its appreciation of the importance of personal power. It openly revered those who were strong and domineering. Indeed, the virtue most cherished by males was "honor," the recognition of one's precedence over others, established chiefly through one's ability to achieve physical, economic, or political dominance. Other virtues were related to how one expressed this domination, for example, by showing courage and "manliness" when it was threatened, and self-control and restraint when it was exercised.

In Roman society, those who were "weaker" were supposed to be subservient to those who were stronger, and women were, by their very nature, weaker than men. Nature itself had set up a kind of pecking order, in which men were to be dominant over women as imperfect and underdeveloped beings, and women accordingly were to be submissive to men. This notion of dominance played itself out in all sorts of relationships, especially the sexual and domestic.

Most people in the Roman world appear to have thought that women were to be sexually dominated by men. This view was sometimes expressed in terms that might strike us as crass; it was widely understood that men were designed to be penetrators while women were designed to be penetrated. Being sexually penetrated was a sign of weakness and submission. This is why same-sex relations between adult males were so frowned upon, not because of some natural repulsion that people felt for homosexual unions (in parts of the ancient world it was common for adult males to have adolescent, and therefore inferior, boys as sex partners), but because such a relation meant that a man was being penetrated and therefore dominated. To be dominated was to lose one's claim to power and therefore one's honor, the principal male virtue.

Women's virtues, on the other hand, derived from their own sphere of influence. Whereas a man's were associated with the public arena of power relations—the forum, the business place, and the military—a woman's were associated with the domestic sphere of the home. To be sure, women were extremely active and overworked and burdened with responsibilities and duties, but these were almost always associated with the household: making clothes, preparing food, having babies, educating children, taking care of personal finances, and the like. Even wealthy women shouldered considerable burdens, having to serve as household managers over family, slaves, and employees, while husbands concerned themselves with public affairs.

Figure 23.2 Painting of a Christian Woman in Prayer, from the Catacomb of Priscilla.

The domestic nature of a woman's virtues generally required her to keep out of the public eye. At least this is what the Roman men who wrote moral essays for women urged them to do. They were not to speak in public debates, they were not to exercise authority over their husbands, and they were not to be involved with other men sexually, since this would mean that one man was dominating the wife of another, calling into question the husband's own power and, consequently, his honor.

For this reason, women who sought to exercise any power or authority over men were thought to be "unnatural." When women did attain levels of authority, as was happening with increasing regularity in the Roman world during the time of the New Testament, they were often viewed suspiciously and maligned for not knowing their place, for not maintaining properly female virtues, and for being sexually aggressive, even if their personal sex lives were totally unknown.

GENDER IDEOLOGY AND THE PAULINE CHURCHES

Our theoretical discussion of the ideology of gender in the Roman world, that is, of the way that people mentally and socially constructed sexual difference, gives us a backdrop for reconsidering the progressive oppression of women in the Pauline churches. Women may have been disproportionately represented in the earliest Christian communities. This at least was a constant claim made by the opponents of Christianity in the second century, who saw the inordinate number of women believers as a fault; remarkably, the defenders of the faith never denied it. The large number of women followers is not surprising given the circumstance that the earliest Christian communities, including those established by Paul, were not set up as public institutions like the Jewish synagogues or the local trade associations, which met in public buildings and had high social visibility. Paul established house churches, gatherings of converts who met in private homes (see box 11.3), and in the Roman world, matters of the household were principally handled by women. Of course, the husband was lord of the house, with ultimate authority over everything from finances to household religion, but since the home was private space instead of public, most men gave their wives relatively free reign within its confines. If Paul's churches met in private homes, that is, in the world where women held some degree of jurisdiction, it is small wonder that women often exercised authority in his churches. It is also small wonder that men often allowed them to do so, for the home was the woman's domain. The heightened possibility for their own involvement is perhaps one reason why so many women were drawn to the religion in the first place.

Why, then, did women's roles come to be curtailed? It may be that as the movement grew and individual churches increased in size, more men came to be involved and the activities within the church took on a more public air. People thoroughly imbued with the ancient ideology of gender naturally found it difficult to avoid injecting into the church the perspectives that they brought with them when they converted. These views were a part of them and were accepted without question as being natural and right. And they could always be justified on other, Christian, grounds. For instance, the Scriptures that these people inherited could be used to justify refusing women the right to exercise authority. The Jewish Bible was itself a product of antiquity, rooted in an Israelite world that advocated an ideology of submission as much as the Roman world did, though in a different way.

As a result of the mounting tensions, some Pauline believers, many of them women, we might suppose, began to urge that the views of sexual relations dominant in their culture were no longer appropriate for those who were "in Christ." In reaction to social pressures exerted on them from all sides, these people urged abstinence from marital relations altogether, arguing for sexual continence and freedom from the constraints imposed upon them by marriage. Moreover, they maintained that since they had been set free from all forms of evil by Christ, they were no longer restricted in what they could do in the public forum; they had just as much right and ability to teach and exercise authority as men.

Unfortunately for them, their views never became fully rooted. Indeed, their ideas may have contained the seed of their own destruction, in a manner of speaking. These celibate Christians obviously could not raise a new generation of believers in their views without producing children to train. With the passing of time, and the dwindling of the apocalyptic hope that had produced a sense of equality in the first place, there appeared to be little chance that the ideas so firmly implanted in people by their upbringing could be changed.

Those who advocated the rights of women to exercise authority in the church came to be widely opposed, and probably not only by men. As is true of women today, women in antiquity were molded by their culture's assumptions about what is right and wrong, natural and unnatural, appropriate and inappropriate. The proponents of the cultural status quo took the message of Paul (and Jesus) in a radically different direction, different not only from those who advocated a high-profile for women in the churches but also from Paul and Jesus themselves. The eschatological fervor that had driven the original proclamation began to wane (notice how it is muted already in the Pastorals), and the church grew in size and strength. More and more it took on a public dimension, with a hierarchy and a structure, a public mission, a public voice, and a concern for public relations. The church, in other words, settled in for the long haul, and the apocalyptic message that had brought women relative freedom from the oppressive constraints of their society took a back seat, taking along with it those who had appealed to its authority to justify their important role in the life of the community.

Women came to be restricted in what they could do in the churches; no longer could they evangelize or teach or exercise authority. These were public activities reserved for the men. The women were to stay at home and protect their modesty, as was "natural" for them; they were to be submissive in all things to their husbands; and they were to bear children and fulfill their functions as the weaker and less perfect members of the human race. The Roman ideology of gender relations became Christianized, and the social implications of Paul's apocalyptic vision became lost except among the outcasts relegated to the margins of his churches, women whose tales have survived only by chance discovery, not by their inclusion in the pages of canonical scripture.

SUGGESTIONS FOR FURTHER READING

Fiorenza, Elizabeth Schüssler. *In Memory of Her: A Feminist Theological Reconstruction of Christian Origins*. New York: Crossroad, 1983. A sophisticated, controversial, and highly influential study of the roles women played in the formation of Christianity by a noted feminist New Testament scholar and historian; for more advanced students.

Kraemer, Ross. *Her Share of the Blessings: Women's Religions among Pagans, Jews, and Christians in the Greco-Roman World*. New York: Oxford University Press, 1992. A superb study of religion as embraced and molded by women in ancient Greek, Roman, Jewish, and Christian cultures.

Lefkowitz, Mary R., and Maureen B. Fant, eds. *Women's Lives in Greece and Rome: A Source Book in Translation*. Baltimore: Johns Hopkins University Press, 1982. An extremely valuable collection of ancient texts illuminating all the major aspects of women's lives in the Greco-Roman world.

Pomeroy, Sarah. *Goddesses, Whores, Wives, and Slaves: Women in Classical Antiquity*. New York: Schocken, 1975. An important study of the changing views and social realities of women in the Greek and Roman worlds.

Torjesen, Karen Jo. *When Women Were Priests: Women's Leadership in the Early Church and the Scandal of their Subordination in the Rise of Christianity*. San Francisco: HarperCollins, 1993. An interesting study of the shifting roles of women in the early Christian churches; particularly suitable for beginning students.

CHAPTER 24

Christians and Jews: Hebrews, Barnabas, and Later Anti-Jewish Literature

Now that we have completed our study of the Gospels, Acts, and the letters attributed to Paul, we can move on to explore the remaining books of the New Testament: the catholic epistles and the Apocalypse of John. The term "catholic" may cause some confusion for modern readers: contrary to what one might think, these books were not written only by or for Roman Catholics. In this context, "catholic" means "universal" or "general"; for this reason, these books are sometimes called the general epistles. Through the Christian ages, they have been thought to address universal problems experienced by Christians everywhere, as opposed to the letters of Paul, which have been thought to address specific congregations about specific problems.

In fact, however, the general epistles are not, strictly speaking, general. We have already seen that three of them—1, 2, and 3 John—do address specific problems of a particular community. Moreover, one of them, 1 John, is not even an epistle. At the same time, one of the fruitful ways to go about studying these books is to situate them in a broader historical context to see how they address problems that Christians generally came to experience during the period in which they were written. Many of these problems have already cropped up in our study; they involve the early Christians' relationships with (a) non-Christian Jews, (b) antagonistic pagans, (c) their own wayward members, and (d) the history of the cosmos itself. In the present chapter we will consider the first of these relationships (the others will be addressed in Chapters 25–28).

Although Jesus and his earliest disciples were Jews, and the authors of the New Testament all understood their movement as springing from Judaism, as time went on, conflicts arose between Jews who believed in Jesus and those who did not. Tensions mounted as Jewish Christians began to convert Gentiles to this new faith and to claim that they too could be heirs of the promises given to Israel in the Hebrew Scriptures, even without adhering to Jewish customs and practices. The social conflicts that ensued created theological difficulties for the emerging Christian communities: if Gentiles did not have to become Jews in order to be Christians, how were they (and their Jewish-Christian brothers and sisters in the church) to understand themselves in relationship to Judaism?

Before seeing how these issues came to be resolved in some of the early Christian writings, we should begin by examining the more general problem of how early Christians came to understand themselves as a social group that was distinct from Judaism. To use a modern sociological term, this involves the problem of Christian "self-definition."

EARLY CHRISTIAN SELF-DEFINITION

Self-definition is a process by which any group of individuals understands itself to be a distinct group. Each of us, of course, belongs to a number of social groups. You are a member of a family, a student at a college or university or professional school, a citizen of a state and a country, perhaps the member of a church, synagogue, or other religious congregation, and possibly a participant in some other academic, religious, or civic group (e.g., a sorority, Campus Crusade, or the Rotary Club). Each of these social networks has ways of understanding and defining itself with respect to both what its members have in common and what makes them different from those who don't belong. These boundaries between insiders and outsiders are part of the group's self-definition.

For some social groups the boundaries are well-defined and rigid; for others they are quite loose. For instance, members of a strict fundamentalist Bible church may have a very firm understanding of who is inside and who is outside the body of the faithful. To belong to this church, you may have to hold certain beliefs without wavering (e.g., a belief in the Bible as the inerrant Word of God and the literal second coming of Christ) and participate in certain practices without fail (e.g., you must be baptized in this particular church, and you must attend church twice on Sunday and prayer meeting on Wednesday evening). Those who do such things are among the "saved" (insiders) and those who don't are among the "damned" (outsiders).

This rigorous form of self-definition stands in sharp contrast to that found, say, in a liberal Presbyterian church, where members know why they are Christian and, in general, what it means to be Presbyterian but do not at all think that they alone are God's chosen or that it would be an irreversible tragedy and unpardonable sin if some of their members were to transfer membership to the Methodist church across the street.

All social groups define themselves by establishing what it means to be a member and how belonging to the group sets a person off from those who do not belong. This has always been the case, as long as human societies have existed. It was certainly true in the early days of Christianity, when one group of Jews understood themselves to be distinct from other Jews (and from everyone else as well) in that they believed that the Messiah had come, that he had died and been raised from the dead, and that they could have a right standing before God by faith in him. These beliefs helped to characterize the group and to distinguish it from all other social groups. Bitter conflicts eventually emerged as the group began to define itself more and more rigidly and as those who were outside of the community grew hostile toward their beliefs and practices (see, for example, our earlier discussion of the Johannine community in Chapter 10). Opposition drove the group yet further inward, as its members began to insist on conversion for admission, practiced distinct initiation rites such as baptism, observed other periodic rituals such as the Lord's Supper, devised distinctive sets of beliefs that were to be confessed by all group members, and condemned those who remained on the outside.

As Christianity developed, it was compelled to define itself not only in relation to the Jewish world from which it emerged but also in relation to the polytheistic world into which it moved and from which it began to draw its greatest number of converts. Sometimes these different aspects of self-definition reinforced each other. Let me point out just one of the many issues involved. As we have seen, Jews were somewhat anomalous in the Greco-Roman world in that *(a)* they maintained that only one God, the God of Israel, was to be worshipped and *(b)* they adhered to ancient practices that had been ordered by this God as part of his Law, for example, the circumcision of males, Sabbath observance, and dietary restrictions (these were among the social boundaries of the Jews as a group in the Roman world). Within Roman society, all other people were expected to participate in the cult to the state gods. Jews were exempt because they were an ancient people with ancient customs that forbade such participation.

Along came the Christians, most of them former pagans, who did not appear to be Jewish to most outsiders: they worked on the Sabbath, they ate pork, and their men weren't circumcised. Yet they claimed to worship the God of the Jews and him alone; in fact, they claimed to be the new people of this God. As a result of their monotheism, they refused to worship the state gods. But they had no ancestral traditions to claim—except the traditions of the Jews, most of which they did not even seem to keep (e.g., circumcision, kosher food laws, and so on). If the gods were angered by those who refused to offer them cult (Jews excepted, given their ancestral traditions), and Christians refused to offer them cult without having any ancestral traditions to fall back on, who would be to blame when the gods sent disaster against the city—an earthquake, famine, epidemic, or the like? You guessed it.

Partially to defend themselves in a world in which nearly everyone knew that a new religion could not possibly be true and in which an exclusivistic cult would certainly not be protected by the state, Christians eventually had to explain how their religion was not recent but venerable with age, as old as Moses and the prophets, the ancient writers of ancient Israel. This act of self-definition was carried out, to some extent at least, for the purpose of public relations, for political gain. If the Christians were the true heirs of the promises of Israel, they had a defense against persecution.

The need for self-defense is just one aspect of the relationship of Christianity to Judaism that drove Christians to develop a sense of group identity. There were other, more internal aspects as well, such as the need for Christians to explain some of the basics of the new faith to converts. How was it that the God who had chosen the Jews to be his people in days of old had now in these recent days chosen a different people, the Christians? How were believers in Jesus related to Jews who did not believe in him? And what was their connection to the Jewish Scriptures?

We have already seen that different Christians answered these questions in different ways—recall our studies of Matthew, Luke-Acts, Galatians, and Ephesians, as well as the views of the Ebionites and Marcion (see box 24.1). The differences are magnified even more when we turn to two other writings produced by early Christians: the canonical Epistle to the Hebrews and the noncanonical Epistle of Barnabas.

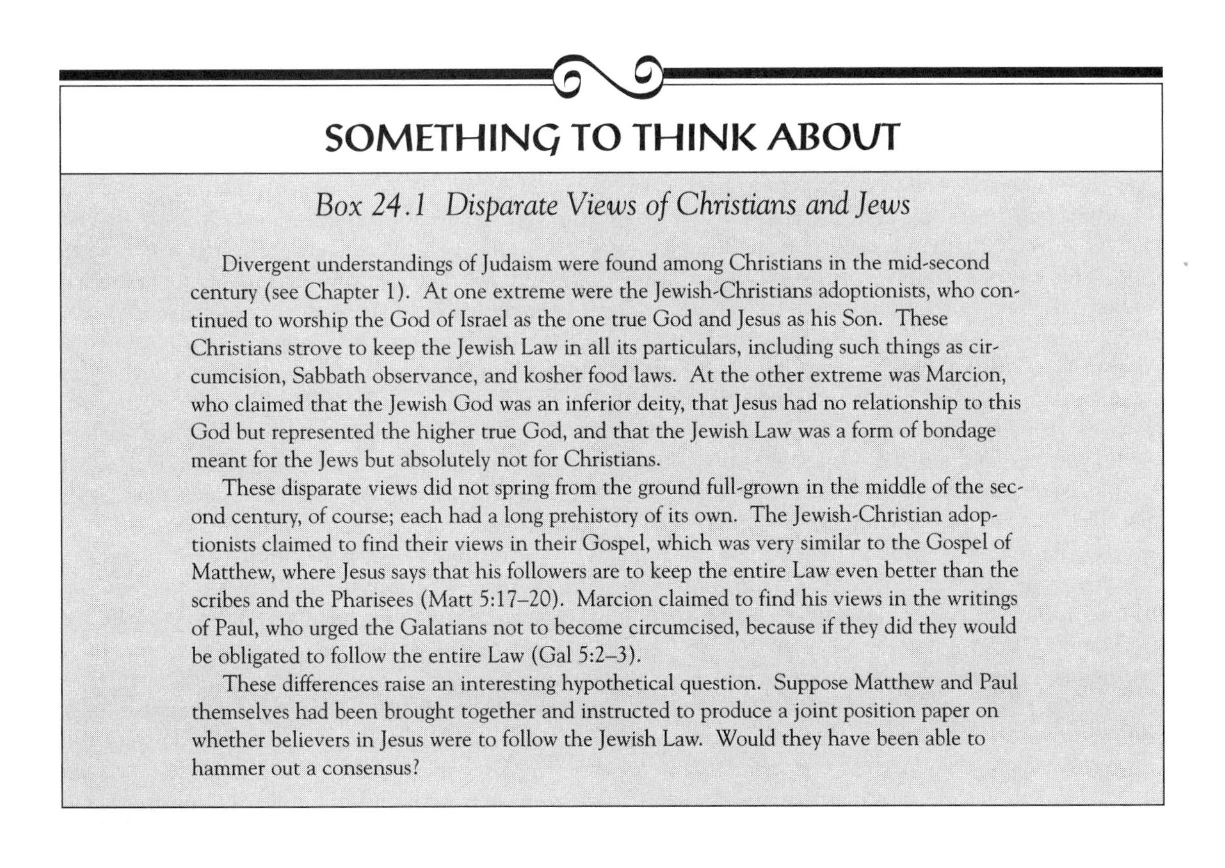

SOMETHING TO THINK ABOUT

Box 24.1 Disparate Views of Christians and Jews

Divergent understandings of Judaism were found among Christians in the mid-second century (see Chapter 1). At one extreme were the Jewish-Christians adoptionists, who continued to worship the God of Israel as the one true God and Jesus as his Son. These Christians strove to keep the Jewish Law in all its particulars, including such things as circumcision, Sabbath observance, and kosher food laws. At the other extreme was Marcion, who claimed that the Jewish God was an inferior deity, that Jesus had no relationship to this God but represented the higher true God, and that the Jewish Law was a form of bondage meant for the Jews but absolutely not for Christians.

These disparate views did not spring from the ground full-grown in the middle of the second century, of course; each had a long prehistory of its own. The Jewish-Christian adoptionists claimed to find their views in their Gospel, which was very similar to the Gospel of Matthew, where Jesus says that his followers are to keep the entire Law even better than the scribes and the Pharisees (Matt 5:17–20). Marcion claimed to find his views in the writings of Paul, who urged the Galatians not to become circumcised, because if they did they would be obligated to follow the entire Law (Gal 5:2–3).

These differences raise an interesting hypothetical question. Suppose Matthew and Paul themselves had been brought together and instructed to produce a joint position paper on whether believers in Jesus were to follow the Jewish Law. Would they have been able to hammer out a consensus?

CONTINUITY AND SUPERIORITY: THE EPISTLE TO THE HEBREWS

The Epistle to the Hebrews portrays the Jewish Law as partial and imperfect, unable to accomplish its task of putting people into a right standing before God. The inadequacy of the old covenant, the book claims, was recognized even by the Old Testament prophets, who predicted that God would establish a new covenant to do what the old one could not. This new covenant was foreshadowed in the legislation of Moses and came to reality only in the work of Jesus. The old has now passed away and believers must cling to the new.

The Book, the Author, and the Audience

Although Hebrews is normally labeled an epistle, this designation is not particularly fitting. Even though the book has an epistolary closing (13:20–25), there is no epistolary prescript. The author names neither himself nor his addressees, nor does he include an opening prayer, benediction, or thanksgiving on their behalf. Moreover, the author describes his book not as a letter but as a "word of exhortation" (13:22). This is a fair summary of the book's contents, leading most scholars to think that it was originally a sermon or homily delivered by a Christian preacher to his congregation. The author may have composed the sermon to be read aloud (most literature in antiquity was, in fact, read publicly), or possibly he wrote it down after it was delivered orally (from notes?). If it did originate as a sermon, then the epistolary closing with its benediction, exhortation, travel plans, final greetings, and farewell (13:20–25) may have simply been tacked on by its author, or by someone else who read the piece, when he sent it to another community. It is particularly intriguing that Timothy is mentioned at the end (13:23). Are we to infer from this that Paul wrote the sermon?

The book does not explicitly claim to be written by Paul; like the New Testament Gospels, it is anonymous. But it came to be included in the canon only after Christians of the third and fourth centuries became convinced that Paul had written it. Modern scholars, however, are unified in recognizing that he did not. The writing style is not Paul's, and the major topics of discussion (e.g., the Old Testament priesthood and the Jewish sacrificial system) are things that Paul scarcely mentions, let alone emphasizes. Moreover, the way this author understands such critical terms as "faith" (11:1) differs markedly from what you find in the writings of the apostle. It is difficult to say, then, who did write the book. A number of names have been proposed over the years, including such early Christian notables as Barnabas, Apollos, and Priscilla. It is safest, though, simply to accept the pronouncement of a famous Christian scholar of the third century, Origen of Alexandria, who said: "As for who has written it, only God knows."

We are in a better position to say something about the book's audience. The author presupposes that they are Christians who had previously undergone some serious persecutions for their faith, including imprisonment and the confiscation of property (10:32–34), although none of them had been martyred (12:4). From antiquity the book has been titled "To the Hebrews," but there is some considerable doubt over whether these persecuted Christians were Jews or Gentiles. For instance, when the author reminds them of the instruction they received upon first coming into the fold, he includes such matters as faith in God, belief in the resurrection of the dead, and eternal judgment (6:1–2). Surely Jews attracted to the Christian religion would already have known about such things. It seems more probable, then, that we are dealing with a group of Gentile converts who had experienced some persecution for their Christian faith, possibly (though not certainly) for reasons similar to those sketched earlier, that is, for refusing to worship state gods without having the Jewish roots that would make this refusal acceptable to local state officials.

The author, then, is writing to demonstrate to them that Christianity is superior to Judaism. Possibly he fears that members of his audience are being tempted to convert away from Christianity to non-Christian Judaism, perhaps to escape persecution. To abandon Christ for Judaism, in his judgment, would be a serious mistake. To do so would be to prefer the foreshadowing of God's salvation to salvation itself and to opt for the imperfect and flawed religion of the Jewish Scriptures rather than its perfect and complete fulfillment in Christ. For this author, Christ does indeed stand in continuity with the religion of the Jews as set forth in their sacred writings; but he is superior to that religion in every way, and those who reject the salvation that he alone can provide are in danger of falling under the wrath of God.

The Overarching Theme of the Sermon: The Superiority of Christ

The superiority of Christ and of the salvation he brings is the constant refrain sounded throughout this homily. Consider the following points that the author stresses.

Christ Is Superior to the Prophets (1:1–3). The Jewish prophets were God's spokespersons in former times, but now he has spoken through his own Son, the perfect image of God himself.

Christ Is Superior to the Angels (1:4–11; 2:5–18). The angels mentioned in the Old Testament are God's messengers par excellence, but Christ is his very Son, exalted to a position of power next to God's heavenly throne. Angels are ministers for those destined for salvation, but Christ is the Son of God whose suffering actually brought this salvation.

Christ Is Superior to Moses (3:1–6). Moses was a servant in "God's house," but Jesus is the Son of the house.

Christ Is Superior to Joshua (4:1–11). Joshua gave the people of Israel peace (or "rest") after the Promised Land had been conquered; but as the Scriptures themselves indicate, the people of Israel could not fully enjoy that peace (or "enter into their rest") because they were disobedient. Christ brings a more perfect peace.

Christ Is Superior to the Jewish Priesthood (4:14–5:10; 7:1–29). Like the Jewish high priests, Jesus was personally acquainted with human weaknesses that require a mediator before God, but unlike them, he was without sin and did not need to offer a sacrifice for himself before representing the people. He is superior to the priests descended from Levi because he is the one promised in the Scriptures as the priest from the line of Melchizedek (Ps 110:4), the mysterious figure whom Abraham, the ancestor of Levi, honored by paying one-tenth of his goods (Gen 14:17–20). For this reason, Levi himself, as represented by his ancestor, was inferior and subservient to Melchizedek and the descendant from his line. If the Levitical priests had been able to make the people of God perfect, God would not have had to promise to send a priest from the line of Melchizedek into the world. Moreover, Christ is superior to these other priests because they are many, but he is one, and unlike them, he needed to offer his sacrifice only once, not repeatedly.

Christ Is Minister of a Superior Covenant (8:1–13). God promised in the Scriptures to bring a new covenant (Jer 31:31–34), thereby showing that the old covenant with the Jews was outmoded and imperfect. Christ is the minister of this new covenant.

Christ Is Minister in a Superior Tabernacle (9:1–28). The earthly tabernacle, where Jewish sacrifices were originally performed, was constructed according to a heavenly model. Unlike the Jewish priests, Christ did not minister in the earthly replica; he brought his sacrifice to heaven, to the real sanctuary, into the presence of God himself.

Christ Makes a Superior Sacrifice (10:1–18). Christ's sacrifice was perfect, unlike those that had to be offered year after year by the Jewish priests. His death brought complete forgiveness of sins; there is no longer any need for sacrifice.

The Method of the Author's Demonstration

Like the author of Matthew, the author of Hebrews bases his understanding of Jesus on the Jewish Scriptures. This may seem somewhat ironic, in view of his insistence that Jesus is superior to anything that the Jewish religion has to offer. But as we have already seen, he was not the only Christian author who used the Jewish Bible to

SOMETHING TO THINK ABOUT

Box 24.2 Divergent Views of Christ in Hebrews

We have seen that sometime during the second century, Christians began to debate whether Jesus was God or man or somehow both. It is fairly easy to see how a book like Hebrews could have been used by all sides in such a debate. On the one hand, there are passages that appear to embrace an exalted view of Christ, more exalted than what is found almost anywhere else in the New Testament. You may have noticed that elsewhere in the New Testament Jesus is rarely, if ever, explicitly called "God" (although he is constantly called "Son of God"). Yet Hebrews 1:8 presents a quotation of the Psalms in which God is said to be speaking to his Son and calls him "God": "But of the Son, [God] says, 'Your throne, O God, is forever and ever.'"

Is this not an unequivocal statement that Christ himself is God? One difficulty is that the Greek of this verse can be translated in different ways. For example, it could also be rendered: "But of the Son [God] says, 'God is your throne forever and ever.'"

Other passages in Hebrews could be used by the opposite side in the later christological debates to show that Jesus was a full flesh-and-blood human. One of the most striking verses is 5:7, which indicates that Jesus went to his death with "loud cries and tears," beseeching God to save him from death, and that he "learned obedience" (meaning that he learned how to obey?) through his suffering. This does not sound like the calm and assured Jesus of some of the Gospel accounts (e.g., Luke and John); here Jesus almost seems to go to the cross kicking and screaming.

Other second- and third-century Christians, of course, could have argued that since Hebrews has both kinds of passages they have to be reconciled in some way, for example, by saying that Jesus started out as a normal human being but became divine at his exaltation (cf. Phil 2:6–10), or that Jesus was at one and the same time both man and God.

How would the author of Hebrews himself have reacted to these debates or reconciled the divergent views that he appears to have written? Regrettably, we will never know.

show that the Judaism he knew was inadequate and passé. The apostle Paul, for example, argued that the Jewish Law itself taught his doctrine of justification by faith apart from the Law. The author of Hebrews takes a different tack. He claims that the Scriptures anticipated a future act of God that would surpass everything that had come before. Somewhat like Matthew, he conceptualizes this anticipation in two different ways, as a prophecy that was to be fulfilled and as a foreshadowing that was to be made real.

Prophecy-Fulfillment. On several occasions the author uses predictions of the Jewish Scriptures to show that God had planned something new and better to supplant the Jewish religion. This new something, of course, would stand in continuity with Judaism; otherwise there would scarcely be any reason for the author to quote the Jewish Bible. As something new, however, it would be superior to that which it had been sent to replace. The clearest expression of the author's view comes in his lengthiest citation of the Old Testament (Jer 31:31–34):

> For if that first covenant had been faultless, there would have been no need to look for a second one. God finds fault with them when he says: "The days are surely coming, says the Lord, when I will establish a new covenant with the house of Israel and with the house of Judah . . . for they did not continue in my covenant, and so I had no concern for them," says the Lord. (8:7–9)

He concludes the citation, which continues for three more verses, by saying that "in speaking of 'a new covenant,' [God] has made the first one obsolete" (8:13). That is to say, the Scriptures predicted that God would establish a new covenant which would make the older religion, as set forth in the Scriptures themselves, invalid. In the author's judgment, the Scriptural prediction has now been fulfilled in Christ.

Shadow-Reality. The author of Hebrews also understands Christ to be superior to the religion of the Jews to the extent that the reality of a thing is superior to its foreshadowing. On two occasions he makes this claim explicit: both the Old Testament tabernacle (8:5) and the Law itself (10:1) were but "shadows" of another reality; on yet other occasions he appears to presuppose this view without explicitly stating it (9:23–24; 13:10–13).

Scholars have long recognized that the terms "shadow" and "reality" were popular philosophical metaphors that had been developed nearly 500 years earlier by Plato. Plato insisted that things appearing to be real are often only shadows of a greater reality. Physical pleasure, for example, has all the appearance of being a superior good; why else would so many people actively pursue it, some of them devoting their entire lives to little else? In itself, though, pleasure is good only in appearance. Witness the hangover, the county jail, and the halfway house. For Plato, the real good is located somewhere outside of bodily pleasure, which is itself, therefore, a mere shadow of reality.

Plato's most famous illustration of this idea is his Allegory of the Cave, found in Book VII of his influential dialogue *The Republic*. Let us suppose, says Socrates, the speaker of the dialogue, that there is a cave in which a number of people are chained together on the floor in such a way as to be unable to see anything except what lies in front of their eyes. These prisoners have always lived this way and so do not realize that they are in a cave or that there are other things in the world to be seen. Some distance behind them, unbeknownst to them, is a low half-wall and beyond that a large fire. Between the half-wall and the fire are people carrying puppets in the shapes of plants and animals and humans. The light from the fire casts the shadows of these objects on the wall of the cave that is before the prisoners' eyes. The prisoners themselves can see only the shadows, and when they hear the voices of those who carry the puppets echoing off the wall before them, they naturally assume that it is the images themselves who are speaking. These shadows are the only phenomena that they experience and they take them to be real—in fact, to be reality in its fullness. For them, these shadows *are* plants, animals, and humans.

What would happen, asks Socrates, if one of these chained persons were set free from his bondage and stood up to look around? He would no doubt be blinded by the bright light; in his terror, he might sit down and beg to be chained again. But if this person's eyes grew accustomed to the light, so that he could see that the images on the wall were actually shadows of puppets, he would then realize how fully his senses had been deceived. What he had taken to be reality were in fact only shadows.

Suppose this person then proceeded to leave the cave and to enter into the light of the sun. A similar sequence of events would no doubt occur. First he would be blinded by the light (in comparison to which the fire in the cave could itself be thought of as only a shadow). Only after his eyes adjusted would he come to see that not even the puppets had been the real thing, but only imperfect representations of real-life plants, animals, and people. No one who came to this kind of realization would choose to return to the cave to spend the rest of his days watching shadows cast on the wall. Once one has experienced reality, there is no turning back.

For the author of Hebrews, Christ is the reality that was foreshadowed in the Jewish Scriptures. As such, he is superior to anything Judaism has to offer. The author, however, is not concerned merely with making a debating point to an impartial audience. He is writing to Christians, and his ultimate goal is quite clear: he wants to convince his readers that there is no turning back to the shadow of Judaism once they have experienced the reality of Christ.

The Goal of the Author's Exposition

Throughout his exposition the author of Hebrews repeatedly exhorts his readers not to fall away from their commitment to Christ. Many of these exhortations are based on the notion that Christ is the reality behind the shadows of the Jewish Scriptures. The Old Testament contains numerous stories of individuals who chose to disobey God. As a rule, the penalties for disobedience were not pretty—being left as rotting carcasses in the wilderness and the like. If this was what happened to people who spurned the imperfect and incomplete revelation of God, asks the author, what gruesome fate awaits those who reject the revelation that is perfect and complete? If rejecting God's servants was bad, what happens to those who reject his Son? The logic of this argument can be easily illustrated: if I was upset when my son played with matches, think how I'd react if he torched the house.

The first exhortation occurs in 2:1–4: "If the message declared through angels was valid, and every transgression or disobedience received a just penalty, how can we escape if we neglect so great a salvation [i.e., provided by Christ]?" The answer: there will be no escape. A similar exhortation appears in 3:7–18: if those who were disobedient to Moses, God's servant, were destroyed in the wilderness, imagine what will happen to those who disobey Jesus, God's Son.

Sometimes these warnings leave less to the imagination, as in the dire and threatening words of 6:1–6, where the author claims there can be no hope of salvation for those who have "fallen away" after having "been enlightened," that is, for those who leave the faith after once having joined. In the author's opinion, such people are "crucifying again the Son of God and . . . holding him up to contempt" (v. 6). So too in chapter 10:

> If we willfully persist in sin after having received the knowledge of the truth, there no longer remains a sacrifice for sins, but a fearful prospect of judgment, and a fury of fire that will consume the adversaries (vv. 26–27). . . . It is a fearful thing to fall into the hands of the living God (v. 29).

Why does the author need to give such vitriolic warnings to people who are members of the congregation? Evidently, some of them were being tempted to fall away. The author does not explicitly state where these people might go after leaving the Christian community, but there can scarcely be any doubt, given everything else that he says about Christ's superiority to non-Christian Judaism. He is afraid that Christians will renounce Christ to join the synagogue, and he's doing everything in his power to stop them.

The author's bottom line is that his readers will inherit the salvation that God has promised only if they remain within the Christian church. And so he exhorts them: "Do not, therefore, abandon that confidence of yours; it brings great reward. For you need endurance, so that when you have done the will of God, you may receive what was promised" (10:35–36). As the Scriptures say, "my righteous one will live by faith" (Hab 2:4, quoted in 10:37). For this author to live by faith appears to mean something different from what it meant for Paul, who also quoted Habakkuk 2:4 (Rom 1:17; Gal 3:11). For the author of Hebrews, faith does not mean a trusting acceptance of Christ's death and resurrection for sins; it means being confident that God will do what he promised. Or in his own more poetic words, "Faith is the assurance of things hoped for, the conviction of things not seen" (11:1).

Chapter 11 recounts the deeds of the faithful from the Jewish Scriptures, those who lived by and acted on their assurance of that which they had not yet experienced. Jesus himself acted in this way (12:1–2). His followers need to emulate his example. Even though they suffer (as he himself did), they need to remain faithful to God's promises so as to reap their future reward. The book ends with a series of exhortations to love one another, to refrain from sexual improprieties, to obey the community's leaders, and to abstain from false teachings, especially those that promote adherence to the laws of Judaism (13:1–18).

The Epistle to the Hebrews and the Problem of Self-Definition

Even though we don't know the full story, we can make some plausible stabs at the social context of this author and his audience. As we have seen, from its earliest days the Christian message was closely tied to the apocalyptic notion that the end of the age was imminent, that the forces of evil were on the rise but God would soon intervene on behalf of his people and vindicate their suffering. With the passing of time and the failure of the end to appear, some believers gave up their confidence in this apocalyptic message. Generally, we don't know what happened to such people. Did some of them return to their former gods? Probably. Did some of them maintain their monotheistic devotion to the God of Israel but jettison their faith in Christ as his messiah and join the local synagogue as Gentile "God-fearers"? No doubt some of them did that as well. The author appears fearful that such a conversion (or return) to Judaism might occur among some members of his community.

We don't know where the author's community was located or when he lived. When he conveys greetings from "those from Italy" (13:24), he could mean either "those of us who are presently living in Italy" or "those who hale from Italy but are presently living with us." Some scholars have thought that his references to priests who continually perform sacrifices indicate that the Temple was still standing when he wrote, and therefore that the book must have been written before 70 C.E. Others have pointed out that later Jewish authors also spoke of the Temple in the present tense long after it was gone and have noted that almost all of the references to the Jewish sacrificial system in the book are drawn from the descriptions in the Old Testament rather than from first-century practice. Moreover, the few explicit references to the community's history suggest a somewhat later date, possibly during the final quarter of the first century. These Christians had earlier suffered persecution but were now experiencing some complacency and possibly some defections.

Whenever he was writing, the anonymous author of Hebrews was concerned to establish appropriate boundaries for his Christian community; that is, he was involved with the problem of Christian self-definition. Even though his community was evidently made up largely of converted polytheists, they understood themselves (or at least the author thought they ought to understand themselves) as the true heirs of the traditions of Israel. They were clearly in conflict with other groups that also claimed these traditions for themselves, in particular, with groups of non-Christian Jews. As we will discuss later in this chapter, non-Christian Jews far outnumbered Christians at this time, and as a rule they found it ludicrous for non-Jews to claim to understand the Jewish religion better than they themselves did.

Nonetheless, the Christian author of Hebrews, whether he himself was Jewish or not, claimed that Christ fulfilled the Old Testament revelation and that his followers were the true people of God. Those outside the Christian faith, whether Jews or Gentiles, could not legitimately claim to be the heirs of the religion espoused by Moses, for that religion looked forward to what was to come. It was but a foreshadowing of the salvation that God had promised in the prophets, a salvation brought in the person of his son Jesus, the messiah. In this sense, the Christian religion was continuous with, but ultimately superior to, the religion of non-Christian Judaism, and Christians were not to yield to the temptation of preferring the foreshadowing of salvation to salvation itself. Those who fell away from their Christian faith would learn firsthand that it is indeed "a fearful thing to fall into the hands of the living God" (10:29).

DISCONTINUITY AND SUPREMACY: THE EPISTLE OF BARNABAS

A somewhat different perspective emerges in the so-called *Epistle of Barnabas*, a book that portrays Judaism as a false religion from the very beginning. According to this author, Jews broke God's covenant as soon as it was made with them; they have never been the people of God and have never understood their own Scriptures. Indeed, the Old Testament is and always has been a Christian book.

Barnabas has traditionally been called an epistle, even though its epistolary opening contains only a greeting; neither its author nor its recipients is named. The second- and third-century Christians who first referred to the book claimed that it had been written by Paul's companion Barnabas (hence its name), but they may have been simply guessing. Indeed, these later authors may have ascribed the book to a companion of the apostle in order to elevate its importance. The earliest writer to mention the book, Clement of Alexandria, includes it among the writings of the New Testament, as do other Christian writers in Egypt through the fourth century. Most scholars, however, date the book to a period long after the real Barnabas's death. Several comments in the text itself suggest a date of around 130 C.E. or so. For instance, the book mentions the destruction of the Temple, which occurred in the year 70 C.E. (16:3), and refers to the possibility of its soon being rebuilt (16:4). That possibility was very much alive during the first decades of the second century, but it more or less evaporated when the emperor Hadrian (132–34 C.E.) had a Roman shrine constructed over the Temple's ruins.

Given the popularity of the epistle in the city of Alexandria, many scholars think that it was written there. Alexandria had a large Jewish population, and the city eventually came to house one of the largest Christian churches in the Empire. Relations between the groups were occasionally tense and sometimes even volatile. Moreover, and more intriguing, we know of Alexandrian Jews who practiced allegorical methods to interpret the Scriptures. One of the most famous of them was the first-century philosopher Philo, whose methods of interpretation are comparable to those used by the second-century Gnostics, many of whom also came from Alexandria. The author of Barnabas, whoever he was, also utilizes an allegorical mode of interpretation, taking the text to mean something other than what a literal reading would suggest, but he uses his allegorical readings not to support Judaism, as Philo did, but to attack it. Barnabas (as I'll continue to call him) understood the Old Testament to be a Christian book that had always been misinterpreted by the Jews, who, in his opinion, foolishly maintained that their religion had been given them by God. He claims that they were misled in this by an evil angel, who persuaded them to take the laws of the Old Testament literally rather than as figurative pointers to Christ and the religion that he was to establish (9:5).

Barnabas himself considers only parts of the Old Testament to be literally true, especially the parts that recount the repeated acts of disobedience by the children of Israel. For him, it is true that when Moses came down from Mount Sinai after receiving the Ten Commandments, he smashed the two tablets of the Law into bits, having seen the idolatry and immorality of the Israelites in the camp below. This act showed that

Figure 24.1 Coin of the emperor Vespasian that commemorates the conquest of Judea by Titus with the inscription "Judea Taken Captive." The fall of Jerusalem was a significant event in the development of Jewish-Christian relations.

God's covenant had been broken, quite literally, by the Jews, a disobedient and immoral people; and once broken, the covenant could never be renewed (4:6–8).

In the author's view, Jews failed to understand the figurative meaning of the Law that was given to Moses. Barnabas devotes most of his energies to driving home this basic point, time and again giving the "true" interpretation of the Jews' Law in opposition to their own literalistic understandings of it. For example, when God spoke of honoring the Sabbath day and keeping it holy, he did not mean that Jews should refrain from work on the seventh day. As unholy people, Barnabas claims, Jews could not possibly keep the day itself holy. God was instead referring to his own act of creation in which he spent six days making the world before resting on the seventh. Moreover, as the Scriptures themselves testify, "with the Lord a day is as a thousand years and a thousand years as one day." The six days of creation, then, refer to a period of six thousand years in which God is actively involved with the world, to be followed by a seventh day of rest, in which he will finally put an end to sin and bring peace on earth once and for all. The injunction to keep the Sabbath day holy is therefore not to be interpreted as a commandment to refrain from work; it is an instruction concerning the future apocalypse in which God's milenial kingdom will come to earth. Only then will there be a completely holy people who can keep "the day" holy (15:1–8).

Jews are also wrong to take the dietary laws of the Old Testament literally. God did not mean that his people were not to eat pork or rabbit or hyena, all of which are proscribed in the Torah. The injunction not to eat pork means not to live like swine, who grunt loudly when hungry and keep silent when full. People are not to treat God in this way, coming to him with loud petitions when they are in need and ignoring him when they are not (10:3). Not to eat rabbit means not to live like those wild creatures, who with every passing year increase their sexual appetites and multiply the number of their sexual partners, propagating at random and even committing incest (10:6). Likewise, not to eat hyena means not to live licentious lives, like those promiscuous animals who were thought to change their gender every year, alternately becoming male and female (10:7).

For Barnabas, the laws of God are meant to induce ethical behavior; they are totally misread if taken literally. This rule also applies to the most distinctive Jewish law of all, the law of circumcision. God did not want his people literally to cut off the foreskins of their baby boys. The sign of circumcision given to Abraham was something quite different; it was the sign that salvation would be given to the world through the cross of Jesus. To justify this interpretation, Barnabas points to the first account of circumcision in the Bible, where Abraham took his 318 servants into the wilderness to rescue his nephew Lot, who had been taken prisoner by an army of invading kings (Gen 17). Prior to going into battle, Abraham had these 318 members of his household circumcised. What is significant for Barnabas is the number 318 itself, a mysterious number that he explains by using the method of interpretation known in ancient Jewish sources as "gematria."

Gematria was a way of interpreting words in light of their numerical value (see box 24.3). In ancient languages, the letters of the alphabet performed double duty as numbers, unlike in English, where we use Roman letters but Arabic numerals. The practice is similar, though, to our occasional use of Roman numerals, in which, for instance, the *I* represents one, *V* is five, and *X* is ten. In the case of both ancient Greek and ancient Hebrew, every letter had a numerical value (so that in Greek, for example, the alpha was one, beta two, gamma three, and so on). For this reason, every word written in these languages had a numerical equivalent (the sum of the numbers represented by its letters). Conversely, every number was represented by a sequence of letters.

In explaining Abraham's circumcision of his 318 servants, Barnabas notes that 318 is represented (in Greek) by the letters tau, iota, and eta (τιη). For him, this number is significant because it clearly shows that circumcision prefigures the Christian religion. The tau (τ), he points out, is made in the shape of the cross (it looks like the English t) and iota (ι) and eta (η) are the first two letters of the name "Jesus" (ιησους) in Greek (9:1–8). The true circumcision is thus not the literal cutting of the flesh of the foreskin. It is the cross of Jesus. Adherence to the cross is what makes a person a member of the people of God. According to Barnabas, this doctrine is found in the text of the Jewish Scriptures themselves in the story of Abraham, the father of circumcision. Barnabas assures his readers that no one had ever heard a more excellent lesson from him (9:9).

This fascinating piece of early Christian writing ends on a different note by describing the "Two Ways" of life: the morally upright way of "light" and the morally perverse way of "darkness." These are paths that all people must choose between, and the author indicates the moral practices and improprieties pertaining to each.

In conclusion, what can we say about Christian self-definition as expressed in the Epistle of Barnabas? Christians here do not, strictly speaking, stand in continuity with historic Judaism. Judaism is a false religion followed by people who do not understand their own Scriptures. This harsh indictment of the Jews serves to differentiate them from the Christians, who are the only true heirs of the promises of God. The Scriptures belong to the Christians, and the Jews have no right to them. As the people of God, on the other hand, the Christians' roots are as ancient as Moses

SOME MORE INFORMATION

Box 24.3 Gematria in Early Christianity

The possibilities of gematria seem almost endless. Since any sequence of letters in Greek or Hebrew "adds up" to a total number, different words can be related to one another by their numerical totals. One second-century Gnostic group, for example, pointed out that the letters in the Greek word for "dove" add up to 801, the same numerical value contained in the Greek letters alpha (worth 1) and omega (worth 800). From this they concluded that the Spirit of God that descended upon Jesus "as a dove" was in fact an element of the divine itself, the "alpha and the omega" (see Rev 1:8), which came into the man Jesus to empower him for his ministry. Other Christians, needless to say, were not convinced.

Some Christian scribes used the numerical value of letters to help them devise abbreviations. In some ancient Greek texts, rather than concluding a prayer with the word "amen," these scribes simply wrote the two Greek letters that represented 99, the numerical value obtained by adding up the letters in amen, thereby saving themselves a second of time and a smidgen of ink.

The use of gematria is important in other early Christian texts, as we will see especially when we try to determine what the author of the Book of Revelation might have meant when he claimed that the number of the Antichrist was 666 (see Chapter 27).

and the prophets. Christians may not appear to be distinct from the rest of the world in the ways that Jews are, but that is only because the Jews have misconstrued their own religion. True religion means accepting the cross of Christ and living a moral upright life as a member of God's covenantal community, the Christian church.

CONCLUSION: THE RISE OF CHRISTIAN ANTI-SEMITISM

To modern ears the anti-Jewish invectives of the *Epistle of Barnabas* sound incendiary. As we know in hindsight, such attacks against the Jewish religion have led to hateful crimes against the Jewish people, some of them of unthinkable audacity. Anyone propounding such inflammatory views in our own day would be subject, quite rightly, to public denunciation and censure.

It is important, though, to understand the *Epistle* of Barnabas in the context of its own day. We do not know exactly when or where Barnabas was writing, although around the year 130 in the city of Alexandria is not a bad guess. In any event, it is safe to say that as a Christian, Barnabas represented a tiny minority of persons within the empire of his day, a marginalized religious sect that had never been heard of by most people and was scorned by most of those who did know something about it.

Demographic estimates from antiquity are extremely problematic, but the best guesses put the population of the Roman empire at the beginning of the second century at around 60 million, with Jews making up something like 7 percent of the total. Christians, on the other hand, would have comprised not even 1 percent of the population. As we saw previously, there may have been more women than men in the earliest Christian churches, and the majority of Christians, both men and women, appear to have come from the lower classes. We have no indication that any Christian in this period came from the very upper echelons of Roman society. Throughout this period churches continued to meet in private homes, so in urban areas there may have been a number of small individual congregations, possibly a large number, spread throughout a city. Church buildings were not to be built for more than a century.

In light of these basic demographics, Christianity was clearly not a massive unified movement with a centralized power base and political clout. On the contrary, it was scattered and poorly financed, with little public presence and less public credence. Most of the people who had heard of Christians did not consider their views acceptable, and they sometimes harassed the local Christian communities as a result. This put Christianity in stark contrast to Judaism, which not only had far greater numbers but also had visible public structures, wide public recognition, and prominent public representatives, some of whom had the ear of the highest officials in the empire, on occasion even of the emperor himself.

How was Christianity to justify its own existence in this world? The people of the Jewish God did not believe in Jesus, the crucified criminal, as the messiah, and anyone in society at large could see that Christians did not practice the ancestral traditions of the Jews, whose God the Christians claimed to serve. If the religion could receive no recognition from Jewish leaders, since it advanced an aberrant set of beliefs and practices, and had no protection from Roman administrators, since it lacked ancestral tradition, what recourse was left to the Christian church?

Christians who were convinced that their faith was not misfounded or misguided struck back at those who rejected and persecuted them. One form that this opposition took was the anti-Jewish literature that began to be written with increasing frequency as more and more literate and outspoken persons were converted to the Christian faith. At one stage, and in some places, the writers of this literature simply tried to claim the promises of Israel for the followers of Jesus; this basic position taken in very different ways by Matthew, Paul, and the author of Hebrews. It was also the position was taken by the Jewish-Christian adoptionists who continued not only to embrace the Jewish Scriptures but also to follow Jewish practices such as circumcision, Sabbath observance, and kosher food laws (see Chapter 1). Early in the second century, however, other Christian authors began to paint their opponents as adherents of a false religion. These Christians denied that their religion had any real continuity with Judaism, though they still claimed continuity with the Old Testament itself. This, in a nutshell, was the position of Barnabas.

Still later in the second century, real intellectuals converted to the Christian faith—philosophers like Justin of Rome and rhetorically sophisticated writers like Tertullian of North Africa. These intellectuals put their literary skills to work both to defend their faith from accusations made by pagans and to attack Jews who failed to recognize its superiority. More highly trained than their predecessors, these authors were frequently impressive in their rhetoric, even though the positions they advanced may sound appalling to modern ears. Both Justin and Tertullian, for example, admitted that circumcision was given as a sign to set Jews apart from all other peoples, but for Justin it was to set them apart for persecution, and for Tertullian it was to show who would not be allowed into the holy city. (Tertullian was writing after the Romans had made it illegal for Jews to live in Jerusalem after the violence of the second Jewish uprising in 132–35 C.E.)

Other authors raised the ante even higher. One of the most eloquent homilies of the second century derives from a Christian orator named Melito, who lived in the city of Sardis in Asia Minor (see box 24.4). His sermon text is the story of the Passover in the book of Exodus and his mode of interpretation is figurative. He sees Jesus as the real Passover lamb, rejected and killed by his own people; even more than this, he was also God himself. The implications, for Melito, are severe: Israel is guilty of murdering its own God. Indeed, Jews who continue to reject Christ are themselves culpable of this hateful deed. With Melito we are clearly at the beginning of a form of anti-Semitic hatred that had not appeared on the stage of human history prior to the advent of Christianity.

We are not yet at the point when anything much could be done about this hatred. Such inflammatory words mean one thing when they come from the pen of a relatively obscure preacher of a weak and powerless minority group within the empire and something quite different when taken to heart by people in positions of authority and power. For Melito and his predecessors, such opposition to the Jews represented an attempt to justify the existence

SOME MORE INFORMATION

Box 24.4 Melito's Passover Sermon

Melito of Sardis died around the year 190 C.E., so his sermon lambasting Jews for the role they played in the death of Jesus must have been written sometime during the middle of the second century. It is thus the first instance that we have of a Christian charging Jews with the crime of "deicide," the murder of God. This charge has been used to justify hateful acts of violence against Jews over the centuries. In part, the rhetorical eloquence with which the charge was sometimes leveled has contributed to the emotional reaction that it has produced. Consider Melito's own gripping, if terrifying, rhetoric:

> This one was murdered. And where was he murdered? In the very center of Jerusalem! Why? Because he had healed their lame and had cleansed their lepers, and had guided their blind with light, and had raised up their dead. For this reason he suffered. . . . (chap. 72)
>
> Why, O Israel, did you do this strange injustice? You dishonored the one who had honored you. You held in contempt the one who held you in esteem. You denied the one who publicly acknowledged you. You renounced the one who proclaimed you his own. You killed the one who made you to live. Why did you do this, O Israel? (chap. 73)
>
> It was necessary for him to suffer, yes, but not by you; it was necessary for him to be dishonored, but not by you; it was necessary for him to be judged, but not by you; it was necessary for him to be crucified, but not by you, not by your right hand, O Israel! (chaps. 75–76)
>
> Therefore, hear and tremble because of him for whom the earth trembled. The one who hung the earth in space, is himself hanged; the one who fixed the heavens in place, is himself impaled; the one who firmly fixed all things, is himself firmly fixed to the tree. The Lord is insulted, God has been murdered, the king of Israel has been destroyed, by the hand of Israel. . . . (chaps. 95-96)

of Christianity in a world that refused to recognize it. These Christians believed that their right to exist hinged on the inadequacies of the religion from which they had originated. If the majority of Jews was right, then Christians (so they understood) were necessarily wrong. Christian survival required a defensive posturing that was spun out in vitriolic tracts designed to mold a Christian identity.

It is doubtful that these Christian counterattacks proved convincing to anyone except those who already believed. To use a modern metaphor, these writers were preaching to the choir. By a few hundred years later, however, this bitter lashing out against a much larger opponent had become the confident attack of the high and mighty against a relatively defenseless minority.

For reasons more or less unrelated to the anti-Jewish writings of the early church, Christianity became the dominant religion of the empire. The shift did not occur overnight. By the beginning of the fourth century, Christians still comprised far less than 10 percent of the empire's population (perhaps some five million people). But in one of the most momentous conversions in history, the Roman emperor Constantine came to profess belief in the Christian God, and from then on everything changed. Constantine not only put an end to official persecution of the church (somewhat before his conversion, in the year 313 C.E.), but he also bestowed special imperial favors upon it. He provided extensive lands, magnificent buildings, and sizable revenues to churches, patronized leaders of the church in Rome and elsewhere, and took an active part in critical matters of Christian doctrine and church administration, for instance, by calling the famous Council of Nicea in 325 C.E., where the orthodox doctrine of Christology was established.

It became not only acceptable but also fashionable and even advisable in some circles to become a Christian. By the end of the fourth century, Christianity was named the official religion of the empire, with something like half of the entire population, some 30 million people, professing belief. This historic upheaval had profound effects for Jewish-Christian relations. In the early part of the second century Christians were a marginalized group that occasionally produced revolutionary and incendiary tractates. By the end of the fourth century the tables had turned, and turned with a vengeance. What had started as the defensive posture of an insignificant and powerless minority group became a view shared by prominent members of the Roman bureaucracy. The official policies of the empire did not actively require or promote the persecution of Jews, but in many instances the Christian governors looked the other way or privately condoned it. Synagogues were burned, properties were confiscated, and Jews were publicly mocked and sometimes subjected to mob violence. Leading the way were Christians, who took the defensive rhetoric of their predecessors in the faith all too literally and acted on it by striving to deprive Jews of their right to exist.

The result is one of the tragic ironies of history. Even though the founder of the Christian religion was a Jew, who lived among the Jewish people, followed the Jewish Law, worshipped in the Jewish synagogue, and selected Jewish followers, even though his Jewish disciples were taught to love their fellow Jews as themselves, and even though after their founder's death they developed a theology, a system of ethics, and a basic view of the world that continued to be rooted in Judaism, understanding themselves in light of the Jewish Scriptures that they believed had been given to the Jewish people by the Jewish God—despite all these things, much of the subsequent history of Christianity involved a falling away from its Jewish roots and a sometimes violent opposition to the Jewish people. In an effort to define themselves in the world, Christians came to deny their ties to the history, religion, and people of the Jews. The tragic effects of that denial remain with us even today.

SUGGESTIONS FOR FURTHER READING

Gager, John. *The Origins of Anti-Semitism*. Oxford: Oxford University Press, 1983. An important study of the rise of anti-Semitism among Christians in the early centuries of the Common Era.

Lindars, Barnabas. *The Theology of the Letter to the Hebrews*. Cambridge: Cambridge University Press, 1991. A clear account of the theological perspective of Hebrews.

Ruether, Rosemary. *Faith and Fratricide: The Theological Roots of Anti-Semitism*. New York: Seabury, 1974. A compelling and controversial study that argues that Christian claims about Jesus, by their very nature, are necessarily anti-Semitic.

Sanders, Jack T. *Schismatics, Sectarians, Dissidents, Deviants: The First One Hundred Years of Jewish-Christian Relations*. Valley Forge, Pa.: Trinity Press International, 1993. A study that evaluates the social interactions of Jews and Christians through the New Testament period up to the middle of the second century.

Sandmel, Samuel. *Anti-Semitism in the New Testament?* Philadelphia: Fortess, 1978. A clear and interesting discussion, from the perspective of a Jewish scholar, of whether parts of the New Testament should be viewed as anti-Semitic.

Setzer, Claudia J. *Jewish Responses to Early Christians*. Philadelphia: Fortress, 1994. A nicely written examination of the verbal and physical reactions that Christians evoked among Jews during the first two centuries of the Common Era.

Tugwell, Simon. *The Apostolic Fathers*. Harrisburg, Pa.: Morehouse, 1990. A clear and straightforward discussion of the background and message of the Apostolic Fathers, including the *Epistle of Barnabas;* ideal for beginning students.

CHAPTER 25

Christians and Pagans: 1 Peter, the Letters of Ignatius, the Martyrdom of Polycarp, and Later Apologetic Literature

We have seen that one area of ongoing concern for the early Christians was their relationship with non-Christian Jews. Sometimes this relationship became tense, leading to wide-open conflict. In some measure, the conflict involved more than Jew against Christian. Once Christians left the protective embrace of the ancestral religion of Judaism, they found themselves open to attack by a pagan society that generally did not respect new religious movements and occasionally feared the wrath of the gods who punished the flagrant neglect of their cult. In this chapter we shift our attention to this other form of early Christian conflict, focusing on the tensions that arose between Christians and pagans in the Roman Empire.

THE PERSECUTION OF THE EARLY CHRISTIANS

Perhaps as a result of too many bad Hollywood movies, many people have a completely erroneous sense of what it meant to be a Christian in the Roman Empire. It is commonly imagined, for example, that Christians were of immediate and important concern to the upper echelons of the Roman administration, who saw the Christian movement as taking the world by storm and felt constrained to stop it by any means necessary, and therefore launched massive and violent persecutions as a kind of counterattack. In this view, the Roman emperor or senate declared the religion illegal and used the troops and the law courts to the fullest extent possible to repress it. As a result, the Christians went into hiding; they met secretly in the catacombs, conversed only in private, and identified one another in public through secret signs such as the symbol of the fish.

This view of Christianity in the Roman Empire may make for an indifferent screenplay, but it is far worse from a historical perspective. In fact, Christianity appears to have made only a scant impact on the empire during the first hundred years of its existence. In none of the documents that have survived from pagan authors of the first century of the Common Era—whether histories or philosophical treatises, travelogues or works of fiction, private correspondence or public inscriptions, legal documents or personal notes—in no pagan document of any kind is either Jesus or Christianity mentioned at all. This was not a religion that was on everybody's minds and inspired terror in the hearts of the Roman administration.

I do not mean to say that no one had ever heard of Christianity. People obviously had heard of it, and many of those who did were not kindly disposed toward it. This included at least one of the first-century emperors, as we will see. But the religion was not of major concern to the rulers of the empire or their underlings. During the second half of the first century it was a minor and insignificant nuisance, a mosquito to be swatted, not a tiger to be tamed.

It was not swatted through an officially enacted empire-wide persecution. Contrary to popular imagination, there was no imperial legislation against Christianity and correspondingly no empire-wide persecution of the Christians until nearly two centuries after the time of Paul. Not until 250 C.E. did an emperor proscribe the religion and urge persecutions on a large scale, and even then there is some question concerning how massive the scale was. In any event, during the first century Christians were not driven underground and forced to communicate in private and to hide in the Roman catacombs from the authorities.

Christians had the same rights and responsibilities as everyone else in the empire. Starting a new cult was not illegal; it happened occasionally throughout the entire Hellenistic-Roman period. Christians had the right to worship whatever God they chose, even the Jewish God. Furthermore, the Roman authorities did not care whether the Christians who worshipped this God lived and acted as Jews. It was certainly not against any law for Christians to believe and proclaim that Jesus himself was divine, as some of them eventually came to do. As we have seen, most people believed that gods could come to earth in human form, sometimes as great philosophers or powerful rulers. Some people thought that the emperor himself was a god. To proclaim one more person divine was neither sacrilegious nor sinister.

Christians were within their legal rights to communicate their faith to others, to meet together in private homes, to participate in their own distinctive cultic practices, and to read their sacred Scriptures. Why, then, were Christians like Paul sometimes put in prison, subjected to corporal punishment, and made to stand trial? If they hadn't broken the law, how was it that Christians were found guilty of crimes and punished by torture and imprisonment? To answer the question, we must first visit the Roman legal system.

Roman civil law was extremely sophisticated and nuanced; indeed, it provided the basis for the systems of civil legislation found in European and North American countries today. Disputes over property rights, contractual obligations, financial liabilities, and marriage arrangements were all hammered out by Roman legislators in careful and precise detail. Roman criminal law, on the other hand, was a different matter altogether. Criminal activities were not strictly defined, and punishments were not prescribed by law. In fact, odd though it may seem, neither the Roman emperor nor the Roman senate passed criminal legislation that was binding on all inhabitants of the provincial realms.

The provinces were ruled by governors who were appointed by either the senate or the emperor (depending on whose jurisdiction the province was under). These governors were drawn from the highest ranking officials of the empire, senators and, occasionally, other aristocrats who were judged capable of handling the rule of an indigenous population. The provincial governors had two main responsibilities: to keep the peace and to collect the taxes. They themselves had more than a little stake in these matters, for the governors received a cut of the tax money they brought in. Moreover, they were granted nearly absolute power to accomplish their objectives. To assist them in their duties, the senate would occasionally pass bills proposing rules of governance; these were not federal laws, however, but more like pieces of official advice. In any situation, the governor was expected to use his best judgment to deal with problems that arose, employing whatever means necessary to maintain public order and maximize revenue collection.

Being able to employ any means necessary gave governors the power of life and death. From a Roman administrative point of view, Pontius Pilate was altogether justified in condemning Jesus to death as a public nuisance. People like Pilate were expected to deal with cases like this with justice where possible and severity where necessary.

This takes us, now, to the minor irritations caused by the Christians and the resultant persecutions that were launched in various localities throughout the early empire. Even though the Christian religion was not illegal, in the strict sense of the term, we know that Christians themselves were frequently involved in socially disruptive and therefore punishable behavior, as can be seen, for instance, in the accounts in Acts. It was the magistrate's job to resolve the situation by following his best judgment, for example, by punishing parties that caused the disturbance.

What kinds of public disturbances did Christians cause? From our earliest sources we learn that Christians considered their communities of faith to be self-contained groups that made exclusive demands on the individual member. People were to leave behind their former associations to join the church. This involved abandoning their earlier religious affiliations and, if necessary, their own families. Christians claimed that their Lord himself had meant to disrupt the normal family lives of his followers. From a historical perspective it is difficult to know whether Jesus actually spoke the words that are attributed to him on this score, but they certainly reflect the realities of the churches that later professed his name:

> Do not think that I have come to bring peace to the earth; I have not come to bring peace, but a sword. For I have come to set a man against his father, and a daughter against her mother, and a daughter-in-law against her mother-in-law; and one's foes will be members of one's own household. Whoever loves father or mother more than me is not worthy of me; and whoever loves son or daughter more than me is not worthy of me. (Matt 10:34–37)

Families were disrupted when one member became a Christian and rejected all family ties in favor of a commitment to the church. Indeed, the Christian church portrayed itself as a convert's new family: believers called one another brother and sister, they had "fathers" and "mothers" in the faith, and God himself was the Father of all.

That this new family of faith was to replace one's real family is evident in such early Christian narratives as Paul's adventures with Thecla, a model convert who left her betrothed to follow the apostle in a life of chastity (see also box 25.1). This religious family opened up new possibilities of life for Christian converts; for those outside, however, the impact was sometimes jarring and disruptive. As you might imagine, the abandoned parents and the men left at the altar were not at all pleased. At least in the apocryphal Acts they sometimes did something about it by stirring up public opinion against the Christians and demanding judgment from the governor.

The early Christian communities apparently were viewed with suspicion and distrust for other reasons as well. As we have already seen, these communities were closed to outsiders. Closed societies are always seen as suspicious by society at large: what exactly are they trying to hide? When word leaked out concerning the Christians' activities, the news did little to allay other people's fears. It was known that Christians often met with their brothers and sisters either after dark or before dawn to hold a "love feast" (their term for the Lord's Supper), a celebration that included ritual kissing (e.g., see Rom 16:16; 1 Pet 5:14). At this meal they ate the body and drank the blood of the Son of God. Rumors began to fly, and if you can imagine the worst you won't be far off the mark. Christians were thought to meet under the cloak of darkness in order to hide their despicable deeds from the world. They engaged in wild sex orgies (the love feasts, where the passionate kiss of peace was just the beginning), they committed communal incest with their "brothers and sisters," and most sinister of all, they performed acts of infanticide and ritual cannibalism (eating the son).

These charges may sound ludicrous to us, but they were widely believed by non-Christians in the second century, as evidenced by the fact that Christian authors repeatedly had to defend themselves against them (see box 18.2). Similar charges were leveled against other groups in antiquity as well; evidently one of the common ways to cast aspersions on an unpopular group was to claim that they held nocturnal orgies and ate babies.

Compounding these problems was the fact that Christians refused to participate in local cults and, even worse, in state cults that honored the Roman gods. This refusal was widely seen as treasonous. These were the gods who protected society, who brought peace and prosperity to the empire through the agency of the emperor, who was himself sometimes considered divine in the provinces where Christianity was most successful. In modern terms, failing to worship these gods was a political statement as much as a religious one, for as we have seen, people in the ancient world didn't separate religion and politics into distinct categories. For them, to spurn the state gods was to repudiate the state.

The earliest Christians were attacked principally for causing public disturbances. This is the consistent testimony of the accounts in Acts and the references in Paul's letters, where followers of

SOMETHING TO THINK ABOUT

Box 25.1
The Christian Disruption of the Family: The Case of Perpetua

Early Christians recognized, and sometimes even celebrated, the fact that adherence to their religion could disrupt family lives. For many of them, the Christian church was a new family that replaced their old, biological family. Nowhere can the disruptive possibilities of Christianity be seen more clearly than in the gripping account of the trial and execution of a Roman matron named Perpetua and her female slave Felicitas. The first part of the report actually derives from a private diary that Perpetua kept while in prison awaiting her fate among the wild beasts of a Roman amphitheatre in North Africa.

Perpetua reports that she had an infant son whom she had given over to the care of her family. In one of the most powerful and pathetic scenes of the account, her father pleads with her to consider the pain she is causing her loved ones by her senseless determination to die a martyr's death:

> And then my father came to me [in prison], worn out with anxiety. He came up to me, that he might cast me down [from the faith], saying, "Have pity my daughter, on my grey hairs. Have pity on your father, if I am worthy to be called a father by you. . . . Have regard to your brothers, have regard to your mother and your aunt, have regard to your son, who will not be able to live after you. Lay aside your courage, and do not bring us to destruction; for none of us will speak in freedom if you should suffer anything." . . . And I grieved over the grey hairs of my father . . . and I comforted him saying, "On that scaffold, whatever God wills shall happen. . . ." And he departed from me in sorrow.
>
> Another day . . . an immense number of people were gathered together. We mount the platform. The rest were interrogated and confessed. Then they came to me and my father immediately appeared with my boy and withdrew me from the step, and said in a supplicating tone, "Have pity on your babe." And Hilarianus the procurator . . . said, "Spare the grey hairs of your father, spare the infancy of your boy, offer sacrifice for the well-being of the emperors." And I replied, "I will not do so." Hilarianus said, "Are you a Christian?" And I replied, "I am a Christian."
>
> And as my father stood there to cast me down from the faith, he was ordered by Hilarianus to be thrown down, and was beaten with rods. . . . The procurator then delivers judgment on all of us, and condemns us to the wild beasts, and we went down cheerfully to the dungeon (*Passion of Perpetua and Felicitas*. 2).

Perpetua and her slave Felicitas, who had herself given birth just days before the event, were thrown to the wild beasts for confessing to be Christians. A detailed and gory account of the incident was recorded by an eyewitness and forms the final portion of the second-century martyrology called *The Passion of Perpetua and Felicitas*.

Christ are sometimes subject to mob violence (e.g., Acts 7:54–60; 13:48–51; 14:19–21, 21:27–36; 1 Thess 2:13–16). At other times they suffer an official punishment by order of a Roman magistrate, as indicated, for instance, by Paul's reference to being beaten three times with rods (2 Cor 11:25; see also Acts 16:22). Outsiders evidently considered the followers of Christ to be public nuisances, not the moral, upright citizens one might have expected them to be.

The negative public image of the early Christians can be deduced from the caustic remarks directed against them by pagan authors of the early second century (see box 13.1). Thus, for example, the Roman historian Tacitus calls Christianity a "pernicious superstition" and claims that Nero could use Christians as scapegoats for the burning of Rome because they were the "hatred of the human race" (*Annals* 15). At about the same time (ca. 115 C.E.), the historian Suetonius described Christians as people who held "to a novel and mischievous superstition" (*Life of Nero* 16). The Roman governor of Bithynia-Pontus, Pliny the Younger, considered the Christians to be "obstinate" and "mad" adherents of a "depraved superstition" and expressed some surprise when he learned that at their community meals they ate ordinary food, possibly because he suspected them of cannibalism (Letter 10 to Trajan). Later authors like the emperor Marcus Aurelius considered Christians to be misguided and hardheaded (*Meditations* XI, 3); the satiricist Lucian portrayed them as irrational, gullible dolts (*Death of Peregrinus*, 11–13).

This widespread disapproval of the Christians lies at the root of the earliest governmental actions against them. The first full-blown episode appears to have been the persecution under Nero. When Nero's enemies blamed him for the fire that leveled a good portion of the city—a blame that he evidently deserved—he decided to use the Christians in Rome as his scapegoats. According to the Roman historian Tacitus, Nero made a public display of Christians, having some of them clothed in animal skins to be eaten by ravenous dogs and others rolled in pitch and set aflame to light his public gardens. Tacitus suggests that Nero could treat the Christians this way with impunity because of the general loathing for them. Nero, however, did not order persecutions of Christians living outside of Rome, and more importantly, he did not punish the Christians of Rome *for being Christians*. He condemned them for arson (even though they were apparently innocent of the charge). Thus, Christians were accused of committing actual crimes.

Nero may have set a precedent. Christians who were already looked upon with suspicion and hatred increasingly came to be seen as a public problem, and governors in the provinces must have known the disdain that the emperor himself had shown for them. The problems mounted with the passing of time, as Christians grew in number and openly refused to worship the state gods. This becomes clear in the second incident of official persecution that we can speak about with some confidence. In 112 C.E., Pliny, the governor of Bithynia-Pontus in Asia Minor, heard complaints about the Christians in his province and put them on trial. Afterward, he wrote to the Emperor Trajan to see whether he had handled the situation properly. The letter still survives. In it Pliny tells the emperor that he arrested those suspected of being Christians and forced them to prove their loyalty to the state by paying homage to the images of the emperor and the state gods by offerings of incense and wine. He executed those who refused.

Pliny had these people executed not because they worshipped the Christian God—they were free to do that—but because they refused to worship the gods that supported the empire of Rome. Also, Pliny did not punish those who were suspected of having formerly been Christians so long as they were willing to worship the Roman gods. This procedure shows that it was not a crime to have been a Christian (since crimes are punished even after someone stops committing them). The crime was being adamant in refusing to worship the state gods. Pliny appears to have recognized that Christians were prevented by their religion from worshipping these gods. For this reason, anyone who persisted in claiming to be a Christian was automatically subject to prosecution.

Trajan gave his full approval to Pliny's procedure in a written reply which also still survives, and governors of other Roman provinces appear to have taken his response to heart. Christians weren't hunted down—Trajan explicitly forbade such a practice—and anonymous accusations were generally disallowed, but when difficulties arose within a community and Christians were thought to be to blame, persecutions erupted, even if for a brief period of time. As the existence of the Christians became more widely known, it became increasingly clear that they were (a) antisocial, in that they did not participate in the normal social life of their communities, (b) sacrilegious, in that they refused to worship the gods, and (c) dangerous, since the gods did not take kindly to communities that harbored those who failed to offer them cult. By the end of the second century, the Christian apologist (literally, "defender" of the faith) Tertullian could complain about the widespread perception that Christians were the source of all disasters brought against the human race by the gods:

> They think the Christians the cause of every public disaster, of every affliction with which the people are visited. If the Tiber rises as high as the city walls, if the Nile does not send its waters up over the fields, if the heavens give no rain, if there is an earthquake, if there is famine or pestilence, straightway the cry is, "Away with the Christians to the lion!" (*Apology*, 40)

Christians, of course, had to devise ways of understanding and reacting to the hatred that confronted them on every side. That is to say, the opposition that Christians faced from the rest of the world drove them to define themselves against it. Sociologists have long recognized that a social group often achieves stronger solidarity and internal bonds of cohesion when faced with an enemy, especially one that is powerful and threatening. Speaking in the most general terms, the opposition and persecution that confronted various early Christian communities strengthened the commitment of their members to one another, as they were compelled to face their adversaries together. It also pushed them to explain to themselves theologically why they, the people of God's special favor, should have to undergo such intense and cruel suffering.

These issues are addressed at length in a number of the early Christian writings, some of which we have already considered. In this chapter we will examine several additional documents that derive from this context—the book of 1 Peter, the letters of Ignatius, and the *Martyrdom of Polycarp*—ranging from the end of the first century to the middle of the second. By coincidence, each of these documents relates in one way or another to Asia Minor, the region where Pliny describes his own persecution of Christians during roughly the same period. By exploring these writings we will acquire further insights into how Christians saw themselves in light of the antagonistic world in which they lived.

CHRISTIANS IN A HOSTILE WORLD: THE LETTER OF 1 PETER

The book of 1 Peter is a kind of circular letter written in the name of the apostle Peter to "the exiles of the Dispersion" in several of the provinces of Asia Minor: "Pontus, Galatia, Cappadocia, Asia, and Bithynia" (1:1). Before considering the question of whether Simon Peter himself actually wrote this letter, we need to learn something about its recipients and their situation.

The Addressees

The author calls his readers "exiles" (1:1) and "aliens" (2:11). Most scholars have understood these to be figurative designations of Christians, whose real home is heaven and who are therefore exiles in this world for the time being. Supporting this interpretation are verses where the author indicates that his readers are in exile only for "a while" (1:17) and that their real allegiance is to their heavenly calling (1:13).

Other scholars, however, have suggested that the addressees really were exiles and aliens in the communities in which they lived, that is, they were persons who had moved to new communities but were not fully integrated into them. In the Roman world, such "resident aliens" stood on the margins of society, with more legal rights, for example, than slaves but fewer than native-born citizens (with respect, for instance, to the ownership of property). As is often the case with people in our own world who are new in town, especially if they are entering a close-knit community whose families have been together for a long time, these outsiders would no doubt have felt a sense of alienation from their social world.

How should we weigh these two options for understanding the addressees of 1 Peter? On the one hand, resident aliens or foreigners would have been prime candidates for membership in the new churches that were being established by the early Christians. First Peter may well have been addressed to such persons. They stood on the margins of society at large but who been welcomed into a new community of faith in which they could enjoy the benefits of warm fellowship and family ties unavailable to them on the outside. Moreover, this new community was not just any social gathering of like-minded individuals; it was "the household of God" (4:17).

At the same time, it is a little difficult to believe that the author of 1 Peter actually thought that resident aliens were the only people who were Christians in the churches that he addressed (were there no citizens?) or that social outcasts would be the only Christians who would be interested in reading his letter. It is probably best, then, not to press the literal meaning of these designations too far. Many of his addressees may have been resident aliens, but surely not all of them were.

One thing that we can say with relative certainty about the addressees is that, whether or not they were foreigners, they were Christian believers undergoing suffering, and this author is trying to tell them how to deal with it. The word for "suffering" occurs more often in this short letter than in any other book of the New Testament, even more than in the much longer works of Luke and Acts combined. Even where the author is not talking directly about how to handle suffering, he appears to be speaking about it indirectly. Throughout the letter, for example, he urges his readers to live moral lives so that those on the outside can see that they are doing nothing wrong and causing nobody any harm. They are to be obedient slaves, submissive wives, and tender husbands, and they are to obey all governing authority and to be devoted subjects of the emperor. These are not simply pieces of moral advice; they are also guidelines for avoiding persecution from suspicious authorities and for putting to shame those who wrongfully cause abuse.

The Context of Persecution

Those recipients who were literally resident aliens would no doubt have been accustomed to feeling ostracized by society at large. These feelings would have been assuaged to some extent once they joined the Christian community. Here they would have found a home for themselves in the "household of God" (4:17). Joining this new family also would have had a downside, however, in the public opposition that the group evoked.

We have seen that the persecution of Christians in Bithynia-Pontus during the governorship of Pliny erupted at the grassroots level. Correspondingly, 1 Peter indicates that Christians are principally opposed by their former colleagues and friends who "are surprised that you no longer join them in the same excesses of dissipation" (4:4). That is to say, the Christian converts have caused a good deal of consternation for those with whom they used to spend their time. There has been a public outcry, apparently by those who felt abandoned by their former friends (and spouses?), and it may have reached the point of mob violence or administrative intervention. Thus the author speaks of "the fiery ordeal that is taking place among you" (4:12).

The Author's Response

Persecution often functions to solidify the ties that bind a social group together, giving the members of the group a greater sense of cohesion and belonging as they realize they are "all in it together." Although the author of 1 Peter was obviously not versed in modern sociological theory, he was clearly attuned to the social dimensions of suffering as they were being experienced in the communities that he addressed. One of his goals was to keep these communities together, which meant keeping individual members from leaving as the pressure from the outside mounted.

He constantly reminds his readers that they acquired a privileged status when they joined God's household; they were specially chosen by God, they were "sanctified by the Spirit," and they were "sprinkled with [Christ's] blood" (1:2). He wants them to remember that they have been brought into this new family by means of a new birth (1:3, 23) and that they are now children of God their Father (1:14, 17), having been purchased by the precious gift of Christ's blood (1:19). They are the chosen people, set apart from the rest of the world, belonging to God alone (2:9). Indeed, they are the place of God's residence, his own temple, where sacrifices are made to God; at the same time they are the holy priests who make these sacrifices (2:4–9). Clearly these believers are special before God and unique in the world. Indeed, to some extent they are suffering *because* they are so distinct. Outsiders can't fathom why the members of God's house behave so differently;

and in their ignorance they lash out at what they don't understand (4:3–5). In this they are driven on by the devil himself, God's cosmic enemy (5:8).

Christians, then, should expect to suffer and should not be surprised when they do (4:12), for just as Christ suffered, so too must his followers (4:13). They must not suffer for doing what is wrong, however, but only for doing what is right. They are therefore to live moral, upright lives (3:14–17; 4:14–15). Moreover, when they suffer in this way, they must be prepared to defend themselves by explaining who they are and what they stand for: "Always be ready to make your defense to anyone who demands from you an accounting for the hope that is in you; yet do it with gentleness and reverence" (3:15–16). By making this kind of defense, Christians will put their enemies to shame (3:17).

Thus the author of 1 Peter is concerned not only to create solidarity in the Christian communities but also, and perhaps primarily, to bring an end to the suffering. He makes precisely this point when he urges his readers to "conduct yourselves honorably among the Gentiles, so that though they malign you as evildoers, they may see your honorable deeds and glorify God" (2:11). His injunctions to moral behavior appear to be designed to win over the skeptical (3:1). In a world in which the Christian community was regarded as antisocial, the believers are to "accept the authority of every human institution, whether of the emperor as supreme or of governors as sent by him to punish those who do wrong and to praise those who do right. For it is God's will that by doing right you should silence the ignorance of the foolish" (3:13–15).

The ultimate reward for those who remain steadfast in suffering will be the salvation that is soon to come (1:1–3, 9). This author has not abandoned the eschatological hope of the earliest Christian communities; he embraces it, confident that God will soon bring the believers' suffering to an end (4:17; 5:10). Who was this author?

The Author of 1 Peter

The book claims, of course, to be written by Peter, the disciple of Jesus, and it suggests that he was writing from the capital of the empire. This is intimated at the close of the letter, where the author says that he has written from "Babylon" (5:13), a code word in early Christianity for Rome, the locus of the evil empire that was opposed to God (see Rev 17:5; 18:2). Peter has been traditionally associated with Rome as its first bishop (i.e., the first Pope; see box 20.1).

Many scholars, however, doubt that Peter wrote this letter. Virtually the only thing that we can say for certain about the disciple Peter is that he was a lower-class fisherman from Galilee (Mark 1:16) who was known to have been illiterate (Acts 4:13). His native tongue was Aramaic. This letter, on the other hand, is written by a highly literate Greek-speaking Christian who is intimately familiar with the Old Testament in its Greek translation and with a range of Greek rhetorical constructions. It is possible of course, that Peter went back to school after Jesus' resurrection, learned Greek, became an accomplished writer, mastered the Greek Old Testament, and moved to Rome before writing this letter, but to most scholars, this seems unlikely.

Some have suggested that the letter was actually produced by Silvanus, who is mentioned in 5:12. This is certainly possible as well, but one might then wonder why Silvanus is named not as the author of the letter but only as its scribe (or carrier). Others have thought that Silvanus penned the letter as it was dictated by Peter and that he put Peter's rough dictation into a more aesthetically pleasing and rhetorically persuasive style of Greek. If so, one would still have difficulty accounting for the detailed interpretations of the Greek Old Testament—and, indeed, for most of the detailed argument—without supposing that Silvanus, rather than Peter, was the real author.

I should point out that there are an extraordinary number of pseudonymous writings forged in Peter's name outside of the New Testament. In addition to the *Gospel of Peter* that we have already discussed, there are three apocalypses attributed to Peter, several "Acts" of Peter, and other Petrine letters. In addition, as we will see, scholars are virtually unanimous in thinking that the book of 2 Peter within the New Testament is pseudonymous as well. On balance, then, it is probably best to regard 1 Peter as yet another example of Christian pseudepigraphy, in which a later author took the name of Jesus' closest disciple to lend authority to his own views.

It is difficult to say, however, when the author would have been writing, or even from where and to whom. If the letter is indeed associated with Asia Minor, as its prescript suggests, it should probably be assigned to the first century, possibly near its end, when persecution was on the rise but the later church hierarchy with a solitary bishop over each church had not yet developed. There is no trace of this hierarchy in the letter, where the churches of Asia Minor appear to be ruled by groups of "elders" (5:1–4). A hierarchy *is* in evidence in this region, however, at the beginning of the second century, especially in the letters of Ignatius.

CHRISTIANS SENTENCED TO DEATH: THE LETTERS OF IGNATIUS

The letters of Ignatius of Antioch are among the most intriguing early Christian writings to be preserved from antiquity, in no small measure because of their unusual historical setting. They are addressed to several churches of Asia Minor that had sent delegations to meet Ignatius as he passed through the region en route to Rome, around the year 110 C.E. This was no vacation jaunt for Ignatius. Convicted of crimes against the state, he was traveling under armed guard to face his death by execution, having been condemned to the wild beasts of the Roman arena because of his Christian faith. Far from shuddering in the face of his coming martyrdom, however, Ignatius embraced it ecstatically; he looked forward to the opportunity to be torn apart and devoured for the sake of Christ. Ignatius was an intriguing personality, to put it mildly. He is seen by some modern readers as the ideal Christian martyr and by others as a case study in pathology. In any event, his status in early proto-orthodox circles is clear, for some Christians of later centuries counted his letters among the sacred Scriptures.

The Historical Background

We know almost nothing about the man Ignatius apart from what can be inferred from his letters. From these we learn that he was the bishop of the church of Antioch, Syria, one of the oldest and largest of the empire. He was obviously educated and gives some evidence of knowing secular Greek literature (e.g., in Ign. *Rom.* 4:1). It could be that as a highly literate convert from the upper classes Ignatius had made inroads into the Christian community in Antioch and eventually rose to the position of bishop.

Ignatius appears to have left the church in a state of turmoil. He intimates that there had been an internal squabble, possibly a struggle for control, and that the matter had been resolved just recently. The side that Ignatius himself backed in the dispute (whatever it was about) had apparently won. Some scholars have supposed that Ignatius himself was the issue. It may be that his authority as bishop had been challenged by other members of the church before he left.

We do not know exactly what happened during the persecution that sent Ignatius to Rome. He does indicate that several other members of the Syrian church had gone before him, apparently also to face execution (Ign. *Rom.* 10:2). It is reasonable to assume that a local outcry had led to the arrest of Christian leaders in Antioch; the situation in that case would be somewhat similar to that which arose about the same time under Pliny in Bithynia-Pontus, just north of where Ignatius passed through Asia Minor. Moreover, since Ignatius was sent to the empire's capital for punishment (possibly to stand trial first), it may be that he and his predecessors were Roman citizens and so had to receive special treatment, unlike native citizens of Antioch, who could have been put on trial and executed on the spot.

Ignatius was accompanied across the land route from Syria to Rome by a group of soldiers whom he likens to ten wild leopards who behave more cruelly when treated kindly (Ign. *Rom.* 5:1). News of his journey evidently preceded him, since local churches sent representatives to visit him at several of his stopping points, possibly to provide him with supplies. In response to this outpouring on his behalf, Ignatius wrote letters to the churches in the cities of Tralles, Magnesia, Ephesus, Philadelphia, and Smyrna. He also wrote a separate letter to the bishop of one of these churches, Polycarp of Smyrna, whom we will meet again momentarily, as well as a letter to the Christian

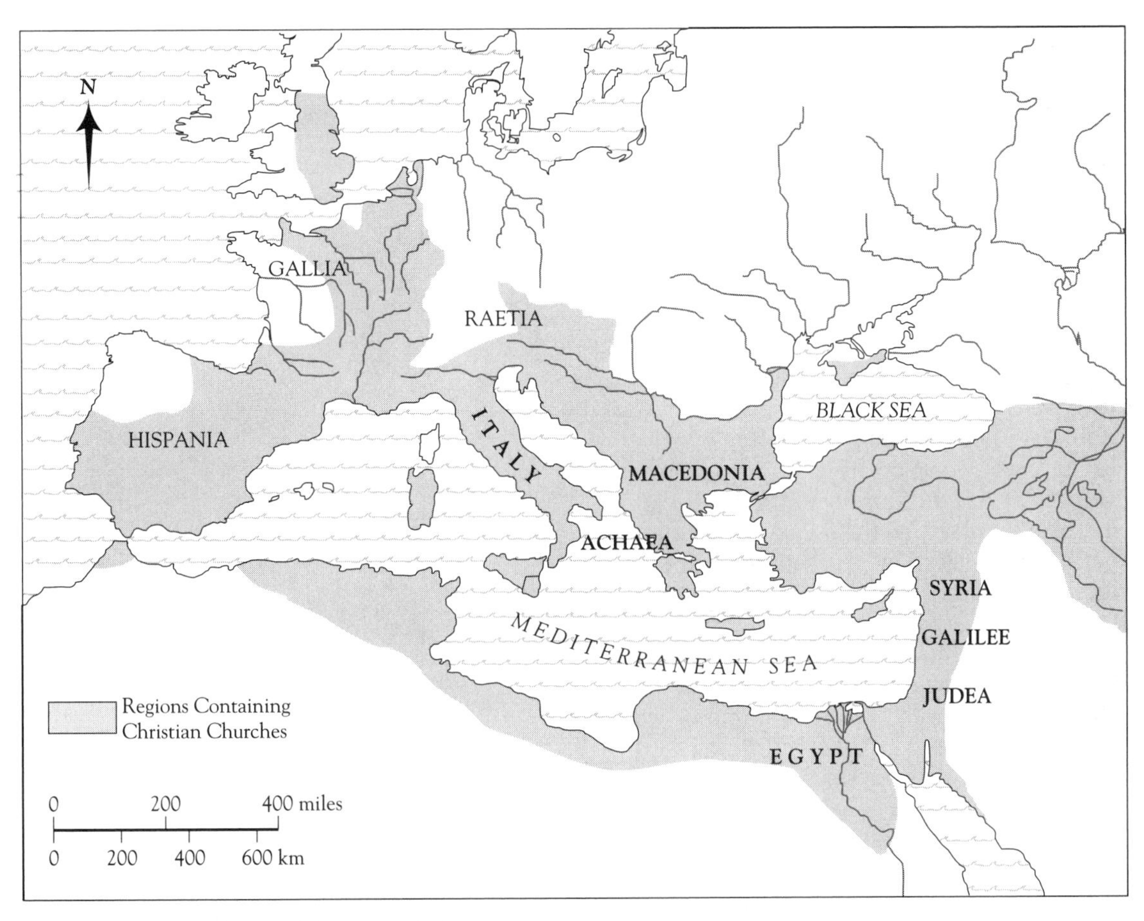

Figure 25.1 The Distribution of Christianity by 300 C.E.

congregation in Rome. These letters were obviously written in relative haste by a man in highly unusual circumstances. Several themes recur throughout them.

The Overarching Themes

The Church's Unity. Ignatius insists that Christian communities throughout the world be unified. That this would be a pressing concern of a proto-orthodox bishop should come as no surprise given the widespread diversity of early Christianity that we have repeatedly observed. Indeed, even Ignatius's own church in Antioch appears to have been internally divided, possibly over Ignatius's own authority as bishop or over the appointment of his successor once he was gone. From other sources we know that there were Gnostic Christians there and possibly also Jewish Christians with adoptionistic views. These various groups may actually have dominated some of the house churches in town and urged their own "candidates" for the post of local bishop. If so, then part of the internal struggle of the community may have involved widely divergent theological perspectives among its leading members.

The Church's Purity. If Ignatius had himself experienced theological controversy in Antioch, this would explain his insistence that the churches of Asia Minor maintain the "pure" doctrine that had been given them by the apostles, and not depart from the truth to embrace heretical speculations.

Ignatius is particularly concerned to combat different kinds of christological heresies, that is, teachings about Jesus that he regarded as false. We have already seen that various New Testament writers living before Ignatius had different views of Jesus. These differences came to be magnified with the passing of time, leading some Christian leaders to declare that only one of them could be right. In this struggle over who was right and who was wrong, some of the parties insisted that Jesus should be seen as a human being chosen by God but not as himself divine. Others claimed that Jesus was actually God and therefore not a flesh-and-blood human being. Still others, including Ignatius himself, maintained that both of these views were right in what they affirmed but wrong in what they denied. For this group, Jesus was both human and divine. The resultant view, at least as Ignatius himself worked it out, was probably meant to sound somewhat paradoxical: Christ was "of flesh, and yet spiritual, born yet unbegotten, God incarnate, genuine life in the midst of death, sprung from Mary as well as God, first subject to suffering then beyond it" (Ign. *Eph.* 7:2).

For Ignatius, the purity of the church depended on this basic confession of faith. Anyone who rejected it was to be rejected from the church. But who was to guarantee that Christians throughout the world would continue to subscribe to it? Who was responsible for the purity of the church? The answer for Ignatius was the single bishop who was to preside over every Christian community, the leader who was to guide the church in the way that it ought to go.

The Church's Leadership. Even more than the Pastoral epistles, the letters of Ignatius stress the importance of the church hierarchy in all matters of doctrine and practice and maintain that the bishop is God's representative on earth, whose rule is law (see box 22.4). No one is permitted to engage in any church activities apart from the bishop and no one is allowed to gainsay his authority. In Ignatius's words: "It is essential to act in no way without the bishop" (Ign. *Trall.* 9:2), "You ought to respect him as you respect the authority of God the Father" (Ign. *Magn.* 3:1), and "We should regard the bishop as the Lord himself" (Ign. *Eph.* 6:1). What better way to bring order out of chaos than to claim that the leaders of the churches, with whom one happens to agree, have been appointed by God himself to run the show.

Ignatius and Christian Persecution

In some respects, the most interesting of Ignatius's writings is the letter to the Romans, where he deals explicitly with his upcoming martyrdom. We might expect that Ignatius would want to find some way to avoid having to pay the ultimate price for his faith, if he could do so without compromising his convictions. Ignatius, however, goes to his death eagerly, longingly. He writes to the Romans in order to urge them not to interfere, for he believes that only by suffering a glorious and bloody martyrdom will he become a true disciple of Christ, only by imitating Christ's own Passion will he be able to "get to God."

Ignatius asks the Roman congregation to "grant me no more than to be a sacrifice to God while there is an altar at hand" (1:2). He wants them to pray for him, not so he might escape his suffering but so he might embrace it: "Pray that I may have strength of soul and body so that I may not only talk [about martyrdom], but really want it" (3:2). Most of all, he does not want them to interfere in the proceedings: "I plead with you, do not do me an unseasonable kindness. Let me be fodder for wild beasts—that is how I can get to God. I am God's wheat and I am being ground by the teeth of wild beasts to make a pure loaf for Christ. I would rather that you fawn on the beasts so that they may be my tomb and no scrap of my body be left" (4:1–2). This longing for death may appear to some modern readers to border on the pathological:

> What a thrill I shall have from the wild beasts that are ready for me! I hope they will make short work of me. I shall coax them on to eat me up at once and not to hold off, as sometimes happens, through fear. And if they are reluctant, I shall force them to it. . . . May nothing seen or unseen begrudge me making my way to Jesus Christ. Come fire, cross, battling with wild beasts, wrenching of bones, mangling of limbs, crushing of my whole body, cruel tortures of the devil—only let me get to Jesus Christ. (5:2–3)

Figure 25.2 Mosaic from a villa in North Africa showing a lion attacking a man. During the persecutions, Christians were sometimes martyred by wild beasts in the arena.

We would be wrong, though, to write Ignatius off as a demented soul who was out of touch with reality. He was very much in touch with reality; it just happened to be a reality that most other people don't see. Ignatius's reality (speaking from his own perspective) was a kingdom that was not of this world, a kingdom that he wanted to obtain with all his heart. The kingdoms of earth meant nothing to him and were clearly run by the forces of evil. One could escape bondage to these forces by letting them do their worst, by allowing them to kill the body so as to free the soul. He believed that by escaping this world he would attain to God. Ignatius was thus one of the first in a long line of Christian martyrs who came to be seen by some of their fellow Christians as people of true faith because they alone were willing to suffer horrible abuses of their bodies for the sake of the kingdom that was not of this world (but see box 25.2).

We lose track of Ignatius after he penned his letters, although later Christian sources indicate that he did indeed face martyrdom in the Roman amphitheater. For an actual depiction of a martyr in the face of death, we have to go elsewhere—but not too far, because the first written account of a Christian martyr happens to be that of Polycarp, the bishop of Smyrna to whom Ignatius wrote on the way to Rome.

CHRISTIANS BEFORE THE TRIBUNAL: THE MARTYRDOM OF POLYCARP

Polycarp appears to have been a relatively young man when he was befriended by Ignatius. His martyrdom occurred some forty-five years later,

SOMETHING TO THINK ABOUT

Box 25.2 An Alternative View of Martyrdom

Most of the surviving Christian writings from antiquity take a positive view of Christian martyrdom, urging Christians to go willingly to their deaths for the faith and to endure all the tortures that humans can devise. By doing so, Christians would imitate the Passion of their Lord, Jesus.

But not everyone agreed. We know from the letters of Pliny and the writings of several Christian authors, for example, that there were large-scale defections from the Christian ranks in times of persecution. Indeed, one of these authors, Tertullian, specifically attacks Christian Gnostic groups for opposing martyrdom. These groups tried to persuade their fellow Christians not to be so foolish as to die for their faith. In their view, Christ died so that his followers would not have to do so. For them, anyone who embraced the need for martyrdom in effect denied that Jesus' death itself was sufficient for salvation (Tertullian *Scorpion's Sting* 1). It appears likely that such people urged Christians to perform the necessary sacrifices to the state gods without actually committing apostasy in their hearts, since God after all was concerned with the heart, not with such meaningless actions as tossing a handful of incense on a burning altar.

If there were competing Christian views of martyrdom, why do most of our surviving texts embody only one of them? The proto-orthodox Christians who won the struggle over whose views were right were quite strong in their insistence that Christians should go to their own deaths willingly, in no small measure because this view was closely related to other theological positions that they took. In particular, the physical sufferings of the Christians served to highlight the reality of Christ's own death, a point of great importance in the debates over docetism and Gnosticism in the second and third centuries. The connection between the virtues of martyrdom and the reality of Christ's death was already made clear in the writings of Ignatius:

> For [Christ] suffered all these things on our account that we might be saved. And he truly suffered, just as he truly raised himself, not as some unbelievers say, that he only appeared to suffer. For they are the ones who are only an appearance. . . . For if these things were done by the Lord in appearance only, then also I am bound only in appearance. And why then have I given myself over to death, to fire, to the sword, to the wild beasts? (Ign. *Smyr.* 2, 4)

around 156 C.E. It is somewhat difficult to gauge his age at that time, since at his trial Polycarp claimed to have served Christ for eighty-six years (*Mart. Pol.* 9:3). If he became a Christian at a very young age, he may thus have been born sometime around 60 or 65 C.E.

The execution of Polycarp and the events leading up to it were recorded by a member of his congregation in Smyrna, in a letter directed to the church in Philomelium in the province of Phrygia in Asia Minor. Even though this "martyrology," or account of a martyrdom, derives from an eyewitness's observation, it cannot be taken as an objective report of what happened to the aged bishop (since any historical document will reflect the subjective views of the person who produced it). There is, for example, a good deal of artistry in this account in that the author goes out of his way to show that Polycarp's martyrdom was "conformable to the gospel" (1:1), that is, similar to Jesus' own martyrdom as described in the early Christian traditions (cf. Ignatius's desire to suffer like Christ).

Thus, in the narrative, Polycarp knows in advance how he is to die (5:2), he is betrayed by his own companions (6:2), the police chief in charge of his arrest is named Herod (6:2), Polycarp refuses to escape arrest but instead prays that "God's will be done" (7:1), he enters the city mounted on a donkey (8:1), and he is put on trial before the Roman tribunal, who tries to have him released, but is opposed by the crowds, especially the Jews among them, who demand Polycarp's death (chaps. 9–13).

In addition to these literary touches, there are several legendary accretions to the account, particularly in the description of Polycarp's execution itself. The Roman governor condemns Polycarp to death by burning. When the executioners build a fire around him, however, he is not touched by the blaze; the flames instead form a kind of chamber around him. His skin does not burn but takes on the appearance of baking bread, and it emits not the stench of charred flesh but the aroma of precious spices. When his enemies behold this miracle, they order the executioner to pierce him with a dagger, but when he does, such a quantity of blood gushes forth that it douses the entire conflagration. A scribe who later copied the story added an even more miraculous detail: a dove flies out from the dagger wound in Polycarp's side (representing his holy spirit?). So died Polycarp, according to the story, the martyr whom God rewarded in death as much as he did in life.

Despite the obviously fictional touches in the account, there are some very interesting historical features as well. We are shown by the narrative, for example, that the only crime Polycarp had committed was siding with the Christians in refusing to worship the state gods. In order to be delivered from the sentence of death, all he was required to do was "swear by the fortune of Caesar" (i.e., do homage to the emperor's divine spirit) and to curse "the atheists," that is, the Christians, who did not acknowledge the gods and were therefore, in the eyes of these pagans, "a-theists" (literally, "not-theists," those who do not accept the gods). Polycarp refused to disavow Christ or his followers and so forced the governor to do his duty to the people by having him executed.

Why would Polycarp not reject his Christian faith, even if just for the moment, in order to avoid a brutal and cruel death? Obviously, we will never know what Polycarp himself thought of the matter, since he never had the chance to tell us, but the author of the account provides an answer, which is no doubt representative of much Christian thinking about suffering for the sake of the faith (but see box 25.2). In speaking of "all the martyrdoms" that Christians had experienced with such boldness (which indicates, of course, that Ignatius and Polycarp were not the only ones known to have died in this way), the anonymous author tells us that

> they despised the tortures of this world, purchasing for themselves in the space of one hour the life eternal. To them the fire of their inhuman tortures was cold; for they set before their eyes escape from the fire that is everlasting and never quenched. (2:2–3)

According to this author, Christian martyrs thought of their future glory rather than their present sufferings and were willing to exchange torment in the present for ecstasy in the hereafter. Moreover, they recognized the reverse side of this commitment: to retreat from their Christian faith to avoid pain now would mean to suffer eternal torment later, in the life to come. Surely it was better to experience agony for an hour than the cruel torments of hell for a million years and beyond.

This view of suffering can tell us something interesting about the direction in which some Christians were heading in their thinking. As we have seen, from the outset Christians had looked to the future; for most of them, it was a future to be brought soon by Christ, when he came in power at his second coming. When this imminent appearance never occurred, many Christians stopped concerning themselves with the salvation of this world and began to reflect on their own salvation from this world. For them, the present life was not the end of the story; indeed, it was only the beginning. After this life came eternity, and no one could afford to let the allures and pleasures of this mortal existence interfere with the true ecstasies of the world to come, which would be granted to those who remain faithful to God and to his Christ.

CHRISTIANS ON THE DEFENSE: THE LATER APOLOGETIC LITERATURE

We have seen in our discussions of Acts and 1 Peter that Christians who were opposed by their non-Christian neighbors and by hostile rulers were bound to make a defense, or apology, for their beliefs and actions. As Christianity spread through the empire in the second century, it eventually came to attract converts not only from among the lower classes but also, occasionally, from the ranks of those who were wealthier, more powerful, and more highly educated. The more intellectually oriented Christians of the second century, of course, were just as prone to persecution for their faith as were their lower-class associates. Some of them reacted to the situation by employing their literary skills to develop an intellectual defense of Christianity, for example, by writing open letters to the emperor to urge him to bring an end to the sporadic persecution of Christians. Some of these Christian thinkers, including such authors as Justin in Rome, Tertullian in North Africa, and Origen in Alexandria, continue to be well known even today (see Chapter 24). While we cannot devote a substantial amount of time to this later apologetic literature in an introduction to the New Testament, we can at the least see how Christian thinkers of the second century followed the lead of the New Testament writers (e.g., the author of 1 Peter) while developing their ideas in new directions as they defended themselves against the charges brought against them.

The Christian apologists claimed that the Christians' beliefs were superior to anything found in the other religions of the empire and that Christians were altogether innocent of the charges of immorality and atheism. To show the superiority of Christianity, the apologists argued that the religion could not have spread so far and wide, and with such speed, if the hand of Providence had not been behind it. They maintained that individual Christians could not display such superhuman bravery in the face of death unless they were supported by the power of God. They insisted that Christ could not have miraculously fulfilled prophecies made hundreds of years before his time in the Hebrew Bible if he himself were not divine and if the religion that he founded did not represent the true interpretation of the traditions of Israel. Indeed, the apologists claimed that their religion was superior precisely because it was so ancient, more ancient than the philosophical traditions stemming from Plato (who lived 800 years after Moses) and even than the religious traditions dependent on Homer (who lived 400 years after Moses).

The antiquity of the Christian religion could also be seen, according to the apologists, in the fact that other (acceptable) religions of the empire had taken over so many of its important beliefs. Thus, belief in a supreme God, in a human as his Son, in the Son of God's virgin birth, in his miracles, resurrection from the dead, and ascent into heaven—all of these things had their parallels in Greek mythology. Why should Christians be punished for beliefs that others subscribed to as well, especially when Christianity, which was older than the oldest Greek myths (since it can be found already in the writings of Moses), was the source for these beliefs?

Finally, the apologists insisted that even if pagans decided to rebel against the truth and reject the true knowledge of God offered by this ancient religion they should at least have the decency to leave it alone. Christians had done nothing to deserve their persecution. Indeed, the apologists claimed, the charges of disrespect for Roman authority and the accusations of flagrant immorality leveled against the Christians were outrageous and unsubstantiated. Christians were the "salt of the earth," the element of society that prevented it from crumbling altogether. They were good citizens and loyal to the state; they were faithful wives, husbands, and slaves; and they were moral and upright members of their communities, who deserved to be thanked rather than punished. Furthermore, argued the apologists, it would be in the authorities' own best interests to leave the Christians in peace, for every attempt to squelch the religion had failed miserably. As often as Christians were persecuted and martyred, other converts flooded in to swell their ranks. To paraphrase Tertullian, "the blood of the martyrs is the seed of the church" (*Apology* 50).

The positions staked out by these Christian apologists may sound completely reasonable to most of us who live in the Western world that emerged from the victory they ultimately won. To most pagans of the time, however, these Christian arguments would have seemed altogether irrelevant. It is not that pagans in the empire were intolerant of diversity. Quite the contrary, as we have seen, pagan religions and their devotees, whether from the lower or upper classes, were as a rule remarkably tolerant. But Christianity was something that many could *not* tolerate precisely because, ironically, the Christians themselves were perceived to be so stubborn and intolerant. Unlike followers of other religions, many Christians claimed that they knew the one and only way, that they alone had the truth. Those who accepted this truth would be blessed by God, those who rejected it did so at their own eternal peril. At the end, believers would be rewarded, nonbelievers damned. Many Christian people, especially the proto-orthodox Christians who ended up dominating the religion, believed that theirs was the only God and that anyone who rejected him would suffer the eternal consequences.

These Christians thus urged non-Christians to live and let live—when it came to their own Christian beliefs—but these very beliefs consigned to the flames of hell all those who did not accept them. This kind of intolerance was intolerable to most pagans.

The apologists' request that the government not get tangled up in the affairs of religion by persecuting aberrant cults may also seem reasonable to us, especially in the United States, where there is a constitutional guarantee of the separation of church and state. For ancient persons, though, such a separation was unheard of and nonsensical. The gods had made the state great and in response the state honored the gods. The gods, after all, did not ask for much—simply the respect and honor that was due their name, shown in such simple acts as the sacrifice of some incense on an altar. Those who refused to make such a sacrifice were obviously obstinate and dangerous—obstinate because so little was involved and dangerous because the gods did not take kindly to those who willfully neglected their cult, or to the communities that housed them. To be sure, the gods themselves were tolerant, but only up to a point, and once offended, they knew full well how to exact retribution. For the state not to promote the worship of the gods—indeed, for the state not to insist upon it—would have been to commit social suicide.

It was the task of the Christian apologists to show that this pagan view was wrong. In one interesting respect, they failed miserably. After the conversion of Constantine, the state did not take the apologists' advice to get out of the business of religion. The Christian emperors promoted religion just as avidly as the pagan emperors had before them, but rather than using the power of the state in support of the Roman gods, they used it to advance the worship of the Christian God. It was not until the Enlightenment that European thinkers came to believe that a separation of church and state would prove beneficial to both. And only when this novel idea entered into the public domain and became a centerpiece of the constitutional charter of the United States did religion and politics come to be seen as two discrete entities, for the first time in the history of Western civilization.

SUGGESTIONS FOR FURTHER READING

Elliott, J. H. *A Home for the Homeless: A Sociological Exegesis of 1 Peter, Its Situation, and Strategy*. Philadelphia: Fortress, 1981. A groundbreaking examination of the communities addressed by 1 Peter from a sociological perspective; for advanced students.

Lane Fox, Robin. *Pagans and Christians*. New York: Alfred A. Knopf, 1987. A long but fascinating and often brilliant discussion of the relationship of pagans and Christians during the first centuries of Christianity; for more advanced students.

Frend, W. H. C. *Martyrdom and Persecution in the Early Church*. Oxford: Blackwell, 1965. The best full-length study of hostilities against Christians during the first three centuries of the Common Era.

Macmullen, Ramsey. *Christianizing the Roman Empire A.D. 100–400*. New Haven, Conn.: Yale University Press, 1984. A concise and insightful discussion of the difficulties Christians encountered, the methods they used, and the success they enjoyed in propagating their religion throughout the Roman world.

Musurillo, H., ed. *The Acts of the Christian Martyrs*. Oxford: Clarendon, 1972. An intriguing collection of twenty-eight accounts of Christian martyrdoms, from eyewitness sources of the second to fourth centuries.

Tugwell, Simon. *The Apostolic Fathers*. Harrisburg, Pa.: Morehouse, 1990. A clear and straightforward discussion of the background and message of the Apostolic Fathers, including the epistles of Ignatius; ideal for beginning students.

Wilken, Robert. *The Christians As the Romans Saw Them*. New Haven. Conn.: Yale University Press, 1984. A popular study of the largely derogatory views of Christians held by several Roman authors; particularly suitable for beginning students.

CHAPTER 26

Christians and Christians: James, the Didache, Polycarp, 1 Clement, Jude, and 2 Peter

Up to this stage in our examination of the general problems of the general epistles, we have explored two areas of social conflict encountered by the early Christians: those involving non-Christian Jews and those involving pagans. We have seen that these areas of conflict affected more than the external aspects of Christianity; they were profoundly related to certain internal dynamics as well. The Jewish opposition to Christianity, for example, compelled Christians to engage in acts of self-definition as they tried to understand themselves in relation to the religion from which they had emerged and to the people who continued to embrace it. Not all Christians agreed on the self-definitions that were devised. Pagan opposition also forced Christians to attend to their public image. Church leaders urged their communities to maintain high ethical standards so as to earn the respect of those who suspected the group's motives and activities. Again, not every Christian agreed on what these ethical standards should entail.

We now to turn from these external forms of conflict to controversies that raged within the Christian communities themselves. The issues affect not only the general epistles; we have already seen numerous instances of internal Christian conflicts in the other writings we have examined. One need only think of Paul's conflicts with the Judaizing Christians in Galatia or with the "superapostles" in Corinth, of the Pastoral epistles and the problems of false teaching that they were written to address, or of the Johannine epistles and their attacks on the secessionists from the community. Indeed, it appears that most of our early Christian authors saw as many enemies inside the church as outside.

Internal conflicts arose in no small measure because Christianity was so remarkably diverse in the first two centuries. From the beginnings of this religious movement, believers who insisted that they had a corner on the truth found some of their most energetic adversaries among those who also claimed to be Christian but who advanced a different point of view or promoted a different kind of lifestyle. As we have already seen, only one basic form of Christianity emerged victorious from these conflicts and thereafter declared itself "orthodox," and every major form of modern Christianity—Catholic, Protestant, Eastern Orthodox—traces its roots to this victory. Indeed, the collection of twenty-seven ancient Christian writings that became the sacred canon of Scripture is itself one of the legacies of this victory. During the period we are exploring in this study, however, no New Testament canon had yet come into being, and Christians were by no means in agreement on some of the most basic questions about what to believe and how to live.

We can see some of the conflicts at work in several of the general epistles of the New Testament as well as in other early Christian writings that happen to survive from roughly the same period of time. In this chapter we will consider some of these writings, following a sequence based more on the content of these books than on their chronology (precise dates are nearly impossible to establish with these writings in any case). As we will see, the major internal conflicts of the early Christian movement involved ethics, leadership, and doctrine. These three areas of concern were not, of course, mutually exclusive. On the contrary, many early Christians believed that bad leaders introduced false teachings that promoted immoral activities. We have already seen this view reflected in the Pastoral epistles and the letters of Ignatius, books that are roughly contemporary with the works we are about to consider: the epistle of James, the *Didache*, the letter of Polycarp to the Philippians, *1 Clement*, Jude, and 2 Peter.

THE EPISTLE OF JAMES

Of all of the writings that we will be examining in the present chapter, James appears to be the least concerned with corrupt leaders or false teachings infiltrating the community (but see 3:1–3). Nonetheless, parts of the letter appear to be directed against aberrant notions advanced by Christians known to the author. In particular, as we have already seen in Chapter 21, it is possible that some Christians had taken Paul's doctrine of justification by faith apart from the works of the Law to mean something that Paul himself did not, namely, that it only mattered what a person believed, not how he or she lived. James (whoever he was) may have written this letter to stake out the opposing position, arguing that true faith will always be manifest in one's life, especially in the ways one treats the poor and the oppressed. To put it in his own words, "a person is justified by works and not by faith alone" (2:24) because "faith without works is dead" (2:26).

The book consists of a series of ethical admonitions to those "who believe in our glorious Lord Jesus Christ" (2:1). One of its most striking features, however, is that Jesus is otherwise scarcely ever mentioned. Apart from the epistolary opening (1:1) and the verse just quoted (2:1), Jesus makes no appearance at all. What is even more intriguing is that, apart from these two verses, almost none of the ideas in the book is uniquely Christian. The various ethical injunctions have numerous parallels, for instance, in non-Christian Jewish writings, and all of the examples of ethical behavior are drawn from stories of the Hebrew Bible (Abraham 2:21, Rahab 2:25, Job 5:11, Elijah 5:17) rather than from the life of Jesus or the activities of his apostles. Even the communities of believers that are addressed appear in Jewish guise—they are described as "the twelve tribes in the Dispersion," and their place of assembly is literally called a "synagogue" (2:2).

For these reasons, some scholars have argued that James is a kind of Jewish book of wisdom (somewhat like the Book of Proverbs but without as many one-liners) with only a thin Christian veneer. According to this opinion, the author took over a piece of Jewish writing and "Christianized" it by adding a couple of references to Jesus.

Not everyone is persuaded by this point of view, however. Many scholars, for example, have observed that a large number of the admonitions in James have close parallels in Matthew's Sermon on the Mount (see the examples in Chapter 21). In addition, portions of the book relate closely to other teachings of Jesus (compare, for instance, 4:13–15 with Jesus' parable of the rich fool in Luke 12:16–21). How then does one account for the general nature of these admonitions, that is, the fact that most of them are not distinctively Christian, *and* for their close similarities to older traditions about Jesus? It may be that the author strung together a number of important ethical admonitions that could be found in a variety of settings, such as Jewish wisdom literature and traditions of Jesus' own teaching, and has applied them to the Christian communities that he is addressing.

James emphasizes that those who have faith need to manifest it in the way they live (1:22–27; 2:14–26). Other recurring themes include the importance of controlling one's "tongue" (i.e., one's speech; 1:26; 3:1–12), the danger of riches for believers (1:9–11; 4:13–17; 5:1–6), and the need to be patient in the midst of suffering (1:2–8, 12–16; 5:7–11). The author, however, is not concerned only with what we might call individual ethics. Near the end of the book he turns to

Figure 26.1 The first page of *The Didache*, a book thought by some ancient Christians to belong among the Christian Scriptures.

address communal activities within the church as well, giving his readers advice about saying prayers, singing psalms, anointing the sick with oil, confessing sins, and restoring those who have strayed from the faith (5:13–16).

THE DIDACHE

The idea that false teachers and fraudulent Christian leaders were abroad is somewhat more prominent in a book of the early second century known as *The Didache of the Twelve Apostles* ("*didache*" literally means "the teaching"). The book was virtually unknown until the end of the nineteenth century, when it was discovered in a monastery library in Constantinople. Since then it has made a tremendous impact on our understanding of the inner life of the early Christian communities. Among other things, it (*a*) preserves our earliest account of how the early Christians practiced their rituals of baptism and the eucharist, (*b*) discloses the kinds of prayers that early Christians said, (*c*) indicates the days on which they fasted, and (*d*) demonstrates the existence of itinerant Christian apostles, prophets, and teachers who roved from town to town, addressing the spiritual needs of the Christian communities in exchange for daily food and shelter.

The first six chapters of the book present a set of ethical admonitions organized according to the doctrine of the "Two Ways," which we have already seen in the *Epistle of Barnabas*. Here, though, rather than being presented as "the Ways of Light and Darkness," the two ways are said to be those "of Life and Death." The broad similarities to Barnabas have led most scholars to think that this portion of the writing was drawn from an earlier source that was more widely available to various Christian authors.

In many respects, the "Way of Life" is more interesting than the "Way of Death." At least the author devotes considerably more space to it—all of chapters 1–4, as opposed to merely chapter 5. Many of the moral exhortations are reminiscent of James: a Christian's words are to be backed up by actions (2:5; cf. James 2:14–26); jealousy and anger are to be avoided, since they lead to murder (3:2; cf. James 4:1–2); believers should associate with the humble and upright rather than the high and mighty (3:8; cf. James 2:5–7); and Christians are not to show favoritism or to turn their backs on the needy (4:3; cf. James 2:1–4) but instead to share their goods with one another (4:8; cf. James 2:14–16).

The Way of Death is described far more tersely; it involves "murders, adulteries, lusts, fornications, thefts, idolatries . . . deceit, arrogance, malice, stubbornness, greediness, filthy talk, jealousy, audacity, haughtiness," and so on (5:1). Once again, the exhortations are not uniquely Christian, in that other moralists of the Greco-Roman world agreed that such activities and attitudes were to be avoided. As a result, some scholars have maintained that this notion of the Two Ways ultimately originated in non-Christian Jewish circles. Still, the various authors who incorporate this source into their writings (Barnabas, the *Didache*, and several later writers) were all Christian. Moreover, just as James has a number of parallels to Matthew's Sermon on the Mount, so too does the *Didache*—even more, in fact, including references to praying for one's enemies, turning the other cheek, and going the extra mile.

Unlike the "Teaching of the Two Ways," the second portion of the *Didache* does not appear to be drawn from an earlier source and may well represent the anonymous author's own composition. It is a kind of "church order" in which instructions are given for various kinds of church activities. For example, Christians are to perform their baptisms in cold running water (i.e., in an outdoor stream) wherever possible, although standing or warm water is permissible where necessary. If none of these options is available, water is to be poured over a person's head three times "in the name of the Father, Son, and Holy Spirit" (chap. 7).

Christians are to fast twice a week, on Wednesdays and Fridays (8:1), not on Mondays and Thursdays since that is when "the hypocrites," presumably non-Christian Jews (cf. Matt 6:16–18), do so. Nor are they to pray "like the hypocrites," but they should repeat the Lord's prayer three times a day (8:2–3; see box 26.1). When they celebrate the Eucharist they are first to bless the cup with a prayer that the author provides and then to bless the broken bread, with another set prayer (9:1–4). This way of celebrating the Lord's Supper by starting with the cup and ending with the bread has long puzzled scholars, since the typical practice of the early Christians appears to be reflected in the New Testament accounts of the Last Supper, where Jesus distributes first the bread and then the cup (e.g., see Mark 14:22–25).

The *Didache* continues by giving extended instructions concerning what to do with the traveling apostles, teachers, and prophets who arrive in town to minister to the community (chaps. 11–13). These three categories of persons appear to overlap. Evidently, problems had arisen because some itinerant Christians were scoundrels who had become traveling preachers solely for financial gain. For this reason, the author insists that visiting prophets not be allowed to have more than two days' room and board at the community's expense, and that they be considered false if they demand money while uttering a pronouncement from God. Moreover, any wandering prophets who disagree with the "doctrines" expressed in this document, or who fail to practice what they preach, are to be rejected as false (11:1–2, 10).

The *Didache* finally gives instructions concerning wandering prophets who decide to settle down within the community. True prophets are to be treated with the highest honor and offered the "first-fruits" of the community's wine, harvest, and livestock, as if they were its chief priests (13:1–3). In addition, the Christian communities are to elect bishops and deacons from among their own ranks to run the affairs of the church (15:1–2).

The concluding chapter of the book provides a kind of apocalyptic discourse, an exhortation to be ready for the imminent end of the world which will be brought by "the Lord coming on the clouds of heaven" (16:7). Given its loose connection with what precedes it, this chapter may have been tacked on to the *Didache* at a later date.

What is the date of the earlier portion of the book (chaps. 1–15)? Scholars have debated the issue for as long as they have known of the document's existence. Part of the dispute centers around the question of the book's unity, that is, whether or not its different parts derived from different times

SOME MORE INFORMATION

Box 26.1 The Development of the Lord's Prayer

The Lord's Prayer is not found in the Gospels of Mark or John. Luke appears to represent the oldest surviving form of the prayer, possibly the form that was original to Q. Matthew's Gospel expands this version by adding some additional petitions. One of the many intriguing features of the *Didache* is that it also presents the Lord's Prayer, but in a slightly different form from what can be found in either of the canonical Gospels. Interestingly, of the three extant versions, the Didache's is closest to the form of the prayer familiar to most Christians today.

Consider first the versions in Luke (Q) and Matthew, side by side:

Luke 11:2–4	**Matthew 6:9–13**
Father, hallowed be your name.	Our Father, who is in heaven, hallowed be your name.
Let your kingdom come.	Let your kingdom come. Let your will be done, even on earth as it is in heaven.
Give us our daily bread every day. And forgive us our sins, for we ourselves have forgiven everyone who is indebted to us.	Give us today our daily bread. And forgive our debts, as we have forgiven those who are our debtors.
And do not lead us into temptation.	And do not lead us into temptation, but deliver us from evil.

The *Didache* agrees almost word for word with Matthew's form of the prayer but tacks on the conclusion "For yours is the power and glory forever." Later scribes who copied Matthew's Gospel supplied a similar ending but added several more words to form the familiar conclusion "For yours is the kingdom and the power and the glory forever, amen."

and places and were combined by someone living later. Recent scholars tend to think that the book was produced by a single author on the basis of earlier sources at his disposal. Its final production may date to around 100 C.E. One reason for choosing some such date is that the document appears to presuppose Christian communities that are not yet highly structured, unlike the proto-orthodox communities that we know about from later in the second century. Moreover, the author knows a wide range of earlier Christian traditions such as those embodied in the Sermon on the Mount, and it appears that his community, somewhat like Matthew's, held views that were widespread in Judaism even though it rejected Judaism as it was currently practiced (thus the references to the "hypocrites"; cf. Matthew 23). It appears then that the document dates from a time when a variety of Christian traditions, possibly even Matthew itself, were in circulation—that is, sometime after the middle third of the first century. Yet it was apparently produced before the mid second century, when the proto-orthodox churches had developed their rigid form of structure.

As for the inner life of the congregation(s) that the author addresses, it appears that they are in the process of developing a strict ethical code (or at least that he hopes they are) and establishing the most important early Christian sacraments and ceremonial practices (baptism, eucharist, set prayers, and days of fasting). They are also experiencing both the benefits and problems of wandering Christian "authorities," some of whom provide useful guidance for the communities while others actively exploit them. In our discussion of the Pastoral epistles we saw how charismatic communities like these ended up solving their problems by establishing clerical hierarchies, creedal statements, and canonical authorities. In many ways the communities of the *Didache* are moving in this direction themselves, as is evident in the appointment of local bishops and deacons, the insistence on conformity to certain views, and the devotion to certain recognized traditions such as those that at an earlier stage had come to be incorporated in the Sermon on the Mount.

POLYCARP'S LETTER TO THE PHILIPPIANS

Problems of morality and church structure are also evidenced in a writing whose historical circumstances are somewhat clearer to us and whose author we have already had occasion to meet. This is the letter written to the church of Philippi by the Bishop of Smyrna, Polycarp, the friend of Ignatius, who like him came to be martyred on behalf of his Christian faith (see Chapter 25 above). You will recall that Polycarp was himself the recipient of a letter from Ignatius around 110 C.E., some forty-five years or so before his own death. Soon after he received this letter, he wrote to the Philippian Christians, evidently in response to their requests on several matters (Pol. *Phil.* 3:1).

One of the things the Philippians had requested was a copy of "the letters of Ignatius, those he sent to us [in Smyrna] and any others which we had by us" (13:2). Polycarp complied with this request, sending his own epistle as a kind of cover letter for the collection. The collection itself would have included the two letters Ignatius wrote from Troas to the Smyrneans and their bishop, and possibly those that he had earlier written while he was actually staying with them in Smyrna: *Ephesians*, *Magnesians*, *Trallians*, and *Romans*. Whether it also included the letter to the Philadelphians (written from Troas) or any of Ignatius's other writings is something that we will probably never know.

Polycarp indicates that both Ignatius and the Philippians had requested that he, or one of his representatives, take letters to the Christian church in Syria (13:1). This was the church over which Ignatius had been bishop prior to his arrest and which had just recently experienced considerable internal turmoil, evidently due to an in-house fight over who would control the church, possibly involving persons with widely disparate theological views. The struggle had ended successfully from Ignatius's perspective. Churches with leaders who embraced views similar to his were requested to send delegations to Antioch showing their support. Polycarp announces his plan to go there personally, if the opportunity presents itself (13:1).

One of the problems that scholars have had in understanding Polycarp's letter is in knowing when he wrote it. Parts of the letter suggest that Ignatius had just passed through town on his way to Rome. Thus, for example, Polycarp asks the Philippians for any news that they have heard about him (13:2). But earlier in the letter Polycarp seems to know that Ignatius has already met his death by martyrdom (9:1). Some scholars have proposed, on these grounds, that chapters 13–14 represent part of a letter written around 110 C.E., soon after Polycarp had seen Ignatius, but that the earlier chapters derive from a letter written some twenty-five years later on the occasion of problems that had arisen in the church of Philippi. As happened with Paul's letters to the Philippians, according to this theory, the letters of Polycarp were later cut up and pasted together to form one larger letter for broader circulation.

Scholars continue to take different sides in this debate. The majority today appear to think that when Polycarp refers to Ignatius as a martyr for the faith in chapter 9, he is indicating what he knows is going to happen once his friend arrives in Rome. If this is the case, then the letter as a whole would not necessarily have been a composite piece but could have been penned at one time, fairly early in the second century.

In any event, whether it was written around the year 110 or some time later, it is clear that the letter was composed in part because Polycarp felt constrained to address the serious internal problems that the Philippian church was experiencing, problems involving both the ethical misconduct of one of its elders and the appearance of false teachers. The problem with the false teachers is somewhat elusive, but it appears that some members of the Philippian congregation, perhaps outsiders who came into their midst, had begun to proclaim a docetic kind of Christology similar to that countered in the letters of Ignatius himself and, from roughly the same time, in the Johannine epistles. This much, at least, can be inferred from Polycarp's castigation of someone whom he calls "the first-born of Satan." This was an epithet that Polycarp later used specifically against the docetist Marcion, according to the testimony of his own student, the proto-orthodox church father Irenaeus. The person Polycarp attacks in his letter to the Philippians is an "antichrist" who denies that there will be a resurrection of the flesh and who accordingly "does not confess that Jesus Christ has come in the flesh" (7:1).

Somewhat less opaque is the problem involving the ethical misconduct of the Philippian elder, a man named Valens, who along with his wife had evidently embezzled funds from the church and been caught red-handed (chap. 11). The Philippians asked Polycarp's advice in the matter and he willingly gives it. He states that the incident should be a lesson to them not to crave worldly goods. With respect to the offending couple themselves, Polycarp advises that they be allowed to repent and return to the good graces of the church. No such kindly treatment is recommended for the unrepentant false teachers.

Apart from these specific issues, Polycarp's letter consists chiefly of general moral exhortations. The Philippians are to love one another and to pray for one another and to give alms whenever possible, their wives are to love their husbands and to educate their children in the fear of God, their widows are to be discreet and devoted to prayer, their deacons are to be moral and upright, their younger men are to avoid passions of the flesh, and so on. Many readers of the letter have found these guidelines somewhat uninspiring, or at least uncreative. Indeed, Polycarp devotes almost the entire letter to quoting or alluding to other early Christian authorities. Rather than formulating views of his own, he has produced a kind of pastiche of earlier traditions.

This circumstance in itself, however, is of considerable interest to the historian of early Christianity. In this short letter of only about five pages in English translation, there are well over 100 quotations of and allusions to other authors. Moreover, only about 10 percent of these are drawn from the pages of the Jewish Scriptures, writings that Polycarp frankly confesses not to know very well (12:1). He does, however, claim to know the letters of Paul (3:2; 11:2–3), and indeed a number of his quotations are drawn from letters ascribed to the apostle, including the Pastoral epistles and Hebrews. In addition, Polycarp displays considerable knowledge of the traditions that are embedded in the Synoptic Gospels, Acts, and 1 Peter (see box 26.2).

In short, despite the fact that he was writing so early in the second century, Polycarp evidences precisely the concerns that will come to dominate proto-orthodox authors of the later second and third centuries, who engaged in the internecine conflicts of their communities and who subscribed to positions that later came to be dominant throughout the church at large. He urges an upright church hierarchy (with respect to the elder Valens), appeals to a doctrinally pure creed (with respect to the docetists), and uses earlier Christian traditions and writings as authoritative for guiding the ongoing life of the church.

1 CLEMENT

Concerns for the leadership of the church are even more central in the epistle known as *1 Clement*, a letter whose single-minded purpose is to address schism in the church of Corinth. Since the second century, the letter has been attributed to a man named Clement, thought to be the third bishop of the church in Rome. The letter itself, however, never mentions Clement; it claims to have been produced by "the church of God, living in exile in Rome, to the church of God, exiled in Corinth"

SOMETHING TO THINK ABOUT

Box 26.2 Polycarp and the Early Christian Tradition

To get an idea of how thoroughly immersed Polycarp was in the Christian tradition, consider the following passage drawn from the fifth chapter of his letter to the Philippians. I have placed possible echoes and citations of earlier Christian writings in parentheses.

> Knowing, then, that God is not mocked (Gal 6:7), we ought to walk worthily (Phil 1:27) of his commandment and glory. Likewise let deacons be blameless before his righteousness as servants of God and Christ, and not of humans; let them not be slanderers, or double-tongued (1 Tim 3:8), or lovers of money (1 Tim 3:3), but let them be temperate in all things, compassionate, careful, walking according to the truth of the Lord, who became the servant of all (Mark 9:35). For if we are pleasing to him in the present world, we will receive also the world that is coming, just as he promised us to raise us from the dead (John 5:21), and that if we conduct ourselves worthily of him, we will also reign with him (2 Tim 2:12) . . . for it is a good thing to be cut off from the desires that are in this world, because every desire wages war against the Spirit (1 Pet 2:11), and neither the sexually immoral nor the effeminate nor men who have sex with men will inherit the kingdom of God (1 Cor 6:9–10).

Here within a half page of text Polycarp repeats phrases found in eight different books that eventually became part of the New Testament. It appears that by the early second century earlier Christian writings had already begun to mold the thoughts and views of proto-orthodox church leaders.

(1:1). That is to say, it was a letter from the Roman Christian community to the Corinthian church. Since the letter was presumably not compiled as a kind of committee project, however, its actual author may well have been the leader of the Roman church.

Unlike most of the other books that we are considering in this chapter, *1 Clement* provides some concrete clues concerning the time of its writing. Its author speaks of the Corinthian church as "ancient" (47:6) and yet maintains that there are still church leaders throughout the world who were hand-picked by the apostles (chap. 44). Taken together these comments may indicate that the author is living sometime near the end of the first century. Corroborating evidence may be found in the author's reference to the martyrdoms of both Peter and Paul as having taken place during an earlier persecution in the city in "our own time" (chap. 5; they are generally thought to have been executed during the reign of Nero) and in his indication that hostilities against the Christians have recently been renewed (1:1; 7:1). For many scholars, these references suggest a time of composition sometime around the year 95 or 96, when the emperor Domitian is thought to have engaged in some local persecution of Christians, although hard evidence of this persecution is scanty.

A later author living in Corinth, a proto-orthodox Christian named Dionysius, indicates that *1 Clement* was used as Scripture by the Corinthian church around 170 C.E. Moreover, other evidence suggests that the book was sometimes considered to be a part of the "New" Testament in some regions of the church. It appears, then, that the book was written near the end of the first century, that it was an immediate success in Corinth (at least among some of the Christians there), and that it was then distributed to other parts of the Christian world where it was also read with favor.

The author (whom I will continue to call Clement for the sake of convenience) has learned of an "abominable and unholy schism" in the Corinthian church (1:1). Evidently the elders of the church had been forcibly deposed from their office, and others had taken their places (3:2–4). We are not told how, exactly, the coup had been staged, that is, whether there had been (a) an actual act of violence (which seems unlikely, since the issue was church leadership, not military or civil government), (b) an election of leaders which the former officers lost, (c) the appearance of charismatic figures in the church who simply won over the hearts and minds of the congregation and thereby assumed de facto positions of authority, or (d) something else. What is clear is that the church in Rome found the circumstance altogether unsettling and wrote a relatively long letter to rectify the situation to its own satisfaction, somewhat as Paul himself had done some forty years earlier.

The letter gives no concrete information concerning who the new leaders were or what they stood for. We do not know, for example, whether they embraced theological positions that Clement found to be untenable, whether they were people whom Clement himself simply didn't like or admire, or whether Christian leaders in Rome opposed a change of church leadership on general principle, perhaps out of fear that if such things could happen abroad, they could happen at home as well. Whatever the real historical situation, *1 Clement* states firmly its primary guideline for church governance, a guideline that is imbued with divine authority and backed by the words of sacred Scripture. The leaders of the Christian churches have been appointed by the apostles, who were chosen by Christ, who was sent from God. Anyone who deposes these leaders is therefore in rebellion against God (chaps. 42–44).

These chapters of *1 Clement* provide one of the earliest surviving expressions of the notion of "apostolic succession," which later came to play such a significant role in the theological controversies of the second century. We have already seen that proto-orthodox Christians used the church hierarchy as a way of controlling theological deviation in their congregations, but the proto-orthodox bishops, elders, and deacons were only as effective as their offices were stable. If there was considerable and repeated turnover in the offices held by the church leaders, as happens today, for example, in the political arena, then there could be no guarantee of a stable agenda and unified outlook—the sine qua non for proto-orthodox Christians wanting to establish their form of belief and practice as dominant throughout Christendom.

This is not, however, the argument that the author of *1 Clement* uses to castigate those who have taken over the leadership of the church in Corinth. Instead, he appeals to Scripture to show that throughout the history of the people of God, envy and strife have always been promoted by sinners who are opposed to the righteous. Thus he maligns the actions of the Corinthian "usurpers" by citing examples of jealousy and rivalry all the way from Cain and Abel up to his own day. He also appeals to the words of the prophets to show that God opposes those who exalt themselves over the ones whom he has chosen. Furthermore, this author does not restrict his citations to the words of the Jewish Scripture but applies the teachings of Jesus and the writings of his apostles to the contemporary situation as well (e.g., chaps. 12, 46). For him, these are just as authoritative as the Old Testament. Here we are on the way to having uniquely Christian authorities—eventually, Christian writings—serve as the ultimate arbiters over all matters of faith and practice.

Toward the end of his letter, Clement offers some practical advice for dealing with the leadership crisis. What has happened in the church Corinth is a disgrace, and those responsible must repent and return the leadership to those who formerly possessed it:

> It is disgraceful, exceedingly disgraceful, and unworthy of your Christian upbringing, to have it reported that because of one or two individuals the solid and ancient Corinthian Church is in revolt against its presbyters. . . . (47:6)

> You who are responsible for the revolt must submit to the presbyters. You must humble your hearts and be disciplined so that you repent. You must learn obedience, and be done with your proud boasting and curb your arrogant tongues. For it is better for you to have an insignificant yet creditable place in Christ's flock than to appear eminent and be excluded from Christ's hope. (57:1–2)

We cannot know for certain how well this letter to the Corinthians was first received. No doubt the former leaders of the church (friends of the leaders in Rome?) welcomed it with open arms, whereas those who had taken over their positions of authority found it more than a little discomfiting. It is possible that they relinquished control of the church, but it is also possible that, even if they did so, the infighting did not come to an immediate end. What is clear is that the Roman position eventually became more widely known and appreciated: the leaders of the churches were thought to owe their position to God himself and could not be opposed without opposing God.

A number of scholars have found it significant that this view was first promoted, so far as we can tell, within the church of Rome, whose bishop was eventually to assume a position of special prominence within all of Christendom. Here in *1 Clement* the Roman leaders exert influence not only over their own congregation but also over a congregation located far away. This Roman influence made itself increasingly felt with the passage of time, until the Roman bishop came to be considered the father of all the bishops and thus the leader of the entire Christian church. It is probably no accident that the form of Christianity that eventually established itself as orthodox in the third and fourth centuries proceeded out of Rome, and that the universal church came to be known

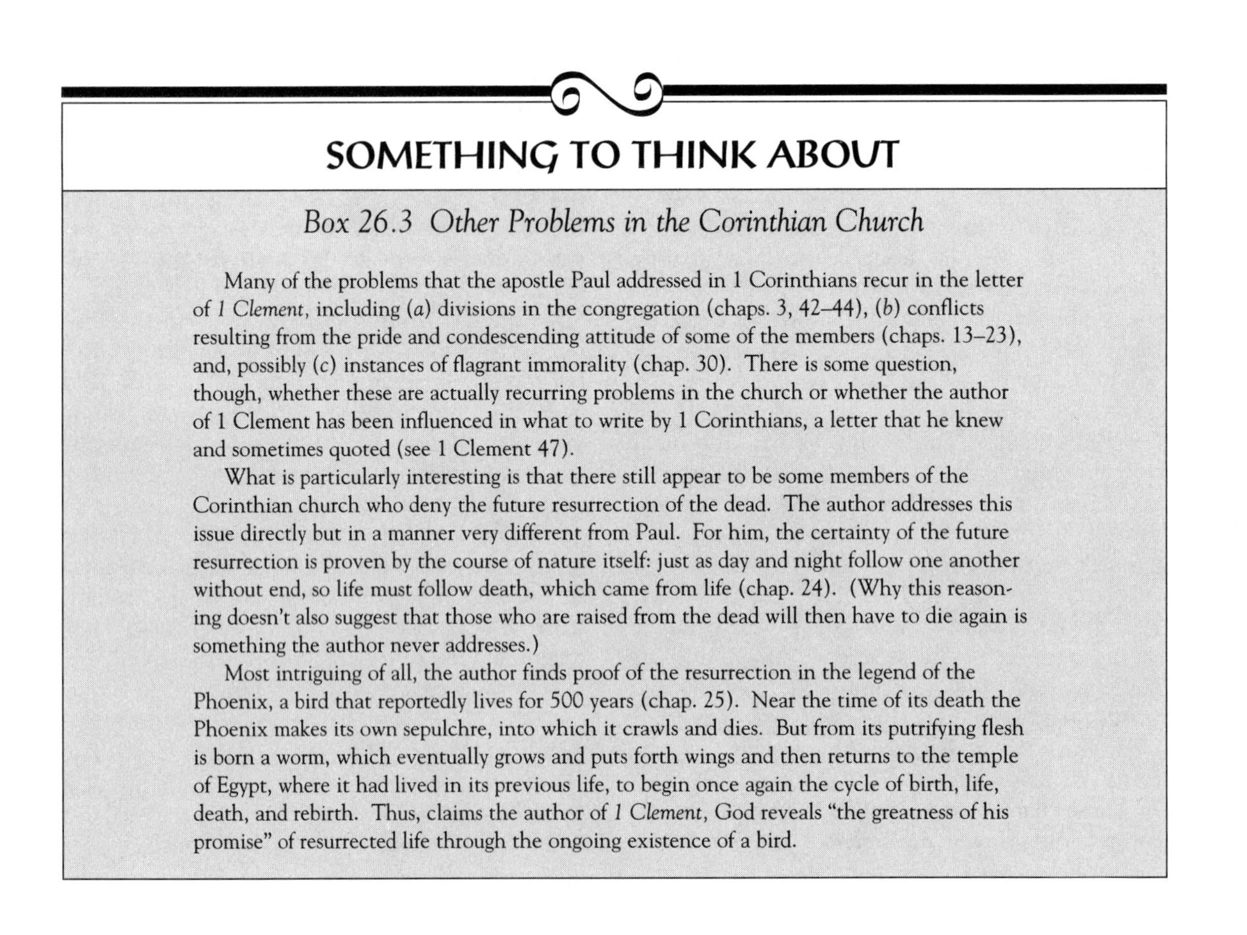

SOMETHING TO THINK ABOUT

Box 26.3 Other Problems in the Corinthian Church

Many of the problems that the apostle Paul addressed in 1 Corinthians recur in the letter of *1 Clement*, including (*a*) divisions in the congregation (chaps. 3, 42–44), (*b*) conflicts resulting from the pride and condescending attitude of some of the members (chaps. 13–23), and, possibly (*c*) instances of flagrant immorality (chap. 30). There is some question, though, whether these are actually recurring problems in the church or whether the author of 1 Clement has been influenced in what to write by 1 Corinthians, a letter that he knew and sometimes quoted (see 1 Clement 47).

What is particularly interesting is that there still appear to be some members of the Corinthian church who deny the future resurrection of the dead. The author addresses this issue directly but in a manner very different from Paul. For him, the certainty of the future resurrection is proven by the course of nature itself: just as day and night follow one another without end, so life must follow death, which came from life (chap. 24). (Why this reasoning doesn't also suggest that those who are raised from the dead will then have to die again is something the author never addresses.)

Most intriguing of all, the author finds proof of the resurrection in the legend of the Phoenix, a bird that reportedly lives for 500 years (chap. 25). Near the time of its death the Phoenix makes its own sepulchre, into which it crawls and dies. But from its putrifying flesh is born a worm, which eventually grows and puts forth wings and then returns to the temple of Egypt, where it had lived in its previous life, to begin once again the cycle of birth, life, death, and rebirth. Thus, claims the author of *1 Clement*, God reveals "the greatness of his promise" of resurrected life through the ongoing existence of a bird.

as the Roman Catholic Church, with the bishop of that church, the Pope, serving as the head of the church throughout the world.

JUDE

This concern for the leadership of the church is addressed in yet another way in a much shorter letter that did have the good fortune to be included in the New Testament, perhaps because the author claimed to be someone of high standing in early Christian circles. The writer of this one-page epistle names himself Jude (literally, Judas), the "brother of James" (v. 1). As you know, there were early traditions that two of Jesus' own brothers were named Jude and James (Mark 6:3). This author, then, is apparently claiming to be related to the great leader of the Jerusalem church, James, and therefore to be a family relation of Jesus himself.

The letter itself gives scant reason for accepting this ascription, and many critical scholars think that it is another example of early Christian pseudepigraphy. Jesus' brother Jude, of course, would have been a lower-class Aramaic-speaking peasant. Indeed, we learn from sources dating to the second century that Jude's family did not attain social prominence and were therefore, presumably, not well educated: his grandsons were known to be uneducated peasant farmers. The author of this book, on the other hand, was someone who was well trained in Greek and was conversant with a wide range of apocryphal Jewish literature. He quotes, for example, from a lost apocryphal account of the angelic battle over Moses' body (v. 9), and he cites the book of 1 Enoch as Scripture (v. 14). Thus, it does not appear to be likely that Jesus' own brother wrote the book.

The book is concerned with false teachers who have invaded the Christian community:

> Beloved. . . I find it necessary to write and appeal to you to contend for the faith that was once for all entrusted to the saints. For certain intruders have stolen in among you, people who long ago were designated for this condemnation as ungodly, who pervert the grace of our God into licentiousness and deny our only Master and Lord, Jesus Christ. (vv. 3–4)

It is hard to know how Christian leaders can be thought to have denied Christ, but it may be that, from the author's perspective, anyone who understands the religion in a way that is significantly different from the way he does is liable to this charge. We saw a similar state of affairs in our study of 1 John. Also, just as the secessionists from the Johannine community were thought to have engaged in immoral and illegal activities because of their false beliefs, so the opponents of Jude are chiefly maligned for their licentious and perverse lifestyles. They are "like irrational animals" (v. 10), they engage in "deeds of ungodliness" (v. 15), they are "grumblers and malcontents, they indulge their own lusts, and they are bombastic in speech" (v. 16). The author likens them to the children of Israel, who after escaping from Egypt reveled in wanton acts of disbelief (adulteries and idolatries), and to the inhabitants of Sodom and Gomorrah, who "indulged in sexual immorality and pursued unnatural lust" (vv. 5–7).

From the historians' point of view it is to be regretted that the author never tells us what these people actually stood for, that is, what they taught and how they lived. Most of the letter is simply filled with invective and name-calling. The author's enemies are "waterless clouds carried along by the winds; autumn trees without fruit, twice dead, uprooted; wild waves of the sea, casting up the foam of their own shame" (vv. 12–13).

It is clear, however, the author feels that his community is in jeopardy from these "worldly people, devoid of the Spirit, who are causing divisions" (v. 19). These false teachers need to realize what happens to those who oppose God and lead his people astray. In the past those who have caused disturbances and promoted immorality among God's people have been confronted with God's judgment. The offenders must take heed and repent, lest they become like the inhabitants of Sodom and Gomorah, serving "as an example by undergoing a punishment of eternal fire" (v. 7).

We do not know exactly when the pseudonymous author produced his account; most modern scholars date it somewhere near the end of the first century. We do know that the book was used as a source some years later by another pseudonymous author, who produced a similarly vitriolic attack on false teachers who promoted immoral behavior among the Christians. This author wrote in the name of the apostle Peter and produced a letter that was in all likelihood the final book of the New Testament to be written, the epistle of 2 Peter.

2 PETER

For a variety of reasons, there is less debate about the authorship of 2 Peter than any other pseudepigraphon in the New Testament. The vast majority of critical scholars agree that whoever wrote the book, it was not Jesus' disciple Simon Peter. As was the case with 1 Peter, this author is a relatively sophisticated and literate Greek-speaking Christian, not an Aramaic-speaking Jewish peasant. At the same time, the writing style of the book is so radically different from that of 1 Peter that linguists are virtually unanimous in thinking that if Simon Peter was responsible for producing the former book, he could not have written this one. Even more to the point, a major portion of this letter has been taken over from the book of Jude and incorporated into chapter 2. If Jude can be dated near the end of the first century, 2 Peter must be somewhat later. Therefore, it could not have been penned by Jesus' companion Peter, who was evidently martyred in Rome sometime around 64 C.E. during the reign of Emperor Nero (see discussion of *1 Clement* above).

This letter, then, should probably be included among the large number of pseudonymous writings in the name of Peter, which include the *Gospel of Peter* that we considered in Chapter 12 and the *Apocalypse of Peter* that we will examine in Chapter 26. In this connection, it is striking that the letter was not widely accepted as Peter's, or even known to exist, for most of the first three Christian centuries. There is not a solitary reference to it until around 220 C.E., and it does not appear to have been widely circulated for at least another century after that. It was no doubt included in the canon because the orthodox fathers of the fourth century accepted the claims of its author to be Peter, and because it served their purposes in opposing those who promote false teaching.

The author goes out of his way to insist that he is none other than Jesus' disciple—a case, perhaps, of protesting too much. Not only does he begin by naming himself "Simeon Peter, a servant and apostle of Jesus Christ" (1:1), but he proceeds to recount his own personal experience with Jesus on the Mount of Transfiguration, where he beheld for himself Jesus' divine glory and heard God's affirmation of his Son in the voice from heaven (1:17; as we will see, the pseudonymous author of the *Apocalypse of Peter* also appeals to his "memory" of this event). He assures his reader that he was there to see these things: "We ourselves heard this voice come from heaven, while we were with him on the holy mountain" (1:18). Why does he choose to parade his credentials in this manner? It is probably to convince his readers that he has no need of "cleverly devised myths" to understand Jesus (1:16) since he knows about him firsthand.

This reference to myths may intimate something about the author's opponents. They may be early Gnostics, who use their creative mythologies and genealogies to support their "unorthodox" points of view, for the author goes on to attack people who provide idiosyncratic interpretations of Scripture—a favorite activity of the Gnostics, according to the proto-orthodox church fathers: "First of all you must understand this, that no prophecy of scripture is a matter of one's own interpretation" (1:21). Moreover, the author's opponents appeal to the writings of the apostle Paul, which by this time are evidently in circulation as a collection and are even being considered as "Scripture"—other indications that the letter was written long after the apostle's death. As we have previously seen, the Gnostics took a particular liking to Paul as an authority for their views.

> So also our beloved brother Paul wrote to you according to the wisdom given him, speaking of this as he does in all his letters. There are some things in them hard to understand, which the ignorant and unstable twist to their own destruction, as they do the other scriptures. (3:16)

Unfortunately, the author of 2 Peter does not set forth the actual views of his opponents but simply enters into invective against them. Much of his attack has simply been borrowed from the

SOME MORE INFORMATION

Box 26.4 Peter, the Smoked Tuna, and the Flying Heretic

Among the pseudepigrapha connected with the apostle Peter, none is more interesting than the apocryphal *Acts of Peter,* a document that details Peter's various confrontations with the heretical magician Simon Magus (cf. Acts 8:14–24). The narrative shows how Peter outperforms the magician by invoking the power of God. Consider the following entertaining account, in which Peter proves the divine authorization of his message by raising a dead tunafish back to life:

> But Peter turned round and saw a smoked tunny-fish hanging in a window; and he took it and said to the people, "If you now see this swimming in the water like a fish, will you be able to believe in him whom I preach?" And they all said with one accord, "Indeed we will believe you!" Now there was a fish-pond near by; so he said, "In thy name, Jesus Christ, in which they still fail to believe" [he said to the tunny] "in the presence of all these be alive and swim like a fish!" And he threw the tunny into the pond, and it came alive and began to swim. And the people saw the fish swimming; and he made it do so not merely for that hour, or it might have been called a delusion, but he made it go on swimming, so that it attracted crowds from all sides and showed that the tunny had become a live fish; so much so that some of the people threw in bread for it, and it ate it all up. And when they saw this, a great number followed him and believed in the Lord." (*Acts of Peter* 5)

In the ultimate showdown between the heretical sorcerer and the man of God, Simon the magician uses his powers to leap into the air and fly like a bird over the temples and hills of Rome. Not to be outdone, Peter calls upon God to smite Simon in midair; God complies, much to the magician's dismay and demise. Unprepared for a crash landing, he plunges to earth and breaks his leg in three places. Seeing what has happened, the crowds rush to stone him to death as an evildoer. And so the true apostle of God triumphs over his enemy, the preacher of heresy.

epistle of Jude. He sees his opponents as "false prophets" (2:1) who engage in acts of flagrant immorality: "They have eyes full of adultery, insatiable for sin. . . . They speak bombastic nonsense, and with licentious desires of the flesh they entice people who have just escaped from those who live in error" (2:14, 18). Moreover, these persons are not outsiders but members of the Christian community who, in the author's judgment, have gone astray to their own destruction:

> For it would have been better for them never to have known the way of righteousness than, after knowing it, to turn back from the holy commandment that was passed on to them. It has happened to them according to the true proverb, "The dog turns back to its own vomit." (2:21–22)

One additional piece of information about these Christian adversaries is that they scoff at the traditional apocalyptic belief that the end of the world is imminent. The author assures his readers that the prophets and Jesus himself, speaking through the apostles, predicted that "in the last days scoffers will come scoffing and indulging their own lusts and saying, 'Where is the promise of his coming? For ever since our ancestors died, all things continue as they were from the beginning of the creation'" (3:3–4).

The author goes on to indicate that the end *is* destined to come. Whereas the world had once been destroyed by water, it is now being preserved for destruction by fire. Indeed, this end seems to be slow in coming only for those who measure time in human terms. For God, however, "one day is like a thousand years, and a thousand years are like one day" (3:8)—meaning, one might suppose, that if the end is still 6,000 years away, it is still coming "soon."

The author emphasizes that the end has been delayed to allow all people adequate time to repent and turn to the truth. But the day of judgment is nonetheless destined to come, and when it does it will appear "like a thief" (3:9). The certainty of this final day should drive people to live "lives of holiness and godliness, waiting for and hastening the coming of the day of God, because of which the heavens will be set ablaze and dissolved, and the elements will melt with fire" (3:11–12).

CONCLUSION: CONFLICTS WITHIN THE EARLY CHRISTIAN COMMUNITIES

In the Christian writings that have survived from the end of the first century and the beginning of the second we get some sense of the state of Christianity at the close of the New Testament period. The Christian communities were by no means unified at this time. Different Christian leaders and teachers were proclaiming different versions of the faith, and many of them were at serious odds with one another. Christians had different views of how to conduct themselves both within the Christian community and within society as a whole. Some Christians were thought to be engaging in wild, immoral activities and to be promoting such ventures in the church.

As historians of the period we should remember that we have only one side of almost every story. There can be no doubt that the "immoral and corrupt heretics" attacked in surviving writings would have had much to say in their own defense. Indeed, they did defend their views and attack their proto-orthodox opponents for propagating error, as we have discovered from the Gnostic writings of the Nag Hammadi library. Regrettably, almost all of the other books produced by advocates of alternative Christian perspectives came to be destroyed on order of their victorious adversaries. Typically, from the ancient world, only the writings of the winners survive.

The authors who later came to be canonized in the New Testament, some of them claiming to be apostles, urged their own versions of the faith, their own leaders, and their own systems of ethics. These authors may not have been in full agreement with one another on every point, but most of their differences came to be smoothed over when their books were later collected into a sacred canon of Scripture and read and interpreted only in light of one another. The proto-orthodox Christians chiefly responsible for this canon of Scripture also advocated a church structure that could trace itself back to Jesus and his apostles. In their conflicts with aberrant forms of the religion, these late-first-century and early-second-century believers thus set the stage for the battles over orthodoxy that were to rage throughout the second and third centuries, as different Christian groups representing different understandings of the faith strove for converts both from the outside (through evangelism) and from within.

SUGGESTIONS FOR FURTHER READING

Bauckham, Richard. *Jude and the Relatives of Jesus in the Early Church*. Edinburgh: T & T Clark, 1990. An interesting study that argues, among other things, that the epistle of Jude was actually written by Jesus' own brother.

Chester, Andrew, and Ralph Martin. *The Theology of the Letters of James, Peter, and Jude*. Cambridge: Cambridge University Press, 1994. A nice discussion of the social context and theological perspectives of these Catholic epistles.

Jefford, Clayton, ed. *The Didache in Context: Essays on Its Text, History, and Transmission*. Leiden: Brill, 1995. A significant collection of essays that cover every major aspect of the study of the Didache; for advanced students.

Tugwell, Simon. *The Apostolic Fathers*. Harrisburg, Pa.: Morehouse, 1990. A clear and straightforward discussion of the background and message of the Apostolic Fathers, including the *Didache*, the epistle of Polycarp, and *1 Clement;* ideal for beginning students.

CHAPTER 27

Christians and the Cosmos: The Revelation of John, The Shepherd of Hermas, and the Apocalypse of Peter

INTRODUCTION: THE END OF THE WORLD AND THE REVELATION OF JOHN

The end of the world was near. So proclaimed Jesus, and some years after him, the apostle Paul. So proclaimed most of the earliest Christians of whom we have any knowledge. The end of time had come, God was about to intervene in history; Christ was soon to return from heaven in judgment on the earth, and people were to repent and prepare for his coming.

With the passing of time, this message lost its appeal in some Christian circles. For the end never did come, and Christians had to reevaluate (or even reject) the earlier traditions that said it would. We have already observed such reevaluations among some of the early Christian authors. We have noticed, for example, how the Gospel of Luke modifies Jesus' predictions so that he no longer claims that the Son of Man will arrive in his disciples' lifetimes. We have also seen that in several later Gospels, such as John and Thomas, Jesus tells no parables concerning the coming kingdom of God. We have also observed that among the Christians in Corinth, Jesus' return and the resurrection of the dead became heated questions, as some believers claimed that the divine plan of redemption had already come to completion and that they were already experiencing the full benefits of salvation. Moreover, we have seen that still other Christians, such as those attacked by the author of 2 Peter, came to mock the idea that Jesus was soon to return from heaven in judgment.

Nonetheless despite the passing of time and the failure of their hopes to materialize, many Christians remained firmly committed to this belief. It stood at the heart of the message proclaimed by the apostle Paul some twenty years after Jesus himself had died, and by the Gospel of Mark some fifteen years after Paul, by the Gospel of Matthew some fifteen years after Mark, and by 2 Peter and the *Didache* some thirty years after Matthew.

The coming of the end was also the fervent conviction of a prophet named John, who lived near the end of the first century. John was a Christian seer who penned a majestic and awe-inspiring account of the end of the world, an account that has spawned endless speculation and debate among those who have continued to await the return of Jesus over the intervening nineteen hundred years. John was not the only Jewish or Christian author to narrate visions of the end of the world. Indeed, the kind of book that he wrote was quite popular among people looking for the heavenly truths that could give meaning to their earthly realities. But none of the other early apocalypses has enjoyed nearly the success of the Apocalypse of John. Indeed, the book of Revelation continues to serve many Christians today as a kind of blueprint of events that are still to transpire in the future, when the history of the world, as they believe, will be brought to a screeching halt.

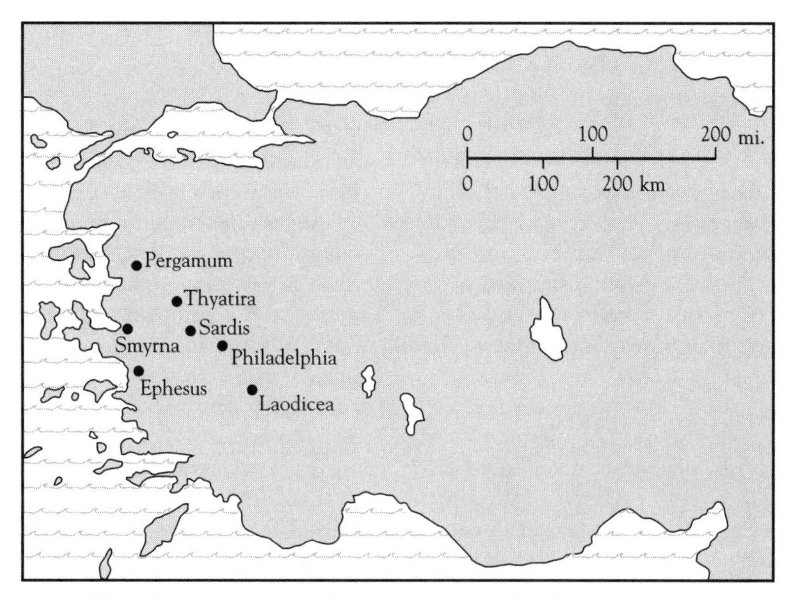

Figure 27.1 The seven churches of Asia Minor addressed in Revelation 2–3.

THE CONTENT AND STRUCTURE OF THE BOOK OF REVELATION

The title of the book comes from its opening words: "The revelation of Jesus Christ, which God gave him to show his servants what must soon take place" (Rev 1:1). The revelation, or apocalypse (from the Greek word for "unveiling" or "revealing") concerns the end of time; it is given by God through Jesus and his angel to "his servant John" (1:1). The author appears to be known to his readers, who are identified as Christians of seven churches in Asia Minor (1:11). He begins to narrate his visionary experiences by describing his extraordinary encounter with the exalted Christ, the "one like a Son of Man" who walks in the midst of seven golden lampstands (1:12–20).

Christ instructs John to "write what you have seen, what is, and what is to take place after this" (1:19). In other words, he is to (*a*) narrate the vision of Christ that he has just had ("what you have seen"), (*b*) describe the present situation of the churches in his day ("what is"), and (*c*) record his visions of the end of time ("what is to take place after this"). The first task is accomplished in chapter 1. The second is undertaken in chapters 2–3. Christ dictates brief letters to each of the seven churches of Asia Minor, describing their situations and urging certain courses of action. These churches are experiencing difficulties: persecutions, false teachings, and apathy. Christ praises those who have done what is right, promising them a reward, but upbraids those who have fallen away, threatening them with judgment.

The third task is accomplished in chapters 4–22, which record John's heavenly vision of the future course of history, down to the end of time. Briefly, the narrative unfolds as follows. The prophet is taken up into heaven through a window in the sky. There he beholds the throne of God, who is eternally worshipped and praised by twenty-four human "elders" and four "living creatures" (angelic beings in the shapes of animals; chap. 4). In the hand of the figure on the throne is a scroll sealed with seven seals, which cannot be broken except by one who is found worthy. This scroll records the future of the earth, and the prophet weeps when he sees that no one can break its seals; but one of the elders informs him that there is one who is worthy. He then sees next to the throne a "Lamb standing as if it had been slaughtered" (5:6). The Lamb, of course, is Christ.

The Lamb takes the scroll from the hand of God, amidst much praise and adoration from the twenty-four elders and the four living creatures, and begins to break its seals (chap. 5). With each broken seal, a major catastrophe strikes the earth: war, famine, death. The sixth seal marks the climax, a disaster of cosmic proportions: the sun turns black, the moon turns red as blood, the stars fall from the sky, and the sky itself disappears. One might think that we have come to the end of all things, the destruction of the universe. But we are only in chapter 6.

The breaking of the seventh seal leads not to a solitary disaster but to a period of silence that is followed by an entirely new set of seven more disasters. Seven angels appear, each with a trumpet. As each one blows his trumpet, further devastations strike the earth: natural disasters on the land and sea and in the sky, the appearance of dread beasts who torture and maim, widespread calamity and unspeakable suffering (chaps. 8–9). The seventh trumpet marks the beginning of the end (11:15), the coming of the antichrist and his false prophet on earth (chaps. 12–13) and the appearance of seven more angels, each with a bowl filled with God's wrath. As the angels pour out their bowls upon the earth, further destruction and agony ensue: loathsome diseases, widespread misery, and death (chaps. 15–16).

The end comes with the destruction of the great "whore of Babylon," the city ultimately responsible for the persecution of the saints (chap. 17). The city is overthrown, to much weeping and wailing on earth but to much rejoicing in heaven (chaps. 18–19). The defeat of the city is followed by a final cosmic battle in which Christ, with his heavenly armies, engages the forces of the antichrist aligned against him (19:11–21). Christ wins a resounding victory. The enemies of God are completely crushed, and the antichrist and his false prophet are thrown into a lake of burning sulfur to be tormented forever.

Satan himself is then imprisoned in a bottomless pit, while Christ and his saints rule on earth for a thousand years. Afterwards, the Devil emerges for a brief time to lead some of the nations astray. Then comes a final judgment, in which all persons are raised from the dead and rewarded for their deeds. Those who have sided with Christ are brought into the eternal kingdom; those who have aligned themselves with the Devil and his antichrist are taken away for eternal torment in the lake of fire. The Devil himself is thrown into the lake, as are finally Hades and Death itself (chap. 20).

The prophet then has a vision of the new heaven and the new earth that God creates for his people. A new Jerusalem descends from heaven, with gates made of pearl and streets paved with gold. This is a beautiful and utopian place where Christ reigns eternal, where there is no fear or darkness, no pain or suffering or evil or death, a place where the good and righteous will dwell forever (chaps. 21–22). The prophet ends his book by emphasizing that his vision is true, and that it will come to fulfillment very soon.

THE BOOK OF REVELATION FROM A HISTORICAL PERSPECTIVE

To most modern readers the Apocalypse of John seems mystical and bizarre, quite unlike anything else that we read. In part, this explains our continual fascination with the book—it is so strange, so unearthly, that its descriptions cannot simply have been dreamt up. Its supernatural feel seems to vindicate its supernatural character.

The historian who approaches the book, however, sees it in a somewhat different light, for this was not the only book of its kind to be written in the ancient world, even if it is the only one that most of us have ever read. Indeed, a number of other apocalypses were produced by ancient Jews and Christians. These works also offer unworldly accounts of happenings in heaven, bizarre descriptions of supranatural events and transcendent realities that impinge on the history of our world, and deeply symbolic visions of the end of time that are given by God through his angels to a human prophet, who writes them down in cryptic and mysterious narratives filled with emphatic claims that they are true and soon to take place.

Some of these other apocalypses still survive, and together they make up a distinct genre of literature. Thus, far from being unique in its own day, the Apocalypse of John followed a number of literary conventions that were well known among Jews and Christians of the ancient world. A historian who wants to understand this one ancient text, then, will situate it in the context of this related literature and explain its important features in light of the literary conventions of the genre.

APOCALYPTIC WORLDVIEWS AND APOCALYPSE GENRE

Apocalypses were written to convey an apocalyptic agenda. Here it is important to be very clear about our terms. Throughout our discussion I have used the term "apocalypticism" to refer to an ancient Jewish and Christian worldview that maintained that there were two fundamental components of reality, good and evil, and that everything in the world was aligned on one side or the other (God versus the Devil, the angels versus the demons, life versus death, and so on; see Chapter 15 above). This dualistic perspective applied to human history: the present age was seen to be evil, controlled by the Devil and his forces, whereas the age to come would be good, controlled by God. According to this view, there was to be a cataclysmic break between these ages, when God would destroy the forces of evil to bring in his kingdom. At that time there would be a judgment of all beings, both living and dead. This judgment was imminent.

Whereas the term "apocalypticism" refers to this worldview, "apocalypse" refers to a genre of literature that embodies it. Everyone who wrote a Jewish or Christian apocalypse was obviously an apocalypticist (i.e., he or she embraced the apocalyptic worldview). The reverse, however, is not true: not every apocalypticist wrote an apocalypse. Thus, neither John the Baptist nor Jesus nor Paul, to take three prominent examples, appears to have written a detailed vision of the heavenly realities. The first Jewish apocalypticist to do so, to our knowledge, was the author of Daniel (around 165 B.C.E.), the second half of which contains several brief apocalypses. Other apocalypses written somewhat later include the noncanonical Jewish works of 1 Enoch, 2 Baruch, and 4 Ezra, and two important Christian apocalypses that we will be exploring later in this chapter: *The Shepherd* of Hermas and the *Apocalypse of Peter*.

These apocalypses differ in important ways. Some of their most obvious differences relate to whether they were written by Jews or Christians, since the apocalyptic drama unfolds differently depending on whether or not Jesus himself is the key to the future. One of the things that all of these books have in common, however, is that they were evidently written in times of distress and suffering. In large measure, apocalypses were books that protested the present order of things and the powers that maintained it; these powers were seen to be inimical to the ways and people of God. These books invariably show that despite the suffering experienced by the people of God, God is ultimately in control and will soon intervene on their behalf. One of the important purposes of these works, then, is to encourage those who are experiencing the forces of evil to hold on and keep the faith. Their suffering is not in vain and it will not last long, for soon they will be vindicated in the glorious climax of history in which God will destroy the forces of evil and exalt those who have remained faithful to him.

Apocalypse as a Genre: General Description

The various Jewish and Christian apocalypses that convey this message share a number of literary features. All of these books are first-person narratives by prophets who have been granted highly symbolic visions or dreams. The visions are usually interpreted by a heavenly being who serves as a mediator. For the most part, the visions serve to explain the realities of earth from the perspective of heaven—realities such as the ultimate meaning of life and the future course of earth's history. These narratives always embody a triumphal movement from the painful existence of life here below to the glorious life up above or from the hardships and sufferings of the present to the vindication and bliss of the future.

There are two major kind of ancient apocalypses. These are not mutually exclusive categories. As you will see, the book of Revelation has aspects of each, although some other apocalypses are of only one type or the other:

1. *Heavenly Journeys*. In this kind of apocalypse, the prophet is taken up into heaven and given a tour of the heavenly realm by an angelic companion, where he beholds symbols and events that have earthly implications. The idea implicit in this kind of apocalypse is that life on earth directly reflects life in heaven; that is, it is somewhat like the earthly shadow of a heavenly reality (cf. our discussion of the Platonic notion of shadow versus reality in Chapter 24).
2. *Historical Sketches*. In this kind of apocalypse, the prophet has a symbolic vision of the future course of history. For example, grotesque beasts might arise out of the sea to wreak havoc on the earth, representing various kingdoms that will come to dominate the people of God (see Daniel 7). The symbolism is often explained to the seer by the heavenly mediator, and through him to the reader.

Apocalypse as a Genre: Specific Literary Features

Despite their wide-ranging differences, the surviving apocalypses typically share specific literary features. The most common of these are the following:

Pseudonymity. Almost all of the ancient apocalypses were written pseudonymously in the name of a famous religious person from the past (the book of Revelation is a rare exception). Among the surviving Jewish apocalypses are some claiming to be written by Moses, Abraham, Enoch, and even Adam. We have Christian apocalypses reputedly from the pens of the prophet Isaiah and the apostles Peter, Paul, and Thomas.

Is there a particular reason for authors of apocalypses to hide their identity behind a pseudonym? We have already seen that pseudonymity can help to secure a hearing for one's views, by lending a kind of authority to one's writing that it otherwise could not hope to enjoy. Nowhere is this kind of authority more important than when one is writing a detailed description of heavenly realities that explain the tragedies and suffering of earthly life. Such visions of transcendent truth are obviously not granted to just anybody. It makes sense, then, that authors of apocalypses typically claimed to be famous persons of the past who were renowned for their religious piety and devotion to God. Only to such as these would God reveal the ultimate truths that could unlock the mysteries of human existence.

The use of a pseudonym made particular sense for apocalypses of the historical sketch type. By pretending to be someone living in the distant past, an author could "predict" the future. A typical ploy, then, was to write in the name of a prophet from ancient times to whom was revealed a number of events that were to take place. When the author then continued to predict what was soon to happen in his own day—the reader didn't know when this was, of course, since the author claimed to be writing from the distant past—he was naturally granted the benefit of the doubt. That is to say, these future events (from the time of the reader) were just as certain to occur as those that had already happened. The prophet had been right about everything else; surely he was also right about what would come next!

The first apocalypse known to use this technique came to be included in the Hebrew Bible. The book of Daniel, allegedly written by the great wise man of the sixth century B.C.E. during the days of the Babylonian captivity, was actually written, in the judgment of almost all critical scholars, sometime during the period of suffering associated with the Maccabean revolt, some 400 years later. No wonder "Daniel" could predict the rise of the Persians and the Greeks, and even more accurately detail events that were to transpire near the time of the Jewish uprising; the author of these "prophecies" lived after they had taken place.

We should probably not pass moral judgment on this kind of literary device, for it is not at all clear that the apocalyptic authors meant to lead people astray by writing under a pseudonym. Rather, they intended to provide comfort and hope for those who were in the throes of terrible suffering.

Bizarre Symbolic Visions. Rarely do apocalypses describe the geography of heaven or the events of the future in straightforward and easily understood terms. Instead, they delight in the mystical and revel in the symbolic. The future is envisioned as a series of wild and grotesque beasts that appear on the face of the earth; there are fantastic spectacles, bizarre images, strange figures, mysterious events. The symbols often confuse not only the reader but also the prophet himself, who sometimes presses the angelic mediator for an interpretation of what he has seen. Sometimes the explanation itself is mysterious and subject to a wide range of interpretations.

Violent Repetitions. Apocalypses often convey the mysteries of the heavenly realm through violent repetitions. By this I do not mean that there is always repeated violence in these texts—although there often is—but that the repetitions themselves are violent in that they violate the literal sense of the narrative. That is, apocalypticists often emphasize their points by producing countless repetitions for effect. If one were to take Revelation's descriptions of future tribulations literally, for example, there would be no way to map them out chronologically on a time line. As we have already seen, at the breaking of the sixth seal, the sun, moon, and stars are destroyed; surely this is the end—no life could possibly go on existing. But life does go on, and we enter into a new phase of sufferings on the earth with the heavenly lights shining in full force.

What we have, then, is a kind of spiral effect in narrative. The catastrophes that it describes cannot be sketched in linear fashion as if one event necessarily occurred after another. One benefit of this kind of repetition is that it allows the author to employ important numbers known to have mystical qualities. In the book of Revelation, for example, there are three major sets of seven disasters sent from heaven, the number three probably symbolizing fullness and perfection and seven symbolizing divinity—as opposed to six, which is one short of seven, and therefore imperfect (see below the number of the Beast, 666).

Triumphalist Movement. By their nature, apocalypses are designed to provide hope for those who suffer and despair. In the end, God will prevail. The present suffering is intense, and that to come will be yet more intense, but ultimately God will triumph over evil and vindicate his people.

Motivational Function. These books exhort their readers to remain faithful to their religious com-

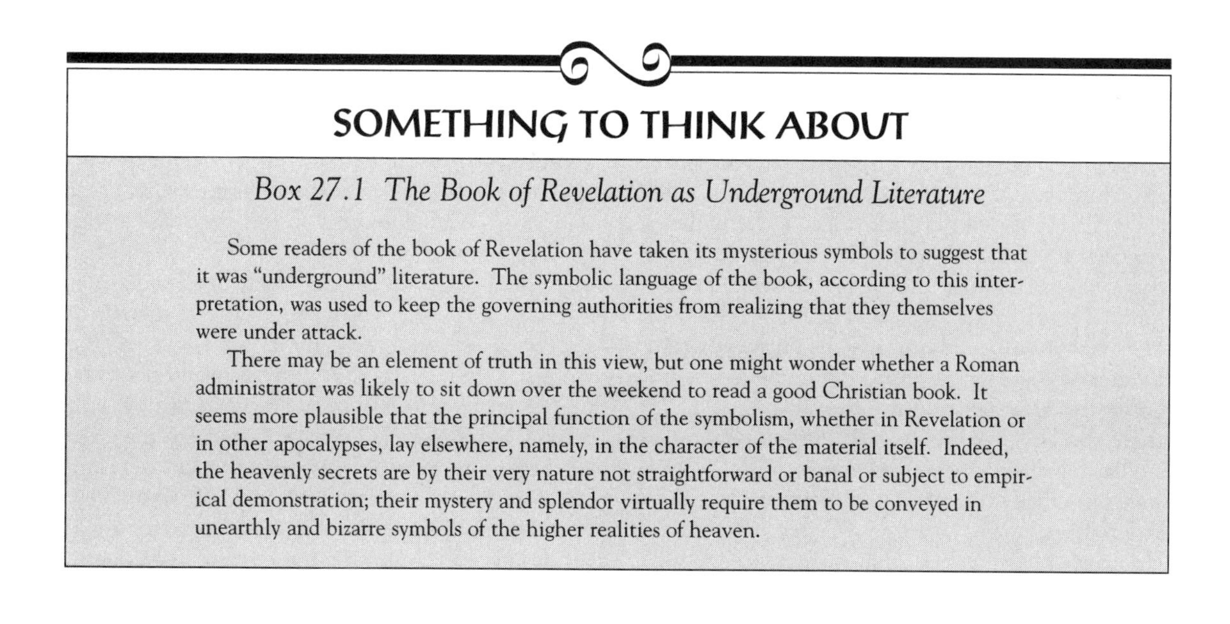

SOMETHING TO THINK ABOUT

Box 27.1 The Book of Revelation as Underground Literature

Some readers of the book of Revelation have taken its mysterious symbols to suggest that it was "underground" literature. The symbolic language of the book, according to this interpretation, was used to keep the governing authorities from realizing that they themselves were under attack.

There may be an element of truth in this view, but one might wonder whether a Roman administrator was likely to sit down over the weekend to read a good Christian book. It seems more plausible that the principal function of the symbolism, whether in Revelation or in other apocalypses, lay elsewhere, namely, in the character of the material itself. Indeed, the heavenly secrets are by their very nature not straightforward or banal or subject to empirical demonstration; their mystery and splendor virtually require them to be conveyed in unearthly and bizarre symbols of the higher realities of heaven.

mitments, to keep true to their faith, and to refuse to give up hope. This point is worth emphasizing: ancient Jewish and Christian apocalypses were written not so much to reveal the precise details of the future as to provide motivation for those who were in danger of growing slack in their commitments and of losing hope in the midst of their suffering. The hope they provided was rooted in the belief that when all was said and done, God was in control of the world and would eventually reward those remained faithful to him.

THE REVELATION OF JOHN IN HISTORICAL CONTEXT

The Book of Revelation is virtually unique among apocalypses in that it does not appear to be pseudonymous. The author simply calls himself John without claiming to be a famous person from the past.

Some Christians of the second and third centuries claimed that this John was none other than Jesus' own disciple, the son of Zebedee. Others rejected this notion and as a result refused to admit the book into the Christian canon of Scripture. (If the author had claimed to be that John, the book would probably have to be considered pseudonymous, for reasons we will see momentarily.) One of the ironies of the New Testament is that the Fourth Gospel, which does not claim to be written by someone named John, is called John, whereas the book of Revelation, which does claim to be written by someone named John, is not called by this name. In any event, it can be stated without reservation that whoever wrote the Gospel did not also write this book. For one thing, the theological emphases are quite distinct. In the Gospel of John there is virtually no concern for the coming end of the age (contrast the Synoptics, with their proclamation of the imminent arrival of the Son of Man); in the book of Revelation the end is nearly the entire concern. Even more importantly, as recognized even by linguists in early Christianity, the writing styles of these two books are completely different. Detailed studies have shown that the author of Revelation was principally literate in a Semitic language, probably Aramaic, and knew Greek as a second language. His Greek is clumsy in places, sometimes even ungrammatical. This is not at all the case with the Gospel of John, which is written in an entirely different style and therefore by a different author.

We have already seen that the Fourth Gospel was probably not written by John the son of Zebedee. Is it possible, then, that the book of Revelation was? The difficulty with this view is that parts of the book could scarcely be explained if it were written by Jesus' own disciple. The author, for example, occasionally mentions "the apostles," but he never indicates that he is one of them (e.g., 21:14). Even more intriguingly, at one point in the narrative the prophet sees twenty-four elders around the throne of God (chap. 4). Most interpreters understand these figures to represent the twelve Jewish Patriarchs and the twelve apostles of Jesus (cf. 21:12, 14); among them, of course, would be the two sons of Zebedee. But the author gives no indication that he is seeing himself! It appears, then, that the book was written by some other Christian named John, a prophet who was known to several of the churches of Asia Minor.

It is difficult to know exactly when he wrote this book. Modern interpreters usually appeal to details in some of the visions to pinpoint a date. For example, the Beast of Babylon in chapter 17, which, as we will see, appears to represent the city of Rome, is said to have seven horns on its head. These represent seven "kings," evidently meaning the rulers of Rome (17:9). Five of these are said to have come and gone and one is currently reigning (17:10). This would presumably mean that the vision was written during the reign of the sixth Roman ruler, but with which ruler should we begin counting—with the dictator Julius Caesar or with his adopted son, the first emperor, Caesar Augustus? And does this vision date the entire book or simply this portion of it?

On the basis of a detailed study of all such clues in the text, most investigators think that parts of the book were written during the 60s of the Common Era, soon after the persecution of the Christians under Nero. If we begin counting with Julius Caesar, Nero happens to have been the sixth ruler of Rome. He was also one of the author's chief enemies. The book was evidently not completed, however, until some thirty years or so later, probably around 95 C.E., during the reign of Domitian.

Somewhat less complicated is the question of the social context of the book. The author describes the Christian churches of Asia Minor in chapters 2–3. They are persecuted, they have false teachers in their midst, and a number of their members have lost their fervor for their faith, possibly because of the passing of time and the hardships imposed upon them as Christians. Elsewhere in the book we read of extensive Christian martyrdoms (6:5) and find hints that the Christian communities that the author addresses are among the poorer classes, who hate the rich and powerful (18:11–20). In particular, John directs his anger against the political institutions of his day, especially the Roman government, which was responsible for the oppression and suffering of the people of God. In his view, this government will not survive, since God was soon going to destroy it.

In short, Christianity as experienced by this author was an oppressed and persecuted religion. Indeed, interpreters have traditionally maintained that John actually wrote the book while in exile from his homeland because of his Christian proclamation (see 1:9). The churches of his world had suffered from economic exploitation and some Christians had been martyred, but God was going to put an end to it all, and he would do so very soon.

In general terms, Revelation corresponds to the basic description of an apocalypse. It is a firsthand account written by a prophet who has been shown a vision of heaven that explains the realities of earth, a vision that is mediated by angels and full of bizarre and mysterious symbolism. The nature of the book is indicated at the outset in the magnificent vision of the exalted Christ that the prophet describes in chapter 1. Here Christ appears as "one like a Son of Man" (cf. Dan 7:13–14, where the phrase describes the cosmic judge of the earth) and is seen walking amidst the seven golden lampstands (i.e., he is present among the seven churches of Asia Minor, 1:20) with seven stars in his hands (i.e., he himself is in control of the guardian angels of these churches and therefore of the churches' own destinies, 1:20). His appearance is symbolic: among other things, he is a king (wearing a long robe with golden sash, 1:12), he is ancient (with white hair, 1:14), he is the cosmic judge (with eyes like fire, 1:14), he is full of splendor (with feet of burnished bronze, 1:15), he is all-powerful (with a voice of many waters, 1:15), he speaks

Figure 27.2 Roman coin showing the son of the emperor Domitian seated on a globe and reaching out to seven stars, with the inscription "To the Divine Caesar." Notice the similarities with the visions found in Revelation, where Christ too is a divine being, the Son of God and ruler of the earth, in whose hand are seven stars (e.g., Rev 1:12–16). Interestingly Revelation was written during the time of Domitian, when this coin was minted.

the word of God (has a two-edged sword coming from his mouth, 1:16), and he is totally overpowering (with a face like the sun, 1:16). The prophet's response to this vision is understandable: he falls down as if dead. But Christ raises him up and commands him to convey both the message of his vision and the truth of what is yet to come. Many other features of the book are also typical of the genre.

Bizarre Symbolism. The symbolic character of John's visions is obvious. Sometimes he himself doesn't understand what he sees and needs an angel to explain it to him (e.g., 17:7). Not everything he says is shrouded in mystery, however. Many of the symbols are not difficult to understand for those who know enough about the Old Testament (e.g., the image of "one like a son of man") or about common images in ancient culture (e.g., eyes of fire). The explanations of other symbols are hinted at in the text. These are among the most interesting features of the book. A few prominent examples will illustrate the process of historical interpretation.

Figure 27.3 Painting of Christ as the alpha and Omega (cf. Rev 21:6; 22:13), from the catacomb of Commodilla.

The Great Whore of Babylon. In chapter 17 the prophet is taken into the wilderness to see "the great whore . . . with whom the kings of the earth have committed fornication" (v. 2). He sees a "woman sitting on a scarlet beast that was full of blasphemous names" (v. 3). The woman is wearing fine clothes and jewels and holds in her hand "a golden cup full of abominations and impurities of her fornication" (v. 4). Across her forehead is written the name "Babylon the great." She is "drunk with the blood of the saints and the blood of the witnesses to Jesus" (v. 6).

An amazing vision. Fortunately, the accompanying angel gives enough of an explanation to enable us to interpret its major points with relative ease (though even so some of the details are a bit puzzling). The beast on which the woman is seated is about to descend to the bottomless pit (v. 8); we learn in 20:2 that Satan is about to be thrown into the pit, so this woman, whoever she is, appears to be supported by the Devil. (This is an important point to observe, for the book of Revelation will sometimes interpret its own symbols for the attentive reader.) Who is the woman herself? The beast has seven heads, and we are told that these are seven mountains on which the woman is seated (v. 9). For those who know enough about the world in which the prophet was writing, this will be the only clue that is needed. For those who don't, the angel makes the matter still clearer in verse 18: "The woman you saw is the great city that rules over the kings of the earth."

The meaning of the vision is now reasonably transparent. The "great city" that ruled the world in

SOMETHING TO THINK ABOUT

Box 27.2 Futuristic Interpretations of the Book of Revelation

One of the most popular ways to interpret the book of Revelation today is to read its symbolic visions as literal descriptions of what is going to transpire in our own day and age. But there are problems with this kind of approach. On the one hand, we should be suspicious of interpretations that are blatantly narcissistic; this way of understanding the book maintains that the entire course of human history has now culminated with us! An even larger problem, though, is that this approach inevitably has to ignore certain features of the text in order to make its interpretations fit.

Consider, as just one example, an interpretation sometimes given of the "locusts" that emerge from the smoke of the bottomless pit in order to wreak havoc on earth in chapter 9. The seer describes the appearance of these dread creatures as follows:

> On their heads were what looked like crowns of gold; their faces were like human faces, their hair like women's hair, and their teeth like lions' teeth; they had scales like iron breastplates, and the noise of their wings was like the noise of many chariots with horses rushing into battle. They have tails like scorpions, with stingers, and in their tails is their power to harm people. . . . (Rev 9:7–10)

According to one futuristic interpretation, these locusts are modern attack helicopters flying forth through the smoke of battle. The seer, living many centuries before the advent of modern warfare, had no way of knowing what these machines really were, and so he described them as best he could. They fly like locusts but are shaped like huge scorpions. The rotors on top appear like crowns, they seem to have human faces as their pilots peer through their windshields, they are draped with camouflage that from a distance looks like hair, they have fierce teeth painted on their fronts, they are made of steel and so appear to have iron breastplates, the beating of their rotors sounds like chariots rushing to battle, and they have machine guns attached to their tails, like scorpions' stingers.

What could be more plausible? The prophet has glimpsed into the future and seen what he could not understand. We, however, living in the age in which his predictions will come to pass, understand them full well.

The problem is that the interpretation simply doesn't work, because it overlooks some of the most important details of the passage. Consider, for example, what these locusts are actually said to do. The text is quite emphatic: they are not allowed to harm any grass or trees, but only people; moreover, and most significantly, they are given the power to torture people for five months, but not to kill them (9:4–5). Those who are attacked by the locusts will long to die but will not be able to do so (9:6). These locusts can't be modern instruments of war designed for mass destruction because they are explicitly said to be unable to destroy *anything*.

The same problems occur with virtually every interpretation of the book that takes its visions as literal descriptions of events that will transpire in our own imminent future. These approaches simply cannot account for the details of the text, which is to say that they don't take the text itself seriously enough. It is more reasonable to interpret the text within its own historical context, not as a literal description of the future of the earth, but as a metaphorical statement of the ultimate sovereignty of God over a world that is plagued by evil.

John's day is obviously Rome, commonly called the city "built on seven hills" (hence the beast's seven heads). This city, which in the vision is supported by the Devil himself, had corrupted the nations (the whore fornicates with the kings of earth), exploited the peoples of earth (she is bedecked in fine clothing and jewelry), and persecuted the Christians (she is drunk with the blood of the martyrs). Why is the whore called Babylon? This symbol too is clear for those who know the Old Testament, where Babylon is portrayed as the archenemy of God, the city whose armies devastated Judah, leveled Jerusalem, and destroyed the Temple in 587 B.C.E. In Revelation, then, "Babylon" is a code name for the city opposed to God—Rome, God's principal enemy. Like Babylon of old, Rome too will be destroyed (v. 16). Indeed, this is the point of much of the entire book.

The Number of the Beast, 666. Somewhat earlier in the book we are given a description of another beast, which bears a remarkable resemblance to the one we have just observed. According to chapter 13, this other beast arises from the sea and has ten horns and many heads. One of its heads receives a mortal wound that is then healed. The entire world follows this beast, which is empowered by the dragon (i.e., the Devil, 12:9). The beast makes war on the saints and conquers them (13:7). It has power over all the nations of earth (13:7–8), exploiting them economically (13:17) and demanding to be worshipped (13:15). The author concludes his description of this mortal enemy of God with a final identifying mark, given for those "with understanding." The number of the beast is 666 (13:18).

Interpreters have offered numerous explanations of this number over the years (probably more than six hundred and sixty-six of them). Most of these interpreters have been concerned to show that the beast has finally arisen in their own day. Rarely are the interpretations put forth as conjectures, of course, but almost always with the confidence of those who have the inside scoop. Just within the past several decades, for example, Christian preachers, televangelists, and authors have suggested such tantalizing and diverse candidates as Adolph Hitler, Mussolini, former Secretary of State Henry Kissinger, and Pope Paul VI!

The author of this book, however, was writing for his own day, not for the twentieth century, and he may have had something specific in mind (see box 27.2). Recall our discussion of the ancient art of interpretation known as gematria (in connection with the *Epistle of Barnabas*). In ancient numeral systems, numbers were written by using letters, and conversely, any combination of letters could yield a numerical total. Anyone conversant with gematria would have understood what the author meant by saying that the number of the beast was 666. He was indicating that this was the numerical value of the person's name. An interesting wrinkle in this matter is that some of the ancient Greek manuscripts of the Book of Revelation give a different number for the beast. In these documents, it is 616 rather than 666.

How can we make sense of all this? The beast is described as God's enemy, who controls the world, exploits its people, and kills the saints. Given the similarities to the beast in chapter 17, we may not be too far afield to assume that the beast may be another image of the Roman Empire. If so, then the heads would presumably be the rulers of the empire, some of whom demand to be worshipped (as did some of the emperors). One of these heads was mortally wounded, but then healed. What might this mean? Historians have long known of a group of ancient Jewish books called the Sybilline Oracles, which predict that one of the most hated of the Roman emperors, Caesar Nero, will return from the dead to wreak havoc on the earth—making him comparable to one who recovers from a death-inflicting wound. This popular belief may have something to do with the number of the beast. It should be recalled that Nero was seen as the archenemy of the Christians, whom he ruthlessly and unjustly persecuted for setting fire to the city of Rome. Could he have been the beast described in Revelation 13?

Intriguingly, when the name "Caesar Nero" is spelled in Hebrew letters ("Nero" becomes "Neron"), their numerical total is 666. More intriguingly still, the name can be spelled in another way, without a final *n* at the end. The *n* is worth 50 in the Hebrew numerical system. When the alternative spelling is employed, the name adds up to 616.

The author of Revelation is not referring to Hitler or Mussolini or anyone else in modern times. His enemy was Rome and its Caesars. It was Rome that had dominated the other nations of earth, exploited their native populations, and oppressed the people of God; it was the Roman emperor who was worshipped as divine and who persecuted Christians and sometimes put them to death. This book is about how God was going to overthrow this emperor and his empire at the end of time (see especially chaps. 18–19) prior to rewarding his saints with the kingdom in a new heavens and a new earth (chaps. 20–22).

Violent Repetitions. The book of Revelation follows the literary convention of using violent repetitions. It is impossible to take the predictions of this book as a linear, chronological sequence of events that are to transpire at the end of time. The universe caves in on itself in chapter 6, but the pain and agony continue for another thirteen chapters! The author has written for effect, compounding the tribulations and intensifying the sufferings of the last times to show how dreadful things are going to be.

Triumphalist Movement. The narrative moves through tragedy to triumph, through despair to hope. The fundamental point of the narrative is to provide assurance that, regardless of how terrifying the situation may become, God is ultimately in control of it all. The suffering of the present is part of God's plan, and he will vindicate his people by destroying their enemies. When he does so, he will establish a new kingdom on earth in which there will be no more pain, suffering, or death, no more persecution or exploitation, no more disease, famine, or war. There will only be Christ and his kingdom of saints.

Imminence. The author emphasizes at the beginning and end of his work that the events he records are going to happen soon (1:1, 3; 22:6, 10, 12, 20). This emphasis may suggest that the people he addresses are presently undergoing considerable suffering (note the pervasive references to persecution, exploitation, and martyrdom). He is writing to provide them with hope that they will not have to suffer long before the end comes and God intervenes in history to make right all that has gone wrong.

Encouragement and Admonition. Ultimately, Revelation is a book about hope. In some respects, the author's timetable matters less than his overarching message that God is sovereign over this world, appearances notwithstanding, and that he will soon bring his people's suffering to a crashing halt. This message is meant to encourage those who are persecuted and weak, but it is also meant to admonish those who are tempted to abandon ship in view of their present distress. John emphasizes that those who depart from the faith will face a severe judgment, indeed, they will experience eternal torment. Believers must therefore hold on and not cave in, they must keep the faith and never abandon hope, for the end is near, and with it comes a fearful judgment for those who have proved faithless but an eternal reward for those who have stayed true.

THE SHEPHERD OF HERMAS

We have already seen that early Christian apocalypses employed a variety of means for revealing the heavenly secrets that can make sense of earthly realities. Neither of the two books we will now examine briefly, for example, includes a detailed sketch of the future course of history.

The first is a book titled *The Shepherd,* written by a Christian named Hermas. Like the book of Revelation, *The Shepherd* is unusual among apocalypses in not being pseudonymous. Hermas was a Christian living during the first half of the second century C.E. in Rome, where his brother was the bishop. His book was well received by the Christians throughout the world and was even included among the writings of the New Testament canon by one of our oldest manuscripts. Eventually, however, the judgment articulated by an anonymous author of the second century was sustained; this author urged that *The Shepherd* not be read as Scripture because it was written "recently" (i.e., it wasn't ancient enough) and because its author was someone who was known to the Roman church, not an apostle (see box 27.3).

SOME MORE INFORMATION

Box 27.3 The Shepherd of Hermas and the Muratorian Canon

The anonymous author who dismisses *The Shepherd* of Hermas because it was penned "recently" by someone who was not an apostle is an otherwise unknown figure whose writing continues to intrigue scholars. In the only fragment of his writing that remains, he briefly discusses the books that he considers to be part of the Christian Scriptures. Unfortunately, the fragment begins in the middle of a sentence, followed by the words, "The third book of the Gospel is that according to Luke. . . ." Evidently, he has just discussed Matthew and Mark (assuming that these were his first two Gospels). He proceeds to describe Luke, John, the letters of Paul, and the other books that he accepts as canonical. The piece ends, as it begins, in midsentence.

The fragment was discovered in the eighteenth century in a library in Milan, Italy, by a scholar named Muratori. For this reason, it is known as the Muratorian Fragment. The fragment itself was written in the eighth century by an unskilled Latin scribe; his grammar is terrible, and he was extremely careless. Scholars debate when and where the original text that the scribe was copying was produced; most believe that it was written during the second half of the second century, in or around Rome. The original language of the document was probably Greek.

The Muratorian canon does not mention the books of Hebrews, James, 1 Peter, 2 Peter, or 3 John, but it does accept as canonical all of the other books of our present New Testament. Interestingly, it also accepts the *Wisdom of Solomon* and, somewhat tentatively, the *Apocalypse of Peter*. Finally, the author explicitly condemns two books that he labels as forgeries concocted by followers of Marcion in the name of Paul: a letter to the Laodiceans and another to the Alexandrians. These are not to be accepted by the Catholic church as canonical writings, the author declares, "for it is not fitting for gall to be mixed with honey."

This fragment is of great interest to the historian of early Christianity, for it reveals a period of Christian history in which a closed canon of Scripture appears to be on the horizon, while being still some distance off.

The book takes its name from the angelic mediator who appears to Hermas in the form of a shepherd. There are other angelic beings here as well, in particular, an old woman who identifies herself as a personification of the Christian church. These various figures communicate visions, commandments, and parables to Hermas, who asks for interpretations of what he sees and hears. His heavenly companions typically consent, sometimes grudgingly.

The book divides itself rather neatly into five visions, twelve sets of commandments (or "mandates"), and ten parables (or "similitudes"). The visions and similitudes are enigmatic and symbolic; they are normally explained to Hermas (and the reader) as having a spiritual significance for Christians living on earth. The mandates are somewhat easier to interpret, consisting of direct exhortations to speak the truth, to give alms, to do good to all, to avoid sexual immorality, drunkenness, gluttony, hypocrisy, malice, and so on.

The entire book, not just the mandates, is driven by an ethical concern. The primary issue involves Christians who have lapsed into sin after being baptized. While a number of early Christians insisted that those who returned to a life of sin after their conversion and baptism had lost their salvation (cf. Heb 6:4–6), this book contends that a second repentance is possible. A person who reverts to sin after being baptized has only one second chance to repent, however. If the second opportunity is squandered, then no hope remains.

This promise of a second repentance may not seem like a particularly apocalyptic message, but it is, because the second repentance will prevent a person from suffering the apocalyptic judgment of God. Moreover, the book contains a number of other features of apocalypses.

1. **First-Person Narrative.** The author speaks of his own personal history and of events that have happened to him.
2. **Mediated Revelations.** He experiences visions that convey the truth that he needs to communicate to his readers. These visions are given through angelic intermediaries and are generally interpreted by them as well.
3. **Transcendent Realities.** The visions provide Hermas with the "heavenly" basis for his "earthly" doctrine. The church and its experiences are not the haphazard accidents of human history. They are rooted in divine reality and are directed by higher powers. In this narrative God works behind the scenes to bring his plan for the church to fruition.
4. **Symbolic Visions.** The visions and similitudes that Hermas portrays are manifestly symbolic and often relate to other visions found in other Jewish and Christian apocalypses. Two instances are the visions of the tower and the monster.

   The Tower. In his third vision, Hermas sees a tower being built in the sea by six young men who are assisted by tens of thousands of others. They use a variety of stones for the tower's construction. Some stones are tailor-made for the task, but some are rotten, others are cracked, and others simply do not fit. Those that can be used are joined together to build the tower while the others are cast aside. The angelic interpreter then explains what all of this means. The tower is the church. It is built in the sea because it comes into being through the waters of baptism. The workers are the holy angels who construct the tower, six of whom are more powerful than the others. The stones represent persons who make up the church. Those that fit perfectly are apostles, bishops, teachers, and deacons who are in perfect harmony with one another. The other usable stones are Christians who have been faithful to God unto death. The stones that are rotten, cracked, or misshapen represent people who can form no part of the tower of God, even though they were formerly stones of potential value (i.e., they at one time claimed to be Christian). These would include people who have been hypocritical in their faith or who have abandoned the truth.

   The vision portrays a social reality and its ultimate point is a moral one. Those who have been cast out of the church because of their hypocrisy or complacency are urged to repent before the tower is completed, for once the job is done, they will have no place among the people of God.

   The Monster. In another important vision, Hermas describes his encounter with a grotesque beast that is symbolic of a spiritual reality (Vision IV). Hermas is passing along the road and to his horror sees a gigantic monster breathing fiery locusts from its mouth and rushing upon him with power enough to destroy a city. Frightened nearly to death, Hermas prays for help and is told simply to pass the beast by. As he does so, the monster lies down meekly and does nothing but flick its tongue in the air. We are told that the beast represents a great persecution to come, which will crush everyone who does not turn to God with all their heart, pure and blameless.

5. **Encouragement and Admonition.** Like the Revelation of John, *The Shepherd* of Hermas ultimately aims to encourage and admonish its readers. Those who have fallen into a life of sin after their baptism are encouraged to repent and turn anew to the life of faith; they can trust that they will be given a second chance. But all believers must know that God's patience with sinners is not without limit, for a day of judgment will come in which the tower of the church will be complete, and those who are outside God's good graces will feel the power of his wrath.

THE APOCALYPSE OF PETER

The last Christian apocalypse for us to consider claims to be a firsthand account of the tortures of hell and the ecstasies of heaven written in the name of Jesus' disciple, Peter. As we have seen, there are a large number of early Christian pseudepigrapha written in Peter's name, one or two of which came to be included in the New Testament. Indeed, among Christian apocalypses alone we know of three that claim his name. One is preserved only in an Arabic translation, another was discovered among the Coptic writings of the Nag Hammadi library, and the third has been known by historians for centuries, although they have had it in their possession only since 1887, when it was found in the tomb of a Christian monk along with the pseudonymous *Gospel of Peter*. It is the third apocalypse that will concern us here, for it is a book that was accepted as canonical Scripture in some churches of the second and third centuries (see box 27.3). Even when it finally came to be excluded from the canon, it continued to make an impact on Christian thought. To our knowledge, this is the first Christian writing to describe a journey through hell and heaven, an account that influenced a large number of successors, including, ultimately, Dante's *Divine Comedy*, one of the great inspirational classics of Western civilization.

The book begins with Peter and the other disciples on the Mount of Olives listening to Jesus deliver his "apocalyptic discourse" (see Mark 13). Peter asks about the coming judgment. Jesus responds by describing the terrifying events that will occur when the world is destroyed by fire at the last judgment. He then details the eternal terrors that await those destined for hell and, more briefly (possibly because they are somewhat less interesting and certainly less graphic), the perpetual blessings of those bound for heaven.

There is some ambiguity over whether Jesus actually takes Peter on a journey through these two abodes of the dead or simply describes them in such vivid detail that it *feels* as if Peter is actually seeing them. There is no ambiguity, however, concerning the respective fates of those destined for one place or the other. In an unsettling way, the horrific punishments of the damned are made to fit their crimes. Those guilty of blasphemy are hanged by their tongues over unquenchable fire, to roast eternally. Men who have committed fornication are forever suspended by their genitals. Those who have committed murder are thrown into a gorge to be perpetually tormented by venomous reptiles and swarming worms. Worshippers of idols are chased by hideous demons and driven off of high cliffs, time and again, for all eternity.

Included among the sinners who suffer eternal torments are those who have engaged in extramarital sex, who have disobeyed their parents, who have given alms but not striven to live righteously, and who have lent out money and demanded compound interest. The blessed, on the other hand, are those who have followed Christ and kept the commandments of God. These will be brought into the eternal kingdom, where they will enjoy the blissful life of heaven forever. The book ends with Peter describing firsthand what he saw on the Mount of Transfiguration, possibly to validate the legitimacy of the rest of his vision (cf. 2 Pet 1:17–18).

The ultimate message of this firsthand description of hellish and heavenly realities is reasonably clear. There is only one way to avoid facing eternal torment for sins: don't sin. Only those who believe in Christ and lead upright moral lives can expect to enter into his eternal kingdom. All others will be damned by God to face unspeakable pain and suffering for all eternity. This message no doubt made a considerable impact on its Christian readers; it was, after all, written by "Peter," the closest disciple to Jesus! Moreover, the message became an essential element in the Christian missionary proclamation as well, providing an incentive for pagans and Jews to turn from their false ways and to worship the one true God who would reward those who came to accept his truth and punish for all eternity those who did not.

SUGGESTIONS FOR FURTHER READING

Aune, David. *The New Testament in Its Literary Environment*. Philadelphia: Westminster, 1987. Includes an insightful discussion of the characteristics of apocalypses and the social world that they presuppose.

Collins, Adela Yarbro. *Crisis and Catharsis: The Power of the Apocalypse*. Philadelphia: Westminster, 1984. A superb introductory discussion of the author, social context, and overarching message of the Apocalypse.

Collins, John J. *Apocalypse: The Morphology of a Genre*. In *Semeia* 14. Missoula, Mont.: Scholars Press 1979. A full-scale investigation of the characteristics of apocalypses; for more advanced students.

Elliott, J. K. *The Apocryphal New Testament: A Collection of Apocryphal Christian Literature in an English Translation*. Oxford: Clarendon, 1993. An excellent one-volume collection of noncanonical works, including such important apocryphal apocalypses as the *Apocalypse of Peter*, in a readable English translation with brief introductions.

Osiek, Carolyn. *The Rich and the Poor in the Shepherd of Hermas*. Washington, D.C.: Catholic Biblical Association of America, 1979. A discussion of the social world of the church in Rome in the early second century, as intimated by comments found in *The Shepherd*.

Pilch, J. *What Are They Saying about the Book of Revelation?* New York: Paulist Press, 1978. A clear overview of modern scholarly insights into major aspects of the book of Revelation.

Pippin, Tina. *Death and Desire: The Rhetoric of Gender in the Apocalypse of John*. Louisville, Ky.: Westminster/John Knox, 1992. An intriguing discussion of the female imagery in the book of Revelation and the social world that it presupposes, written from a feminist perspective; for advanced students.

Tugwell, Simon. *The Apostolic Fathers*. Harrisburg, Pa.: Morehouse, 1990. A clear and straightforward discussion of the background and message of the Apostolic Fathers, including *The Shepherd* of Hermas; ideal for beginning students.

CHAPTER 28

Epilogue: Do We Have the Original New Testament?

We have now come to the conclusion of our study of the earliest Christian writings. Our investigations have taken us over a remarkable range of materials—historical movements, social groups, and literary works from before the days of Jesus through the early decades of the second century. In some ways, of course, we have only scratched the surface of this extraordinary segment of human history and the literature that emerged out of it. In this epilogue to our study I will make some brief comments on the fate of the earliest Christian writings after they were produced and pose a question that has probably never occurred to most people: Do we have the original New Testament? The answer may surprise you.

THE MANUSCRIPTS OF THE NEW TESTAMENT

We do not have the original copies of any of the books of the New Testament or of any of the other Christian writings that we have examined in our study (or indeed of any literary text from the ancient world). The originals were lost or destroyed long ago, and all that we have are copies. For the most part, these copies were made hundreds of years after the originals, from other copies, rather than the originals.

Let me explain the situation by giving a solitary example of how things worked. When the Thessalonians received Paul's first letter, someone in the community must have copied it by hand, one word at a time. The copy itself was then copied, possibly in Thessalonica, possibly in another community to which a copy was taken or sent. This copy of the copy was also copied, as were later copies, until before long there were a large number of different copies of the letter circulating in different communities throughout the Mediterranean, all made by hand at a pace that would seem outrageously slow to us who are accustomed to the world of photocopiers, word processors, electronic mail, and desktop publishing.

In this process of recopying the document by hand, the original was eventually thrown away or burned or otherwise destroyed. Perhaps it had been read so much that it simply wore out. In any case, the early Christians saw no need to preserve it as the "original" text, since they had copies of the letter. Possibly, they did not fully appreciate what happens to a text that is copied and recopied by hand, especially by scribes who are not trained professionals but simply literate persons with the time and money to do the job. Copyists, even if they are skilled specialists, inevitably make mistakes. (Anyone who doubts this should copy a long document by hand and see how well he or she does.) Moreover, whenever a copyist makes a copy from a document that has already been copied, the mistakes begin to multiply; scribes not only introduce their own mistakes, but they also reproduce mistakes found in the copy being copied.

We do not have the original copy of 1 Thessalonians or of any other New Testament book. Nor do we have copies made directly from the originals, copies made from the copies of the originals, or copies made from the copies of the first copies. Our earliest manuscripts (i.e., handwritten copies) of Paul's letters date from around 200 C.E., that is, nearly 150 years after he wrote them. The earliest full manuscripts of the Gospels come from about the same time, although we have some fragments of manuscripts that date earlier. One credit-card-sized fragment of John discovered in a trash heap in Egypt is usually dated to the first part of the second century. Even our relatively full manuscripts from around the year 200 are not preserved intact, however. Pages and entire books were lost from them before they were discovered in modern times. Indeed, it is not until the fourth century, nearly 300 years after the New Testament was written, that we begin to find complete manuscripts of all of its books.

After the fourth or fifth century, copies of the New Testament become far more common. Indeed, if we count up all of the New Testament manuscripts that have been discovered, the number is impressive. We presently know of nearly 5,400 Greek copies of all or part of the New Testament, ranging from tiny scraps of a verse or two that could fit in the palm of your hand to massive tomes containing all twenty-seven books bound together. These copies range in date, roughly, from the second century down to the invention of the printing press in the fifteenth century. As a result, the New Testament is preserved in far more manuscripts than any other book from antiquity. There are, for example, fewer than 700 copies of Homer's *Iliad*, fewer than 350 copies of the plays of Euripides, and only one copy of the first six books of the *Annals* of Tacitus.

What is unsettling for those who want to know the original text is not the number of New Testament manuscripts but the dates of these manuscripts and the differences among them. Of course, we would expect the New Testament to be copied in the Middle Ages more frequently than Homer or Euripides or Tacitus; the trained copyists throughout the Western world at the time were Christian scribes, frequently monks, who for the most part were preparing copies of texts for religious purposes. Still, the fact that we have thousands of New Testament manuscripts that were made during the Middle Ages, many of them nearly a thousand years after Paul and his companions had passed off the face of the earth, does not mean that we can rest assured that we know what the original text said. For if we have very few early copies, in fact, scarcely any, how can we know that the text was not changed significantly *before* the New Testament began to be reproduced in such large quantities?

It is not simply a matter of scholarly speculation to say that the words of the New Testament were changed in the process of copying. We know they were changed because we can compare all 5,400 copies with one another. What is striking is that when we do so, we find that no two of these copies (except the smallest fragments) agree in all of their wording. There can be only one reason for this: the scribes who copied the texts changed them. Nobody knows for certain how often they changed them, because no one has yet been able to count all of the differences among the surviving manuscripts. Some estimates put the number at around 200,000, others at around 300,000. Perhaps it is simplest to express the figure in comparative terms: there are more differences among our manuscripts than there are words in the New Testament.

In spite of these remarkable differences, scholars are convinced that we can reconstruct the original words of the New Testament with reasonable (though probably not 100 percent) accuracy. The first step in doing so is to categorize the kinds of changes that scribes made in their texts.

CHANGES IN THE NEW TESTAMENT TEXT

The vast majority of all changes found in our New Testament manuscripts are careless mistakes that are easily recognized and corrected. Christian scribes were fully human and often made mistakes simply because they were tired or inattentive or, sometimes, inept. Indeed, the single most common mistake in our manuscripts is misspelled words (which are significant for little more than showing us that most scribes in antiquity could spell no better than people can today). In addition, we have numerous manuscripts in which scribes have left out entire words, verses, or even pages of a book, presumably by accident. Sometimes scribes rearranged the words on the page, for example, by leaving out a word and then sticking it in later in the sentence. Sometimes they found a marginal note scribbled by an earlier scribe and thought that it was to be included in the text, so they inserted it as an additional verse. These kinds of accidental changes were facilitated, in part, by the fact that ancient scribes did not use punctuation and paragraph divisions or even use spaces between words. Occasionally, as you might imagine, the correct interpretation of a sentence depends on how the words are to be separated (lastnightatdinnerwesawabundanceonthetable.)

Other kinds of changes are both more important and harder for modern scholars to detect. These are changes that scribes appear to have made in their texts intentionally. I say that they "appear" to have made such changes intentionally simply because the scribes are no longer around for us to interview about their intentions. But some of the changes in our manuscripts can scarcely be attributed to fatigue, carelessness, or ineptitude; they instead suggest intention and forethought.

It is difficult to know what might have motivated a scribe to change his text, but it often appears to have been some kind of problem in the text itself that he found disturbing. Sometimes, for example, scribes ran across a statement that appeared to be mistaken, as in Mark 1:2, where a citation from the book of Malachi is quoted as coming from Isaiah. At other times, scribes thought that a passage they were copying contradicted another one, as in Mark 2:25 (cf. 1 Sam. 21:1–7, where Ahimelech, not Abiathar, is said to have been the high priest when David entered the Temple). At still other times, scribes thought that a passage was grammatically incorrect or inelegant. In all such cases, scribes appear to have had little compunction about changing the text (both Mark 1:2 and 2:25 were commonly altered, for example). Some scribes, however, did not make such changes. As a result, a verse found in some manuscripts will appear to embody a mistake, a contradiction, or an awkward construction, but in others it will be worded differently in a way that avoids the problem. Scholars have to decide, then, which form of the verse was probably original and which represents the change made by a scribe.

One of the most common kinds of intentional changes involved the "harmonization" of one text to another, that is, changing a passage in one book to make it conform to a similar passage in another. This kind of change is particularly common in the Synoptic Gospels, since these three books tell so many of the same stories in slightly (or significantly) different ways. By harmonizing such accounts, scribes made them identical. One of the most famous instances of harmonization involves the Lord's Prayer, which is much shorter in Luke 11 than it is in Matthew 6 (see box 26.1). Some scribes, however, smoothed out the differences by adding the petitions found only in Matthew to the prayer found in Luke. This longer form is the way in which the prayer is still recited today. It was also the way it was recited by ancient scribes, which no doubt is what led some of them to add the petitions to Luke's version of the prayer in the first place, rather than, say, to delete them from Matthew.

Among the most interesting changes that were intentionally made by scribes are those that involve Christian doctrines. It appears that in the second and third centuries, when the earliest manuscripts were being produced and when the doctrinal controversies were in full swing (see Chapter 1), scribes would sometimes modify their texts to make them *say* what they were supposed to mean (in the scribes' opinion). For example, when Jesus remains behind in the Temple as a twelve-year-old boy and is discovered there by his mother after a long search, she upbraids him by saying, "Child, why have you treated us like this? Look, your father and I have been searching for you in great anxiety" (Luke 2:48). Some scribes have changed the text so that Jesus' mother no longer says "*your father* and I have been searching for you" but instead says "*we* have been searching for you." The reason for the change should be obvious: proto-orthodox scribes believed that Joseph was not Jesus' father because Jesus was born of a virgin. A second example occurs in Jesus' apocalyptic discourse, where he tells his disciples that no one knows the precise moment when the end will come: "About that day and hour no one knows, neither the angels of heaven nor the Son, but only the Father" (Matt 24:36). Interestingly, though, a large number of scribes changed the text by eliminating the phrase "nor the Son." Again, the rea-

Figure 28.1 The first chapter of the book of Hebrews in one of the oldest and best surviving manuscripts of the New Testament, the fourth-century Codex Vaticanus. Notice the marginal note between the first and second columns. A corrector to the text had erased a word in verse 3 and substituted another word in its place; a second corrector came along, erased the correction, reinserted the original word, and wrote a note in the margin to castigate the first corrector. The note reads, "Fool and knave, leave the old reading, don't change it!"

son is not hard to find. If Jesus truly was divine, as medieval scribes believed he was, then he must have been all-knowing and therefore must have known when the end was to come. Scribes solved the theological problem of Jesus' ignorance by removing any reference to it from the text.

As you can see, some textual changes can be important for textual interpretation. Consider several other important examples. The earliest manuscripts of the Gospel of Mark end at 16:8 with the report that the women fled Jesus' empty tomb in fear and told no one what they had seen. Later manuscripts append an additional twelve verses in which the resurrected Jesus appears before his disciples and delivers a remarkable speech in which he says, among other things, that those who believe in him will be able to handle venomous snakes and drink deadly poison without suffering harm. Are these verses original, or did scribes add them to a text that otherwise seemed to end too abruptly? Most scholars think the Gospel originally ended at 16:8 (I will explain their basic reasoning in coming to decisions like this below).

Did the author of the Fourth Gospel write the famous story of the woman taken in adultery or was this a later addition to the Gospel by a well-meaning scribe? The story is found in many of our later manuscripts between chapters 7 and 8 but not in any of the earliest ones; moreover, the writing style is significantly different from the rest of the Gospel. Almost all scholars acknowledge that the story was in fact added to manuscripts of John's Gospel many years after it had first been circulated.

Did the voice at Jesus' baptism in the Gospel of Luke originally declare, "You are my beloved son, in whom I am well pleased," which are exactly the words found in Mark's account, or did it proclaim, "You are my son, today I have begotten you," as the text is worded by some of the earliest witnesses (Luke 3:23)? This latter statement, a quotation of Psalm 2, would have been acceptable to second-century Christians who denied that Jesus had always been divine: for them, he was "adopted" to be God's son at his baptism. And it may also have been the original text of Luke, which was changed by proto-orthodox scribes who rejected the adoptionists' view of Jesus' baptism. This makes better sense than thinking that a scribe changed Luke's text (by having the voice quote Psalm 2) to make it sound *more* adoptionistic than it already was.

A final example involves the famous words of 1 John 5:7–8, the only passage in the entire Bible that explicitly affirms the Christian doctrine of the trinity—that the Godhead consists of three persons but that "these three are one." Even though the passage is part of the Latin Bible and found its way into the King James Version, it does not occur in any Greek manuscript of the New Testament prior to the fourteenth century. To my knowledge, there is no textual scholar who thinks that the passage was originally found in 1 John.

THE CRITERIA FOR ESTABLISHING THE ORIGINAL TEXT

There is a subdiscipline within New Testament scholarship called "textual criticism," which seeks to establish the original text of the New Testament based on the surviving manuscript evidence. It is a complex task but one that can be extremely intriguing—something like reading a detective story in which a few clues have to be pieced together in order to decide "whodunit." When there are different forms of the text, that is, when a verse is worded in different ways in the surviving manuscripts, the question has to be asked, which manuscripts represent the text of the autograph (the technical term for the original) and which ones represent changes of the text? Inevitably, a choice has to be made between one form of wording and another, and the choice can sometimes make a significant difference in how a document is interpreted. Since it is better to make an intelligent choice based on evidence than simply to guess, critics have developed certain principles for deciding which form of a text is more likely to be the original.

1. *The Number of Witnesses That Support a Reading*. In addition to the nearly 5,400 Greek manuscripts of the New Testament, we have tens of thousands of New Testament manuscripts in other languages into which it was eventually translated (especially Latin but also Syriac, Coptic, Armenian, and others). Moreover, there are dozens of ancient Christian authors from different times and places who quoted the New Testament. By collecting their quotations, we can reconstruct what their own manuscripts probably looked like.

   Given this abundance of evidence, one might suppose that a fairly obvious criterion for deciding which reading is original is to count the witnesses in support of each (different) reading and to accept the one that is most abundantly attested. Suppose, for example, that for a given verse there are 500 witnesses that have one form of wording and only six that have a variant form. All other things being equal, one might suspect that the six represent a mistake.

The problem, though, is that all other things are rarely equal. If the six witnesses, for example, all derive from the third and fourth centuries, whereas the 500 are all later, from the 5th to the 15th centuries, then the six may preserve an earlier form of text that came to be changed to the satisfaction of later scribes. Thus, simply counting the witnesses that support a certain form of the text is generally recognized as a rather unreliable method for reconstructing the original text.

2. *The Age of the Witnesses*. The form of the text that is supported by the oldest witnesses is more likely to be original that a different form found only in later manuscripts, even if these are more numerous. Most scholars recognize that this principle is better than simply counting the manuscripts, but it too can be problematic. For example, it is possible for a sixth-century manuscript to preserve an older form of the text than, say, a fourth-century one. This could happen if the sixth-century manuscript had been produced from a copy that was made in the second century, whereas the fourth-century manuscript derived from one made in the third.
3. *The Quality of the Manuscripts*. In a court of law, the testimony of some witnesses carries more weight than that of others. If there are two witnesses with contradictory testimony, and one is known to be a habitual liar, drunkard, and thief whereas the other is an upstanding member of the community, most juries will have little difficulty deciding whom to believe. A similar situation occurs with manuscripts. Some are obviously full of errors, for instance, when their scribe was routinely inattentive or inept, and others appear to be on the whole trustworthy. The best manuscripts are those that do not regularly preserve forms of the text that are obviously in error.
4. *The Geographical Spread of the Manuscripts*. An even more useful criterion involves the geographical distribution of the different forms of the text, especially among the earliest manuscripts. Suppose our manuscripts support two different forms of a passage, one found only among manuscripts produced in a specific geographical area (say, southern Italy), the other found in witnesses spread throughout the Mediterranean (say, Northern Africa, Alexandria, Syria, Asia Minor, Gaul, and Spain). In this case, the former is more likely to be a local variation reproduced by scribes of the region, whereas the other is more likely to be older since it was more widely known.

The foregoing criteria often have a cumulative effect in helping scholars decide what the original text was. If one form of reading, for example, is found in geographically diverse witnesses that are early and of generally high quality, then there is a good chance that it is original. This judgment has to be borne out, though, by two other factors

5. *The Difficulty of the Reading*. Scholars have found this criterion to be extraordinarily useful. We have seen that scribes sometimes eliminated possible contradictions and discrepancies, harmonized stories, and changed doctrinally questionable statements. Therefore, when we have two forms of a text, one that would have been troubling to scribes—for example, one that is possibly contradictory to another passage or grammatically inelegant or theologically problematic—and one that would not have been as troubling, it is the former form of the text, the one that is more "difficult," that is more likely to be original. That is, since scribes were far more likely to have corrected problems than to have created them, the comparatively smooth, consistent, harmonious, and orthodox readings are more likely to have been created by scribes. Our earliest manuscripts, interestingly enough, are the ones that tend to preserve the more difficult readings.
6. *Conformity with the Author's Own Language, Style, and Theology*. With the preceding criterion we were interested in determining which form of a passage could be most easily attributed to

Figure 28.2 P^{52}, a fragment of the Gospel of John discovered in a trash heap in the sands of Egypt. This credit-card sized scrap is the earliest surviving manuscript of the New Testament, dating from around 125–150 C.E. Both front and back are pictured here.

scribes who copied the text. With our sixth and final criterion we are interested in seeing which form of a passage would be easiest to ascribe to the author who originally produced the text in light of its vocabulary, writing style, and theology. If two forms of a passage are preserved among the New Testament manuscripts and one of them contains words, grammatical constructions, and theological ideas that never occur in the author's writings elsewhere (or that conflict with his other writings), then that form of the text is less likely to be original than the other.

All of these criteria need to be applied to any particular passage in order to decide which reading preserved among the manuscripts is likely to be original. In many instances, the arguments coalesce, so that the earliest and best manuscripts also support the reading that is most difficult and that conforms most closely with the author's own language and style. When this happens, we can be relatively certain that we have uncovered the earliest available form of the text.

CONCLUSION: THE ORIGINAL TEXT OF THE NEW TESTAMENT

At one time or another, you may have heard someone claim that the New Testament can be trusted because it is the best attested book from the ancient world, that because there are more manuscripts of the New Testament than of any other book, we should have no doubts concerning the truth of its message. Given what we have seen in this chapter, it should be clear why this line of reasoning is faulty. It is true, of course, that the New Testament is abundantly attested in manuscripts produced through the ages, but most of these manuscripts are many centuries removed from the originals, and none of them is perfectly accurate. They all contain mistakes—altogether, many thousands of mistakes. It is not an easy task to reconstruct the original words of the New Testament.

Moreover, even if scholars have by and large succeeded in reconstructing the New Testament, this, in itself, does not mean that we can have no doubt about the truth of its message. It simply means that we can be reasonably certain of what the New Testament authors actually said, just as we can be reasonably certain what Plato and Euripides and Josephus and Suetonius all said. Whether or not any of these ancient authors said anything that was *true* is another question, one we cannot answer simply by appealing to the number of surviving manuscripts that preserve their writings.

Since this has been a historical introduction to the New Testament rather than a theological one, we have not entered into this question of the truth claims of the New Testament. Historians are no more qualified to answer questions of ultimate truth than anyone else. If historians do answer such questions, they do so not in their capacity as historians but in their capacity as believers or philosophers or theologians (or skeptics). What the historian can say as a historian, however, is that the early Christian truth claims have been handed down from one generation to the next, not only orally but also through written texts that have inspired hope and faith in believers and, sometimes, hatred and fear in their enemies.

That these texts were often changed in the process of their transmission is a useful lesson for us. Only rarely have the documents of the New Testament been read out of historical interest, pure and simple. For those who read, heard, and transcribed these texts over the centuries, before the invention of "history" as a modern discipline, they preserved a living faith, one that could continue to change and grow and that now continues to change and grow. Whereas it may be the task of a believer to decide where to stand within this great living entity of the Christian tradition, it is the task of the historian, Christian or otherwise, to study the tradition and to read its texts, to see whence it came, to learn how it changed, and to understand how it became what it did.

SUGGESTIONS FOR FURTHER READING

Aland, Kurt, and Barbara Aland. *The Text of the New Testament: An Introduction to the Critical Editions and to the Theory and Practice of Modern Textual Criticism*. 2d ed. Trans. Erroll F. Rhodes. Grand Rapids, Mich.: Eerdmans; Leiden: Brill, 1989. An invaluable introduction to the various editions of the Greek New Testament and other aspects of textual criticism; most suitable for students who know Greek.

Ehrman, Bart D. *The Orthodox Corruption of Scripture: The Effect of Early Christological Controversies on the Text of the New Testament*. New York: Oxford University Press, 1993. A study of the ways scribes were influenced by doctrinal disputes in the early church and of how they modified their texts of the New Testament to make them conform more closely with their own views; for advanced students.

Ehrman, Bart D., and Michael W. Holmes. *The Text of the New Testament in Contemporary Research: Essays on the "Status Questionis."* Grand Rapids, Mich.: Eerdmans, 1995. A collection of important essays that provide up-to-date discussions and bibliographies of every important aspect of New Testament textual criticism; for advanced students.

Epp, Eldon Jay, and Gordon D. Fee. *Studies in the Theory and Method of New Testament Textual Criticism*. Grand Rapids, Mich.: Eerdmans, 1993. A significant collection of essays on important aspects of textual criticism by two of the leading scholars in the field; for advanced students.

Greenlee, J. Harold. *An Introduction to New Testament Textual Criticism*. 2d ed. Peabody: Mass.: Hendrickson, 1995. A fine introduction to the problems and methods involved in reconstructing the original text of the New Testament.

Metzger, Bruce M. *The Text of the New Testament: Its Transmission, Corruption, and Restoration*. 3d ed. New York: Oxford University Press, 1992. All in all, the best introduction to the history, data, and methods of New Testament textual criticism. Portions of the book require a basic knowledge of Greek.

GLOSSARY OF TERMS

Adoptionism: The view that Jesus was not divine, but a flesh-and-blood human being who had been *adopted* at baptism to be God's son.

Alexander the Great: The great military leader of Macedonia (356–323 B.C.E.) whose armies conquered much of the eastern Mediterranean and who was responsible for the spread of Greek culture (Hellenism) throughout the lands he conquered.

Antiochus Epiphanes: The Syrian monarch who attempted to force the Jews of Palestine to adopt Greek culture, leading to the Maccabean revolt in 167 B.C.E.

Antitheses: Literally, "contrary statements," used as a technical term to designate six sayings of Jesus in the Sermon on the Mount (Matt 5:21–48), in which he states a Jewish law ("You have heard it said . . .") and then sets his own interpretation over it ("But I say to you . . .").

Apocalypse: A literary genre in which an author, usually pseudonymous, reports symbolic dreams or visions, given or interpreted through an angelic mediator, which reveal the heavenly mysteries that can make sense of earthly realities.

Apocalypticism: A worldview held by many ancient Jews and Christians that maintained that the present age is controlled by forces of evil, but that these will be destroyed at the end of time when God intervenes in history to bring in his kingdom, an event thought to be imminent.

Apocrypha: A Greek term meaning, literally, "hidden things," used of books on the fringe of the Jewish or Christian canons of Scripture. The Jewish Apocrypha comprises books found in the Septuagint but not in the Hebrew Bible, including 1 and 2 Maccabees and 4 Ezra.

Apollonius: A pagan philosopher and holy man of the first century C.E., reported to do miracles and to deliver divinely inspired teachings, a man believed by some of his followers to be a son of God.

Apology: A reasoned explanation and justification of one's beliefs and/or practices, from a Greek word meaning "defense."

Apostle: Generally, one who is commissioned to perform a task, from a Greek word meaning "sent"; in early Christianity, the term was used to designate special emissaries of the faith who were understood to be representatives of Christ. *See also* Disciple.

Apostolic Fathers: A collection of noncanonical writings penned by proto-orthodox Christians of the second century who were traditionally thought to have been followers of the apostles; some of these works were considered Scripture in parts of the early church.

Augurs: A group of pagan priests in Rome who could interpret the will of the gods by "taking the auspices." *See also* Auspicy.

Auspicy: A form of divination in which specially appointed priests could determine the will of the gods by observing the flight patterns or eating habits of birds. *See also* Divination.

Autograph: The original manuscript of a literary text, from a Greek word meaning "the writing itself."

B.C.E. / C.E.: Abbreviations for "before the common era" and the "common era" respectively, used as exact equivalents of the Christian designations "before Christ" (B.C.) and "anno domini" (A.D., a Latin phrase meaning, "year of our Lord").

Beatitudes: A Latin word meaning, literally, "blessings," used as a technical term for the sayings of Jesus that begin the Sermon on the Mount (e.g., "Blessed are the poor in spirit . . .," Matt 5:3–12).

Canon: From a Greek word meaning "ruler" or "straight edge." The term came to designate any recognized collection of texts; the canon of the New Testament is thus the collection of books that Christians accept as authoritative.

Catholic: From a Greek word meaning "universal" or "general," used of the New Testament epistles James, 1 and 2 Peter, 1, 2, and 3 John, Jude, and sometimes Hebrews (the "Catholic" epistles) to differentiate them from the letters of Paul.

Christ. *See* Messiah.

Christology: Any teaching about the nature of Christ. *See also* Adoptionism; Docetism.

Cosmos: The Greek word for "world."

Covenant: An agreement or treaty between two social or political parties that have come to terms; used by ancient Jews in reference to the pact that God made to protect and preserve them as his chosen people in exchange for their devotion and adherence to his law.

Cult: Shortened form of *cultus deorum*, a Latin phrase that literally means "care of the gods," generally used of any set of religious practices of worship. In pagan religions, these normally involved acts of sacrifice and prayer.

Cynics: Greco–Roman philosophers, commonly portrayed as street preachers who harangued their audiences and urged them to find true freedom by being liberated from all social conventions. The Cynics' decision to live "according to nature" with none of the niceties of life led their opponents to call them "dogs" (in Greek, *cynes*).

Daimonia: Category of divine beings in the Greco–Roman world. Daimonia were widely thought to be less powerful than the gods but far more powerful than humans and capable of influencing human lives.

Dead Sea Scrolls: Ancient Jewish writings discovered in several caves near the northwest edge of the Dead Sea, widely thought to have been produced by a group of apocalyptically minded Essenes who lived in a monastic-like community from Maccabean times through the Jewish War of 66–70 C.E. *See also* Essenes; Qumran.

Demeter: The Greek and Roman goddess of grain, worshipped in a prominent mystery cult in Eleusis, Greece. *See also* Persephone.

Deutero–Pauline Epistles: The letters of Ephesians, Colossians, and 2 Thessalonians, which have a "secondary" (Deutero) standing in the Pauline corpus because scholars debate whether they were written by Paul.

Diaspora: Greek for "dispersion," a term that refers to the dispersion of Jews away from Palestine into other parts of the Mediterranean, beginning with the Babylonian conquests in the sixth century B.C.E.

Disciple: A follower, one who is "taught" (as opposed to an apostle, one who is "sent" as an emissary).

Divination: Any practice used to ascertain the will of the gods. *See also* Auspicy; Extispicy.

Docetism: The view that Jesus was not a human being but only appeared to be, from a Greek word meaning "to seem" or "to appear."

Ebionites: A group of second-century Adoptionists who maintained Jewish practices and Jewish forms of worship.

Egyptian, The: A Jewish apocalyptic prophet of the first century C.E. who predicted the destruction of the walls of Jerusalem, mentioned by Josephus.

Equestrian: The second-highest socioeconomic class of ancient Rome (below Senator), comprising wealthy aristocrats.

Essenes: An apocalyptic and ascetic Jewish sect started during the Maccabean period, members of which are generally thought to have produced the Dead Sea Scrolls.

Extispicy: A form of divination in Greek and Roman religions in which a specially appointed priest (haruspex) would examine the entrails of a sacrificed animal to determine whether it had been accepted by the gods.

Fourth Philosophy: A group of Jews that Josephus mentions but leaves unnamed, characterized by their insistence on violent opposition to the foreign domination of the Promised Land. *See also* Sicarii; Zealots.

Gamaliel: A famous rabbi of first-century C.E. Judaism.

Gematria: Jewish method of interpreting a word on the basis of the numerical value of its letters (in both Greek and Hebrew, the letters of the alphabet also serve as numerals.)

Genius: A man's guardian spirit (that of a woman was called Iuno).

Gentile: A Jewish designation for a non-Jew.

Gnosticism: A group of ancient religions, some of them closely related to Christianity, that maintained that elements of the divine had become entrapped in this evil world of matter and could be released only when they acquired the secret *gnosis* (Greek for "knowledge") of who they were and of how they could escape. Gnosis was generally thought to be brought by an emissary of the divine realm.

Greco–Roman World: The lands (and culture) around the Mediterranean from the time of Alexander the Great to the Emperor Constantine, roughly 300 B.C.E. to 300 C.E. (*see also* Box 2 in Chapter 2).

Haruspex: In Roman religion, a specially trained priest skilled in the practice of extispicy.

Hasmoneans: An alternative name for the Maccabeans, the family of Jewish priests that began the revolt against Syria in 167 B.C.E. and that ruled Israel prior to the Roman conquest of 63 B.C.E.

Hellenization: The spread of Greek language and culture (Hellenism) throughout the Mediterranean, starting with the conquests of Alexander the Great.

Heresy: Any worldview or set of beliefs deemed by those in power to be deviant, from a Greek word meaning "choice" (because "heretics" have "chosen" to deviate from the "truth".) *See also* Orthodoxy.

High Priest: Prior to 70 C.E., the highest-ranking authority in Judaism when there was no Jewish king, in charge of the operation of the Jerusalem Temple and its priests. *See also* Sadducees; Sanhedrin.

Historiography: The literary reconstruction of historical events; the writing of history; and the study and analysis of historical narrative.

Isis: Egyptian goddess worshipped in mystery cults throughout the Roman world.

Judas Maccabeus: Jewish patriot who led the family responsible for spearheading the Maccabean revolt. *See also* Hasmoneans.

Maccabean Revolt: The Jewish uprising against the Syrians and their king, Antiochus Epiphanes, starting in 167 B.C.E., in protest against the forced imposition of Hellenistic culture and the proscription of Jewish practices such as circumcision. *See also* Hasmoneans.

Manuscript: Any handwritten copy of a literary text.

Marcion: A second-century Christian scholar and evangelist, later labeled a heretic for his docetic Christology and his belief in two Gods—the harsh legalistic God of the Jews and the merciful loving God of Jesus—views that he claimed to have found in the writings of Paul.

Markan Priority: The view that Mark was the first of the Synoptic Gospels to be written and was one of the sources used by Matthew and Luke.

Messiah: From a Hebrew word that literally means "anointed one," translated into Greek as *Christos*, from which derives our English word *Christ*. In the first century C.E., there was a wide range of expectations about whom this anointed one might be, some Jews anticipating a future warrior king like David, others a cosmic redeemer from heaven, others an authoritative priest, and still others a powerful spokesperson from God like Moses.

Mishnah: A collection of oral traditions passed on by generations of Jewish rabbis who saw themselves as the descendants of the Pharisees, finally put into writing around 200 C.E.. *See also* Talmud.

Mithras: A Persian deity worshipped in a mystery cult spread throughout the Roman world.

Mystery Cults: A group of Greco–Roman religions that focused on the devotees' individual needs both in this life and in life after death, so named because their initiation rituals and cultic practices involved the disclosure of hidden things that were to be kept secret from outsiders.

Nag Hammadi: Village in upper (southern) Egypt, near the place where a collection of Gnostic writings, including the *Gospel of Thomas*, were discovered in 1945.

Oracle: A sacred place where the gods answered questions brought by their worshippers to the resident holy person—a priest or, more commonly, a priestess—who would often deliver the divine response out of a trance-like state; the term can also refer to the divine answer itself.

Orthodoxy: From the Greek, literally meaning "right opinion"; a term used to designate a worldview or set of beliefs acknowledged to be true by the majority of those in power. *See also* Heresy.

Paganism: Any of the polytheistic religions of the Greco–Roman world, an umbrella term for ancient Mediterranean religions other than Judaism and Christianity.

Parousia: A Greek word meaning "presence" or "coming," used as a technical term to refer to the Second Coming of Jesus in judgment at the end of time.

Passion: From a Greek word that means "suffering," used as a technical term to refer to the traditions of Jesus' last days, up to and including his crucifixion (hence the "Passion narrative").

Passover: The most important and widely celebrated annual festival of Jews in Roman times, commemorating the exodus from Egypt.

Pastoral Epistles: New Testament letters that Paul allegedly wrote to two pastors, Timothy (1 and 2 Timothy) and Titus, concerning their pastoral duties.

Pauline Corpus: All of the letters of the New Testament that claim to be written by Paul, including the Deutero-Pauline and Pastoral Epistles.

Pentateuch: Literally, the "five scrolls" in Greek, a term used to designate the first five books of the Hebrew Bible, also known as the Torah or the Law of Moses.

Pentecost: A Jewish agricultural festival, celebrated fifty days after the feast of the Passover, from the Greek word for fifty (*pentakosia*).

Persephone: Daughter of the Greek goddess Demeter, reported to have been abducted to the underworld by Hades but allowed to return to life every year to be reunited temporarily with her grieving mother; also known as Kore.

Pharisees: A Jewish sect, which may have originated during the Maccabean period, that emphasized strict adherence to the purity laws set forth in the Torah. *See also* Mishnah.

Prescript: The formal beginning of an epistle, normally including the names of the sender and addressees, a greeting, and often a prayer or wish for good health.

Proto-orthodox Christianity: A form of Christianity endorsed by some Christians of the second and third centuries (including the Apostolic Fathers), which promoted doctrines that were declared "orthodox" in the fourth and later centuries by the victorious Christian party, in opposition to such groups as the Ebionites, the Marcionites, and the Gnostics.

Pseudepigrapha: From the Greek, literally meaning "false writings" and commonly referring to ancient noncanonical Jewish and Christian literary texts, many of which were written pseudonymously.

Pseudonymity: The practice of writing under a fictitious name, evident in a large number of pagan, Jewish, and Christian writings from antiquity.

Q Source: The source used by both Matthew and Luke for the stories they share, principally sayings, that are not found in Mark; from the German word *Quelle*, "source." The document no longer exists, but can be reconstructed on the basis of Matthew and Luke.

Qumran: Place near the northwest shore of the Dead Sea, where the Dead Sea Scrolls were discovered in 1946, evidently home to the group of Essenes who had used the Scrolls as part of their library.

Redaction criticism: The study of how authors modified or edited (i.e., redacted) their sources in view of their own vested interests and concerns.

Rhetoric: The art of persuasion; in the Greco-Roman world, this involved training in the construction and analysis of argumentation and was the principal subject of higher education.

Roman Empire: All of the lands conquered by Rome and ruled, ultimately, by the Roman emperor, starting with Caesar Augustus in 27 B.C.E.; prior to that, Rome was a republic ruled by the Senate (*see also* Box 5 in Chapter 2).

Sadducees: A Jewish party associated with the Temple cult and the Jewish priests who ran it, comprising principally the Jewish aristocracy in Judea. The party leader, the High Priest, served as the highest ranking local official and chief liaison with the Roman governor.

Sanhedrin: A council of Jewish leaders headed by the High Priest, which played an advisory role in matters of religious and civil policy.

Scribes, Christian: Literate Christians responsible for copying sacred scripture.

Scribes, Jewish: Highly educated experts in Jewish Law (and possibly its copyists) during the Greco–Roman period.

Senators: The highest-ranking members of the Roman aristocracy, comprising the wealthiest men of Rome, responsible for governing the vast Roman bureaucracy during the republic and still active and highly visible during the time of the empire.

Septuagint: The translation of the Hebrew Scriptures into Greek, so named because of a tradition that seventy (Latin: *septuaginta*) Jewish scholars had produced it.

Sicarii: A Latin term meaning, literally, "daggermen," a designation for a group of first-century Jews responsible for the assassination of Jewish aristocrats thought to have collaborated with the Romans. *See also* Fourth Philosophy.

Signs Source: A document, which no longer survives, thought by many scholars to have been used as one of the sources of Jesus' ministry in the Fourth Gospel; it reputedly narrated a number of the miraculous deeds of Jesus.

Son of God: In most Greco–Roman circles, the designation of a person born to a god, able to perform miraculous deeds and/or to convey superhuman teachings; in Jewish circles, the designation of persons chosen to stand in a special relationship with the God of Israel, including the ancient Jewish Kings.

Son of Man: A term whose meaning is much disputed among modern scholars, used in some ancient apocalyptic texts to refer to a cosmic judge sent from heaven at the end of time.

Stoics: Greco–Roman philosophers who urged people to understand the way the world worked and to live in accordance with it, letting nothing outside of themselves affect their internal state of well-being.

Synagogue: Jewish place of worship and prayer, from a Greek word that literally means "being brought together."

Synoptic Gospels: The Gospels of Matthew, Mark, and Luke, which narrate so many of the same stories that they can be placed side by side in parallel columns and so "be seen together" (the literal meaning of "synoptic").

Synoptic Problem: The problem of explaining the similarities and differences between the three Synoptic Gospels. *See also* Markan Priority; Q Source.

Talmud: The great collection of ancient Jewish traditions that comprises the Mishnah and the later commentaries on the Mishnah, called the Gemarah. There are two collections of the Talmud, one made in Palestine during the early fifth century C.E. and the other in Babylon perhaps a century later. The Babylonian Talmud is generally considered the more authoritative.

Textual Criticism: An academic discipline that seeks to establish the original wording of a text based on the surviving manuscripts.

Thecla: A (legendary) female disciple of Paul whose adventures are narrated in the novel-like work of the second century, *The Acts of Paul and Thecla*.

Theudas: (1) A first-century Jewish apocalyptic prophet (mentioned by Josephus) who predicted the parting of the Jordan River and, evidently, the reconquest of the Promised Land by the chosen people. (2) An early Gnostic Christian, allegedly the disciple of Paul and the teacher of Valentinus.

Torah: A Hebrew word that means "guidance" or "direction," but that is usually translated "law." As a technical term it designates *either* the Law of God given to Moses *or* the first five books of the Jewish Bible that Moses was traditionally thought to have written—Genesis, Exodus, Leviticus, Numbers, and Deuteronomy.

Tradition: Any doctrine, idea, practice, or custom that has been handed down from one person to another.

Two Ways: The doctrine found in the *Didache* and the *Epistle of Barnabas*, that people must choose between two ways of living, the way of life (or light) and the way of death (or darkness).

Undisputed Pauline epistles: Romans, 1 and 2 Corinthians, Galatians, Philippians, 1 Thessalonians, and Philemon—letters that scholars overwhelmingly judge to be have been written by Paul. *See also* Deutero–Pauline epistles; Pastoral Epistles.

Valentinus: Second-century Gnostic Christian who traced his intellectual lineage through his teacher Theudas back to the apostle Paul.

Zealots: A group of Galilean Jews who fled to Jerusalem during the uprising against Rome in 66–70 C.E., who overthrew the reigning aristocracy in the city and urged violent resistance to the bitter end. *See also* Fourth Philosophy.

INDEX

Abraham, 80–81, 141, 253, 290–91, 307–8, 316, 361
Acts, apocryphal, 368. *See also* Thecla
Acts of the Apostles, 8–9, 98, 115–32, 243–46
Adoptionists, 3–4, 322, 353, 363, 418
Aeons, 165–66
Afterlife, 24, 168, 208, 278
Akiba, Rabbi, 67
Alexander the Great, 19, 20–21, 53, 204
Angel worship, 326
Antichrist, 324–25, 159–60, 389, 400, 408
Anti-Judaism. *See* Anti-Semitism
Antiochus Epiphanes, 204, 215–16
Antipas, Herod, 205, 219
Anti-Semitism, 93–94, 170, 359–65
Antitheses, 88–89
Apocalypse
  as a genre, 215, 400–4
  of John. *See* Revelation of John
Apocalypticism, 86, 152, 180–81, 215–17, 248, 251–53, 262–63, 343–44, 395–96, 400. *See also* End of time
Apocrypha, Jewish, 89, 109, 302
Apollonius of Tyana, 17–18, 24, 54, 199–200
Apollos, 275
Apology, 99, 127–28, 131–32, 380–81
Apostles, epistle of, 181
Apostles' Creed, 338
Apostolic constitutions, 323
Apostolic fathers, 10. *See also individual names*
Apostolic succession, 337, 391
Aquila, 272
Aramaisms, criterion of, 193
Archelaus, 101, 214
Artemis, 262
Asclepius, 43, 199
Asia Minor, seven churches of, 399–400
Associations, trade, 37, 263–64
Assyrians, 203
Athanasius, 11
Atonement, Jesus' death as, 67–68, 110, 126, 131, 238
Augurs, 26
Augustine, 299
Auspicy, 26
Autographs, 418–21

Babylonians, 203
Baptism, Christian, 278, 305–7, 327–28, 352, 386, 411
Bar Kosiba, Simon, 67
Barnabas, epistle of, 12, 359–62
Beatitudes, 87, 113, 227
Beloved disciple, 146, 149–50, 152–53. *See also* Gospels, authors of
Bethlehem, 82–84, 101–2
Biography, Greco-Roman, 52–55, 80, 100, 116, 134
Birth narratives, 100–2. *See also* Jesus, birth of
Bloody sweat, Jesus', 108
Brothers of Jesus, 192, 312, 315–16, 393

Caesar Augustus, 18, 27, 102, 404
Caiaphas, 228, 230
Caligula (emperor), 26, 213
Calvin, John, 299
Canon
  New Testament, 3–13, 323, 339, 354, 359, 383, 390, 396, 404, 410, 412
  Old Testament, 2–3
Cephas, 287–88
Cerinthus, 181
Charismatic communities, 337
Chief priests, 61, 223
Chloe, 275
Christ, as a title. *See* Messiah
Christ hymn, of Philippians, 294–95
Christianity
  beginning of, 233–34
  spread of, 41–44, 362–65, 380–81
Circumcision, 4, 35, 285–92, 361. *See also* Law, Jewish
Claudius (emperor), 187, 300
Clement of Alexandria, 359
Clergy, Christian, 337–38, 376, 391–93
Codex Sinaiticus, 12, 330

Collection, Paul's, 283, 301
Colossians, 325–29
Comparative Method, 96–113, 137–41
Confucius, 89
Constantine, 364–65, 381
Contextual credibility, criterion of, 195–96
Contextual Method, 156–61, 247
Corinth, 271–72
Covenant, Jewish, 34–35, 215, 218, 253, 286, 355–57, 359–60
Creed, Christian, 338
Criteria for reconstructing the life of Jesus, 192–97
Crucifixion, 46–49, 67–69, 109, 175–76, 188, 195, 219, 230–31, 235–37, 249, 252, 273–74, 278, 313, 364
Cult
  Jewish, 35–37, 90–91, 162, 207–8. *See also* Temple, Jewish
  pagan, 20–31, 162, 260–62, 352–54, 368–70
Cumanus, 212
Cynic philosophy, 220, 259–60, 343

Daimonia, 22, 33
Daniel, book of, 215, 225–26, 236–37, 321, 401–2
Dante, 412
David, King, 57, 80–82, 104–5
Day of atonement, 36, 68
Dead Sea Scrolls, 208–11, 215. *See also* Essenes
Deicide, accusations of, 364
Demeter, 30
Deutero-Pauline epistles, 242–43, 323–32
Devil. *See* Satan
Diaspora, 33, 37
Diatesseron, 173
Diatribe, 301
Didache, 385–88
Didymus Judas Thomas, 177–78
Dionysius of Corinth, 390
Dionysus, 30
Diotrephes, 161–62
Dissimilarity, criterion of, 193–95
Divination, 25–26
Docetists, 5, 159, 167, 174–75, 322, 378, 389
Domitian (emperor), 390, 404–5
Dualism, 164–65, 216
Dura-Europas, 161–62

Ebionites, 171–73, 238–39
  Gospel of, 173
Egyptian, The, 213, 218
End of time, 112, 120–21, 131, 152, 399–409. *See also* Apocalypticism
Epaphras, 129, 296, 325
Epaphroditus, 293–96
Ephesians, 329–32
Epicureanism, 30–31, 343
Epistles. *See* Letters
Equestrians, 27
Essenes, 206, 208–10, 221–22, 248. *See also* Dead Sea Scrolls
Eucharist, 279, 334, 352, 385–86
Euodia, 293, 342
Euripides, 239, 415
Extispicy, 24–26

Faith
  and history, 13–14, 198–202
  and works. *See* James, epistle of
  Paul's teaching on, 252–53, 289–92, 302–7, 339. *See also* Justification, Paul's doctrine of
Farewell discourse, 135, 144–45
Fasting, in early Christianity, 386
Felicitas, 369
First Clement, 300, 310, 389–93
First Corinthians, 271–80, 392
First Enoch, 226, 393, 401
First Thessalonians, 263–69
First Timothy, 333. *See also* Pastoral epistles
Forgery. *See* Pseudepigrapha
Four-source hypothesis, 73–78
Fourth Ezra, 226, 401
Fourth Philosophy, 210–11
Fronto, 268
Fulfillment citation, 82–83
Funeral societies, 37, 263–64

Gaius, church leader, 156, 161
Galatia, 285
Galatians, 285–92

Galen, 322
Gamaliel, 247–48
Gematria, 361–62, 408
Gender, ideologies of, 347–50
General epistles, 154–62
Genius, 23, 26
Gentile Christianity, 70, 92, 94–95, 111, 120, 131–32, 253–55, 285–92, 301–10
Gnosticism, 5–6, 10, 161–69, 171, 176–81, 318, 333, 336–38, 342, 360, 362, 378, 394–96. *See also* Nag Hammadi Library
Gods, Roman, 20–24, 43, 266, 352–54, 368–71, 380–81. *See also* Greco-Roman religion
Golden Rule, 89–90
Gospel harmonies, 173
Gospels
  authors of, 44, 49–50. *See also individual Gospels*
  dates, 40–41
  differences among, 44–49
  genre of, 51–55
  as historical sources, 190–91
  Jewish-Christian, 172–74
  non-canonical, 163–64, 171–84. *See also individual names*
Governors, Roman, 266, 367–68. *See also* Pliny; Pontius Pilate
Greco-Roman religion, 18–31, 258. *See also* Gods, Roman
Greco-Roman world (history, society), 16–17, 19, 27, 32, 258–62

Habakkuk, commentary on, 209–10
Hagar, 291
Hanina ben Dosa, 38, 58, 200
Haruspex, 25–26
Hasmoneans. *See* Maccabeans
Head coverings, 278, 344–45
Hebrew Bible, 2–3, 10–11, 166, 354–62
Hebrews, epistle to the, 354–59
Hebrews, Gospel of, 174
Hellenization, 19, 204
Heracleon, 169
Heracles, 24, 178, 238
Heresy, 3–7. *See also* Adoptionists; Docetists; Gnosticism; Proto-orthodox Christianity
Hermas. *See* Shepherd of Hermas
Herodians, 60
Herodotus, 89
Herod the Great, 83–84, 204–5, 343
Hesiod, 22
Hierarchy, Church. *See* Clergy
High priest, Jewish, 36, 61, 68, 204, 228
Hillel, Rabbi, 89
History, as a genre, 98, 100, 115–17
Holy of Holies, 36, 68
Homer, 22, 415
Honi the Circle Drawer, 38, 58
House Churches, 162

"I am" sayings, 141
Idols, 36, 246
Ignatius, 159, 337–38, 374–78, 388–89
Independent attestation, criterion of, 192–93
Infancy Gospels, 181–82
Insurrections, Jewish. *See* Jewish war
Irenaeus, 164–65, 168, 181, 337, 389
Isaac, 291
Ishmael, 291
Isis, 30
Isocrates, 89

James, Apocryphon of, 181
James, brother of Jesus, 188, 289, 315–16, 320, 393
James, epistle of, 315–17, 331, 384–85
James, Gospel of, 183
Jerusalem, 212–15, 221–23, 360
  apostles, 288–91. *See also* Twelve, The
  Council, 125
Jesus
  as apocalyptic prophet, 217–31
  baptism of, 3, 6, 58, 83, 85–86, 105, 173, 218–19, 238, 418. *See also* John the Baptist
  betrayal of. *See* Judas Iscariot
  birth of, 4–5, 80–83, 101–3, 107, 178, 183, 338
  as cynic philosopher, 220
  death of, 228–31. *See also* Atonement, Jesus' death as; Crucifixion
  deeds of, 219–24
  as divine, 4–5, 67, 147–49, 356
  genealogy of, 80–81, 104–5
  historical, canonical sources for. *See* Criteria for reconstructing the life of Jesus

historical, non-canonical sources for, 186–90
as human, 4, 147–49, 356
as Lord, 235–36
as Messiah. *See* Messiah
as Prophet, 107–11
return from heaven, 262, 267–69, 280, 294, 323–25, 327, 331
as Son of God, 56–69, 235–39
as Son of Man, 235–39
teachings of, 224–28
trial of, 47–49, 67, 93–94, 175, 229–31
Jewish scriptures. *See* Hebrew Bible
Jewish war, 36, 70, 188, 210, 214–15
Jews, in the Fourth Gospel, 136, 151
Johannine community, 149–52, 158–60, 169
Johannine epistles, 154–62, 389
John, and the Synoptics, 137–41
John, Apocryphon of, 181
John, Gospel of, 133–53
John, son of Zebedee, 49, 152–53, 181, 404
John, sources of, 142–45
John the Baptist, 58, 85–86, 173, 192, 194–95, 218–19
Josephus, 115, 188–89, 205–15
Judaism, 32–38. *See also* Law, Jewish
Judas, son of Hezekiah, 214
Judas Iscariot, 122–23, 195, 229–30
Jude, epistle of, 393–94
Judicial model, Paul's, 304–7. *See also* Justification, Paul's doctrine of
Julius Caesar, 18, 27, 212, 404
Justification, Paul's doctrine of, 252–53, 289–92, 304–7, 316–17, 384
Justin Martyr, 363, 380

Kingdom of God, 216–17, 224–31. *See also* Apocalypticism
King of the Jews, 84, 229–30
Kore. *See* Persephone

L (Luke's special source), 77–78
Law
Jewish, 2, 4, 34–35, 37, 87–91, 205–11, 326, 329, 331–32, 352, 354–62. *See also* Circumcision; Paul, view of the Law
oral, 207, 248
Roman, 367–68
Lazarus, 138–39
Letters, in the Greco-Roman World, 154–56
Literacy, 45
Literary-historical method, 72, 86, 134–36
Lord's Prayer, 386–87, 416
Lord's Supper. *See* Eucharist
Love
Jesus' teaching on, 227–28, 368
Paul's teaching on, 278–79, 292, 294–96, 308
Lucian of Samosata, 370
Luke, author of, 129–32. *See also* Gospels, authors of
Luke, Gospel of, 96–114
Luther, Martin, 299, 316

M (Matthew's special source), 77–78
Maccabeans, 109, 204–5, 209, 211–12, 215–16, 239, 402
Magi, 84, 101
Magic, 28–29
Magnificat, 107
Manuscripts, New Testament, 10, 12, 54, 104, 108, 111–12, 239, 329–30, 408, 414–21
Marcion, 4–5, 84, 159, 167, 171, 174, 353, 389, 410
Marcus Aurelius, 268, 370
Mark, Gospel of, 56–71
Markan priority, 73–75
Martyrdom, Christian, 126, 310, 369, 374–79, 390, 394, 405
Martyrs, Jewish, 109, 238
Matthew, Gospel of, 79–95, 353
Maximilla, 342
Melanchton, Philip, 299
Melchizedek, 355
Melito of Sardis, 363–64
Messiah, 56–59, 209, 226, 230, 235–39, 249
Messianic secret, 64, 85–86
Miracle, problem of, 198–202
Mishnah, 188, 207, 248
Mithras, 30
Models for salvation, Paul's, 304–9
Montanus, 342
Moses, 34–35, 46–47, 83–84, 86–87, 141, 291, 355, 360. *See also* Covenant, Jewish; Law, Jewish
Muratorian Fragment, 410
Mystery cults, 28–31
Mythology, Greek, 22–23, 30, 178, 380

Nag Hammadi Library, 10, 163–64, 177, 181, 396. *See also* Gnosticism
Nazareans, Gospel of, 172–73
Nero (emperor), 187, 370, 390, 394, 404, 408
New Testament. *See* Canon, New Testament
Nicea, Council of, 364–65
Nicodemus, 137, 196
Novels, Greco-Roman, 117

Old Testament. *See* Hebrew Bible
Onesimus, 296–98
Oracles, 26
Origen, 354, 380
Orthodoxy. *See* Proto-orthodoxy Christianity
Oxyrhynchus, 155, 180

Paganism. *See* Greco-Roman religion
Palestine, history of, 203–15. *See also* Maccabeans
Parables, 59, 91, 93, 137–38, 181–82, 192, 223. *See also* Jesus, teachings of
Parousia. *See* Jesus, return from heaven
Participationist model, Paul's, 305–7
Passover, 45–49, 148, 212–13, 223, 228–29, 363–64
Pastoral epistles, 243–44, 333–40, 346–48
Paul
- in Acts, 121, 127–28, 243–46, 250
- as apocalypticist, 248–53, 262–63, 267–69, 276–80, 284–85, 294, 298, 305, 314, 324–25, 350
- conversion of, 119, 244–45, 249–51, 287, 332
- death of, 310, 390
- difficulties in studying, 242–47
- and the historical Jesus, 189–90, 273, 311–15. *See also* Paul, view of Christ
- life of, 241–56, 287–88
- missionary journeys of, 119, 255
- missionary message, 260–63, 302–09
- missionary tactics, 257–60
- nature of his letters, 246–47, 255
- opponents of, 277–78, 285–89. *See also* Superapostles in Corinth
- view of Christ, 252, 262, 273–74, 276–77, 304–7. *See also* Paul, and the historical Jesus
- view of the Law, 244–45, 248–49, 252–53, 285–92, 301–10, 316–17, 353

Pentateuch, 34. *See also* Law, Jewish
Pentecost, day of, 119, 125–26
Perpetua, 369
Persecution, of Christians, 70, 99–100, 122, 126–27, 249, 263–66, 323–24, 354, 366–81, 390, 405–08. *See also* Nero (emperor)
Persecution, of Jews. *See* Anti-semitism
Persephone, 30
Pesher, 209–10
Peter
- Acts of, 373, 395
- Apocalypse of, 394, 410, 412
- First epistle of, 371–74
- Gospel of, 174–76, 189, 373, 394, 412
- Second epistle of, 11, 373, 394–96
- Simon, 49, 62–65, 121–27, 146, 288, 300, 373, 390, 394, 412

Pharisees, 60–61, 90–92, 206–7, 227, 248–49, 344
Philemon, 296–98, 329
Philip, Gospel of, 195–96
Philippi, 292–93
Philippians, 292–96
Philo, 82, 360
Philosophers and philosophy, Greco-Roman, 30–31, 220, 258–60, 272–73, 343
Philostratus. *See* Apollonius of Tyana
Phoebe, 300, 341
Phoenix, 392
Plato, platonism, 30–31, 277, 357
Plautus, 178
Pliny, 186, 295, 370–71, 378
Plutarch, 53
Polycarp, 377–79, 388–90
Pompey, 204
Pontius Pilate, 47, 93–94, 99, 175–76, 187, 205, 208, 213, 219, 223–24, 228–31, 367
Priests, Jewish, 36, 207–8, 223, 355. *See also* Chief priests; High priest
Priests, Roman, 23–25
Prisca (Paul's companion). *See* Priscilla
Prisca (the Montanist), 342
Priscilla, 273
Prologue, Johannine, 134, 136–37, 143, 145, 149, 159
Prophets, Jewish, 213–15, 219
Proto-orthodox Christianity, 6–7, 162–69, 181, 322, 336–39, 374–75, 378, 381, 387–96, 416–17
Psalms in early Christianity, 123, 236–37

Pseudepigrapha, 242–43, 246, 320–39, 373, 393–96, 402, 412
Pseudonymity. *See* Pseudepigrapha
Pyramid, divine, 22–24
Pythagorus, 321

Q source, 73–78, 180–81, 387
Quirinius, 102
Qumran. *See* Dead Sea Scrolls

Redaction criticism, 72, 78, 80–94, 96–97, 142–45
Resident aliens, 371–72
Resurrection
in apocalypticism, 217, 249, 276–79
of believers, 152, 267, 327–28, 331, 392. *See also* Resurrection, in apocalypticism
of Jesus, 68–69, 126, 128, 175–76, 180, 234–36, 251–53, 275–78, 304–9
Revelation discourses, 181
Revelation of John, 398–409
Roman emperor, 24–27, 213. *See also* Caesar Augustus; Nero; Roman Empire; Titus; Trajan; Vespasian
Roman Empire, 27, 32, 211–15, 366–71, 406–8
Romans, Paul's letter to, 300–11
Rome. *See* Roman Empire; Romans, Paul's letter to

Sabbath, 35, 206–7, 360–61. *See also* Law, Jewish
Sacrifices
Jewish, 35–37, 248, 253, 290, 354–55
Pagan, 23–26. *See also* Gods, Roman
Sadducees, 60, 206–08, 223, 228–30
Samaritans, 120
Sanhedrin, 66–67, 147–48, 207, 228–30
Satan, 216, 282–84, 400, 406, 408
Scribes
Christian, 415–21. *See also* Manuscripts, New Testament
Jewish, 60
Secessionists, 158–60, 169
Second Baruch, 401
Second Corinthians, 280–85
Second Thessalonians, 323–25
Second Timothy, 333–34. *See also* Pastoral epistles
Sects, Jewish, 60–61, 205–11
Self-definition, 351–53, 358–59, 361
Seneca, 242, 246, 332
Senate and Senators, Roman, 26, 27, 367
Septuagint, 33, 57–58, 248
Serapion (bishop of Antioch), 174–75
Sermon on the Mount, 77, 83, 86–91, 384, 386–88
Sex rituals, in early Christianity, 168–69, 267–68, 368, 380
Shepherd of Hermas, 11, 300, 409–11
Sicarii, 211
Signs, in the Fourth Gospel, 135, 138–41, 144–45, 150, 172, 192
Signs source. *See* Signs, in the Fourth Gospel
Silvanus, 244, 258, 269, 272, 275, 282, 323, 373
Simon Magus, 181, 395
Sin, in Paul, 304–7
Sinners. *See* Tax collectors and sinners
666, 408
Slavery, in early Christianity, 272, 275, 296–98, 344–45
Socio-historical method, 133, 146–53, 263–65
Socrates. *See* Plato, platonism
Son of God
Jewish, 38, 58–59, 66
pagan, 17–18, 23–24, 58, 178
Son of Man, 66, 112, 142, 193, 225–28, 230. *See also* Jesus, as Son of Man
Sophia, 166–67
Sosthenes, 276
Speeches, in Acts, 122–29, 245–46
Speeches, in ancient writings, 116, 122
Spiritual gifts, 278–79
Stoicism, 30–31, 220, 259–60
Suetonius, 187, 300, 370
Suffering servant, 237
Superapostles in Corinth, 281–84
Superstition, 187, 370
Sybilline oracles, 408
Symbolism, apocalyptic, 403–9
Synagogues, 37, 106, 149–52, 162, 196, 258, 263
Synoptic problem, 72–78
Syntyche, 293, 342

Tabernacle, 355, 357
Tacitus, 187, 370, 415

Talmud, 188, 207
Tatian, 173
Tax collectors and sinners, 223–24, 227
Taxes, 36, 211–12
Temple, Jewish, 33, 35–37, 46–47, 60–61, 68, 102–4, 110, 128, 203–4, 207–8, 211–15, 221–23, 227–30, 359–60
Tertullian, 163, 174, 242, 321, 337, 363, 371, 378, 380
Textual criticism, 418–21. *See also* Manuscripts, New Testament
Thecla, 317–18, 342, 345, 368
Thematic approach, 117–32, 136–37
Theophilus, 99–100, 118
Thessalonica, 257–58
Theudas (Christian gnostic), 318
Theudas (Jewish prophet), 213, 218
Third Corinthians, 321–22, 332
Thomas
  Gospel of, 77, 177–82, 189, 196, 217
  infancy Gospel of, 183
Thucydides, 116
Timothy, 244, 258, 265, 269, 272, 275, 282, 296, 323, 327, 354. *See also* Pastoral epistles
Titus (companion of Paul), 281–84, 288–89, 317. *See also* Pastoral epistles
Titus (emperor), 214, 360
Titus, letter to, 334. *See also* Pastoral epistles
Titus, Pseudo-, 317
Tobit, 89
Torah. *See* Law, Jewish
Traditions, oral, about Jesus, 40–50
Trajan (emperor), 186, 370–71
Twelve, The, 124, 223, 343. *See also* Jerusalem, apostles
Two Ways, 361, 385–86

Undisputed Pauline epistles, 243
Union with Christ. *See* Participationist model, Paul's

Valens, 389
Valentinus, 319
Vespasian (emperor), 23, 188, 200, 360

War scroll, 209–10
Wisdom of Solomon, 410
Women, in early Christianity, 341–50

Zealots, 211, 214–15